SELECTED SOLUTIONS MANUAL

JOSEPH TOPICH
Virginia Commonwealth University

RUTH TOPICH
Virginia Commonwealth University

GENERAL
CHEMISTRY
ATOMS FIRST

Second Edition

JOHN E. McMURRY
ROBERT C. FAY

PEARSON

Boston   Columbus   Indianapolis   New York   San Francisco   Upper Saddle River
Amsterdam   Cape Town   Dubai   London   Madrid   Milan   Munich   Paris   Montréal   Toronto
Delhi   Mexico City   São Paulo   Sydney   Hong Kong   Seoul   Singapore   Taipei   Tokyo

Editor in Chief: Adam Jaworski

Senior Acquisitions Editor: Terry Haugen

Senior Marketing Manager: Jonathan Cottrell

Project Editor: Jessica Moro

Assistant Editor: Erin Kneuer

Managing Editor, Chemistry and Geosciences: Gina M. Cheselka

Senior Project Manager: Beth Sweeten

Operations Specialist: Jeffrey Sargent

Supplement Cover Designer: Seventeenth Street Studios

Cover Image Credit: *Molecular Chair*, Design by Antonio Pio Saracino—www.antoniopiosaracino.com

Credits and acknowledgments borrowed from other sources and reproduced, with permission, in this textbook appear on the appropriate page within the text.

Many of the designations used by manufacturers and sellers to distinguish their products are claimed as trademarks. Where those designations appear in this book, and the publisher was aware of a trademark claim, the designations have been printed in initial caps or all caps.

2 3 4 5 6 7 8 9 10—EBM—17 16 15 14 13

www.pearsonhighered.com   ISBN-10: 0-321-81332-4; ISBN-13: 978-0-321-81332-9

Contents

Preface

Chemistry is the study of the composition and properties of matter along with the changes that matter undergoes. The principles of chemistry are learned through a combination of different experiences. You hear and read about them in your lecture and textbook. You encounter them first hand in the laboratory, and most importantly, you use these chemistry principles to solve chemistry problems.

Critical thinking and problem solving are your keys to success in chemistry! **GENERAL CHEMISTRY Atoms First, 2/e** by McMurry and Fay contains thousands of problems to help you develop your critical thinking and problem solving skills. Develop your own problem solving strategy. Read each problem carefully. Summarize all the information contained in the problem. Identify what the problem is asking you to do. With your knowledge of chemistry principles, identify connections between the information provided in the problem and the solution that you are seeking. Set up and attempt to solve the problem. Look at your answer. Is it reasonable? Are the units correct? Then, and only then, compare your answer with the *Selected Solutions Manual*.

The *Selected Solutions Manual* to accompany **GENERAL CHEMISTRY Atoms First, 2/e** by McMurry and Fay contains the solutions to all in-chapter, and even numbered conceptual and end-of-chapter problems.

I have worked to ensure that the solutions in this manual are as error free as possible. Solutions have been double-checked and in many cases triple-checked. Small differences in numerical answers between student results and those in the *Selected Solutions Manual* may result because of rounding and significant figure differences. Also recognize that there is, in many cases, more than one acceptable setup for a problem.

I would like to thank John McMurry and Robert Fay for the opportunity to contribute to their **GENERAL CHEMISTRY Atoms First, 2/e** package. I also want to thank them for their helpful comments as I worked on this solutions manual. I want to thank my wife Ruth for her help in preparing this selected solutions manual. I also want to acknowledge and thank the Pearson Education staff. Finally, I want to thank my wife, Ruth, again and our daughter, Judy, both in a very special way for their constant encouragement and support as I worked on this project.

Joseph Topich
Department of Chemistry
Virginia Commonwealth University

0 Chemical Tools: Experimentation and Measurement

0.1   (a) The decimal point must be shifted ten places to the right so the exponent is –10. The result is 3.72×10^{-10} m.

       (b) The decimal point must be shifted eleven places to the left so the exponent is 11. The result is 1.5×10^{11} m.

0.2   (a) microgram   (b) decimeter   (c) picosecond
       (d) kiloampere   (e) millimole

0.3 $°C = \dfrac{5}{9} \times (°F - 32) = \dfrac{5}{9} \times (98.6 - 32) = 37.0\ °C$

       $K = °C + 273.15 = 37.0 + 273.15 = 310.2\ K$

0.4   (a) $K = °C + 273.15 = -78 + 273.15 = 195.15\ K = 195\ K$

       (b) $°F = (\dfrac{9}{5} \times °C) + 32 = (\dfrac{9}{5} \times 158) + 32 = 316.4\ °F = 316\ °F$

       (c) $°C = K - 273.15 = 375 - 273.15 = 101.85\ °C = 102\ °C$

           $°F = (\dfrac{9}{5} \times °C) + 32 = (\dfrac{9}{5} \times 101.85) + 32 = 215.33\ °F = 215\ °F$

0.5 $d = \dfrac{m}{V} = \dfrac{27.43\ g}{12.40\ cm^3} = 2.212\ g/cm^3$

0.6 $\text{volume} = 9.37\ g \times \dfrac{1\ mL}{1.483\ g} = 6.32\ mL$

0.7   (a) $540\ Cal \times \dfrac{1000\ cal}{1\ Cal} \times \dfrac{4.184\ J}{1\ cal} \times \dfrac{1\ kJ}{1000\ J} = 2259\ kJ = 2300\ kJ$

       (b) $100\ \text{watts} = 100\ J/s$

           $\text{time} = 2259\ kJ \times \dfrac{1000\ J}{1\ kJ} \times \dfrac{1\ s}{100\ J} \times \dfrac{1\ min}{60\ s} \times \dfrac{1\ h}{60\ min} = 6.275\ h = 6.3\ h$

0.8   The actual mass of the bottle and the acetone = 38.0015 g + 0.7791 g = 38.7806 g. The measured values are 38.7798 g, 38.7795 g, and 38.7801 g. These values are both close to each other and close to the actual mass. Therefore the results are both precise and accurate.

0.9   (a) 76.600 kJ has 5 significant figures because zeros at the end of a number and after the decimal point are always significant.

(b) 4.502 00 x 10^3 g has 6 significant figures because zeros in the middle of a number are significant and zeros at the end of a number and after the decimal point are always significant.

(c) 3000 nm has 1, 2, 3, or 4 significant figures because zeros at the end of a number and before the decimal point may or may not be significant.

(d) 0.003 00 mL has 3 significant figures because zeros at the beginning of a number are not significant and zeros at the end of a number and after the decimal point are always significant.

(e) 18 students has an infinite number of significant figures because this is an exact number.

(f) 3 x 10^{-5} g has 1 significant figure.

(g) 47.60 mL has 4 significant figures because a zero at the end of a number and after the decimal point is always significant.

(h) 2070 mi has 3 or 4 significant figures because a zero in the middle of a number is significant and a zero at the end of a number and before the decimal point may or may not be significant.

0.10   (a) Darts are clustered together (good precision) but are away from the bullseye (poor accuracy).

(b) Darts are clustered together (good precision) and hit the bullseye (good accuracy).

(c) Darts are scattered (poor precision) and are away from the bullseye (poor accuracy).

0.11   (a) Because the digit to be dropped (the second 4) is less than 5, round down. The result is 3.774 L.

(b) Because the digit to be dropped (0) is less than 5, round down. The result is 255 K.

(c) Because the digit to be dropped is equal to 5 with nothing following, round down. The result is 55.26 kg.

(d) Because the digit to be dropped (1) is less than 5, round down. The first zero is significant because it is in the middle of the number. The second zero is significant because a zero at the end of a number and after the decimal point is always significant. The result is 906.40 kJ.

0.12   (a)

	24.567	g	This result should be expressed with 3 decimal places.
+	0.044 78	g	Because the digit to be dropped (7) is greater than 5, round
	24.611 78	g	up. The result is 24.612 g (5 significant figures).

(b)   4.6742 g / 0.003 71 L = 1259.89 g/L

0.003 71 has only 3 significant figures so the result of the division should have only 3 significant figures. Because the digit to be dropped (first 9) is greater than 5, round up. The result is 1260 g/L (3 significant figures), or 1.26 x 10^3 g/L.

(c)

	0.378	mL	This result should be expressed with 1 decimal place.
+	42.3	mL	Because the digit to be dropped (9) is greater than 5, round
−	1.5833	mL	up. The result is 41.1 mL (3 significant figures).
	41.0947	mL	

0.13   The level of the liquid in the thermometer is just past halfway between the 32 °C and 33 °C marks on the thermometer. The temperature is 32.6°C (3 significant figures).

0.14   (a) Calculation: $°F = (\frac{9}{5} \times °C) + 32 = (\frac{9}{5} \times 1064) + 32 = 1947\ °F$

Ballpark estimate: $°F \approx 2 \times °C$ if °C is large. The melting point of gold $\approx$ 2000 °F.
(b) $r = d/2 = 3 \times 10^{-6}$ m $= 3 \times 10^{-4}$ cm; h $= 2 \times 10^{-6}$ m $= 2 \times 10^{-4}$ cm
Calculation: volume $= \pi r^2 h = (3.1416)(3 \times 10^{-4}\ cm)^2(2 \times 10^{-4}\ cm) = 6 \times 10^{-11}\ cm^3$
Ballpark estimate: volume $= \pi r^2 h \approx 3r^2 h \approx 3(3 \times 10^{-4}\ cm)^2(2 \times 10^{-4}\ cm) \approx 5 \times 10^{-11}\ cm^3$

0.15   1 carat $= 200$ mg $= 200 \times 10^{-3}$ g $= 0.200$ g

Mass of Hope Diamond in grams $= 44.4$ carats $\times\ \dfrac{0.200\ g}{1\ carat} = 8.88$ g

1 ounce $= 28.35$ g

Mass of Hope Diamond in ounces $= 8.88$ g $\times\ \dfrac{1\ ounce}{28.35\ g} = 0.313$ ounces

0.16   Volume of Hope Diamond $= 8.88$ g $\times\ \dfrac{1\ cm^3}{3.52\ g} = 2.52\ cm^3$

C atoms in Hope Diamond $= 8.88$ g $\times\ \dfrac{5.014 \times 10^{21}\ C\ atoms}{0.1000\ g} = 4.45 \times 10^{23}$ C atoms

0.17   600 mg $= 600 \times 10^{-3}$ g

mass of person in kg $= 175$ lb $\times\ \dfrac{1.00\ kg}{2.2046\ lb} = 79.4$ kg

LD_{50} for 79.4 kg person $= 79.4$ kg $\times\ \dfrac{600 \times 10^{-3}\ g}{kg} = 47.6$ g

47.6 g oxalic acid $= 0.5\%$ x (mass of dry rhubarb leaves)
47.6 g oxalic acid $= 0.005$ x (mass of dry rhubarb leaves)

mass of dry rhubarb leaves $= \dfrac{47.6\ g}{0.005} \times \dfrac{1\ kg}{1000\ g} \times \dfrac{2.2046\ lb}{1\ kg} = 21.0$ lb

Undried rhubarb leaves are 95% water and 5% dry rhubarb leaves
21.0 lb $= 5\%$ x (mass of undried rhubarb leaves)
21.0 lb $= 0.05$ x (mass of undried rhubarb leaves)

mass of undried rhubarb leaves $= \dfrac{21.0\ lb}{0.05} = 420$ lb

Conceptual Problems

0.18   For balance (a), the mass of the red block is greater than the mass of the green block. The volume of the red block is less than the volume of the green block.

Because density $=\ \dfrac{mass}{volume}$, the red block is more dense.

For balance (b), the mass of the green block is greater than the mass of the red block. The volume of both blocks is the same. Because $density = \dfrac{mass}{volume}$, the green block is more dense.

0.20

(a)

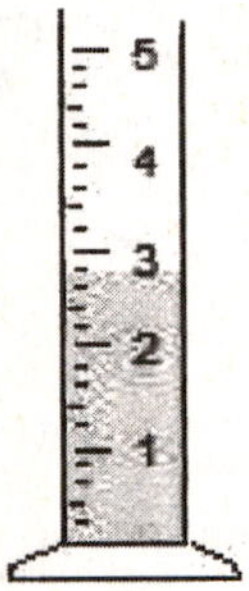

(b)

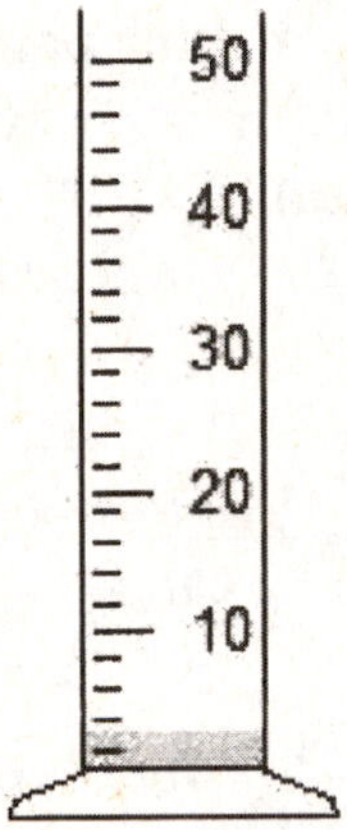

The 5 mL graduated cylinder is marked every 0.2 mL and can be read to $\pm\,0.02$ mL. The 50 mL graduated cylinder is marked every 2 mL and can only be read to $\pm\,0.2$ mL. The 5 mL graduated cylinder will give more accurate measurements.

Section Problems
Units, Significant Figures, and Rounding (Sections 0.2, 0.9, 0.10)

0.22   Mass measures the amount of matter in an object, whereas weight measures the pull of gravity on an object by the earth or other celestial body.

0.24   (a) kilogram, kg   (b) meter, m   (c) kelvin, K   (d) cubic meter, m^3
(e) joule, $(kg \cdot m^2)\,/\,s^2$ (f) kg/m^3 or g/cm^3

0.26   A Celsius degree is larger than a Fahrenheit degree by a factor of $\dfrac{9}{5}$.

0.28   The volume of a cubic decimeter (dm^3) and a liter (L) are the same.

0.30   (a) and (b) are exact numbers because they are both definitions.
(c) and (d) are not exact numbers because they result from measurements.

0.32   cL is centiliter $(10^{-2}\,L)$

0.34   (a) Convert cm to km and compare the two quantities.
$5.63 \times 10^6 \text{ cm} \times \dfrac{1 \text{ m}}{100 \text{ cm}} \times \dfrac{1 \text{ km}}{1,000 \text{ m}} = 5.63 \times 10^1 \text{ km}$
6.02×10^1 km is larger.
(b) Convert µs to ms and compare the two quantities.
$46 \text{ µs} \times \dfrac{1 \text{ s}}{1 \times 10^6 \text{ µs}} \times \dfrac{1000 \text{ ms}}{1 \text{ s}} = 4.6 \times 10^{-2} \text{ ms}$
46 µs is larger.

(c) Convert g to kg and compare the two quantities.

$$200,098 \text{ g} \times \frac{1 \text{ kg}}{1000 \text{ g}} = 20.0098 \times 10^1 \text{ kg}$$

200,098 g is larger.

0.36 $1 \text{ mg} = 1 \times 10^{-3} \text{ g}$ and $1 \text{ pg} = 1 \times 10^{-12} \text{ g}$

$$\frac{1 \times 10^{-3} \text{ g}}{1 \text{ mg}} \times \frac{1 \text{ pg}}{1 \times 10^{-12} \text{ g}} = 1 \times 10^9 \text{ pg/mg}$$

$35 \text{ ng} = 35 \times 10^{-9} \text{ g}$ $\dfrac{35 \times 10^{-9} \text{ g}}{35 \text{ ng}} \times \dfrac{1 \text{ pg}}{1 \times 10^{-12} \text{ g}} = 3.5 \times 10^4 \text{ pg/35 ng}$

0.38   (a) $5 \text{ pm} = 5 \times 10^{-12} \text{ m}$

$$5 \times 10^{-12} \text{ m} \times \frac{100 \text{ cm}}{1 \text{ m}} = 5 \times 10^{-10} \text{ cm}$$

$$5 \times 10^{-12} \text{ m} \times \frac{1 \text{ nm}}{1 \times 10^{-9} \text{ m}} = 5 \times 10^{-3} \text{ nm}$$

(b) $8.5 \text{ cm}^3 \times \left(\dfrac{1 \text{ m}}{100 \text{ cm}}\right)^3 = 8.5 \times 10^{-6} \text{ m}^3$

$8.5 \text{ cm}^3 \times \left(\dfrac{10 \text{ mm}}{1 \text{ cm}}\right)^3 = 8.5 \times 10^3 \text{ mm}^3$

(c) $65.2 \text{ mg} \times \dfrac{1 \times 10^{-3} \text{ g}}{1 \text{ mg}} = 0.0652 \text{ g}$

$65.2 \text{ mg} \times \dfrac{1 \times 10^{-3} \text{ g}}{1 \text{ mg}} \times \dfrac{1 \text{ pg}}{1 \times 10^{-12} \text{ g}} = 6.52 \times 10^{10} \text{ pg}$

0.40   (a) 35.0445 g has 6 significant figures because zeros in the middle of a number are significant.
(b) 59.0001 cm has 6 significant figures because zeros in the middle of a number are significant.
(c) 0.030 03 kg has 4 significant figures because zeros at the beginning of a number are not significant and zeros in the middle of a number are significant.
(d) 0.004 50 m has 3 significant figures because zeros at the beginning of a number are not significant and zeros at the end of a number and after the decimal point are always significant.
(e) 67,000 m² has 2, 3, 4, or 5 significant figures because zeros at the end of a number and before the decimal point may or may not be significant.
(f) 3.8200×10^3 L has 5 significant figures because zeros at the end of a number and after the decimal point are always significant.

0.42   To convert 3,666,500 m³ to scientific notation, move the decimal point 6 places to the left and include an exponent of 10^6. The result is 3.6665×10^6 m³.

0.44   (a) To convert 453.32 mg to scientific notation, move the decimal point 2 places to the left and include an exponent of 10^2. The result is 4.5332×10^2 mg.
(b) To convert 0.000 042 1 mL to scientific notation, move the decimal point 5 places to the right and include an exponent of 10^{-5}. The result is 4.21×10^{-5} mL.
(c) To convert 667,000 g to scientific notation, move the decimal point 5 places to the left and include an exponent of 10^5. The result is 6.67×10^5 g.

0.46   (a) Because the digit to be dropped (0) is less than 5, round down. The result is 3.567×10^4 or 35,670 m (4 significant figures).
Because the digit to be dropped (the second 6) is greater than 5, round up. The result is 35,670.1 m (6 significant figures).
(b) Because the digit to be dropped is 5, round up. The result is 69 g (2 significant figures).
Because the digit to be dropped (0) is less than 5, round down. The result is 68.5 g (3 significant figures).
(c) Because the digit to be dropped is 5, round up. The result is 5.00×10^3 cm (3 significant figures).
(d) Because the digit to be dropped is 5, round up. The result is 2.3099×10^{-4} kg (5 significant figures).

0.48   (a) $4.884 \times 2.05 = 10.012$
The result should contain only 3 significant figures because 2.05 contains 3 significant figures (the smaller number of significant figures of the two). Because the digit to be dropped (1) is less than 5, round down. The result is 10.0.
(b) $94.61 / 3.7 = 25.57$
The result should contain only 2 significant figures because 3.7 contains 2 significant figures (the smaller number of significant figures of the two). Because the digit to be dropped (second 5) is 5 with nonzero digits following, round up. The result is 26.
(c) $3.7 / 94.61 = 0.0391$
The result should contain only 2 significant figures because 3.7 contains 2 significant figures (the smaller number of significant figures of the two). Because the digit to be dropped (1) is less than 5, round down. The result is 0.039.

(d)
$$
\begin{array}{r}
5502.3 \\
24 \\
+\quad 0.01 \\
\hline
5526.31
\end{array}
$$
This result should be expressed with no decimal places. Because the digit to be dropped (3) is less than 5, round down. The result is 5526.

(e)
$$
\begin{array}{r}
86.3 \\
+\ 1.42 \\
-\ 0.09 \\
\hline
87.63
\end{array}
$$
This result should be expressed with only 1 decimal place. Because the digit to be dropped (3) is less than 5, round down. The result is 87.6.

(f) $5.7 \times 2.31 = 13.167$
The result should contain only 2 significant figures because 5.7 contains 2 significant figures (the smaller number of significant figures of the two). Because the digit to be dropped (second 1) is less than 5, round down. The result is 13.

0.50 1 mile = 1.6093 km; the time is 1 h, 5 min, and 26.6 s.

Convert the time to seconds and then hours.

$$\text{time} = \left(1 \text{ hr} \times \frac{60 \text{ min}}{1 \text{ h}} \times \frac{60 \text{ s}}{1 \text{ min}} \right) + \left(5 \text{ min} \times \frac{60 \text{ s}}{1 \text{ min}} \right) + 26.6 \text{ s} = 3926.6 \text{ s}$$

$$\text{time} = 3926.6 \text{ s} \times \frac{1 \text{ min}}{60 \text{ s}} \times \frac{1 \text{ h}}{60 \text{ min}} = 1.0907 \text{ h}$$

Convert meters to miles.

$$20{,}000 \text{ m} \times \frac{1 \text{ km}}{1000 \text{ m}} \times \frac{1 \text{ mi}}{1.6093 \text{ km}} = 12.4278 \text{ mi}$$

$$\text{average speed} = \frac{12.4278 \text{ mi}}{1.0907 \text{ h}} = 11.394 \text{ mi/h}$$

Temperature (Section 0.5)

0.52 $^\circ\text{F} = (\frac{9}{5} \times {^\circ\text{C}}) + 32$

$^\circ\text{F} = (\frac{9}{5} \times 39.9 {^\circ\text{C}}) + 32 = 103.8 \, ^\circ\text{F}$ \qquad (goat)

$^\circ\text{F} = (\frac{9}{5} \times 22.2 {^\circ\text{C}}) + 32 = 72.0 \, ^\circ\text{F}$ \qquad (Australian spiny anteater)

0.54 $^\circ\text{C} = \frac{5}{9} \times (^\circ\text{F} - 32) = \frac{5}{9} \times (6192 - 32) = 3422 \, ^\circ\text{C}$

$K = {^\circ\text{C}} + 273.15 = 3422 + 273.15 = 3695.15 \text{ K or } 3695 \text{ K}$

0.56

Ethanol boiling point	78.5 °C	173.3 °F	200 °E
Ethanol melting point	−117.3 °C	−179.1 °F	0 °E

(a) $\dfrac{200 \, ^\circ\text{E}}{[78.5 \, ^\circ\text{C} - (-117.3 \, ^\circ\text{C})]} = \dfrac{200 \, ^\circ\text{E}}{195.8 \, ^\circ\text{C}} = 1.021 \, ^\circ\text{E/}^\circ\text{C}$

(b) $\dfrac{200 \, ^\circ\text{E}}{[173.3 \, ^\circ\text{F} - (-179.1 \, ^\circ\text{F})]} = \dfrac{200 \, ^\circ\text{E}}{352.4 \, ^\circ\text{F}} = 0.5675 \, ^\circ\text{E/}^\circ\text{F}$

(c) $^\circ\text{E} = \dfrac{200}{195.8} \times (^\circ\text{C} + 117.3)$

$\text{H}_2\text{O melting point} = 0 ^\circ\text{C}; \quad ^\circ\text{E} = \dfrac{200}{195.8} \times (0 + 117.3) = 119.8 \, ^\circ\text{E}$

$\text{H}_2\text{O boiling point} = 100 ^\circ\text{C}; \quad ^\circ\text{E} = \dfrac{200}{195.8} \times (100 + 117.3) = 222.0 \, ^\circ\text{E}$

(d) $^\circ\text{E} = \dfrac{200}{352.4} \times (^\circ\text{F} + 179.1) = \dfrac{200}{352.4} \times (98.6 + 179.1) = 157.6 \, ^\circ\text{E}$

(e) $^\circ\text{F} = \left(^\circ\text{E} \times \dfrac{352.4}{200} \right) - 179.1 = \left(130 \times \dfrac{352.4}{200} \right) - 179.1 = 50.0 \, ^\circ\text{F}$

Because the outside temperature is 50.0°F, I would wear a sweater or light jacket.

Chapter 0 – Chemical Tools: Experimentation and Measurement

Density (Section 0.7)

0.58   mass = 10.5 g/cm³ x $\dfrac{1 \text{ kg}}{1000 \text{ g}}$ x $\left(\dfrac{100 \text{ cm}}{1 \text{ m}}\right)^3$ x $(0.62 \text{ m})^3$ = 2500 kg

0.60   250 mg x $\dfrac{1 \times 10^{-3} \text{ g}}{1 \text{ mg}}$ = 0.25 g;   V = 0.25 g x $\dfrac{1 \text{ cm}^3}{1.40 \text{ g}}$ = 0.18 cm³

       500 lb x $\dfrac{453.59 \text{ g}}{1 \text{ lb}}$ = 226,795 g;   V = 226,795 g x $\dfrac{1 \text{ cm}^3}{1.40 \text{ g}}$ = 161,996 cm³ = 162,000 cm³

0.62   d = $\dfrac{m}{V}$ = $\dfrac{220.9 \text{ g}}{(0.50 \times 1.55 \times 25.00) \text{ cm}^3}$ = 11.4 $\dfrac{\text{g}}{\text{cm}^3}$ = 11 $\dfrac{\text{g}}{\text{cm}^3}$

Energy (Section 0.8)

0.64   1 oz = 28.35 g

       energy = 0.450 oz x $\dfrac{28.35 \text{ g}}{1 \text{ oz}}$ x $\dfrac{2498 \text{ kJ}}{45.0 \text{ g}}$ x $\dfrac{1 \text{ kcal}}{4.184 \text{ kJ}}$ = 169 kcal

Unit Conversions (Section 0.11)

0.66   (a) 0.25 lb x $\dfrac{453.59 \text{ g}}{1 \text{ lb}}$ = 113.4 g = 110 g

       (b) 1454 ft x $\dfrac{12 \text{ in.}}{1 \text{ ft}}$ x $\dfrac{2.54 \text{ cm}}{1 \text{ in.}}$ x $\dfrac{1 \text{ m}}{100 \text{ cm}}$ = 443.2 m

       (c) 2,941,526 mi² x $\left(\dfrac{1.6093 \text{ km}}{1 \text{ mi}}\right)^2$ x $\left(\dfrac{1000 \text{ m}}{1 \text{ km}}\right)^2$ = 7.6181 x 10¹² m²

0.68   (a) 1 acre-ft x $\dfrac{1 \text{ mi}^2}{640 \text{ acres}}$ x $\left(\dfrac{5280 \text{ ft}}{1 \text{ mi}}\right)^2$ = 43,560 ft³

       (b) 116 mi³ x $\left(\dfrac{5280 \text{ ft}}{1 \text{ mi}}\right)^3$ x $\dfrac{1 \text{ acre-ft}}{43,560 \text{ ft}^3}$ = 3.92 x 10⁸ acre-ft

0.70   (a) $\dfrac{200 \text{ mg}}{100 \text{ mL}}$ x $\dfrac{1000 \text{ mL}}{1 \text{ L}}$ = 2000 mg/L

       (b) $\dfrac{200 \text{ mg}}{100 \text{ mL}}$ x $\dfrac{1 \times 10^{-3} \text{ g}}{1 \text{ mg}}$ x $\dfrac{1 \text{ μg}}{1 \times 10^{-6} \text{ g}}$ = 2000 μg/mL

       (c) $\dfrac{200 \text{ mg}}{100 \text{ mL}}$ x $\dfrac{1 \times 10^{-3} \text{ g}}{1 \text{ mg}}$ x $\dfrac{1000 \text{ mL}}{1 \text{ L}}$ = 2 g/L

(d) $\dfrac{200\ \text{mg}}{100\ \text{mL}} \times \dfrac{1 \times 10^{-3}\ \text{g}}{1\ \text{mg}} \times \dfrac{1000\ \text{mL}}{1\ \text{L}} \times \dfrac{1\ \text{ng}}{1 \times 10^{-9}\ \text{g}} \times \dfrac{1 \times 10^{-6}\ \text{L}}{1\ \mu\text{L}} = 2000\ \text{ng}/\mu\text{L}$

(e) $2\ \text{g/L} \times 5\ \text{L} = 10\ \text{g}$

0.72 $\quad 55\ \dfrac{\text{mi}}{\text{h}} \times \dfrac{5280\ \text{ft}}{1\ \text{mi}} \times \dfrac{12\ \text{in.}}{1\ \text{ft}} \times \dfrac{2.54\ \text{cm}}{1\ \text{in.}} \times \dfrac{1\ \text{h}}{3600\ \text{s}} \times \dfrac{2.5 \times 10^{-4}\ \text{s}}{1\ \text{shake}} = 0.61\ \dfrac{\text{cm}}{\text{shake}}$

Chapter Problems

0.74 $\quad d = \dfrac{m}{V} = \dfrac{8.763\ \text{g}}{(28.76 - 25.00)\ \text{mL}} = \dfrac{8.763\ \text{g}}{3.76\ \text{mL}} = 2.331\ \dfrac{\text{g}}{\text{cm}^3} = 2.33\ \dfrac{\text{g}}{\text{cm}^3}$

0.76 $\quad$ NaCl melting point = 1074 K

$\quad$ °C = K − 273.15 = 1074 − 273.15 = 800.85 °C = 801 °C

$\quad$ °F = $(\dfrac{9}{5} \times$ °C$) + 32 = (\dfrac{9}{5} \times 800.85) + 32 = 1473.53$ °F = 1474 °F

$\quad$ NaCl boiling point = 1686 K

$\quad$ °C = K − 273.15 = 1686 − 273.15 = 1412.85 °C = 1413 °C

$\quad$ °F = $(\dfrac{9}{5} \times$ °C$) + 32 = (\dfrac{9}{5} \times 1412.85) + 32 = 2575.13$ °F = 2575 °F

0.78 $\quad V = 112.5\ \text{g} \times \dfrac{1\ \text{mL}}{1.4832\ \text{g}} = 75.85\ \text{mL}$

0.80 $\quad$ (a) density = $\dfrac{1\ \text{lb}}{1\ \text{pint}} \times \dfrac{8\ \text{pints}}{1\ \text{gal}} \times \dfrac{1\ \text{gal}}{3.7854\ \text{L}} \times \dfrac{453.59\ \text{g}}{1\ \text{lb}} \times \dfrac{1\ \text{L}}{1000\ \text{mL}} = 0.958\ 61\ \text{g/mL}$

$\quad$ (b) area in m^2 =

$\quad 1\ \text{acre} \times \dfrac{1\ \text{mi}^2}{640\ \text{acres}} \times \left(\dfrac{5280\ \text{ft}}{1\ \text{mi}}\right)^2 \times \left(\dfrac{12\ \text{in.}}{1\ \text{ft}}\right)^2 \times \left(\dfrac{2.54\ \text{cm}}{1\ \text{in.}}\right)^2 \times \left(\dfrac{1\ \text{m}}{100\ \text{cm}}\right)^2 = 4047\ \text{m}^2$

$\quad$ (c) mass of wood =

$\quad 1\ \text{cord} \times \dfrac{128\ \text{ft}^3}{1\ \text{cord}} \times \left(\dfrac{12\ \text{in.}}{1\ \text{ft}}\right)^3 \times \left(\dfrac{2.54\ \text{cm}}{1\ \text{in.}}\right)^3 \times \dfrac{0.40\ \text{g}}{1\ \text{cm}^3} \times \dfrac{1\ \text{kg}}{1000\ \text{g}} = 1450\ \text{kg} = 1400\ \text{kg}$

$\quad$ (d) mass of oil =

$\quad 1\ \text{barrel} \times \dfrac{42\ \text{gal}}{1\ \text{barrel}} \times \dfrac{3.7854\ \text{L}}{1\ \text{gal}} \times \dfrac{1000\ \text{mL}}{1\ \text{L}} \times \dfrac{0.85\ \text{g}}{1\ \text{mL}} \times \dfrac{1\ \text{kg}}{1000\ \text{g}} = 135.1\ \text{kg} = 140\ \text{kg}$

$\quad$ (e) fat Calories =

$\quad 0.5\ \text{gal} \times \dfrac{32\ \text{servings}}{1\ \text{gal}} \times \dfrac{165\ \text{Calories}}{1\ \text{serving}} \times \dfrac{30.0\ \text{Cal from fat}}{100\ \text{Cal total}} = 792\ \text{Cal from fat}$

0.82 $\quad$ (a) number of Hershey's Kisses =

$\quad 2.0\ \text{lb} \times \dfrac{453.59\ \text{g}}{1\ \text{lb}} \times \dfrac{1\ \text{serving}}{41\ \text{g}} \times \dfrac{9\ \text{kisses}}{1\ \text{serving}} = 199\ \text{kisses} = 200\ \text{kisses}$

(b) Hershey's Kiss volume = $\dfrac{41 \text{ g}}{1 \text{ serving}}$ x $\dfrac{1 \text{ serving}}{9 \text{ kisses}}$ x $\dfrac{1 \text{ mL}}{1.4 \text{ g}}$ = 3.254 mL = 3.3 mL

(c) Calories/Hershey's Kiss = $\dfrac{230 \text{ Cal}}{1 \text{ serving}}$ x $\dfrac{1 \text{ serving}}{9 \text{ kisses}}$ = 25.55 Cal/kiss = 26 Cal/kiss

(d) % fat Calories =

$\dfrac{13 \text{ g fat}}{1 \text{ serving}}$ x $\dfrac{9 \text{ Cal from fat}}{1 \text{ g fat}}$ x $\dfrac{1 \text{ serving}}{230 \text{ Cal total}}$ x 100% = 51% Calories from fat

0.84 $^\circ$C = $\dfrac{5}{9}$ x ($^\circ$F – 32);   Set $^\circ$C = $^\circ$F: $^\circ$C = $\dfrac{5}{9}$ x ($^\circ$C – 32)

Solve for $^\circ$C: $^\circ$C x $\dfrac{9}{5}$ = $^\circ$C – 32

($^\circ$C x $\dfrac{9}{5}$) – $^\circ$C = –32

$^\circ$C x $\dfrac{4}{5}$ = –32

$^\circ$C = $\dfrac{5}{4}$(–32) = –40 $^\circ$C

The Celsius and Fahrenheit scales "cross" at –40 $^\circ$C (–40 $^\circ$F).

0.86   1 lb = 453.59 g

rat mass = 0.75 lb x $\dfrac{453.59 \text{ g}}{1 \text{ lb}}$ x $\dfrac{1 \text{ kg}}{1000 \text{ g}}$ = 0.34 kg

LD_{50} = 1.1 g/kg
mass of aspirin required = 0.34 kg x 1.1 g/kg = 0.37 g

aspirin tablets = 0.37 g x $\dfrac{1000 \text{ mg}}{1 \text{ g}}$ x $\dfrac{1 \text{ tablet}}{81 \text{ mg}}$ = 4.6 aspirin tablets = 5 aspirin tablets

0.88   Convert 8 min, 25 s to s.   8 min x $\dfrac{60 \text{ s}}{1 \text{ min}}$ + 25 s = 505 s

Convert 293.2 K to $^\circ$F:

293.2 – 273.15 = 20.05 $^\circ$C and $^\circ$F = ($\dfrac{9}{5}$ x 20.05) + 32 = 68.09 $^\circ$F

Final temperature = 68.09 $^\circ$F + 505 s x $\dfrac{3.0 \ ^\circ\text{F}}{60 \text{ s}}$ = 93.34 $^\circ$F

$^\circ$C = $\dfrac{5}{9}$ x (93.34 – 32) = 34.1 $^\circ$C

0.90   Average brass density = (0.670)(8.92 g/cm³) + (0.330)(7.14 g/cm³) = 8.333 g/cm³

length = 1.62 in. x $\dfrac{2.54 \text{ cm}}{1 \text{ in.}}$ = 4.115 cm diameter = 0.514 in. x $\dfrac{2.54\ cm}{1\ in.}$ = 1.306 cm

volume = $\pi r^2 h$ = (3.1416)[(1.306 cm)/2]^2(4.115 cm) = 5.512 cm^3

mass = 5.512 cm^3 x $\dfrac{8.333\ g}{1\ cm^3}$ = 45.9 g

0.92   (a)   Ga density = $\dfrac{0.2133\ lb}{1\ in.^3}$ x $\dfrac{453.59\ g}{1\ lb}$ x $\dfrac{1\ in.^3}{(2.54\ cm)^3}$ = 5.904 g/cm^3

     (b)   Ga boiling point   2204 °C   1000 °G
          Ga melting point   29.78 °C   0 °G

$\dfrac{1000\ °G - 0\ °G}{2204\ °C - 29.78\ °C}$ = $\dfrac{1000\ °G}{2174.22\ °C}$ = 0.4599 °G/°C

°G = 0.4599 x (°C − 29.78)
°G = 0.4599 x (801 − 29.78) = 355 °G
The melting point of sodium chloride (NaCl) on the gallium scale is 355 °G.

1 The Structure and Stability of Atoms

1.1   (a) Cd   (b) Sb   (c) Am

1.2   (a) silver   (b) rhodium   (c) rhenium   (d) cesium   (e) argon   (f) arsenic

1.3   (a) Ti, metal   (b) Te, semimetal   (c) Se, nonmetal
      (d) Sc, metal   (e) At, semimetal   (f) Ar, nonmetal

1.4   The three "coinage metals" are copper (Cu), silver (Ag), and gold (Au).

1.5   First, find the S:O ratio in each compound.
Substance A: S:O mass ratio = (6.00 g S) / (5.99 g O) = 1.00
Substance B: S:O mass ratio = (8.60 g S) / (12.88 g O) = 0.668

$$\frac{\text{S:O mass ratio in substance A}}{\text{S:O mass ratio in substance B}} = \frac{1.00}{0.668} = 1.50 = \frac{3}{2}$$

1.6 $0.005 \text{ mm} \times \dfrac{1 \text{ cm}}{10 \text{ mm}} \times \dfrac{1 \text{ Au atom}}{2.9 \times 10^{-8} \text{ cm}} = 2 \times 10^{4} \text{ Au atoms}$

1.7 $1 \times 10^{19} \text{ C atoms} \times \dfrac{1.5 \times 10^{-10} \text{ m}}{\text{C atom}} \times \dfrac{1 \text{ km}}{1000 \text{ m}} \times \dfrac{1 \text{ time}}{40{,}075 \text{ km}} = 37.4 \text{ times} \approx 40 \text{ times}$

1.8 $^{75}_{34}\text{Se}$ has 34 protons, 34 electrons, and (75 − 34) = 41 neutrons.

1.9 $^{35}_{17}\text{Cl}$ has (35 − 17) = 18 neutrons. $^{37}_{17}\text{Cl}$ has (37 − 17) = 20 neutrons.

1.10   The element with 47 protons is Ag. The mass number is the sum of the protons and the neutrons, 47 + 62 = 109. The isotope symbol is $^{109}_{47}\text{Ag}$.

1.11   atomic mass = (0.6915 x 62.93) + (0.3085 x 64.93) = 63.55

1.12 $2.15 \text{ g} \times \dfrac{1 \text{ amu}}{1.6605 \times 10^{-24} \text{ g}} \times \dfrac{1 \text{ Cu}}{63.55 \text{ amu}} = 2.04 \times 10^{22} \text{ Cu atoms}$

1.13   (a)   g Ti = 1.505 mol Ti x $\dfrac{47.867 \text{ g Ti}}{1 \text{ mol Ti}}$ = 72.04 g Ti

(b)   g Na = 0.337 mol Na x $\dfrac{22.989\ 770 \text{ g Na}}{1 \text{ mol Na}}$ = 7.75 g Na

(c)   g U = 2.583 mol U x $\dfrac{238.028\ 91 \text{ g U}}{1 \text{ mol U}}$ = 614.8 g U

1.14   (a)   mol Ti = 11.51 g Ti x $\dfrac{1 \text{ mol Ti}}{47.867 \text{ g Ti}}$ = 0.2405 mol Ti

(b)   mol Na = 29.127 g Na x $\dfrac{1 \text{ mol Na}}{22.989\ 770 \text{ g Na}}$ = 1.2670 mol Na

(c)   mol U = 1.477 kg x $\dfrac{1000 \text{ g}}{1 \text{ kg}}$ x $\dfrac{1 \text{ mol U}}{238.028\ 91 \text{ g U}}$ = 6.205 mol U

1.15   (a)   In beta emission, the mass number is unchanged, and the atomic number increases by one. $^{106}_{44}\text{Ru} \rightarrow\ ^{0}_{-1}\text{e} +\ ^{106}_{45}\text{Rh}$

(b)   In alpha emission, the mass number decreases by four, and the atomic number decreases by two. $^{189}_{83}\text{Bi} \rightarrow\ ^{4}_{2}\text{He} +\ ^{185}_{81}\text{Tl}$

(c)   In electron capture, the mass number is unchanged, and the atomic number decreases by one. $^{204}_{84}\text{Po} +\ ^{0}_{-1}\text{e} \rightarrow\ ^{204}_{83}\text{Bi}$

1.16   The mass number decreases by four, and the atomic number decreases by two. This is characteristic of alpha emission. $^{214}_{90}\text{Th} \rightarrow\ ^{210}_{88}\text{Ra} +\ ^{4}_{2}\text{He}$

1.17 $^{148}_{69}\text{Tm}$ decays to $^{148}_{68}\text{Er}$ by either positron emission or electron capture.

1.18   H and He

1.19   (a) $^{1}_{1}\text{H} +\ ^{1}_{1}\text{H} \rightarrow\ ^{2}_{1}\text{H} +\ ^{0}_{1}\text{n}$

(b) $^{1}_{1}\text{H} +\ ^{2}_{1}\text{H} \rightarrow\ ^{3}_{2}\text{He}$

Conceptual Problems

1.20

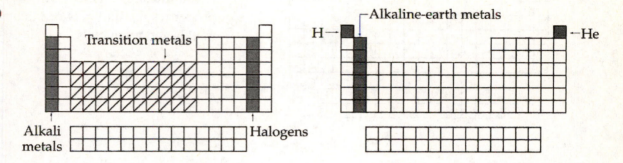

1.22 red – gas; blue – 42;
green – lithium, sodium, potassium or rubidium are possible answers

1.24 Drawing (a) represents a collection of SO_2 units. Drawing (d) represents a mixture of S atoms and O_2 units.

1.26 Figures (b) and (d) illustrate the law of multiple proportions. The ⬭⬭●/⬭● mass ratio is 2.

1.28 Figures (b) and (c) both contain two protons but different numbers of neutrons. They are isotopes of the same element. Figure (a) contains only one proton. It is a different element than (b) and (c).

Section Problems
Elements and the Periodic Table (Sections 1.1–1.3)

1.30 118 elements are presently known. About 90 elements occur naturally.

1.32 There are 18 groups in the periodic table. They are labeled as follows:
1A, 2A, 3B, 4B, 5B, 6B, 7B, 8B (3 groups), 1B, 2B, 3A, 4A, 5A, 6A, 7A, 8A

1.34

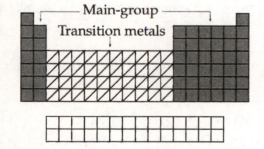

1.36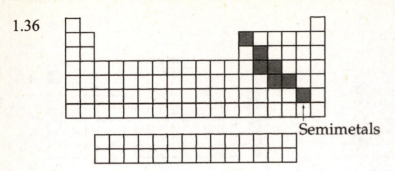

Semimetals

A semimetal is an element with properties that fall between those of metals and nonmetals.

1.38 Li, Na, K, Rb, and Cs

1.40 F, Cl, Br, and I

1.42 An element that is a soft, silver-colored solid that reacts violently with water and is a good conductor of electricity has the characteristics of a metal.

1.44 All match in groups 2A and 7A.

1.46 (a) gadolinium, Gd (b) germanium, Ge (c) technetium, Tc (d) arsenic, As

1.48 (a) Te, tellurium (b) Re, rhenium (c) Be, beryllium (d) Ar, argon
(e) Pu, plutonium

1.50 (a) Tin is Sn, Ti is titanium. (b) Manganese is Mn, Mg is magnesium.
(c) Potassium is K, Po is polonium.
(d) The symbol for helium is He. The second letter is lowercase.

Atomic Theory (Sections 1.4 and 1.5)

1.52 The law of mass conservation in terms of Dalton's atomic theory states that chemical reactions only rearrange the way that atoms are combined; the atoms themselves are not changed.
The law of definite proportions in terms of Dalton's atomic theory states that the chemical combination of elements to make different substances occurs when atoms join together in small, whole-number ratios.

1.54 In any chemical reaction, the combined mass of the final products equals the combined mass of the starting reactants.
mass of reactants = mass of products
mass of reactants = mass of Hg + mass of O_2 = 114.0 g + 12.8 g = 126.8 g
mass of products = mass of HgO + mass of left over O_2 = 123.1 g + mass of left over O_2
126.8 g = 123.1 g + mass of left over O_2
mass of left over O_2 = 126.8 g - 123.1 g = 3.7 g

1.56   First, find the C:H ratio in each compound.

Benzene: C:H mass ratio = (4.61 g C) / (0.39 g H) = 12

Ethane: C:H mass ratio (4.00 g C) / (1.00 g H) = 4.00

Ethylene: C:H mass ratio = (4.29 g C) / (0.71 g H) = 6.0

$$\frac{\text{C:H mass ratio in benzene}}{\text{C:H mass ratio in ethane}} = \frac{12}{4.00} = \frac{3}{1}$$

$$\frac{\text{C:H mass ratio in benzene}}{\text{C:H mass ratio in ethylene}} = \frac{12}{6.0} = \frac{2}{1}$$

$$\frac{\text{C:H mass ratio in ethylene}}{\text{C:H mass ratio in ethane}} = \frac{6.0}{4.00} = \frac{3}{2}$$

1.58   (a) For benzene:

$$4.61 \text{ g} \times \frac{1 \text{ amu}}{1.6605 \times 10^{-24} \text{ g}} \times \frac{1 \text{ C atom}}{12.011 \text{ amu}} = 2.31 \times 10^{23} \text{ C atoms}$$

$$0.39 \text{ g} \times \frac{1 \text{ amu}}{1.6605 \times 10^{-24} \text{ g}} \times \frac{1 \text{ H atom}}{1.008 \text{ amu}} = 2.3 \times 10^{23} \text{ H atoms}$$

$$\frac{\text{C}}{\text{H}} = \frac{2.31 \times 10^{23} \text{ C atoms}}{2.3 \times 10^{23} \text{ H atoms}} = \frac{1 \text{ C}}{1 \text{ H}} \qquad \text{A possible formula for benzene is CH.}$$

For ethane:

$$4.00 \text{ g} \times \frac{1 \text{ amu}}{1.6605 \times 10^{-24} \text{ g}} \times \frac{1 \text{ C atom}}{12.011 \text{ amu}} = 2.01 \times 10^{23} \text{ C atoms}$$

$$1.00 \text{ g} \times \frac{1 \text{ amu}}{1.6605 \times 10^{-24} \text{ g}} \times \frac{1 \text{ H atom}}{1.008 \text{ amu}} = 5.97 \times 10^{23} \text{ H atoms}$$

$$\frac{\text{C}}{\text{H}} = \frac{2.01 \times 10^{23} \text{ C atoms}}{5.97 \times 10^{23} \text{ H atoms}} = \frac{1 \text{ C}}{3 \text{ H}} \qquad \text{A possible formula for ethane is CH}_3.$$

For ethylene:

$$4.29 \text{ g} \times \frac{1 \text{ amu}}{1.6605 \times 10^{-24} \text{ g}} \times \frac{1 \text{ C atom}}{12.011 \text{ amu}} = 2.15 \times 10^{23} \text{ C atoms}$$

$$0.71 \text{ g} \times \frac{1 \text{ amu}}{1.6605 \times 10^{-24} \text{ g}} \times \frac{1 \text{ H atom}}{1.008 \text{ amu}} = 4.2 \times 10^{23} \text{ H atoms}$$

$$\frac{\text{C}}{\text{H}} = \frac{2.15 \times 10^{23} \text{ C atoms}}{4.2 \times 10^{23} \text{ H atoms}} = \frac{1 \text{ C}}{2 \text{ H}} \qquad \text{A possible formula for ethylene is CH}_2.$$

(b) The results in part (a) give the smallest whole-number ratio of C to H for benzene, ethane, and ethylene, and these ratios are consistent with their modern formulas.

1.60   The mass of 6.02×10^{23} atoms is its atomic mass expressed in grams. If the atomic mass of an element is X, then 6.02×10^{23} atoms of this element weighs X grams.

1.62 $$\text{mass} = \frac{x \text{ g}}{6.02 \times 10^{23} \text{ atoms}} \times 3.17 \times 10^{20} \text{ atoms} = (x) \times 5.27 \times 10^{-4} \text{ g}$$

1.64   Assume a 1.00 g sample of the binary compound of zinc and sulfur.
        0.671 x 1.00 g = 0.671 g Zn;   0.329 x 1.00 g = 0.329 g S

$$0.671 \text{ g x } \frac{1 \text{ amu}}{1.6605 \text{ x } 10^{-24} \text{ g}} \text{ x } \frac{1 \text{ Zn atom}}{65.39 \text{ amu}} = 6.18 \text{ x } 10^{21} \text{ Zn atoms}$$

$$0.329 \text{ g x } \frac{1 \text{ amu}}{1.6605 \text{ x } 10^{-24} \text{ g}} \text{ x } \frac{1 \text{ S atom}}{32.066 \text{ amu}} = 6.18 \text{ x } 10^{21} \text{ S atoms}$$

$$\frac{\text{Zn}}{\text{S}} = \frac{6.18 \text{ x } 10^{21} \text{ Zn atoms}}{6.18 \text{ x } 10^{21} \text{ S atoms}} = \frac{1}{1}$$

1.66   For the "other" compound: C:H mass ratio = (32.0 g C) / (8.0 g H) = 4
        The "other" compound is not methane because the methane C:H mass ratio is 3.

$$\frac{\text{C:H mass ratio in "other"}}{\text{C:H mass ratio in methane}} = \frac{4}{3}$$

Elements and Atoms (Sections 1.6–1.9)

1.68   The atomic number is equal to the number of protons.
        The mass number is equal to the sum of the number of protons and the number of neutrons.

1.70   The subscript giving the atomic number of an atom is often left off of an isotope symbol because one can readily look up the atomic number in the periodic table.

1.72   An element's atomic mass is the weighted average of the isotopic masses of the element's naturally occurring isotopes. The atomic mass for Cu (63.546) must fall between the masses of its two isotopes. If one isotope is ^{65}Cu, the other isotope must be ^{63}Cu, and not ^{66}Cu. If the other isotope was ^{66}Cu, the atomic mass for Cu would be greater than 65.

1.74   (a) carbon, C   (b) argon, Ar   (c) vanadium, V

1.76   (a) $^{220}_{86}$Rn   (b) $^{210}_{84}$Po   (c) $^{197}_{79}$Au

1.78   (a) $^{15}_{7}$N, 7 protons, 7 electrons, (15 – 7) = 8 neutrons

        (b) $^{60}_{27}$Co, 27 protons, 27 electrons, (60 – 27) = 33 neutrons

        (c) $^{131}_{53}$I, 53 protons, 53 electrons, (131 – 53) = 78 neutrons

        (d) $^{142}_{58}$Ce, 58 protons, 58 electrons, (142 – 58) = 84 neutrons

1.80   (a) $^{24}_{12}$Mg, magnesium   (b) $^{58}_{28}$Ni, nickel

        (c) $^{104}_{46}$Pd, palladium   (d) $^{183}_{74}$W, tungsten

1.82 $^{12}_{5}$C, the atomic number for carbon is 6, not 5.

 $^{33}_{35}$Br, the mass number must be greater than the atomic number.

 $^{11}_{5}$Bo, the element symbol for boron is B.

1.84 $(0.199 \times 10.0129) + (0.801 \times 11.009\ 31) = 10.8$ for B

1.86 $24.305 = (0.7899 \times 23.985) + (0.1000 \times 24.986) + (0.1101 \times Z)$
Solve for Z. $Z = 25.982$ for ^{26}Mg.

Nuclear Reactions and Radioactivity (Sections 1.10–1.11)

1.88 Positron emission is the conversion of a proton in the nucleus into a neutron plus an ejected positron.
Electron capture is the process in which a proton in the nucleus captures an inner-shell electron and is thereby converted into a neutron.

1.90 In beta emission a neutron is converted to a proton and the atomic number increases. In positron emission a proton is converted to a neutron and the atomic number decreases.

1.92 (a) $^{126}_{50}$Sn $\rightarrow$ $^{0}_{-1}$e $+$ $^{126}_{51}$Sb   (b) $^{210}_{88}$Ra $\rightarrow$ $^{4}_{2}$He $+$ $^{206}_{86}$Rn

 (c) $^{77}_{37}$Rb $\rightarrow$ $^{0}_{1}$e $+$ $^{77}_{36}$Kr   (d) $^{76}_{36}$Kr $+$ $^{0}_{-1}$e $\rightarrow$ $^{76}_{35}$Br

1.94 (a) $^{188}_{80}$Hg $\rightarrow$ $^{188}_{79}$Au $+$ $^{0}_{1}$e   (b) $^{218}_{85}$At $\rightarrow$ $^{214}_{83}$Bi $+$ $^{4}_{2}$He

 (c) $^{234}_{90}$Th $\rightarrow$ $^{234}_{91}$Pa $+$ $^{0}_{-1}$e

1.96 (a) $^{162}_{75}$Re $\rightarrow$ $^{158}_{73}$Ta $+$ $^{4}_{2}$He   (b) $^{138}_{62}$Sm $+$ $^{0}_{-1}$e $\rightarrow$ $^{138}_{61}$Pm

 (c) $^{188}_{74}$W $\rightarrow$ $^{188}_{75}$Re $+$ $^{0}_{-1}$e   (d) $^{165}_{73}$Ta $\rightarrow$ $^{165}_{72}$Hf $+$ $^{0}_{1}$e

1.98 $^{241}_{95}$Am $\rightarrow$ $^{237}_{93}$Np $+$ $^{4}_{2}$He

 $^{237}_{93}$Np $\rightarrow$ $^{233}_{91}$Pa $+$ $^{4}_{2}$He

 $^{233}_{91}$Pa $\rightarrow$ $^{233}_{92}$U $+$ $^{0}_{-1}$e

 $^{233}_{92}$U $\rightarrow$ $^{229}_{90}$Th $+$ $^{4}_{2}$He

 $^{229}_{90}$Th $\rightarrow$ $^{225}_{88}$Ra $+$ $^{4}_{2}$He

 $^{225}_{88}$Ra $\rightarrow$ $^{225}_{89}$Ac $+$ $^{0}_{-1}$e

 $^{225}_{89}$Ac $\rightarrow$ $^{221}_{87}$Fr $+$ $^{4}_{2}$He

 $^{221}_{87}$Fr $\rightarrow$ $^{217}_{85}$At $+$ $^{4}_{2}$He

 $^{217}_{85}$At $\rightarrow$ $^{213}_{83}$Bi $+$ $^{4}_{2}$He

$$^{213}_{83}\text{Bi} \rightarrow {}^{213}_{84}\text{Po} + {}^{0}_{-1}\text{e}$$

$$^{213}_{84}\text{Po} \rightarrow {}^{209}_{82}\text{Pb} + {}^{4}_{2}\text{He}$$

$$^{209}_{82}\text{Pb} \rightarrow {}^{209}_{83}\text{Bi} + {}^{0}_{-1}\text{e}$$

1.100 Each alpha emission decreases the mass number by four and the atomic number by two. Each beta emission increases the atomic number by one.

$$^{232}_{90}\text{Th} \rightarrow {}^{208}_{82}\text{Pb}$$

Number of α emissions = $\dfrac{\text{Th mass number} - \text{Pb mass number}}{4}$

$$= \dfrac{232 - 208}{4} = 6 \text{ α emissions}$$

The atomic number decreases by 12 as a result of 6 alpha emissions. The resulting atomic number is $(90 - 12) = 78$.
Number of β emissions = Pb atomic number − 78 = 82 − 78 = 4 β emissions

Chapter Problems

1.102 atomic mass = $(0.205 \times 69.924) + (0.274 \times 71.922) + (0.078 \times 72.923) + (0.365 \times 73.921)$
$+ (0.078 \times 75.921) = 72.6$

1.104 For NH_3, $(2.34 \text{ g N})\left(\dfrac{3 \times 1.0079 \text{ amu H}}{14.0067 \text{ amu N}}\right) = 0.505 \text{ g H}$

For N_2H_4, $(2.34 \text{ g N})\left(\dfrac{4 \times 1.0079 \text{ amu H}}{2 \times 14.0067 \text{ amu N}}\right) = 0.337 \text{ g H}$

1.106 (a) I   (b) Kr

1.108 $\dfrac{12.0000}{15.9994} = \dfrac{X}{16.0000}$;   X = 12.0005 for ^{12}C prior to 1961.

1.110 Molecular weight = $(8 \times 12.011) + (9 \times 1.0079) + (1 \times 14.0067) + (2 \times 15.9994) = 151.165$

1.112 $^{100}_{43}\text{Tc} \rightarrow {}^{0}_{1}\text{e} + {}^{100}_{42}\text{Mo}$ (positron emission)

$^{100}_{43}\text{Tc} + {}^{0}_{-1}\text{e} \rightarrow {}^{100}_{42}\text{Mo}$ (electron capture)

1.114 (a) Arrange the droplet charges in increasing order.
2.21×10^{-16} C
4.42×10^{-16} C
4.98×10^{-16} C
6.64×10^{-16} C
7.74×10^{-16} C
Find the smallest charge difference between droplet charges.

$(4.42 - 2.21) \times 10^{-16}$ C $= 2.21 \times 10^{-16}$ C
$(4.98 - 4.42) \times 10^{-16}$ C $= 0.56 \times 10^{-16}$ C
$(6.64 - 4.98) \times 10^{-16}$ C $= 1.66 \times 10^{-16}$ C
$(7.74 - 6.64) \times 10^{-16}$ C $= 1.10 \times 10^{-16}$ C

The smallest difference (0.56×10^{-16} C) is the approximate value for the charge on one blorvek. To get a more accurate value, divide the droplet charges by the 0.56×10^{-16} C to determine the total number of blorveks.

2.21×10^{-16} C$/0.56 \times 10^{-16}$ C $= 4$
4.42×10^{-16} C$/0.56 \times 10^{-16}$ C $= 8$
4.98×10^{-16} C$/0.56 \times 10^{-16}$ C $= 9$
6.64×10^{-16} C$/0.56 \times 10^{-16}$ C $= 12$
7.74×10^{-16} C$/0.56 \times 10^{-16}$ C $= 14$

There are $(4 + 8 + 9 + 12 + 14) = 47$ blorveks on the 5 droplets with a total charge of $(2.21 + 4.42 + 4.98 + 6.64 + 7.74) \times 10^{-16}$ C $= 25.99 \times 10^{-16}$ C.

The charge on one blorvek is $\dfrac{25.99 \times 10^{-16}\ \text{C}}{47\ \text{blorveks}} = 5.53 \times 10^{-17}$ C/blorvek

(b) Again arrange the droplet charges in increasing order.

2.21×10^{-16} C
4.42×10^{-16} C
4.98×10^{-16} C
5.81×10^{-16} C
6.64×10^{-16} C
7.74×10^{-16} C

Find the smallest charge difference between droplet charges.

$(4.42 - 2.21) \times 10^{-16}$ C $= 2.21 \times 10^{-16}$ C
$(4.98 - 4.42) \times 10^{-16}$ C $= 0.56 \times 10^{-16}$ C
$(5.81 - 4.98) \times 10^{-16}$ C $= 0.83 \times 10^{-16}$ C
$(6.64 - 5.81) \times 10^{-16}$ C $= 0.83 \times 10^{-16}$ C
$(7.74 - 6.64) \times 10^{-16}$ C $= 1.10 \times 10^{-16}$ C

The new charge difference (0.83×10^{-16} C) is not an integer multiple of 5.53×10^{-17} C which means the charge on the blorvek must be smaller than 5.53×10^{-17} C.

The difference between 0.83×10^{-16} C and 0.56×10^{-16} C $= 0.27 \times 10^{-16}$ C is the approximate value for the charge on one blorvek.

To get a more accurate value, divide the droplet charges by the 0.27×10^{-16} C to determine the total number of blorveks.

2.21×10^{-16} C$/0.27 \times 10^{-16}$ C $= 8$
4.42×10^{-16} C$/0.27 \times 10^{-16}$ C $= 16$
4.98×10^{-16} C$/0.27 \times 10^{-16}$ C $= 18$
5.81×10^{-16} C$/0.27 \times 10^{-16}$ C $= 21$
6.64×10^{-16} C$/0.27 \times 10^{-16}$ C $= 24$
7.74×10^{-16} C$/0.27 \times 10^{-16}$ C $= 28$

There are $(8 + 16 + 18 + 21 + 24 + 28) = 115$ blorveks on the 6 droplets with a total charge of $(2.21 + 4.42 + 4.98 + 5.81 + 6.64 + 7.74) \times 10^{-16}$ C $= 31.80 \times 10^{-16}$ C.

The charge on one blorvek is $\dfrac{31.80 \times 10^{-16}\ \text{C}}{115\ \text{blorveks}} = 2.77 \times 10^{-17}$ C/blorvek

2 Periodicity and the Electronic Structure of Atoms

2.1   Gamma ray $\nu = \dfrac{c}{\lambda} = \dfrac{3.00 \times 10^8 \text{ m/s}}{3.56 \times 10^{-11} \text{ m}} = 8.43 \times 10^{18} \text{ s}^{-1} = 8.43 \times 10^{18} \text{ Hz}$

     Radar wave $\nu = \dfrac{c}{\lambda} = \dfrac{3.00 \times 10^8 \text{ m/s}}{10.3 \times 10^{-2} \text{ m}} = 2.91 \times 10^9 \text{ s}^{-1} = 2.91 \times 10^9 \text{ Hz}$

2.2 $\nu = 102.5 \text{ MHz} = 102.5 \times 10^6 \text{ Hz} = 102.5 \times 10^6 \text{ s}^{-1}$

     $\lambda = \dfrac{c}{\nu} = \dfrac{3.00 \times 10^8 \text{ m/s}}{102.5 \times 10^6 \text{ s}^{-1}} = 2.93 \text{ m}$

     $\nu = 9.55 \times 10^{17} \text{ Hz} = 9.55 \times 10^{17} \text{ s}^{-1}$

     $\lambda = \dfrac{c}{\nu} = \dfrac{3.00 \times 10^8 \text{ m/s}}{9.55 \times 10^{17} \text{ s}^{-1}} = 3.14 \times 10^{-10} \text{ m}$

2.3   The wave with the shorter wavelength (b) has the higher frequency. The wave with the larger amplitude (b) represents the more intense beam of light. The wave with the shorter wavelength (b) represents blue light. The wave with the longer wavelength (a) represents red light.

2.4 $m = 2$; $R_\infty = 1.097 \times 10^{-2} \text{ nm}^{-1}$

     $\dfrac{1}{\lambda} = R_\infty \left[\dfrac{1}{m^2} - \dfrac{1}{n^2} \right]$; $\dfrac{1}{\lambda} = R_\infty \left[\dfrac{1}{2^2} - \dfrac{1}{7^2} \right]$; $\dfrac{1}{\lambda} = 2.519 \times 10^{-3} \text{ nm}^{-1}$; $\lambda = 397.0 \text{ nm}$

2.5 $m = 3$; $R_\infty = 1.097 \times 10^{-2} \text{ nm}^{-1}$

     $\dfrac{1}{\lambda} = R_\infty \left[\dfrac{1}{m^2} - \dfrac{1}{n^2} \right]$; $\dfrac{1}{\lambda} = R_\infty \left[\dfrac{1}{3^2} - \dfrac{1}{4^2} \right]$; $\dfrac{1}{\lambda} = 5.333 \times 10^{-4} \text{ nm}^{-1}$; $\lambda = 1875 \text{ nm}$

2.6 $m = 3$; $R_\infty = 1.097 \times 10^{-2} \text{ nm}^{-1}$

     $\dfrac{1}{\lambda} = R_\infty \left[\dfrac{1}{m^2} - \dfrac{1}{n^2} \right]$; $\dfrac{1}{\lambda} = R_\infty \left[\dfrac{1}{3^2} - \dfrac{1}{\infty^2} \right]$; $\dfrac{1}{\lambda} = 1.219 \times 10^{-3} \text{ nm}^{-1}$; $\lambda = 820.4 \text{ nm}$

2.7 $\lambda = 91.2 \text{ nm} = 91.2 \times 10^{-9} \text{ m}$

     $\nu = \dfrac{c}{\lambda} = \dfrac{3.00 \times 10^8 \text{ m/s}}{91.2 \times 10^{-9} \text{ m}} = 3.29 \times 10^{15} \text{ s}^{-1}$

     $E = h\nu = (6.626 \times 10^{-34} \text{ J·s})(3.29 \times 10^{15} \text{ s}^{-1}) = 2.18 \times 10^{-18} \text{ J/photon}$

     $E = (2.18 \times 10^{-18} \text{ J/photon})(6.022 \times 10^{23} \text{ photons/mol}) = 1.31 \times 10^6 \text{ J/mol} = 1310 \text{ kJ/mol}$

2.8   IR, $\lambda = 1.55 \times 10^{-6}$ m

$$E = \frac{hc}{\lambda} = (6.626 \times 10^{-34} \text{ J·s})\left(\frac{3.00 \times 10^{8} \text{ m/s}}{1.55 \times 10^{-6} \text{ m}}\right)(6.022 \times 10^{23} \text{ /mol})$$

$E = 7.72 \times 10^{4}$ J/mol = 77.2 kJ/mol

UV, $\lambda = 250$ nm = 250×10^{-9} m

$$E = \frac{hc}{\lambda} = (6.626 \times 10^{-34} \text{ J·s})\left(\frac{3.00 \times 10^{8} \text{ m/s}}{250 \times 10^{-9} \text{ m}}\right)(6.022 \times 10^{23} \text{ /mol})$$

$E = 4.79 \times 10^{5}$ J/mol = 479 kJ/mol

X ray, $\lambda = 5.49$ nm = 5.49×10^{-9} m

$$E = \frac{hc}{\lambda} = (6.626 \times 10^{-34} \text{ J·s})\left(\frac{3.00 \times 10^{8} \text{ m/s}}{5.49 \times 10^{-9} \text{ m}}\right)(6.022 \times 10^{23} \text{ /mol})$$

$E = 2.18 \times 10^{7}$ J/mol = 2.18×10^{4} kJ/mol

2.9 $\lambda = 390$ nm = 390×10^{-9} m = 3.90×10^{-7} m

$$E = \frac{hc}{\lambda} = (6.626 \times 10^{-34} \text{ J·s})\left(\frac{3.00 \times 10^{8} \text{ m/s}}{3.90 \times 10^{-7} \text{ m}}\right)(6.022 \times 10^{23} \text{/mol})$$

$E = 3.07 \times 10^{5}$ J/mol = 3.07×10^{2} kJ/mol = 307 kJ/mol
Photons with $\lambda = 390$ nm do not have sufficient energy to eject electrons from a sample of zinc.

2.10 $\nu = 1.22 \times 10^{15}$ Hz = 1.22×10^{15} s^{-1}
$E = (6.626 \times 10^{-34}$ J·s$)(1.22 \times 10^{15}$ s$^{-1})(6.022 \times 10^{23}$/mol$) = 4.87 \times 10^{5}$ J/mol = 487 kJ/mol
Light with $\nu = 1.22 \times 10^{15}$ Hz is in the ultraviolet region of the electromagnetic spectrum.

2.11 $\lambda = \dfrac{h}{mv} = \dfrac{6.626 \times 10^{-34} \text{ kg m}^{2} \text{ s}^{-1}}{(1150 \text{ kg})(24.6 \text{ m/s})} = 2.34 \times 10^{-38}$ m
This wavelength is shorter than the diameter of an atom.

2.12

n	l	m_l		Orbital	No. of Orbitals
5	0	0		5s	1
	1	−1, 0, +1		5p	3
	2	−2, −1, 0, +1, +2		5d	5
	3	−3, −2, −1, 0, +1, +2, +3		5f	7
	4	−4, −3, −2, −1, 0, +1, +2, +3, +4		5g	9

There are 25 possible orbitals in the fifth shell.

2.13   (a) 2p   (b) 4f   (c) 3d

2.14 (a) 3s orbital: $n = 3$, $l = 0$, $m_l = 0$
(b) 2p orbital: $n = 2$, $l = 1$, $m_l = -1, 0, +1$
(c) 4d orbital: $n = 4$, $l = 2$, $m_l = -2, -1, 0, +1, +2$

2.15 The g orbitals have four nodal planes.

2.16 The figure represents a d orbital, $n = 4$ and $l = 2$.

2.17 $m = 1$, $n = \infty$; $R_\infty = 1.097 \times 10^{-2}$ nm^{-1}

$$\frac{1}{\lambda} = R_\infty \left[\frac{1}{m^2} - \frac{1}{n^2} \right]; \quad \frac{1}{\lambda} = R_\infty \left[\frac{1}{1^2} - \frac{1}{\infty^2} \right]; \quad \frac{1}{\lambda} = R_\infty \left[\frac{1}{1} \right] = 1.097 \times 10^{-2} \text{ nm}^{-1}; \lambda = 91.2 \text{ nm}$$

$$E = (6.626 \times 10^{-34} \text{ J} \cdot \text{s}) \left(\frac{3.00 \times 10^8 \text{ m/s}}{91.2 \times 10^{-9} \text{ m}} \right) (6.022 \times 10^{23}/\text{mol})$$

$$E = 1.31 \times 10^6 \text{ J/mol} = 1.31 \times 10^3 \text{ kJ/mol}$$

2.18 Cr, Cu, Nb, Mo, Ru, Rh, Pd, Ag, La, Ce, Gd, Pt, Au, Ac, Th, Pa, U, Np, Cm, Ds, Rg

2.19 (a) Ti, $1s^2 2s^2 2p^6 3s^2 3p^6 4s^2 3d^2$ or [Ar] $4s^2 3d^2$
[Ar] $\underline{\uparrow\downarrow}$ $\underline{\uparrow}$ $\underline{\uparrow}$ $\underline{}$ $\underline{}$ $\underline{}$
   4s   3d
(b) Zn, $1s^2 2s^2 2p^6 3s^2 3p^6 4s^2 3d^{10}$ or [Ar] $4s^2 3d^{10}$
[Ar] $\underline{\uparrow\downarrow}$ $\underline{\uparrow\downarrow}$ $\underline{\uparrow\downarrow}$ $\underline{\uparrow\downarrow}$ $\underline{\uparrow\downarrow}$ $\underline{\uparrow\downarrow}$
   4s   3d
(c) Sn, $1s^2 2s^2 2p^6 3s^2 3p^6 4s^2 3d^{10} 4p^6 5s^2 4d^{10} 5p^2$ or [Kr] $5s^2 4d^{10} 5p^2$
[Kr] $\underline{\uparrow\downarrow}$ $\underline{\uparrow\downarrow}$ $\underline{\uparrow\downarrow}$ $\underline{\uparrow\downarrow}$ $\underline{\uparrow\downarrow}$ $\underline{\uparrow\downarrow}$ $\underline{\uparrow}$ $\underline{\uparrow}$ $\underline{}$
   5s   4d   5p
(d) Pb, [Xe] $6s^2 4f^{14} 5d^{10} 6p^2$

2.20 For Na$^+$, $1s^2 2s^2 2p^6$; for Cl$^-$, $1s^2 2s^2 2p^6 3s^2 3p^6$

2.21 The ground-state electron configuration contains 28 electrons. The atom is Ni.

2.22 (a) Ba; atoms get larger as you go down a group.
(b) Hf; atoms get smaller as you go across a period.
(c) Sn; atoms get larger as you go down a group.
(d) Lu; atoms get smaller as you go across a period.

2.23 Excited mercury atoms in a fluorescent bulb emit photons, some in the visible but most in the ultraviolet region. Visible photons contribute to light we can see; ultraviolet photons are invisible to our eyes. To utilize this ultraviolet energy, fluorescent bulbs are coated on the inside with a phosphor that absorbs ultraviolet photons and re-emits the energy as visible light.

Conceptual Problems

2.24

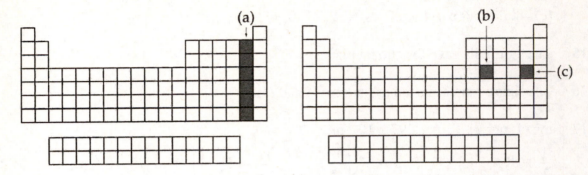

2.26   The green element, molybdenum, has an anomalous electron configuration. Its predicted electron configuration is [Ar] $5s^2 4d^4$. Its anomalous electron configuration is [Ar] $5s^1 4d^5$ because of the resulting half-filled d-orbitals.

2.28   [Ar] $4s^2 3d^{10} 4p^1$ is Ga.

2.30   Ca and Br are in the same period, with Br to the far right of Ca. Ca is larger than Br. Sr is directly below Ca in the same group, and is larger than Ca. The result is Sr (215 pm) > Ca (197 pm) > Br (114 pm).

Section Problems
Radiant Energy and Atomic Spectra (Sections 2.1–2.3)

2.32   Violet has the higher frequency and energy. Red has the higher wavelength.

2.34 1.15×10^{-7} m = 115×10^{-9} m = 115 nm = UV
2.0×10^{-6} m = 2000×10^{-9} m = 2000 nm = IR
The visible region (380 to 780 nm) is completely within this range. The ultraviolet and infrared regions are partially in this range.

2.36 $\lambda = \dfrac{c}{\nu} = \dfrac{3.00 \times 10^8 \text{ m/s}}{5.5 \times 10^{15} \text{ s}^{-1}} = 5.5 \times 10^{-8}$ m

2.38   (a) ν = 99.5 MHz = 99.5×10^6 s^{-1}
$E = h\nu = (6.626 \times 10^{-34} \text{ J·s})(99.5 \times 10^6 \text{ s}^{-1})(6.022 \times 10^{23} \text{ /mol})$
$E = 3.97 \times 10^{-2}$ J/mol = 3.97×10^{-5} kJ/mol
ν = 1150 kHz = 1150×10^3 s^{-1}
$E = h\nu = (6.626 \times 10^{-34} \text{ J·s})(1150 \times 10^3 \text{ s}^{-1})(6.022 \times 10^{23} \text{ /mol})$
$E = 4.589 \times 10^{-4}$ J/mol = 4.589×10^{-7} kJ/mol
The FM radio wave (99.5 MHz) has the higher energy.
(b) $\lambda = 3.44 \times 10^{-9}$ m

$E = \dfrac{hc}{\lambda} = (6.626 \times 10^{-34} \text{ J·s})\left(\dfrac{3.00 \times 10^8 \text{ m/s}}{3.44 \times 10^{-9} \text{ m}} \right)(6.022 \times 10^{23} \text{/mol})$

$E = 3.48 \times 10^7$ J/mol $= 3.48 \times 10^4$ kJ/mol

$\lambda = 6.71 \times 10^{-2}$ m

$$E = \frac{hc}{\lambda} = (6.626 \times 10^{-34} \text{ J·s})\left(\frac{3.00 \times 10^8 \text{ m/s}}{6.71 \times 10^{-2} \text{ m}}\right)(6.022 \times 10^{23}/\text{mol})$$

$E = 1.78$ J/mol $= 1.78 \times 10^{-3}$ kJ/mol

The X ray ($\lambda = 3.44 \times 10^{-9}$ m) has the higher energy.

2.40   (a) $\lambda = \dfrac{c}{v} = \dfrac{3.00 \times 10^8 \text{ m/s}}{825 \times 10^6 \text{ s}^{-1}} \times \dfrac{100 \text{ cm}}{1 \text{ m}} = 36.4$ cm

(b) $\lambda = \dfrac{c}{v} = \dfrac{3.00 \times 10^8 \text{ m/s}}{875 \times 10^6 \text{ s}^{-1}} \times \dfrac{100 \text{ cm}}{1 \text{ m}} = 34.3$ cm

2.42   (a) $E = 90.5$ kJ/mol $\times \dfrac{1000 \text{ J}}{1 \text{ kJ}} \times \dfrac{1 \text{ mol}}{6.02 \times 10^{23}} = 1.50 \times 10^{-19}$ J

$v = \dfrac{E}{h} = \dfrac{1.50 \times 10^{-19} \text{ J}}{6.626 \times 10^{-34} \text{ J·s}} = 2.27 \times 10^{14}$ s^{-1}

$\lambda = \dfrac{c}{v} = \dfrac{3.00 \times 10^8 \text{ m/s}}{2.27 \times 10^{14} \text{ s}^{-1}} = 1.32 \times 10^{-6}$ m $= 1320 \times 10^{-9}$ m $= 1320$ nm, near IR

(b) $E = 8.05 \times 10^{-4}$ kJ/mol $\times \dfrac{1000 \text{ J}}{1 \text{ kJ}} \times \dfrac{1 \text{ mol}}{6.02 \times 10^{23}} = 1.34 \times 10^{-24}$ J

$v = \dfrac{E}{h} = \dfrac{1.34 \times 10^{-24} \text{ J}}{6.626 \times 10^{-34} \text{ J·s}} = 2.02 \times 10^9$ s^{-1}

$\lambda = \dfrac{c}{v} = \dfrac{3.00 \times 10^8 \text{ m/s}}{2.02 \times 10^9 \text{ s}^{-1}} = 0.149$ m, radio wave

(c) $E = 1.83 \times 10^3$ kJ/mol $\times \dfrac{1000 \text{ J}}{1 \text{ kJ}} \times \dfrac{1 \text{ mol}}{6.02 \times 10^{23}} = 3.04 \times 10^{-18}$ J

$v = \dfrac{E}{h} = \dfrac{3.04 \times 10^{-18} \text{ J}}{6.626 \times 10^{-34} \text{ J·s}} = 4.59 \times 10^{15}$ s^{-1}

$\lambda = \dfrac{c}{v} = \dfrac{3.00 \times 10^8 \text{ m/s}}{4.59 \times 10^{15} \text{ s}^{-1}} = 6.54 \times 10^{-8}$ m $= 65.4 \times 10^{-9}$ m $= 65.4$ nm, UV

2.44   (a) $\lambda = \dfrac{c}{v} = \dfrac{3.00 \times 10^8 \text{ m/s}}{3.85 \times 10^{14} \text{ s}^{-1}} = 7.79 \times 10^{-7}$ m $= 779 \times 10^{-9}$ m $= 779$ nm

$E = hv = (6.626 \times 10^{-34} \text{ J·s})(3.85 \times 10^{14} \text{ s}^{-1}) = 2.55 \times 10^{-19}$ J

(b) $\lambda = \dfrac{c}{v} = \dfrac{3.00 \times 10^8 \text{ m/s}}{4.62 \times 10^{14} \text{ s}^{-1}} = 6.49 \times 10^{-7}$ m $= 649 \times 10^{-9}$ m $= 649$ nm

$E = hv = (6.626 \times 10^{-34} \text{ J·s})(4.62 \times 10^{14} \text{ s}^{-1}) = 3.06 \times 10^{-19}$ J

(c) $\lambda = \dfrac{c}{v} = \dfrac{3.00 \times 10^8 \text{ m/s}}{7.41 \times 10^{14} \text{ s}^{-1}} = 4.05 \times 10^{-7} \text{ m} = 405 \times 10^{-9} \text{ m} = 405 \text{ nm}$

$E = hv = (6.626 \times 10^{-34} \text{ J·s})(7.41 \times 10^{14} \text{ s}^{-1}) = 4.91 \times 10^{-19} \text{ J}$

2.46   For n = 3; λ = 656.3 nm = 656.3 × 10⁻⁹ m

$E = \dfrac{hc}{\lambda} = (6.626 \times 10^{-34} \text{ J·s})\left(\dfrac{2.998 \times 10^8 \text{ m/s}}{656.3 \times 10^{-9} \text{ m}}\right)\left(\dfrac{1 \text{ kJ}}{1000 \text{ J}}\right)(6.022 \times 10^{23}/\text{mol})$

$E = 182.3 \text{ kJ/mol}$

For n = 4; λ = 486.1 nm = 486.1 × 10⁻⁹ m

$E = \dfrac{hc}{\lambda} = (6.626 \times 10^{-34} \text{ J·s})\left(\dfrac{2.998 \times 10^8 \text{ m/s}}{486.1 \times 10^{-9} \text{ m}}\right)\left(\dfrac{1 \text{ kJ}}{1000 \text{ J}}\right)(6.022 \times 10^{23}/\text{mol})$

$E = 246.1 \text{ kJ/mol}$

For n = 5; λ = 434.0 nm = 434.0 × 10⁻⁹ m

$E = \dfrac{hc}{\lambda} = (6.626 \times 10^{-34} \text{ J·s})\left(\dfrac{2.998 \times 10^8 \text{ m/s}}{434.0 \times 10^{-9} \text{ m}}\right)\left(\dfrac{1 \text{ kJ}}{1000 \text{ J}}\right)(6.022 \times 10^{23}/\text{mol})$

$E = 275.6 \text{ kJ/mol}$

2.48 $E = (436 \text{ kJ/mol})\left(\dfrac{1000 \text{ J}}{1 \text{ kJ}}\right)\left(\dfrac{1 \text{ mol}}{6.022 \times 10^{23} \text{ photon}}\right) = 7.24 \times 10^{-19} \text{ J/photon}$

$v = \dfrac{E}{h} = \dfrac{7.24 \times 10^{-19} \text{ J}}{6.626 \times 10^{-34} \text{ J·s}} = 1.09 \times 10^{15} \text{ s}^{-1} = 1.09 \times 10^{15} \text{ Hz}$

Particles and Waves (Sections 2.4 and 2.5)

2.50 $\lambda = \dfrac{h}{mv} = \dfrac{6.626 \times 10^{-34} \text{ kg m}^2 \text{ s}^{-1}}{(9.11 \times 10^{-31} \text{ kg})(0.99 \times 3.00 \times 10^8 \text{ m/s})} = 2.45 \times 10^{-12} \text{ m, } \gamma \text{ ray}$

2.52   156 km/h = 156 × 10³ m/3600 s = 43.3 m/s; 145 g = 0.145 kg

$\lambda = \dfrac{h}{mv} = \dfrac{6.626 \times 10^{-34} \text{ kg m}^2 \text{ s}^{-1}}{(0.145 \text{ kg})(43.3 \text{ m/s})} = 1.06 \times 10^{-34} \text{ m}$

2.54   145 g = 0.145 kg; 0.500 nm = 0.500 × 10⁻⁹ m

$v = \dfrac{h}{m\lambda} = \dfrac{6.626 \times 10^{-34} \text{ kg m}^2 \text{ s}^{-1}}{(0.145 \text{ kg})(0.500 \times 10^{-9} \text{ m})} = 9.14 \times 10^{-24} \text{ m/s}$

2.56   0.68 g = 0.68 × 10⁻³ kg

$(\Delta x)(\Delta mv) \geq \dfrac{h}{4\pi}; \quad \Delta x \geq \dfrac{h}{4\pi(\Delta mv)} = \dfrac{6.626 \times 10^{-34} \text{ kg m}^2 \text{ s}^{-1}}{4\pi(0.68 \times 10^{-3} \text{ kg})(0.1 \text{ m/s})} = 8 \times 10^{-31} \text{ m}$

Orbitals and Quantum Mechanics (Sections 2.6–2.9)

2.58   n is the principal quantum number. The size and energy level of an orbital depends on n. l is the angular-momentum quantum number. l defines the three-dimensional shape of an orbital. m_l is the magnetic quantum number. m_l defines the spatial orientation of an orbital. m_s is the spin quantum number. m_s indicates the spin of the electron and can have either of two values, $+\frac{1}{2}$ or $-\frac{1}{2}$.

2.60   The probability of finding the electron drops off rapidly as distance from the nucleus increases, although it never drops to zero, even at large distances. As a result, there is no definite boundary or size for an orbital. However, we usually imagine the boundary surface of an orbital enclosing the volume where an electron spends 95% of its time.

2.62   Part of the electron-nucleus attraction is canceled by the electron-electron repulsion, an effect we describe by saying that the electrons are shielded from the nucleus by the other electrons. The net nuclear charge actually felt by an electron is called the effective nuclear charge, Z_{eff}, and is often substantially lower than the actual nuclear charge, Z_{actual}.
$$Z_{eff} = Z_{actual} - \text{electron shielding}$$

2.64   (a) 4s $n = 4$; $l = 0$; $m_l = 0$; $m_s = \pm\frac{1}{2}$
(b) 3p $n = 3$; $l = 1$; $m_l = -1, 0, +1$; $m_s = \pm\frac{1}{2}$
(c) 5f $n = 5$; $l = 3$; $m_l = -3, -2, -1, 0, +1, +2, +3$; $m_s = \pm\frac{1}{2}$
(d) 5d $n = 5$; $l = 2$; $m_l = -2, -1, 0, +1, +2$; $m_s = \pm\frac{1}{2}$

2.66   (a) is not allowed because for $l = 0$, $m_l = 0$ only.
(b) is allowed.
(c) is not allowed because for $n = 4$, $l = 0, 1, 2,$ or 3 only.

2.68   For $n = 5$, the maximum number of electrons will occur when the 5g orbital is filled:
[Rn] $7s^2 5f^{14} 6d^{10} 7p^6 8s^2 5g^{18} = 138$ electrons

2.70   From problem 2.47, for $n = \infty$, $\lambda = 364.6$ nm $= 364.6 \times 10^{-9}$ m
$$E = \frac{hc}{\lambda} = (6.626 \times 10^{-34} \text{ J·s})\left(\frac{2.998 \times 10^8 \text{ m/s}}{364.6 \times 10^{-9} \text{ m}}\right)\left(\frac{1 \text{ kJ}}{1000 \text{ J}}\right)(6.022 \times 10^{23}/\text{mol})$$
$$E = 328.1 \text{ kJ/mol}$$

2.72 $\lambda = 330$ nm $= 330 \times 10^{-9}$ m
$$E = \frac{hc}{\lambda} = (6.626 \times 10^{-34} \text{ J·s})\left(\frac{3.00 \times 10^8 \text{ m/s}}{330 \times 10^{-9} \text{ m}}\right)\left(\frac{1 \text{ kJ}}{1000 \text{ J}}\right)(6.022 \times 10^{23}/\text{mol})$$
$$E = 363 \text{ kJ/mol}$$

Electron Configurations (Sections 2.10–2.13)

2.74   The number of elements in successive periods of the periodic table increases by the progression 2, 8, 18, 32 because the principal quantum number n increases by 1 from one period to the next. As the principal quantum number increases, the number of orbitals in a shell increases. The progression of elements parallels the number of electrons in a particular shell. The number of electrons per shell is $2n^2$ where n is the shell number.

2.76   (a) 5d   (b) 4s   (c) 6s

2.78   (a) 3d after 4s   (b) 4p after 3d   (c) 6d after 5f   (d) 6s after 5p

2.80   (a) Ti, $Z = 22$ $1s^2\,2s^2\,2p^6\,3s^2\,3p^6\,4s^2\,3d^2$
         (b) Ru, $Z = 44$ $1s^2\,2s^2\,2p^6\,3s^2\,3p^6\,4s^2\,3d^{10}\,4p^6\,5s^2\,4d^6$
         (c) Sn, $Z = 50$ $1s^2\,2s^2\,2p^6\,3s^2\,3p^6\,4s^2\,3d^{10}\,4p^6\,5s^2\,4d^{10}\,5p^2$
         (d) Sr, $Z = 38$ $1s^2\,2s^2\,2p^6\,3s^2\,3p^6\,4s^2\,3d^{10}\,4p^6\,5s^2$
         (e) Se, $Z = 34$ $1s^2\,2s^2\,2p^6\,3s^2\,3p^6\,4s^2\,3d^{10}\,4p^4$

2.82   (a) Rb, $Z = 37$ [Kr]   ↑
                                       5s

         (b) W, $Z = 74$ [Xe]   ↑↓   ↑↓ ↑↓ ↑↓ ↑↓ ↑↓ ↑↓ ↑↓   ↑ ↑ ↑ ↑ _
                                  6s   4f   5d

         (c) Ge, $Z = 32$ [Ar]   ↑↓   ↑↓ ↑↓ ↑↓ ↑↓ ↑↓   ↑ ↑ _
                                  4s   3d   4p

         (d) Zr, $Z = 40$ [Kr]   ↑↓   ↑ ↑ _ _ _
                                  5s   4d

2.84   4s > 4d > 4f

2.86   (a) O $1s^2\,2s^2\,2p^4$ ↑↓ ↑ ↑   2 unpaired e^-
                                          2p

         (b) Si $1s^2\,2s^2\,2p^6\,3s^2\,3p^2$ ↑ ↑ _   2 unpaired e^-
                                          3p

         (c) K   [Ar] $4s^1$ 1 unpaired e^-

         (d) As   [Ar] $4s^2\,3d^{10}\,4p^3$ ↑ ↑ ↑   3 unpaired e^-
                                          4p

2.88   Order of orbital filling:
         1s→2s→2p→3s→3p→4s→3d→4p→5s→4d→5p→6s→4f→5d→6p→7s→5f→6d→7p→8s→5g
         $Z = 121$

Electron Configurations and Periodic Properties (Section 2.14)

2.90   Atomic radii increase down a group because the electron shells are farther away from the nucleus.

2.92   F < O < S

2.94   Mg has a higher ionization energy than Na because Mg has a higher Z_{eff} and a smaller size.

2.96 $Z = 116$ [Rn] $7s^2\,5f^{14}\,6d^{10}\,7p^4$

Chapter Problems

2.98 $m = 2$; $n = 3$; $R = 1.097 \times 10^{-2}$ nm^{-1}

$$\frac{1}{\lambda} = Z^2 R\left[\frac{1}{m^2} - \frac{1}{n^2}\right]; \quad \frac{1}{\lambda} = (2^2)R\left[\frac{1}{2^2} - \frac{1}{3^2}\right] = 6.094 \times 10^{-3} \text{ nm}^{-1}$$

$\lambda = 164$ nm

2.100 $m = 2$; $R_\infty = 1.097 \times 10^{-2}$ nm^{-1}

$$\frac{1}{\lambda} = R_\infty\left[\frac{1}{m^2} - \frac{1}{n^2}\right]; \quad \frac{1}{\lambda} = R_\infty\left[\frac{1}{2^2} - \frac{1}{6^2}\right] = 2.438 \times 10^{-3} \text{ nm}^{-1}$$

$\lambda = 410.2$ nm $= 410.2 \times 10^{-9}$ m

$$E = \frac{hc}{\lambda} = (6.626 \times 10^{-34} \text{ J·s})\left(\frac{2.998 \times 10^8 \text{ m/s}}{410.2 \times 10^{-9} \text{ m}}\right)\left(\frac{1 \text{ kJ}}{1000 \text{ J}}\right)(6.022 \times 10^{23}/\text{mol})$$

$E = 291.6$ kJ/mol

2.102 $m = 5$, $n = \infty$; $R_\infty = 1.097 \times 10^{-2}$ nm^{-1}

$$\frac{1}{\lambda} = R_\infty\left[\frac{1}{5^2} - \frac{1}{\infty^2}\right] = R_\infty\left[\frac{1}{25}\right] = 4.388 \times 10^{-4} \text{ nm}^{-1}; \quad \lambda = 2279 \text{ nm}$$

2.104   (a) $E = h\nu = (6.626 \times 10^{-34} \text{ J·s})(3.79 \times 10^{11} \text{ s}^{-1})\left(\frac{1 \text{ kJ}}{1000 \text{ J}}\right)(6.022 \times 10^{23}/\text{mol}) = 0.151$ kJ/mol

     (b) $E = h\nu = (6.626 \times 10^{-34} \text{ J·s})(5.45 \times 10^4 \text{ s}^{-1})\left(\frac{1 \text{ kJ}}{1000 \text{ J}}\right)(6.022 \times 10^{23}/\text{mol}) = 2.17 \times 10^{-8}$ kJ/mol

     (c) $E = h\nu = (6.626 \times 10^{-34} \text{ J·s})\left(\frac{3.00 \times 10^8 \text{ m/s}}{4.11 \times 10^{-5} \text{ m}}\right)\left(\frac{1 \text{ kJ}}{1000 \text{ J}}\right)(6.022 \times 10^{23}/\text{mol}) = 2.91$ kJ/mol

2.106   (a)   Ra   [Rn] $7s^2$ [Rn]   ⇅
                                           7s

          (b)   Sc   [Ar] $4s^2\,3d^1$ [Ar]   ⇅   ↑ _ _ _ _
                                        4s   3d

          (c)   Lr   [Rn] $7s^2\,5f^{14}\,6d^1$ [Rn]   ⇅   ⇅ ⇅ ⇅ ⇅ ⇅ ⇅ ⇅   ↑ _ _ _ _
                                        7s   5f   6d

          (d)   B   [He] $2s^2\,2p^1$ [He]   ⇅   ↑ _ _
                                        2s   2p

          (e)   Te   [Kr] $5s^2\,4d^{10}\,5p^4$ [Kr]   ⇅   ⇅ ⇅ ⇅ ⇅ ⇅   ⇅ ↑ ↑
                                        5s   4d   5p

2.108 $E = (188 \text{ kJ/mol})\left(\dfrac{1000 \text{ J}}{1 \text{ kJ}}\right)\left(\dfrac{1 \text{ mol}}{6.022 \times 10^{23} \text{ photon}}\right) = 3.12 \times 10^{-19}$ J/photon

$E = \dfrac{hc}{\lambda}, \quad \lambda = \dfrac{hc}{E} = \dfrac{(6.626 \times 10^{-34} \text{ J·s})(3.00 \times 10^{8} \text{ m/s})}{3.12 \times 10^{-19} \text{ J}} = 6.37 \times 10^{-7}$ m $= 637$ nm

2.110 (a) Sr, Z = 38   [Kr]   ⇅
    5s

(b) Cd, Z = 48   [Kr]   ⇅   ⇅ ⇅ ⇅ ⇅ ⇅
    5s   4d

(c) Z = 22, Ti   [Ar]   ⇅   ↑ ↑ __ __ __
    4s   3d

(d) Z = 34, Se   [Ar]   ⇅   ⇅ ⇅ ⇅ ⇅ ⇅   ⇅ ↑ ↑
    4s   3d   4p

2.112 (a) $\lambda = \dfrac{c}{\nu} = \dfrac{3.00 \times 10^{8} \text{ m/s}}{2.9 \times 10^{18} \text{ s}^{-1}} = 1.0 \times 10^{-10}$ m

(b) $E = h\nu = (6.626 \times 10^{-34} \text{ J·s})(2.9 \times 10^{18} \text{ s}^{-1})\left(\dfrac{1 \text{ kJ}}{1000 \text{ J}}\right)(6.022 \times 10^{23}/\text{mol})$

$E = 1.2 \times 10^{6}$ kJ/mol

(c) X rays

2.114 For K, $Z_{\text{eff}} = \sqrt{\dfrac{(418.8 \text{ kJ/mol})(4^{2})}{1312 \text{ kJ/mol}}} = 2.26$

For Kr, $Z_{\text{eff}} = \sqrt{\dfrac{(1350.7 \text{ kJ/mol})(4^{2})}{1312 \text{ kJ/mol}}} = 4.06$

2.116 $q = (350 \text{ g})(4.184 \text{ J/g·°C})(95 \text{ °C} - 20 \text{ °C}) = 109{,}830$ J

$\lambda = 15.0 \text{ cm} = 15.0 \times 10^{-2}$ m

$E = (6.626 \times 10^{-34} \text{ J·s})\left(\dfrac{3.00 \times 10^{8} \text{ m/s}}{15.0 \times 10^{-2} \text{ m}}\right) = 1.33 \times 10^{-24}$ J/photon

number of photons $= \dfrac{109{,}830 \text{ J}}{1.33 \times 10^{-24} \text{ J/photon}} = 8.3 \times 10^{28}$ photons

2.118 $48.2 \text{ nm} = 48.2 \times 10^{-9}$ m

$E(\text{photon}) = 6.626 \times 10^{-34} \text{ J·s} \times \dfrac{3.00 \times 10^{8} \text{ m/s}}{48.2 \times 10^{-9} \text{ m}} \times \dfrac{1 \text{ kJ}}{1000 \text{ J}} \times \dfrac{6.022 \times 10^{23}}{\text{mol}} = 2.48 \times 10^{3}$ kJ/mol

$E_K = E(\text{electron}) = \tfrac{1}{2}(9.109 \times 10^{-31} \text{ kg})(2.371 \times 10^{6} \text{ m/s})^{2}\left(\dfrac{1 \text{ kJ}}{1000 \text{ J}}\right)\left(\dfrac{6.022 \times 10^{23}}{\text{mol}}\right)$

$E_K = 1.54 \times 10^{3}$ kJ/mol

$E(\text{photon}) = E_i + E_K; \quad E_i = E(\text{photon}) - E_K = (2.48 \times 10^{3}) - (1.54 \times 10^{3}) = 940$ kJ/mol

2.120 Substitute the equation for the orbit radius, r, into the equation for the energy level, E, to

get $E = \dfrac{-Ze^2}{2\left(\dfrac{n^2 a_o}{Z}\right)} = \dfrac{-Z^2 e^2}{2a_o n^2}$

Let E_1 be the energy of an electron in a lower orbit and E_2 the energy of an electron in a higher orbit. The difference between the two energy levels is

$$\Delta E = E_2 - E_1 = \frac{-Z^2 e^2}{2a_o n_2^2} - \frac{-Z^2 e^2}{2a_o n_1^2} = \frac{-Z^2 e^2}{2a_o n_2^2} + \frac{Z^2 e^2}{2a_o n_1^2} = \frac{Z^2 e^2}{2a_o n_1^2} - \frac{Z^2 e^2}{2a_o n_2^2}$$

$$\Delta E = \frac{Z^2 e^2}{2a_o}\left[\frac{1}{n_1^2} - \frac{1}{n_2^2}\right]$$

Because Z, e, and a_o are constants, this equation shows that ΔE is proportional to

$\left[\dfrac{1}{n_1^2} - \dfrac{1}{n_2^2}\right]$ where n_1 and n_2 are integers with $n_2 > n_1$.

This is similar to the Balmer-Rydberg equation where $1/\lambda$ or ν for the emission spectra of

atoms is proportional to $\left[\dfrac{1}{m^2} - \dfrac{1}{n^2}\right]$ where m and n are integers with $n > m$.

2.122 (a) 3d, $n = 3$, $l = 2$
   (b) 2p, $n = 2$, $l = 1$, $m_l = -1, 0, +1$
      3p, $n = 3$, $l = 1$, $m_l = -1, 0, +1$
      3d, $n = 3$, $l = 2$, $m_l = -2, -1, 0, +1, +2$
   (c) N, $1s^2 2s^2 2p^3$ so the 3s, 3p, and 3d orbitals are empty.
   (d) C, $1s^2 2s^2 2p^2$ so the 1s and 2s orbitals are filled.
   (e) Be, $1s^2 2s^2$ so the 2s orbital contains the outermost electrons.
   (f) 2p and 3p ($\uparrow$ $\uparrow$ __) and 3d ($\uparrow$ $\uparrow$ __ __ __).

Multiconcept Problems

2.124 (a) $E = h\nu$; $\quad \nu = \dfrac{E}{h} = \dfrac{7.21 \times 10^{-19}\ \text{J}}{6.626 \times 10^{-34}\ \text{J·s}} = 1.09 \times 10^{15}\ \text{s}^{-1}$

   (b) $E(\text{photon}) = E_i + E_K$; from (a), $E_i = 7.21 \times 10^{-19}$ J

   $E(\text{photon}) = \dfrac{hc}{\lambda} = (6.626 \times 10^{-34}\ \text{J·s})\left(\dfrac{3.00 \times 10^8\ \text{m/s}}{2.50 \times 10^{-7}\ \text{m}}\right) = 7.95 \times 10^{-19}$ J

   $E_K = E(\text{photon}) - E_i = (7.95 \times 10^{-19}\ \text{J}) - (7.21 \times 10^{-19}\ \text{J}) = 7.4 \times 10^{-20}$ J
   Calculate the electron velocity from the kinetic energy, E_K.
   $E_K = 7.4 \times 10^{-20}\ \text{J} = 7.4 \times 10^{-20}\ \text{kg·m}^2/\text{s}^2 = \tfrac{1}{2}mv^2 = \tfrac{1}{2}(9.109 \times 10^{-31}\ \text{kg})v^2$

$$v = \sqrt{\frac{2 \times (7.4 \times 10^{-20}\ \text{kg}\cdot\text{m}^2/\text{s}^2)}{9.109 \times 10^{-31}\ \text{kg}}} = 4.0 \times 10^5\ \text{m/s}$$

$$\text{deBroglie wavelength} = \frac{h}{mv} = \frac{6.626 \times 10^{-34}\ \text{kg}\cdot\text{m}^2/\text{s}}{(9.109 \times 10^{-31}\ \text{kg})(4.0 \times 10^5\ \text{m/s})} = 1.8 \times 10^{-9}\ \text{m} = 1.8\ \text{nm}$$

2.126 (a) 5f subshell: $n = 5$, $l = 3$, $m_l = -3, -2, -1, 0, +1, +2, +3$

3d subshell: $n = 3$, $l = 2$, $m_l = -2, -1, 0, +1, +2$

(b) In the H atom the subshells in a particular energy level are all degenerate, i.e., all have the same energy. Therefore, you only need to consider the principal quantum number, n, to calculate the wavelength emitted for an electron that drops from the 5f to the 3d subshell.

$m = 3$, $n = 5$; $R_\infty = 1.097 \times 10^{-2}\ \text{nm}^{-1}$

$$\frac{1}{\lambda} = R_\infty \left[\frac{1}{m^2} - \frac{1}{n^2} \right]; \quad \frac{1}{\lambda} = R_\infty \left[\frac{1}{3^2} - \frac{1}{5^2} \right]; \quad \frac{1}{\lambda} = 7.801 \times 10^{-4}\ \text{nm}^{-1}; \quad \lambda = 1282\ \text{nm}$$

(c) $m = 3$, $n = \infty$; $R_\infty = 1.097 \times 10^{-2}\ \text{nm}^{-1}$

$$\frac{1}{\lambda} = R_\infty \left[\frac{1}{m^2} - \frac{1}{n^2} \right]; \quad \frac{1}{\lambda} = R_\infty \left[\frac{1}{3^2} - \frac{1}{\infty^2} \right]; \quad \frac{1}{\lambda} = R_\infty \left[\frac{1}{3^2} \right] = 1.219 \times 10^{-3}\ \text{nm}^{-1}; \quad \lambda = 820.4\ \text{nm}$$

$$E = (6.626 \times 10^{-34}\ \text{J}\cdot\text{s}) \left(\frac{3.00 \times 10^8\ \text{m/s}}{820.4 \times 10^{-9}\ \text{m}} \right) (6.022 \times 10^{23}/\text{mol}) = 1.46 \times 10^5\ \text{J/mol} = 146\ \text{kJ/mol}$$

Atoms and Ionic Bonds

3.1 (a) LiBr is composed of a metal (Li) and nonmetal (Br) and is ionic.
(b) $SiCl_4$ is composed of only nonmetals and is molecular.
(c) BF_3 is composed of only nonmetals and is molecular.
(d) CaO is composed of a metal (Ca) and nonmetal (O) and is ionic.

3.2 Figure (a) most likely represents an ionic compound because there are no discrete molecules, only a regular array of two different chemical species (ions). Figure (b) most likely represents a molecular compound because discrete molecules are present.

3.3 (a) CsF, cesium fluoride   (b) K_2O, potassium oxide   (c) CuO, copper(II) oxide
(d) BaS, barium sulfide   (e) $BeBr_2$, beryllium bromide

3.4 (a) vanadium(III) chloride, VCl_3 (b) manganese(IV) oxide, MnO_2
(c) copper(II) sulfide, CuS   (d) aluminum oxide, Al_2O_3

3.5 red – potassium sulfide, K_2S; green – strontium iodide, SrI_2; blue – gallium oxide, Ga_2O_3

3.6 (a) $Ca(ClO)_2$, calcium hypochlorite
(b) $Ag_2S_2O_3$, silver(I) thiosulfate or silver thiosulfate
(c) NaH_2PO_4, sodium dihydrogen phosphate   (d) $Sn(NO_3)_2$, tin(II) nitrate
(e) $Pb(CH_3CO_2)_4$, lead(IV) acetate   (f) $(NH_4)_2SO_4$, ammonium sulfate

3.7 (a) lithium phosphate, Li_3PO_4 (b) magnesium hydrogen sulfate, $Mg(HSO_4)_2$
(c) manganese(II) nitrate, $Mn(NO_3)_2$ (d) chromium(III) sulfate, $Cr_2(SO_4)_3$

3.8 Drawing 1 represents ionic compounds with one cation and two anions. Only (c) $CaCl_2$ is consistent with drawing 1.
Drawing 2 represents ionic compounds with one cation and one anion. Both (a) LiBr and (b) $NaNO_2$ are consistent with drawing 2.

3.9 (a) Ra^{2+} [Rn]   (b) Y^{3+} [Kr]   (c) Ti^{4+} [Ar] (d) N^{3-} [Ne]
Each ion has the ground-state electron configuration of the noble gas closest to it in the periodic table.

3.10 The neutral atom contains 30 e^- and is Zn. The ion is Zn^{2+}.

3.11 (a) O^{2-}; decrease in effective nuclear charge and an increase in electron-electron repulsions lead to the larger anion.
(b) S; atoms get larger as you go down a group.

(c) Fe; in Fe^{3+} electrons are removed from a larger valence shell and there is an increase in effective nuclear charge leading to the smaller cation.

(d) H^-; decrease in effective nuclear charge and an increase in electron-electron repulsions lead to the larger anion.

3.12 K^+ is smaller than neutral K because the ion has one less electron. K^+ and Cl^- are isoelectronic, but K^+ is smaller than Cl^- because of its higher effective nuclear charge. K is larger than Cl^- because K has one additional electron and that electron begins the next shell (period). $\qquad$ K^+, r = 133 pm; Cl^-, r = 184 pm; K, r = 227 pm

3.13 (a) Br $\qquad$ (b) S $\qquad$ (c) Se $\qquad$ (d) Ne

3.14 (a) Be $\quad$ $1s^2 2s^2$; $\qquad$ N $\quad$ $1s^2 2s^2 2p^3$
Be would have the larger third ionization energy because this electron would come from the 1s orbital.
(b) Ga $[Ar]\, 4s^2 3d^{10} 4p^1$; $\qquad$ Ge $\quad$ $[Ar]\, 4s^2 3d^{10} 4p^2$
Ga would have the larger fourth ionization energy because this electron would come from the 3d orbitals. Ge would be losing a 4s electron.

3.15 (b) Cl has the highest E_{i1} and smallest E_{i4}.

3.16 Ca (red) would have the largest third ionization energy of the three because the electron being removed is from a filled valence shell. For Al (green) and Kr (blue), the electron being removed is from a partially filled valence shell. The third ionization energy for Kr would be larger than that for Al because the electron being removed from Kr is coming out of a 4p orbital while the electron being removed from Al makes Al isoelectronic with Ne. In addition, Z_{eff} is larger for Kr than for Al. The ease of losing its third electron is Al < Kr < Ca.

3.17 Cr $\quad$ $[Ar]\, 4s^1 3d^5$; $\qquad$ Mn $\quad$ $[Ar]\, 4s^2 3d^5$; $\qquad$ Fe $\quad$ $[Ar]\, 4s^2 3d^6$
Cr can accept an electron into a 4s orbital. The 4s orbital is lower in energy than a 3d orbital. Both Mn and Fe accept the added electron into a 3d orbital that contains an electron, but Mn has a lower value of Z_{eff}. Therefore, Mn has a less negative E_{ea} than either Cr or Fe.

3.18 The least favorable E_{ea} is for Kr (red) because it is a noble gas with a filled set of 4p orbitals. The most favorable E_{ea} is for Ge (blue) because the 4p orbitals would become half filled. In addition, Z_{eff} is larger for Ge than it is for K (green).

3.19 (a) Rb would lose one electron and adopt the Kr noble gas configuration.
(b) Ba would lose two electrons and adopt the Xe noble gas configuration.
(c) Ga would lose three electrons and adopt an Ar-like noble gas configuration (note that Ga^{3+} has ten 3d electrons in addition to the two 3s and six 3p electrons).
(d) F would gain one electron and adopt the Ne noble gas configuration.

3.20 Group 6A elements will gain 2 electrons.

3.21
$$K(s) \rightarrow K(g) \qquad\qquad +89.2 \ \text{kJ/mol}$$
$$K(g) \rightarrow K^+(g) + e^- \qquad +418.8 \ \text{kJ/mol}$$
$$\frac{1}{2}\,[F_2(g) \rightarrow 2\,F(g)] \qquad +79 \ \text{kJ/mol}$$
$$F(g) + e^- \rightarrow F^-(g) \qquad -328 \ \text{kJ/mol}$$
$$K^+(g) + F^-(g) \rightarrow KF(s) \qquad \underline{-821} \ \text{kJ/mol}$$
$$\text{Sum} = -562 \ \text{kJ/mol for } K(s) + \tfrac{1}{2}\,F_2(g) \rightarrow KF(s)$$

3.22   (a) KCl has the higher lattice energy because of the smaller K^+.
(b) CaF_2 has the higher lattice energy because of the smaller Ca^{2+}.
(c) CaO has the higher lattice energy because of the higher charge on both the cation and anion.

3.23   The anions are larger than the cations. Cl^- is larger than O^{2-} because it is below it in the periodic table. Therefore, (a) is NaCl and (b) is MgO. Because of the higher ion charge and shorter cation–anion distance, MgO has the larger lattice energy.

3.24   (a) O^{2-} \qquad (b) O_2^{2-} \qquad (c) O_2^-

3.25   (a) $2\,Cs(s) + 2\,H_2O(l) \rightarrow 2\,Cs^+(aq) + 2\,OH^-(aq) + H_2(g)$
(b) $Rb(s) + O_2(g) \rightarrow RbO_2(s)$
(c) $2\,K(s) + 2\,NH_3(g) \rightarrow 2\,KNH_2(s) + H_2(g)$

3.26   (a) $Be(s) + Br_2(l) \rightarrow BeBr_2(s)$
(b) $Sr(s) + 2\,H_2O(l) \rightarrow Sr(OH)_2(aq) + H_2(g)$
(c) $2\,Mg(s) + O_2(g) \rightarrow 2\,MgO(s)$

3.27 $Mg(s) + S(s) \rightarrow MgS(s)$
In MgS, the oxidation number of S is -2.

3.28   Only about 10% of current world salt production comes from evaporation of seawater. Most salt is obtained by mining the vast deposits of <u>halite</u>, or <u>rock salt</u>, formed by evaporation of ancient inland seas. These salt beds can be up to hundreds of meters thick and may occur anywhere from a few meters to thousands of meters below Earth's surface.

Conceptual Problems

3.30   (a) shows an extended array, which represents an ionic compound.
(b) shows discrete units, which represent a covalent compound.

3.32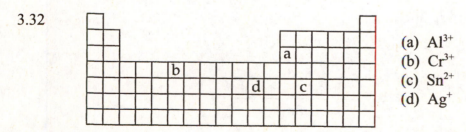
(a) Al^{3+}
(b) Cr^{3+}
(c) Sn^{2+}
(d) Ag^+

3.34 (a) I_2 (b) Na (c) NaCl (d) Cl_2

3.36 Green, CBr_4; Blue, SrF_2; Red, PbS or PbS_2

Section Problems
Naming Ionic Compounds (Section 3.2)

3.38 (a) potassium chloride, KCl (b) tin(II) bromide, $SnBr_2$ (c) calcium oxide, CaO
      (d) barium chloride, $BaCl_2$ (e) aluminum hydride, AlH_3

3.40 (a) Ba^{2+}, barium ion (b) Cs^+, cesium ion (c) V^{3+}, vanadium(III) ion
      (d) HCO_3^-, hydrogen carbonate ion (e) NH_4^+, ammonium ion (f) Ni^{2+}, nickel(II) ion
      (g) NO_2^-, nitrite ion (h) ClO_2^-, chlorite ion
      (i) Mn^{2+}, manganese(II) ion (j) ClO_4^-, perchlorate ion

3.42 (a) $CaBr_2$ (b) $CaSO_4$ (c) $Al_2(SO_4)_3$

3.44 (a) $CaCl_2$ (b) CaO (c) CaS

3.46 (a) sulfite ion, SO_3^{2-} (b) phosphate ion, PO_4^{3-} (c) zirconium(IV) ion, Zr^{4+}
      (d) chromate ion, CrO_4^{2-} (e) acetate ion, $CH_3CO_2^-$ (f) thiosulfate ion, $S_2O_3^{2-}$

3.48 (a) Na^+ and SO_4^{2-}; therefore the formula is Na_2SO_4
      (b) Ba^{2+} and PO_4^{3-}; therefore the formula is $Ba_3(PO_4)_2$
      (c) Ga^{3+} and SO_4^{2-}; therefore the formula is $Ga_2(SO_4)_3$

Ions, Ionization Energy, and Electron Affinity (Sections 3.1, 3.3–3.7)

3.50 A covalent bond results when two atoms share several (usually two) of their electrons.
      An ionic bond results from a complete transfer of one or more electrons from one atom to
      another. The C–H bonds in methane (CH_4) are covalent bonds. The bond in NaCl
      (Na^+Cl^-) is an ionic bond.

3.52 (a) Be^{2+}, 4 protons and 2 electrons (b) Rb^+, 37 protons and 36 electrons
      (c) Se^{2-}, 34 protons and 36 electrons (d) Au^{3+}, 79 protons and 76 electrons

3.54 (a) La^{3+}, [Xe] (b) Ag^+, [Kr] $4d^{10}$ (c) Sn^{2+}, [Kr] $5s^2 4d^{10}$

3.56 Ca^{2+}, [Ar]; Ti^{2+}, [Ar] $3d^2$

3.58 Cr^{2+} [Ar] $3d^4$ $\underset{3d}{\uparrow\ \uparrow\ \uparrow\ \uparrow\ \_}$

      Fe^{2+} [Ar] $3d^6$ $\underset{3d}{\uparrow\downarrow\ \uparrow\ \uparrow\ \uparrow\ \uparrow}$

3.60   The largest E_{i1} are found in Group 8A because of the largest values of Z_{eff}.
        The smallest E_{i1} are found in Group 1A because of the smallest values of Z_{eff}.

3.62   (a) K   [Ar] $4s^1$ Ca   [Ar] $4s^2$
        Ca has the smaller second ionization energy because it is easier to remove the second 4s
        valence electron in Ca than it is to remove the second electron in K from the filled 3p orbitals.

        (b) Ca   [Ar] $4s^2$ Ga   [Ar] $4s^2 3d^{10} 4p^1$
        Ca has the larger third ionization energy because it is more difficult to remove the third
        electron in Ca from the filled 3p orbitals than it is to remove the third electron (second 4s
        valence electron) from Ga.

3.64   (a) $1s^2 2s^2 2p^6 3s^2 3p^3$ is P   (b) $1s^2 2s^2 2p^6 3s^2 3p^6$ is Ar   (c) $1s^2 2s^2 2p^6 3s^2 3p^6 4s^2$ is Ca
        Ar has the highest E_{i2}. Ar has a higher Z_{eff} than P. The 4s electrons in Ca are easier to remove
        than any 3p electrons.
        Ar has the lowest E_{i7}. It is difficult to remove 3p electrons from Ca, and it is difficult to
        remove 2p electrons from P.

3.66   The likely second row element is boron because it has three valence electrons. The large
        fourth ionization energy is from an inner shell 1s electron.

3.68   Using Figure 3.10 as a reference:

	Lowest E_{i1}	Highest E_{i1}
(a)	K	Li
(b)	B	Cl
(c)	Ca	Cl

3.70   The relationship between the electron affinity of a univalent cation and the ionization energy
        of the neutral atom is that they have the same magnitude but opposite signs.

3.72 Na^+ has a more negative electron affinity than either Na or Cl because of its positive charge.

3.74   Energy is usually released when an electron is added to a neutral atom but absorbed when
        an electron is removed from a neutral atom because of the positive Z_{eff}.

3.76   The electron-electron repulsion is large and Z_{eff} is low.

Ionic Bonds and Lattice Energy (Sections 3.9 and 3.10)

3.78 $MgCl_2$ > LiCl > KCl > KBr

3.80   Li → Li^+ + e^- +520 kJ/mol
        Br + e^- → Br^- −325 kJ/mol
                                     +195 kJ/mol

3.82 $Li(s) \rightarrow Li(g)$ +159.4 kJ/mol
      $Li(s) \rightarrow Li(g) + e^-$ +520   kJ/mol
      $\frac{1}{2} [Br_2(l) \rightarrow Br_2(g)]$ +15.4 kJ/mol
      $\frac{1}{2} [Br_2(g) \rightarrow 2 Br(g)]$ +112   kJ/mol
      $Br(g) + e^- \rightarrow Br^-(g)$ –325   kJ/mol
      $Li^+(g) + Br^-(g) \rightarrow LiBr(s)$ <u>–807   kJ/mol</u>
      Sum = –325 kJ/mol for $Li(s) + \frac{1}{2} Br_2(l) \rightarrow LiBr(s)$

3.84 $Na(s) \rightarrow Na(g)$ +107.3 kJ/mol
      $Na(g) \rightarrow Na^+(g) + e^-$ +495.8 kJ/mol
      $\frac{1}{2} [H_2(g) \rightarrow 2 H(g)]$ $\frac{1}{2}$(+435.9) kJ/mol
      $H(g) + e^- \rightarrow H^-(g)$ –72.8 kJ/mol
      $Na^+(g) + H^-(g) \rightarrow NaH(s)$ <u>–U</u>
      Sum = –60 kJ/mol for $Na(s) + \frac{1}{2} H_2(g) \rightarrow NaH(s)$
      $-U = -60 - 107.3 - 495.8 - 435.9/2 + 72.8 = -808$ kJ/mol; $U = 808$ kJ/mol

3.86 $Cs(s) \rightarrow Cs(g)$ +76.1 kJ/mol
      $Cs(g) \rightarrow Cs^+(g) + e^-$ +375.7 kJ/mol
      $\frac{1}{2} [F_2(g) \rightarrow 2 F(g)]$ +79   kJ/mol
      $F(g) + e^- \rightarrow F^-(g)$ –328   kJ/mol
      $Cs^+(g) + F^-(g) \rightarrow CsF(s)$ <u>–740   kJ/mol</u>
      Sum = –537   kJ/mol for $Cs(s) + \frac{1}{2} F_2(g) \rightarrow CsF(s)$

3.88 $Ca(s) \rightarrow Ca(g)$ +178.2 kJ/mol
      $Ca(g) \rightarrow Ca^+(g) + e^-$ +589.8 kJ/mol
      $\frac{1}{2} [Cl_2(g) \rightarrow 2 Cl(g)]$ +121.5 kJ/mol
      $Cl(g) + e^- \rightarrow Cl^-(g)$ –348.6 kJ/mol
      $Ca^+(g) + Cl^-(g) \rightarrow CaCl(s)$ <u>–717   kJ/mol</u>
      Sum = –176   kJ/mol for $Ca(s) + \frac{1}{2} Cl_2(g) \rightarrow CaCl(s)$

3.90

$Na(g) \rightarrow Na^+(g) + e^-$
495.8 kJ/mol

$\frac{1}{2} H_2(g) \rightarrow H(g)$
218 kJ/mol

$Na(s) \rightarrow Na(g)$
107.3 kJ/mol

$Na(s) + \frac{1}{2} H_2(g) \rightarrow NaH(s)$
– 60 kJ/mol

$H(g) + e^- \rightarrow H^-(g)$
–72.8 kJ/mol

$Na^+(g) + H^-(g) \rightarrow NaH(s)$
–808 kJ/mol

Main-Group Chemistry (Sections 3.11–3.14)

3.92   (a) At is in Group 7A. The trend going down the group is gas → liquid → solid. At, being at the bottom of the group, should be a solid.
(b) At is likely to react with Na just like the other halogens, yielding NaAt.

3.94   Group 1A and 2A metals react by losing one and two electrons, respectively. As you go down each group, the valence electrons are farther from the nucleus and more easily removed. This trend parallels chemical reactivity.

3.96   (a) $2 K(s) + 2 H_2O(l) → 2 K^+(aq) + 2 OH^-(aq) + H_2(g)$
(b) $2 K(s) + 2 NH_3(g) → 2 KNH_2(s) + H_2(g)$
(c) $2 K(s) + Br_2(l) → 2 KBr(s)$ (d) $K(s) + O_2(g) → KO_2(s)$

3.98   (a) $Cl_2(g) + H_2(g) → 2 HCl(g)$ (b) $Cl_2(g) + Ar(g) → N. R.$
(c) $Cl_2(g) + 2 Rb(s) → 2 RbCl(s)$

Chapter Problems

3.100 Cu^{2+} has fewer electrons and a larger effective nuclear charge; therefore it has the smaller ionic radius.

3.102 $Mg(s) → Mg(g)$ +147.7 kJ/mol
$Mg(g) → Mg^+(g) + e^-$ +737.7 kJ/mol
$½ F_2(g) → F(g)$ +79   kJ/mol
$F(g) + e^- → F^-(g)$ −328   kJ/mol
$Mg^+(g) + F^-(g) → MgF(s)$ −930   kJ/mol
                Sum = −294   kJ/mol for $Mg(s) + ½ F_2(g) → MgF(s)$

$Mg(s) → Mg(g)$ +147.7 kJ/mol
$Mg(g) → Mg^+(g) + e^-$ +737.7 kJ/mol
$Mg^+(g) → Mg^{2+}(g) + e^-$ +1450.7 kJ/mol
$F_2(g) → 2 F(g)$ +158   kJ/mol
$2[F(g) + e^- → F^-(g)]$ 2(−328)   kJ/mol
$Mg^{2+}(g) + 2 F^-(g) → MgF_2(s)$ −2952   kJ/mol
                Sum = −1114   kJ/mol for $Mg(s) + F_2(g) → MgF_2(s)$
In the reaction of magnesium with fluorine, MgF_2 will form because the overall energy for the formation of MgF_2 is much more negative than for the formation of MgF.

3.104 $Na(s) → Na(g)$ +107.3 kJ/mol
$Na(g) + e^- → Na^-(g)$ −52.9 kJ/mol
$½[Cl_2(g) → 2 Cl(g)]$ +122   kJ/mol
$Cl(g) → Cl^+(g) + e^-$ +1251   kJ/mol
$Na^-(g) + Cl^+(g) → ClNa(s)$ −787   kJ/mol
                Sum = +640   kJ/mol for $Na(s) + ½ Cl_2(g) → Cl^+Na^-(s)$
The formation of Cl^+Na^- from its elements is not favored because the net energy change is positive whereas it is negative for the formation of Na^+Cl^-.

3.106 (a) $2 Li(s) + 2 H_2O(l) \rightarrow 2 Li^+(aq) + 2 OH^-(aq) + H_2(g)$
(b) $2 Li(s) + 2 NH_3(g) \rightarrow 2 LiNH_2(s) + H_2(g)$ (c) $2 Li(s) + Br_2(l) \rightarrow 2 LiBr(s)$
(d) $6 Li(s) + N_2(g) \rightarrow 2 Li_3N(s)$ (e) $4 Li(s) + O_2(g) \rightarrow 2 Li_2O(s)$

3.108 (a) Mg^{2+} and Cl^-, $MgCl_2$, magnesium chloride   (b) Ca^{2+} and O^{2-}, CaO, calcium oxide
(c) Li^+ and N^{3-}, Li_3N, lithium nitride   (d) Al^{3+} and O^{2-}, Al_2O_3, aluminum oxide

3.110 When moving diagonally down and right on the periodic table, the increase in atomic radius caused by going to a larger shell is offset by a decrease caused by a higher Z_{eff}. Thus, there is little net change in the charge density.

3.112

$Mg(s) \rightarrow Mg(g)$	+147.7 kJ/mol
$Mg(g) \rightarrow Mg^+(g) + e^-$	+738 kJ/mol
$Mg^+(g) \rightarrow Mg^{2+}(g) + e^-$	+1451 kJ/mol
$\frac{1}{2}[O_2(g) \rightarrow 2 O(g)]$	+249.2 kJ/mol
$O(g) + e^- \rightarrow O^-(g)$	−141.0 kJ/mol
$O^-(g) + e^- \rightarrow O^{2-}(g)$	E_{ea2}
$Mg^{2+}(g) + O^{2-}(g) \rightarrow MgO(s)$	−3791 kJ/mol
$Mg(s) + \frac{1}{2}O_2(g) \rightarrow MgO(s)$	−601.7 kJ/mol

$147.7 + 738 + 1451 + 249.2 - 141.0 + E_{ea2} - 3791 = -601.7$
$E_{ea2} = -147.7 - 738 - 1451 - 249.2 + 141.0 + 3791 - 601.7 = +744$ kJ/mol
Because E_{ea2} is positive, O^{2-} is not stable in the gas phase. It is stable in MgO because of the large lattice energy that results from the +2 and −2 charge of the ions and their small size.

3.114 (a) The more negative the E_{ea}, the greater the tendency of the atom to accept an electron, and the more stable the anion that results. Be, N, O, and F are all second row elements. F has the most negative E_{ea} of the group because the anion that forms, F^-, has a complete octet of electrons and its nucleus has the highest effective nuclear charge.
(b) Se^{2-} and Rb^+ are below O^{2-} and F^- in the periodic table and are the larger of the four. Se^{2-} and Rb^+ are isoelectronic, but Rb^+ has the higher effective nuclear charge so it is smaller. Therefore Se^{2-} is the largest of the four ions.

3.116

$Cr(s) \rightarrow Cr(g)$	+397 kJ/mol
$Cr(g) \rightarrow Cr^+(g)$	+652 kJ/mol
$Cr^+(g) \rightarrow Cr^{2+}(g)$	+1588 kJ/mol
$Cr^{2+}(g) \rightarrow Cr^{3+}(g)$	+2882 kJ/mol
$\frac{1}{2}(I_2(s) \rightarrow I_2(g))$	+62/2 kJ/mol
$\frac{1}{2}(I_2(g) \rightarrow 2 I(g))$	+151/2 kJ/mol
$I(g) + e^- \rightarrow I^-(g)$	−295 kJ/mol
$Cl_2(g) \rightarrow 2 Cl(g)$	+243 kJ/mol
$2(Cl(g) + e^- \rightarrow Cl^-(g))$	2(−349) kJ/mol
$Cr^{3+}(g) + 2 Cl^-(g) + I^-(g) \rightarrow CrCl_2I(s)$	−U
$Cr(s) + Cl_2(g) + \frac{1}{2} I_2(g) \rightarrow CrCl_2I(s)$	− 420 kJ/mol

$-U = -420 - 397 - 652 - 1588 - 2882 - 62/2 - 151/2 + 295 - 243 + 2(349) = -5295.5$ kJ/mol
$U = 5295$ kJ/mol

Multiconcept Problems

3.118 (a) Fe   [Ar] $4s^2 3d^6$
           Fe^{2+} [Ar] $3d^6$
           Fe^{3+} [Ar] $3d^5$

(b) A 3d electron is removed on going from Fe^{2+} to Fe^{3+}. For the 3d electron, $n = 3$ and $l = 2$.

(c) $E(\text{J/photon}) = 2952 \text{ kJ/mol} \times \dfrac{1 \text{ mol photons}}{6.022 \times 10^{23} \text{ photons}} \times \dfrac{1000 \text{ J}}{1 \text{ kJ}} = 4.90 \times 10^{-18} \text{ J/photon}$

$$E = \frac{hc}{\lambda}$$

$$\lambda = \frac{hc}{E} = \frac{(6.626 \times 10^{-34} \text{ J} \cdot \text{s})(3.00 \times 10^8 \text{ m/s})}{4.90 \times 10^{-18} \text{ J}} = 4.06 \times 10^{-8} \text{ m} = 40.6 \times 10^{-9} \text{ m} = 40.6 \text{ nm}$$

(d) Ru is directly below Fe in the periodic table and the two metals have similar electron configurations. The electron removed from Ru to go from Ru^{2+} to Ru^{3+} is a 4d electron. The electron with the higher principal quantum number, $n = 4$, is farther from the nucleus, less tightly held, and requires less energy to remove.

4   Atoms and Covalent Bonds

4.1

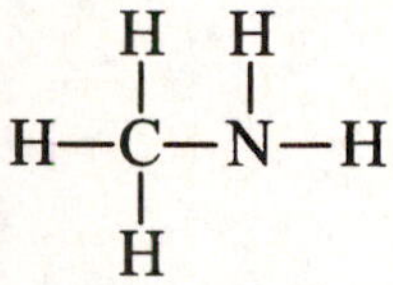

$$H-\overset{\overset{\displaystyle H}{|}}{\underset{\underset{\displaystyle H}{|}}{C}}-\overset{\overset{\displaystyle H}{|}}{N}-H$$

4.2 $C_5H_{11}NO_2S$

4.3   adrenaline, $C_9H_{13}NO_3$

4.4   (a) $SiCl_4$ chlorine $EN = 3.0$
                    silicon $\underline{EN = 1.8}$
                             $\Delta EN = 1.2$ The Si–Cl bond is polar covalent.

    (b) $CsBr$ bromine $EN = 2.8$
                    cesium $\underline{EN = 0.7}$
                             $\Delta EN = 2.1$ The Cs^+Br^- bond is ionic.

    (c) $FeBr_3$ bromine $EN = 2.8$
                     iron $\underline{EN = 1.8}$
                             $\Delta EN = 1.0$ The Fe–Br bond is polar covalent.

    (d) CH_4 carbon $EN = 2.5$
                    hydrogen $\underline{EN = 2.1}$
                             $\Delta EN = 0.4$ The C–H bond is polar covalent.

4.5   (a) CCl_4 chlorine $EN = 3.0$
                    carbon $\underline{EN = 2.5}$
                             $\Delta EN = 0.5$

    (b) $BaCl_2$ chlorine $EN = 3.0$
                    barium $\underline{EN = 0.9}$
                             $\Delta EN = 2.1$

    (c) $TiCl_3$ chlorine $EN = 3.0$
                    titanium $\underline{EN = 1.5}$
                             $\Delta EN = 1.5$

    (d) Cl_2O oxygen $EN = 3.5$
                    chlorine $\underline{EN = 3.0}$
                             $\Delta EN = 0.5$

Increasing ionic character: $CCl_4 \sim ClO_2 < TiCl_3 < BaCl_2$

4.6   H is positively polarized (blue). O is negatively polarized (red). This is consistent with the electronegativity values for O (3.5) and H (2.1). The more negatively polarized atom should be the one with the larger electronegativity.

4.7   (a) NCl_3, nitrogen trichloride   (b) P_4O_6, tetraphosphorus hexoxide
        (c) S_2F_2, disulfur difluoride   (d) SeO_2, selenium dioxide

4.8   (a) disulfur dichloride, S_2Cl_2 (b) iodine monochloride, ICl
        (c) nitrogen triiodide, NI_3

4.9   (a) PCl_5, phosphorus pentachloride   (b) N_2O, dinitrogen monoxide

4.10   (a)   H:S̈:H   (b)   H
                                     :C̈l:C̈:C̈l:
                                        :C̈l:

4.11

        H:Ö:H + H⁺ ⟶ [H:Ö:H with H on top]⁺
                              hydronium ion

4.12   (a)   H H H   (b) H—Ö—Ö—H   (c)   H
              H—C—C—C—H   H—C—N̈—H
               H H H   H H

        (d)   H H   (e) H—C≡C—H   (f)   :C̈l:
              H—C=C—H   :C̈l—C=Ö:

4.13   H H   H   H
       H—C—C—Ö—H and H—C—Ö—C—H
        H H   H   H

4.14   Molecular formula: $C_4H_5N_3O$;

4.15   :C≡O:

4.16 (a) 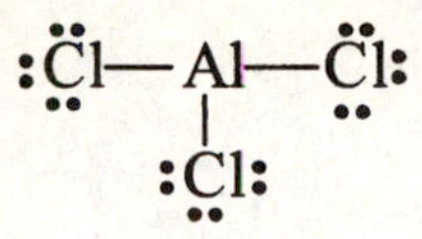 (b) 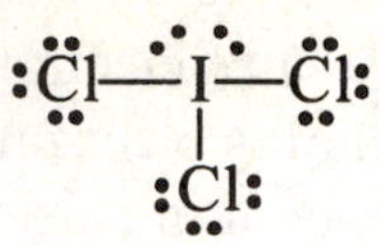 (c) 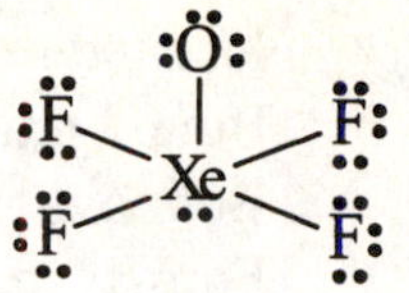 (d) 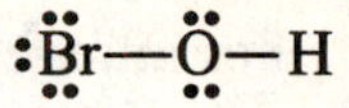

4.17 (a) 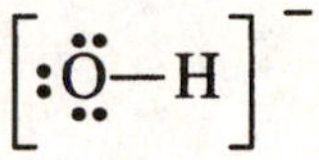 (b) 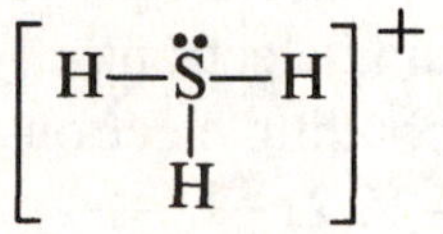 (c) 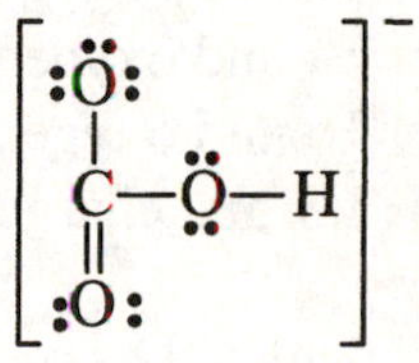 (d) 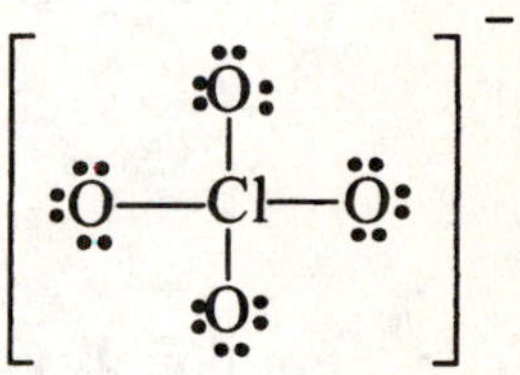

4.18 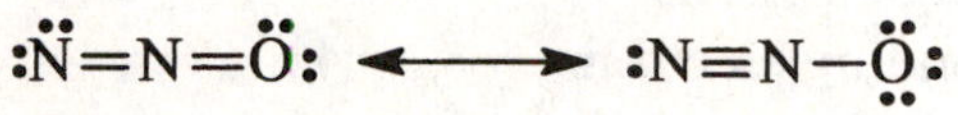

4.19 (a) 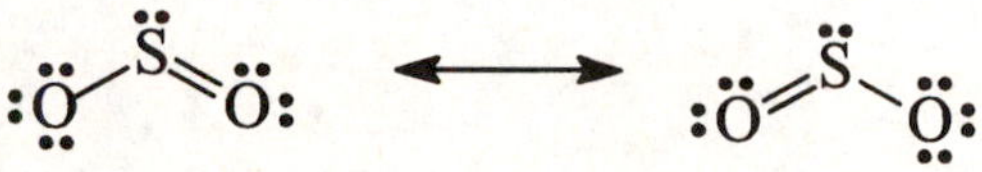

(b) 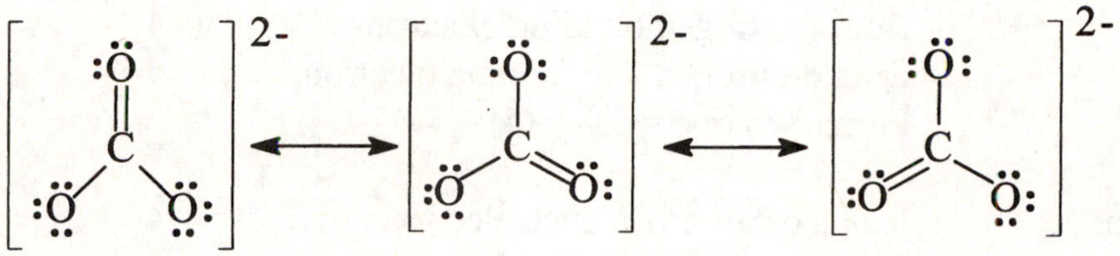

(c) 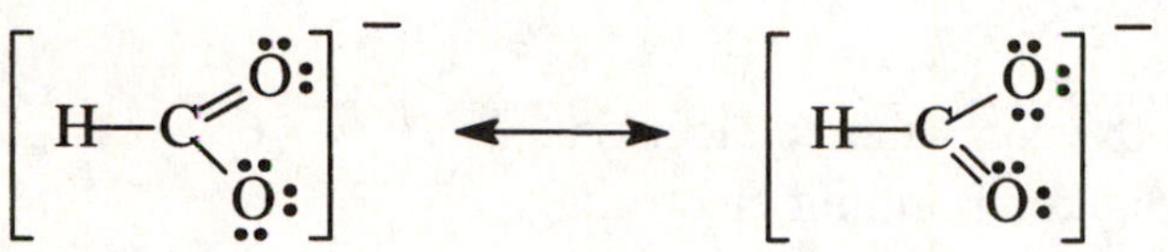

(d) 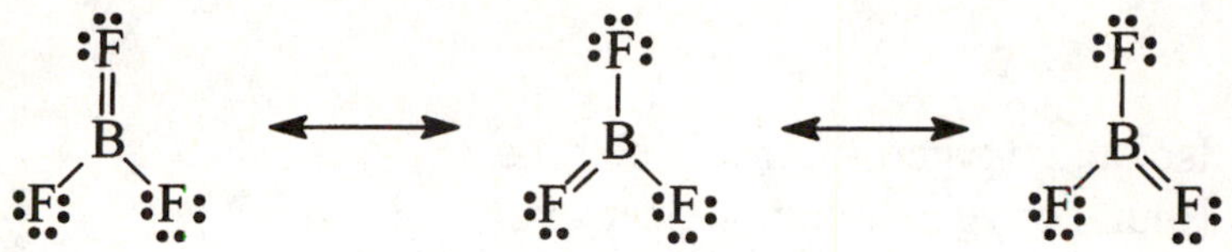

4.20 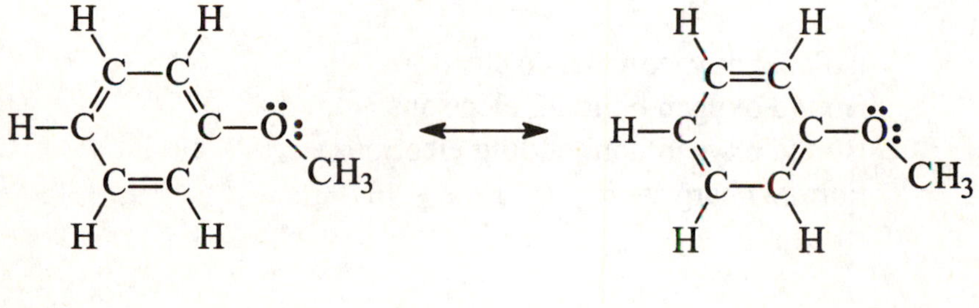

4.21   For nitrogen:

Isolated nitrogen valence electrons	5
Bound nitrogen bonding electrons	8
Bound nitrogen nonbonding electrons	0

Formal charge = $5 - \frac{1}{2}(8) - 0 = +1$

For singly bound oxygen:

Isolated oxygen valence electrons	6
Bound oxygen bonding electrons	2
Bound oxygen nonbonding electrons	6

Formal charge = $6 - \frac{1}{2}(2) - 6 = -1$

For doubly bound oxygen:

Isolated oxygen valence electrons	6
Bound oxygen bonding electrons	4
Bound oxygen nonbonding electrons	4

Formal charge = $6 - \frac{1}{2}(4) - 4 = 0$

4.22   (a) $\left[:\ddot{N}{=}C{=}\ddot{O}: \right]^{-}$

For nitrogen:

Isolated nitrogen valence electrons	5
Bound nitrogen bonding electrons	4
Bound nitrogen nonbonding electrons	4

Formal charge = $5 - \frac{1}{2}(4) - 4 = -1$

For carbon:

Isolated carbon valence electrons	4
Bound carbon bonding electrons	8
Bound carbon nonbonding electrons	0

Formal charge = $4 - \frac{1}{2}(8) - 0 = 0$

For oxygen:

Isolated oxygen valence electrons	6
Bound oxygen bonding electrons	4
Bound oxygen nonbonding electrons	4

Formal charge = $6 - \frac{1}{2}(4) - 4 = 0$

(b) $:\ddot{O}{-}\ddot{O}{=}\ddot{O}:$

For left oxygen:

Isolated oxygen valence electrons	6
Bound oxygen bonding electrons	2
Bound oxygen nonbonding electrons	6

Formal charge = $6 - \frac{1}{2}(2) - 6 = -1$

For central oxygen:

Isolated oxygen valence electrons	6
Bound oxygen bonding electrons	6
Bound oxygen nonbonding electrons	2

Formal charge = $6 - \frac{1}{2}(6) - 2 = +1$

For right oxygen:	Isolated oxygen valence electrons	6
	Bound oxygen bonding electrons	4
	Bound oxygen nonbonding electrons	4
	Formal charge = $6 - \frac{1}{2}(4) - 4 = 0$	

4.23   Radicals are substances with an odd number of electrons and at least one of the electrons is unpaired in a half-filled orbital. Radicals are usually very reactive, rapidly undergoing some kind of reaction that allows them to pair their electrons and form a more stable product.

4.24   Common radicals in the human body are nitric oxide (NO), superoxide ion (O_2^-), and the hydroxyl radical (OH), and melanin.

Nitric oxide is a signaling molecule that is released in the human body by cells lining the interior surface of blood vessels. This release of NO causes the muscles surrounding the vessels to relax, thereby dilating the blood vessels and increasing blood flow.

Superoxide ion and hydroxyl are oxygen-containing radicals that contribute both to the development of many diseases, including stroke and heart attack, and to normal human aging.

Melanin is the biological pigment responsible for skin and hair color. It is one of the very few known stable radicals and functions in the body to absorb harmful ultraviolet rays, thereby protecting the skin from sun damage that might otherwise cause cancer.

4.25

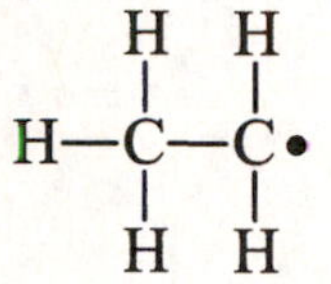

Conceptual Problems

4.26   As the electrostatic potential maps are drawn, the Li and Cl are at the tops of each map. The red area is for a negatively polarized region (associated with Cl). The blue area is for a positively polarized region (associated with Li). Map (a) is for CH_3Cl and Map (b) is for CH_3Li.

4.28 $C_8H_9NO_2$

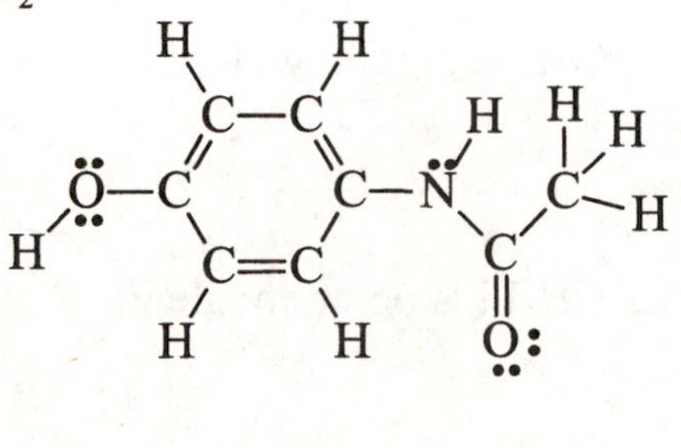

4.30

Section Problems
Electronegativity and Polar Covalent Bonds (Section 4.4)

4.32   Electronegativity increases from left to right across a period and decreases down a group.

4.34   K < Li < Mg < Pb < C < Br

4.36   (a) HF   fluorine   EN = 4.0
                         hydrogen   EN = 2.1
                                        ΔEN = 1.9   HF is polar covalent.

         (b) HI   iodine   EN = 2.5
                          hydrogen   EN = 2.1
                                         ΔEN = 0.4   HI is polar covalent.

         (c) $PdCl_2$ chlorine   EN = 3.0
                             palladium   EN = 2.2
                                            ΔEN = 0.8 $PdCl_2$ is polar covalent.

         (d) BBr_3 bromine   EN = 2.8
                            boron   EN = 2.0
                                           ΔEN = 0.8 BBr_3 is polar covalent.

         (e) NaOH $Na^+ - OH^-$ is ionic
              OH^- oxygen   EN = 3.5
                            hydrogen   EN = 2.1
                                           ΔEN = 1.4 OH^- is polar covalent.

         (f) CH_3Li lithium   EN = 1.0
                            carbon   EN = 2.5
                                           ΔEN = 1.5 CH_3Li is polar covalent.

4.38 (a) $\overset{\delta-}{C} - \overset{\delta+}{H}$ $\overset{\delta+}{C} - \overset{\delta-}{Cl}$ (b) $\overset{\delta-}{Si} - \overset{\delta+}{Li}$ $\overset{\delta+}{Si} - \overset{\delta-}{Cl}$

(c) $N - Cl$ $\overset{\delta-}{N} - \overset{\delta+}{Mg}$

4.40 (a) MgO, $BaCl_2$, $LiBr$ (b) P_4 (c) $CdBr_2$, BrF_3, NF_3, $POCl_3$

4.42 (a) $CaCl_2$ (b) PCl_3

4.44 $H-\overset{..}{N}=\overset{..}{N}-H$

$\overset{H}{\underset{H}{\diagdown}}\overset{}{N}-N\overset{H}{\underset{H}{\diagup}}$

N_2H_2 has the stronger N–N bond because of the higher bond order.

Naming Molecular Compounds (Section 4.5)

4.46 (a) Phosphorus trichloride (b) Dinitrogen trioxide (c) Tetraphosphorus heptoxide
(d) Bromine trifluoride (e) Nitrogen trichloride (f) Tetraphosphorus hexoxide
(g) Disulfur difluoride (h) Selenium dioxide

4.48 (a) S_2Cl_2 (b) ICl (c) NI_3 (d) Cl_2O (e) ClO_3 (f) S_4N_4

Electron-Dot Structures (Sections 4.6–4.8)

4.50 The octet rule states that main-group elements tend to react so that they attain a noble gas electron configuration with filled s and p sublevels (8 electrons) in their valence electron shells. The transition metals are characterized by partially filled d orbitals that can be used to expand their valence shell beyond the normal octet of electrons.

4.52 (a)

$$:\overset{..}{\underset{..}{Br}}-\overset{\overset{\displaystyle :\overset{..}{\underset{..}{Br}}:}{|}}{\underset{\underset{\displaystyle :\overset{..}{\underset{..}{Br}}:}{|}}{C}}-\overset{..}{\underset{..}{Br}}:$$

(b)

$$:\overset{..}{\underset{..}{Cl}}-\overset{\overset{\displaystyle}{|}}{\underset{\underset{\displaystyle :\overset{..}{\underset{..}{Cl}}:}{|}}{N}}-\overset{..}{\underset{..}{Cl}}:$$

(c)

$$H-\overset{\overset{\displaystyle H}{|}}{\underset{\underset{\displaystyle H}{|}}{C}}-\overset{\overset{\displaystyle H}{|}}{\underset{\underset{\displaystyle H}{|}}{C}}-\overset{..}{\underset{..}{C}}\overset{..}{l}:$$

(d)

$$\left[:\overset{..}{\underset{..}{F}}-\overset{\overset{\displaystyle :\overset{..}{F}:}{|}}{\underset{\underset{\displaystyle :\overset{..}{\underset{..}{F}}:}{|}}{B}}-\overset{..}{\underset{..}{F}}:\right]^{-}$$

(e) $\left[:\overset{..}{\underset{..}{O}}-\overset{..}{\underset{..}{O}}:\right]^{2-}$

(f) $\left[:N\equiv O:\right]^{+}$

4.54

$$H-\overset{..}{\underset{..}{O}}-\overset{\overset{\displaystyle :\overset{..}{O}}{||}}{C}-\overset{\overset{\displaystyle \overset{..}{O}:}{||}}{C}-\overset{..}{\underset{..}{O}}-H$$

4.56 (a) The anion has 32 valence electrons. Each Cl has seven valence electrons (28 total). The minus one charge on the anion accounts for one valence electron. This leaves three valence electrons for X. X is Al.

(b) The cation has eight valence electrons. Each H has one valence electron (4 total). X is left with four valence electrons. Since this is a cation, one valence electron was removed from X. X has five valence electrons. X is P.

4.58 (a)

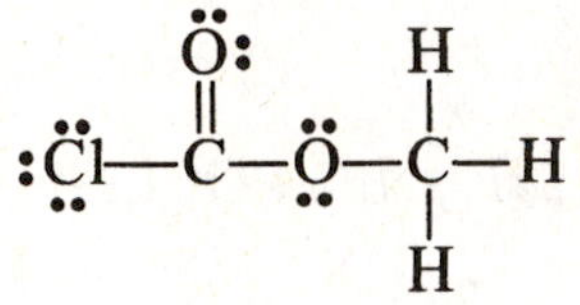

$$\text{:Cl}-\overset{\overset{\displaystyle \ddot{O}:}{\|}}{C}-\ddot{O}-\overset{\overset{\displaystyle H}{|}}{\underset{\underset{\displaystyle H}{|}}{C}}-H$$

(b)

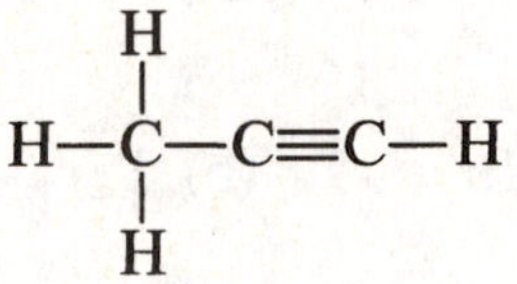

$$H-\overset{\overset{\displaystyle H}{|}}{\underset{\underset{\displaystyle H}{|}}{C}}-C\equiv C-H$$

4.60

$$\text{[structure]}$$

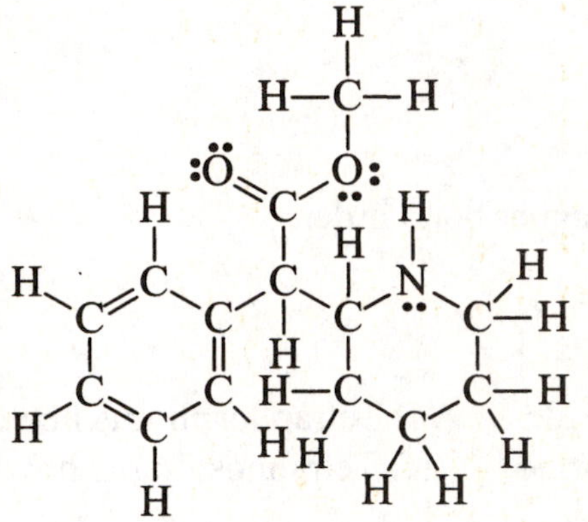

Resonance and Formal Charge (Sections 4.9 and 4.10)

4.62 (a) $H-N\equiv N-\ddot{N}: \longleftrightarrow H-\ddot{N}=N=\ddot{N}: \longleftrightarrow H-\ddot{N}-N\equiv N:$

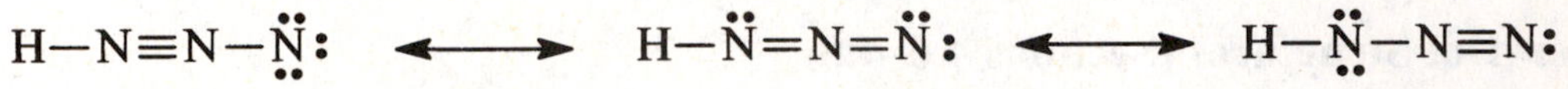

(b) [resonance structures of SO₃]

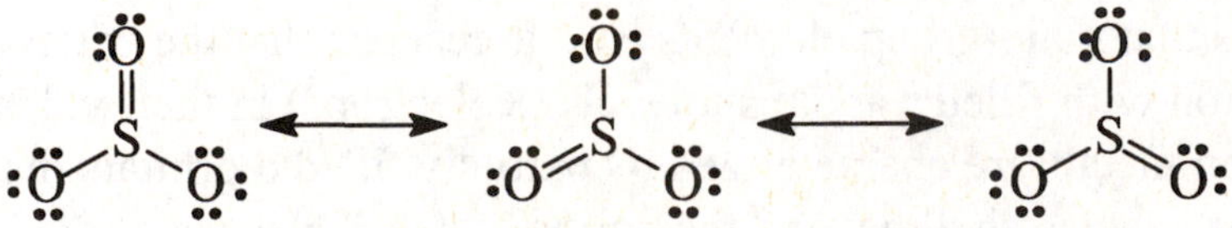

(c) $\left[:N\equiv C-\ddot{S}:\right]^{-} \longleftrightarrow \left[:\ddot{N}=C=\ddot{S}:\right]^{-} \longleftrightarrow \left[:\ddot{N}-C\equiv S:\right]^{-}$

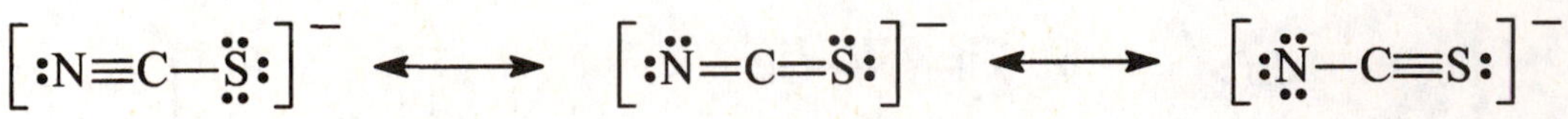

4.64 (a) yes (b) no (c) yes (d) yes

4.66 $:C\equiv O:$

For carbon:	Isolated carbon valence electrons	4
	Bound carbon bonding electrons	6
	Bound carbon nonbonding electrons	2
	Formal charge = 4 – ½(6) – 2 = –1	

For oxygen:	Isolated oxygen valence electrons	6
	Bound oxygen bonding electrons	6
	Bound oxygen nonbonding electrons	2
	Formal charge = 6 – ½(6) – 2 = +1	

4.68

$$\left[:\overset{..}{\underset{..}{O}}-\overset{..}{\underset{..}{Cl}}-\overset{..}{\underset{..}{O}}: \right]^{-}$$

For both oxygens:	Isolated oxygen valence electrons	6
	Bound oxygen bonding electrons	2
	Bound oxygen nonbonding electrons	6
	Formal charge = $6 - \frac{1}{2}(2) - 6 = -1$	

For chlorine:	Isolated chlorine valence electrons	7
	Bound chlorine bonding electrons	4
	Bound chlorine nonbonding electrons	4
	Formal charge = $7 - \frac{1}{2}(4) - 4 = +1$	

$$\left[:\overset{..}{\underset{..}{O}}-\overset{..}{\underset{..}{Cl}}=\overset{..}{\underset{..}{O}} \right]^{-}$$

For left oxygen:	Isolated oxygen valence electrons	6
	Bound oxygen bonding electrons	2
	Bound oxygen nonbonding electrons	6
	Formal charge = $6 - \frac{1}{2}(2) - 6 = -1$	

For right oxygen:	Isolated oxygen valence electrons	6
	Bound oxygen bonding electrons	4
	Bound oxygen nonbonding electrons	4
	Formal charge = $6 - \frac{1}{2}(4) - 4 = 0$	

For chlorine:	Isolated chlorine valence electrons	7
	Bound chlorine bonding electrons	6
	Bound chlorine nonbonding electrons	4
	Formal charge = $7 - \frac{1}{2}(6) - 4 = 0$	

4.70 (a)

$$\overset{H}{\underset{H}{>}}C=N=\overset{..}{N}:$$

For hydrogen:	Isolated hydrogen valence electrons	1
	Bound hydrogen bonding electrons	2
	Bound hydrogen nonbonding electrons	0
	Formal charge = $1 - \frac{1}{2}(2) - 0 = 0$	

For nitrogen: (central)	Isolated nitrogen valence electrons	5
	Bound nitrogen bonding electrons	8
	Bound nitrogen nonbonding electrons	0
	Formal charge = $5 - \frac{1}{2}(8) - 0 = +1$	

For nitrogen: (terminal)	Isolated nitrogen valence electrons	5
	Bound nitrogen bonding electrons	4
	Bound nitrogen nonbonding electrons	4
	Formal charge = $5 - \frac{1}{2}(4) - 4 = -1$	

For carbon:	Isolated carbon valence electrons	4
	Bound carbon bonding electrons	8
	Bound carbon nonbonding electrons	0
	Formal charge = $4 - \frac{1}{2}(8) - 0 = 0$

(b)

$$\begin{array}{c} H \\ \diagdown \\ C-\ddot{N}=\ddot{N}: \\ \diagup \\ H \end{array}$$

For hydrogen:	Isolated hydrogen valence electrons	1
	Bound hydrogen bonding electrons	2
	Bound hydrogen nonbonding electrons	0
	Formal charge = $1 - \frac{1}{2}(2) - 0 = 0$

For nitrogen: (central)	Isolated nitrogen valence electrons	5
	Bound nitrogen bonding electrons	6
	Bound nitrogen nonbonding electrons	2
	Formal charge = $5 - \frac{1}{2}(6) - 2 = 0$

For nitrogen: (terminal)	Isolated nitrogen valence electrons	5
	Bound nitrogen bonding electrons	4
	Bound nitrogen nonbonding electrons	4
	Formal charge = $5 - \frac{1}{2}(4) - 4 = -1$

For carbon:	Isolated carbon valence electrons	4
	Bound carbon bonding electrons	6
	Bound carbon nonbonding electrons	0
	Formal charge = $4 - \frac{1}{2}(6) - 0 = +1$

Structure (a) is more important because of the octet of electrons around carbon.

4.72

$$:\overset{..}{\underset{..}{Cl}}-\overset{\overset{:\overset{..}{Cl}:}{|}}{\underset{\underset{:\overset{..}{Cl}:}{|}}{C}}-\overset{+}{N}-\overset{..}{\underset{..}{O}}:^{-} \quad \longleftrightarrow \quad :\overset{..}{\underset{..}{Cl}}-\overset{\overset{:\overset{..}{Cl}:\overset{..}{O}:^{-}}{|}}{\underset{\underset{:\overset{..}{Cl}:}{|}}{C}}-\overset{+}{N}=\overset{..}{\underset{..}{O}}:$$

For chlorine:	Isolated chlorine valence electrons	7
	Bound chlorine bonding electrons	2
	Bound chlorine nonbonding electrons	6
	Formal charge = $7 - \frac{1}{2}(2) - 6 = 0$

For carbon:	Isolated carbon valence electrons	4
	Bound carbon bonding electrons	8
	Bound carbon nonbonding electrons	0
	Formal charge = $4 - \frac{1}{2}(8) - 0 = 0$

For nitrogen:

Isolated nitrogen valence electrons		5
Bound nitrogen bonding electrons		8
Bound nitrogen nonbonding electrons		0

Formal charge = $5 - \frac{1}{2}(8) - 0 = +1$

For oxygen:
(double bonded)

Isolated oxygen valence electrons		6
Bound oxygen bonding electrons		4
Bound oxygen nonbonding electrons		4

Formal charge = $6 - \frac{1}{2}(4) - 4 = 0$

For oxygen:
(single bonded)

Isolated oxygen valence electrons		6
Bound oxygen bonding electrons		2
Bound oxygen nonbonding electrons		6

Formal charge = $6 - \frac{1}{2}(2) - 6 = -1$

Chapter Problems

4.74   reactants

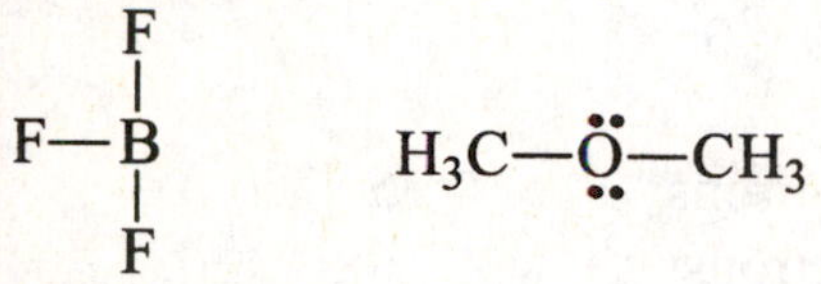

For boron:

Isolated boron valence electrons		3
Bound boron bonding electrons		6
Bound boron nonbonding electrons		0

Formal charge = $3 - \frac{1}{2}(6) - 0 = 0$

For oxygen:

Isolated oxygen valence electrons		6
Bound oxygen bonding electrons		4
Bound oxygen nonbonding electrons		4

Formal charge = $6 - \frac{1}{2}(4) - 4 = 0$

product

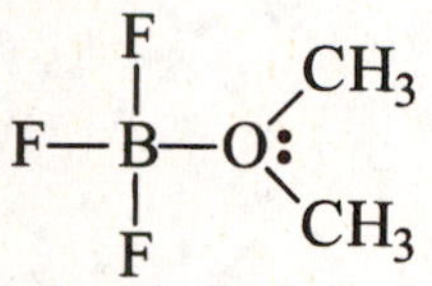

For boron:

Isolated boron valence electrons		3
Bound boron bonding electrons		8
Bound boron nonbonding electrons		0

Formal charge = $3 - \frac{1}{2}(8) - 0 = -1$

For oxygen:

Isolated oxygen valence electrons		6
Bound oxygen bonding electrons		6
Bound oxygen nonbonding electrons		2

Formal charge = $6 - \frac{1}{2}(6) - 2 = +1$

4.76

$$H-\underset{\underset{H}{|}}{\overset{\overset{H}{|}}{C}}-\ddot{N}=C=\ddot{O}: \quad\longleftrightarrow\quad H-\underset{\underset{H}{|}}{\overset{\overset{H}{|}}{C}}-\overset{+}{N}\equiv C-\overset{-}{\ddot{O}}:$$

For $\quad H-\underset{\underset{H}{|}}{\overset{\overset{H}{|}}{C}}-\ddot{N}=C=\ddot{O}:$

For hydrogen:	Isolated hydrogen valence electrons	1
	Bound hydrogen bonding electrons	2
	Bound hydrogen nonbonding electrons	0
	Formal charge = $1 - \frac{1}{2}(2) - 0 = 0$	
For carbon: (left)	Isolated carbon valence electrons	4
	Bound carbon bonding electrons	8
	Bound carbon nonbonding electrons	0
	Formal charge = $4 - \frac{1}{2}(8) - 0 = 0$	
For nitrogen:	Isolated nitrogen valence electrons	5
	Bound nitrogen bonding electrons	6
	Bound nitrogen nonbonding electrons	2
	Formal charge = $5 - \frac{1}{2}(6) - 2 = 0$	
For carbon: (right)	Isolated carbon valence electrons	4
	Bound carbon bonding electrons	8
	Bound carbon nonbonding electrons	0
	Formal charge = $4 - \frac{1}{2}(8) - 0 = 0$	
For oxygen:	Isolated oxygen valence electrons	6
	Bound oxygen bonding electrons	4
	Bound oxygen nonbonding electrons	4
	Formal charge = $6 - \frac{1}{2}(4) - 4 = 0$	

For $\quad H-\underset{\underset{H}{|}}{\overset{\overset{H}{|}}{C}}-\overset{+}{N}\equiv C-\overset{-}{\ddot{O}}:$

For hydrogen:	Isolated hydrogen valence electrons	1
	Bound hydrogen bonding electrons	2
	Bound hydrogen nonbonding electrons	0
	Formal charge = $1 - \frac{1}{2}(2) - 0 = 0$	

For carbon: (left)	Isolated carbon valence electrons	4
	Bound carbon bonding electrons	8
	Bound carbon nonbonding electrons	0
	Formal charge = $4 - \frac{1}{2}(8) - 0 = 0$	

For nitrogen:	Isolated nitrogen valence electrons	5
	Bound nitrogen bonding electrons	8
	Bound nitrogen nonbonding electrons	0
	Formal charge = $5 - \frac{1}{2}(8) - 0 = +1$	

For carbon: (right)	Isolated carbon valence electrons	4
	Bound carbon bonding electrons	8
	Bound carbon nonbonding electrons	0
	Formal charge = $4 - \frac{1}{2}(8) - 0 = 0$	

For oxygen:	Isolated oxygen valence electrons	6
	Bound oxygen bonding electrons	2
	Bound oxygen nonbonding electrons	6
	Formal charge = $6 - \frac{1}{2}(2) - 6 = -1$	

4.78 $CH_4(g) + Cl_2(g) \rightarrow CH_3Cl(g) + HCl(g)$

Energy change = D (Reactant bonds) – D (Product bonds)

Energy change = $[4\,D_{C-H} + D_{Cl-Cl}] - [3\,D_{C-H} + D_{C-Cl} + D_{H-Cl}]$

Energy change = $[(4\text{ mol})(410\text{ kJ/mol}) + (1\text{ mol})(243\text{ kJ/mol}]$

      $- [(3\text{ mol})(410\text{ kJ/mol}) + (1\text{ mol})(330\text{ kJ/mol}) + (1\text{ mol})(432\text{ kJ/mol})] = -109\text{ kJ}$

4.80

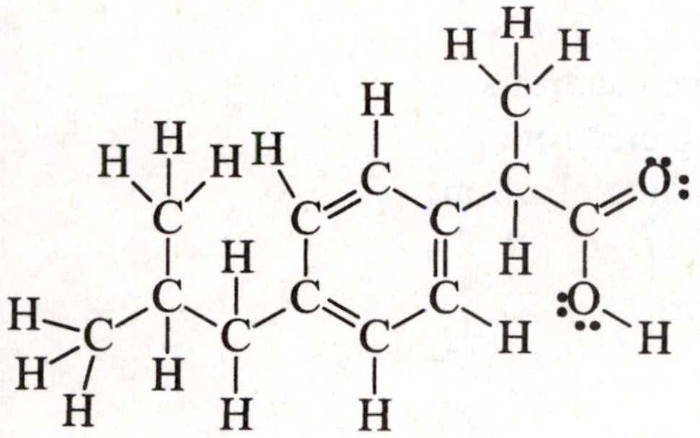

4.82

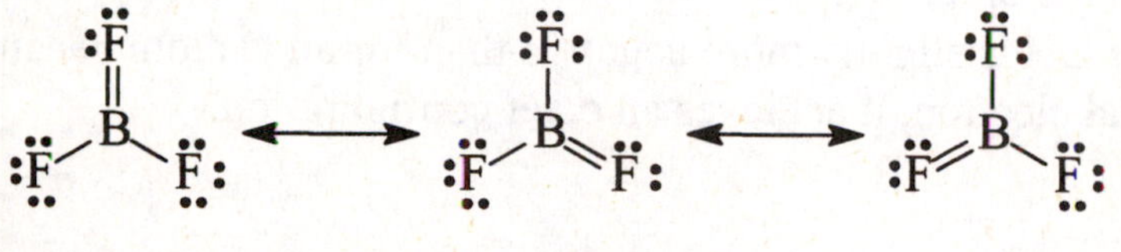

BF$_3$ is a resonance hybrid of the three structures shown here.

4.84   (a) Magnesium fluoride   (b) Phosphorous tribromide

   (c) Germanium tetrachloride

$$:\overset{\cdot\cdot}{\underset{\cdot\cdot}{Cl}}-Ge-\overset{\cdot\cdot}{\underset{\cdot\cdot}{Cl}}:$$

$$:\overset{\cdot\cdot}{\underset{\cdot\cdot}{Cl}}:$$

$$:\overset{\cdot\cdot}{\underset{\cdot\cdot}{Cl}}:$$

   (d) Calcium oxide

Multiconcept Problems

4.86   (a) Assume a 100.0 g sample. From the percent composition data, a 100.0 g sample contains 47.5 g S and 52.5 g Cl.

$$47.5 \text{ g S} \times \frac{1 \text{ mol S}}{32.065 \text{ g S}} = 1.48 \text{ mol S}$$

$$52.5 \text{ g Cl} \times \frac{1 \text{ mol Cl}}{35.453 \text{ g Cl}} = 1.48 \text{ mol Cl}$$

Because the two mole quantities are the same, the empirical formula is SCl.

$$:\overset{\cdot\cdot}{\underset{\cdot\cdot}{Cl}}-\overset{\cdot\cdot}{\underset{\cdot\cdot}{S}}-\overset{\cdot\cdot}{\underset{\cdot\cdot}{S}}-\overset{\cdot\cdot}{\underset{\cdot\cdot}{Cl}}:$$

For sulfur:	Isolated sulfur valence electrons	6
	Bound sulfur bonding electrons	4
	Bound sulfur nonbonding electrons	4
	Formal charge = 6 – ½(4) – 4 = 0	

For chlorine:	Isolated chlorine valence electrons	7
	Bound chlorine bonding electrons	2
	Bound chlorine nonbonding electrons	6
	Formal charge = 7 – ½(2) – 6 = 0	

4.88   (a)

$$\left[:\overset{\cdot\cdot}{\underset{\cdot\cdot}{O}}-H \right]^{-}$$

$$:\overset{\cdot\cdot}{O}-H$$

(b) The oxygen in OH has a half-filled 2p orbital that can accept the additional electron. For a 2p orbital, n = 2 and l = 1.

(c) The electron affinity for OH is slightly more negative than for an O atom because when OH gains an additional electron, it achieves an octet configuration.

5

Covalent Bonds and Molecular Structure

5.1

		Number of Bonded Atoms	Number of Lone Pairs	Shape
(a)	O_3	2	1	bent
(b)	H_3O^+	3	1	trigonal pyramidal
(c)	XeF_2	2	3	linear
(d)	PF_6^-	6	0	octahedral
(e)	$XeOF_4$	5	1	square pyramidal
(f)	AlH_4^-	4	0	tetrahedral
(g)	BF_4^-	4	0	tetrahedral
(h)	$SiCl_4$	4	0	tetrahedral
(i)	ICl_4^-	4	2	square planar
(j)	$AlCl_3$	3	0	trigonal planar

5.2

5.3   Benzene is a planar hexagon.

5.4   (a) tetrahedral   (b) seesaw

5.5   Each C is sp^3 hybridized. The C–C bonds are formed by the overlap of one singly occupied sp^3 hybrid orbital from each C. The C–H bonds are formed by the overlap of one singly occupied sp^3 orbital on C with a singly occupied H 1s orbital.

5.6

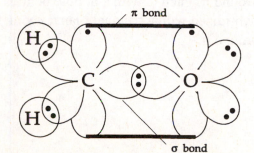

The carbon in formaldehyde is sp^2 hybridized.

5.7

The two Cs with four single bonds are sp³ hybridized. The C with a double bond is sp² hybridized. The C–C bonds are formed by the overlap of one singly occupied sp³ or sp² hybrid orbital from each C. The C–H bonds are formed by the overlap of one singly occupied sp³ orbital on C with a singly occupied H 1s orbital.

5.8

In HCN the carbon is sp hybridized.

5.9

In CO₂ the carbon is sp hybridized.

5.10

In Cl₂CO the carbon is sp2 hybridized.

5.11 (a) sp²   (b) sp   (c) sp³

5.12 (a), (b), (c)

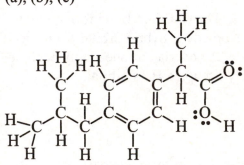

Each C with four single bonds is sp³ hybridized with a tetrahedral geometry around it. Each C with a double bond is sp² hybridized with a trigonal planar geometry around it.

5.13   For He_2^+ σ^*_{1s} $\uparrow$

　　　　　　　　　σ_{1s} $\uparrow\downarrow$

He_2^+ Bond order = $\dfrac{\left(\begin{array}{c}\text{number of}\\\text{bonding electrons}\end{array}\right) - \left(\begin{array}{c}\text{number of}\\\text{antibonding electrons}\end{array}\right)}{2} = \dfrac{2-1}{2} = 1/2$

He_2^+ should be stable with a bond order of 1/2.

5.14   For B_2

σ^*_{2p}　　—

π^*_{2p}　—　—

σ_{2p}　　—

π_{2p}　$\uparrow$　$\uparrow$

σ^*_{2s}　$\uparrow\downarrow$

σ_{2s}　$\uparrow\downarrow$

B_2 Bond order = $\dfrac{\left(\begin{array}{c}\text{number of}\\\text{bonding electrons}\end{array}\right) - \left(\begin{array}{c}\text{number of}\\\text{antibonding electrons}\end{array}\right)}{2} = \dfrac{4-2}{2} = 1$

B_2 is paramagnetic because it has two unpaired electrons in the π_{2p} molecular orbitals.

For C_2

σ^*_{2p}　　—

π^*_{2p}　—　—

σ_{2p}　　—

π_{2p}　$\uparrow\downarrow$　$\uparrow\downarrow$

σ^*_{2s}　$\uparrow\downarrow$

σ_{2s}　$\uparrow\downarrow$

C_2 Bond order = $\dfrac{6-2}{2} = 2$;　　C_2 is diamagnetic because all electrons are paired.

5.15

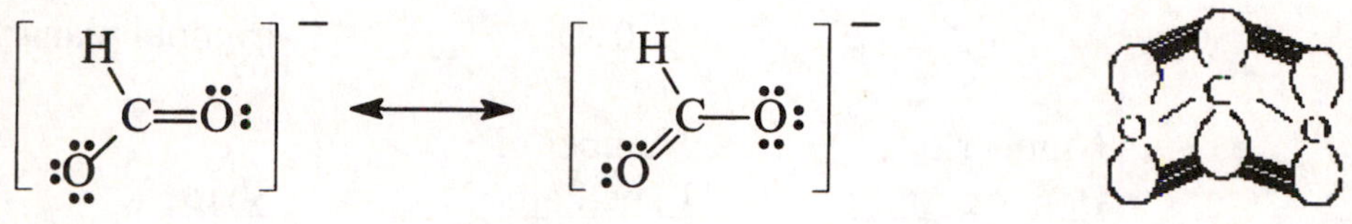

5.16   Molecular shape is important because a biologically active molecule is recognized and fits in a receptor molecule in the same way that a hand fits in a glove.

5.17   The mirror image of molecule (a) has the same shape as (a) and is identical to it in all respects, so there is no handedness associated with it. The mirror image of molecule (b) is different than (b) so there is a handedness to this molecule.

Conceptual Problems

5.18   (a) square pyramidal   (b) trigonal pyramidal
         (c) square planar   (d) trigonal planar

5.20   Molecular model (c) does not have a tetrahedral central atom. It is square planar.

5.22

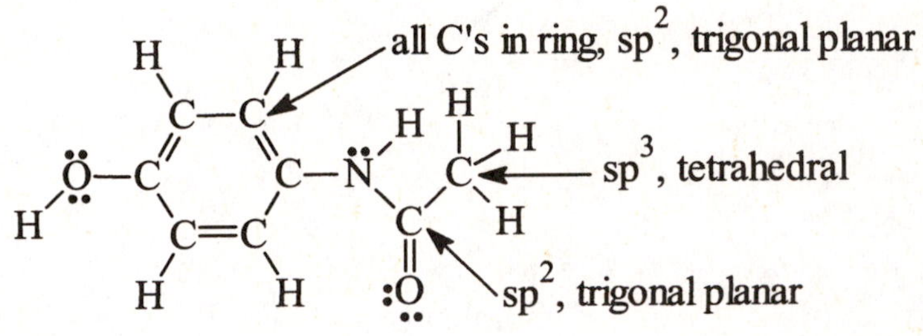

all C's in ring, sp^2, trigonal planar

sp^3, tetrahedral

sp^2, trigonal planar

Section Problems
The VSEPR Model (Section 5.1)

5.24   From data in Table 5.1:
       (a) trigonal planar   (b) trigonal bipyramidal   (c) linear   (d) octahedral

5.26   From data in Table 5.1:
       (a) tetrahedral, 4   (b) octahedral, 6   (c) bent, 3 or 4
       (d) linear, 2 or 5   (e) square pyramidal, 6   (f) trigonal pyramidal, 4

5.28

		Number of Bonded Atoms	Number of Lone Pairs	Shape
(a)	H_2Se	2	2	bent
(b)	$TiCl_4$	4	0	tetrahedral
(c)	O_3	2	1	bent
(d)	GaH_3	3	0	trigonal planar

5.30

		Number of Bonded Atoms	Number of Lone Pairs	Shape
(a)	SbF_5	5	0	trigonal bipyramidal
(b)	IF_4^+	4	1	seesaw
(c)	SeO_3^{2-}	3	1	trigonal pyramidal
(d)	CrO_4^{2-}	4	0	tetrahedral

5.32

		Number of Bonded Atoms	Number of Lone Pairs	Shape
(a)	PO_4^{3-}	4	0	tetrahedral
(b)	MnO_4^-	4	0	tetrahedral
(c)	SO_4^{2-}	4	0	tetrahedral
(d)	SO_3^{2-}	3	1	trigonal pyramidal

(e) ClO_4^-	4	0	tetrahedral
(f) SCN^-	2	0	linear

(C is the central atom)

5.34 (a) In SF_2 the sulfur is bound to two fluorines and contains two lone pairs of electrons. SF_2 is bent and the F–S–F bond angle is approximately 109°.
(b) In N_2H_2 each nitrogen is bound to the other nitrogen and one hydrogen. Each nitrogen has one lone pair of electrons. The H–N–N bond angle is approximately 120°.
(c) In KrF_4 the krypton is bound to four fluorines and contains two lone pairs of electrons. KrF_4 is square planar, and the F–Kr–F bond angle is 90°.
(d) In NOCl the nitrogen is bound to one oxygen and one chlorine and contains one lone pair of electrons. NOCl is bent, and the Cl–N–O bond angle is approximately 120°.

5.36

$H-C_a-H$	~120°
$H-C_a-C_b$	~120°
$C_a-C_b-C_c$	~120°

C_b-C_c-N	180°
C_a-C_b-H	~120°
$H-C_b-C_c$	~120°

5.38

The bond angles are 109.5° around carbon and 90° around S.

5.40 All six carbons in cyclohexane are bonded to two other carbons and two hydrogens (i.e., four charge clouds). The geometry about each carbon is tetrahedral with a C–C–C bond angle of approximately 109°. Because the geometry about each carbon is tetrahedral, the cyclohexane ring cannot be flat.

Valence Bond Theory (Sections 5.2–5.4)

5.42 In a π bond, the shared electrons occupy a region above and below a line connecting the two nuclei. A σ bond has its shared electrons located along the axis between the two nuclei.

5.44 See Table 5.2. (a) sp (b) sp^2 (c) sp^3

5.46 (a) sp^2 (b) sp^2 (c) sp^3 (d) sp^2

5.48

Carbons a, b, and d are sp^2 hybridized and carbon c is sp^3 hybridized.

The bond angles around carbons a, b, and d are ~120°. The bond angles around carbon c are ~109°. The terminal H-O-C bond angles are ~109°.

5.50 Each C with four single bonds is sp^3 hybridized. Each C with a double bond is sp^2 hybridized.

Molecular Orbital Theory (Sections 5.5 and 5.6)

5.52 Electrons in an atomic orbital are localized on a single atom. Electrons in a molecular orbital are delocalized over two or more atoms.

5.54

	O_2^+	O_2	O_2^-
σ^*_{2p}	—	—	—
π^*_{2p}	↑ —	↑ ↑	↑↓ ↑
π_{2p}	↑↓ ↑↓	↑↓ ↑↓	↑↓ ↑↓
σ_{2p}	↑↓	↑↓	↑↓
σ^*_{2s}	↑↓	↑↓	↑↓
σ_{2s}	↑↓	↑↓	↑↓

$$\text{Bond order} = \frac{\left(\begin{array}{c}\text{number of}\\ \text{bonding electrons}\end{array}\right) - \left(\begin{array}{c}\text{number of}\\ \text{antibonding electrons}\end{array}\right)}{2}$$

O_2^+ bond order $= \frac{8-3}{2} = 2.5$ O_2 bond order $= \frac{8-4}{2} = 2$

O_2^- bond order $= \frac{8-5}{2} = 1.5$

All are stable with bond orders between 1.5 and 2.5. All have unpaired electrons.

5.56

	C_2	C_2^-
σ^*_{2p}	—	—
π^*_{2p}	— —	— —
σ_{2p}	—	↑
π_{2p}	↑↓ ↑↓	↑↓ ↑↓
σ^*_{2s}	↑↓	↑↓
σ_{2s}	↑↓	↑↓

$$\text{Bond order} = \frac{\left(\begin{array}{c}\text{number of}\\ \text{bonding electrons}\end{array}\right) - \left(\begin{array}{c}\text{number of}\\ \text{antibonding electrons}\end{array}\right)}{2}$$

(a) C_2 bond order $= \frac{6-2}{2} = 2$

(b) Add one electron because it will go into a bonding molecular orbital.

(c) C_2^- bond order $= \frac{7-2}{2} = 2.5$

5.58

	(a) C_2^{2-}	(b) C_2^{2+}	(c) F_2^-	(d) Cl_2	(e) Li_2^+
σ^*_{2p}	—	—	↑	—	—
π^*_{2p}	— —	— —	↑↓ ↑↓	↑↓ ↑↓	— —
σ_{2p}	↑↓	—	↑↓	—	—
π_{2p}	↑↓ ↑↓	↑ ↑	↑↓ ↑↓	↑↓ ↑↓	— ↑
σ^*_{2s}	↑↓	↑↓	↑↓	↑↓	↑
σ_{2s}	↑↓	↑↓	↑↓	↑↓	↑↓

(a) C_2^{2-}
diamagnetic

(b) C_2^{2+}
paramagnetic

(c) F_2^-
paramagnetic

(d) Cl_2
diamagnetic

(e) Li_2^+
paramagnetic

5.60

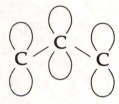 p orbitals in allyl cation

 allyl cation showing only the σ bonds (each C is sp² hybridized)

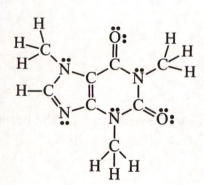 delocalized MO model for π bonding in the allyl cation

Chapter Problems

5.62

Each C with four single bonds is sp³ hybridized. Each C with a double bond is sp² hybridized.

5.64   Every carbon is sp² hybridized. There are 18 σ bonds and 5 π bonds.

5.66   The triple-bonded carbon atoms are sp hybridized. The theoretical bond angle for C–C≡C is 180°. Benzyne is so reactive because the C–C≡C bond angle is closer to 120° and is very strained.

5.68 C_2^{2-}

σ^*_{2p} ___

π^*_{2p} ___ ___

σ_{2p} ↑↓

π_{2p} ↑↓ ↑↓

σ^*_{2s} ↑↓

σ_{2s} ↑↓

Bond order $= \dfrac{\left(\begin{array}{c}\text{number of}\\\text{bonding electrons}\end{array}\right) - \left(\begin{array}{c}\text{number of}\\\text{antibonding electrons}\end{array}\right)}{2}$

C_2^{2-} bond order $= \dfrac{8-2}{2} = 3$; there is a triple bond between the two carbons.

5.70   (a)

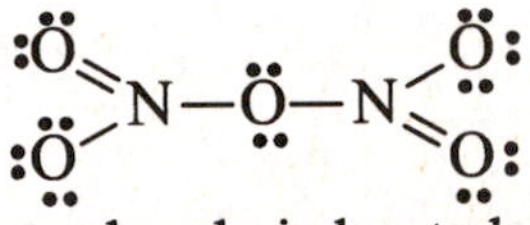

The molecule is bent about the central O and trigonal planar about each N.

(b) The central O is sp^3 hybridized. The N's are sp^2 hybridized.

(c)

(i) 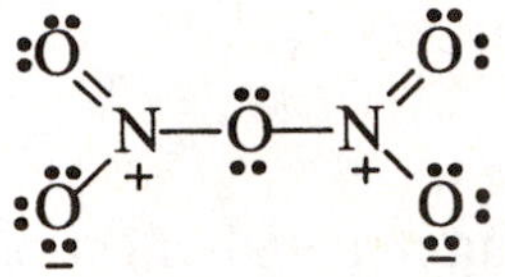   (ii) 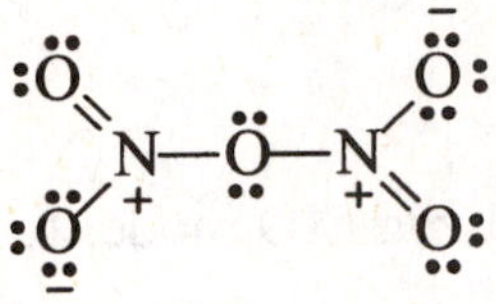   (iii)

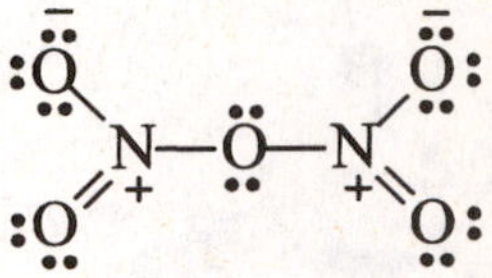

(iv) 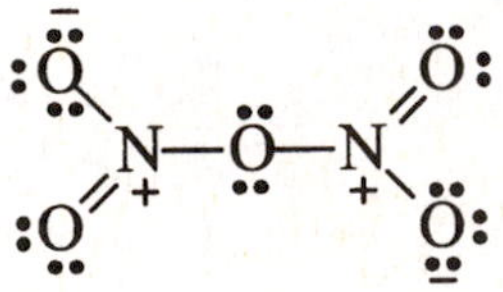   (v) 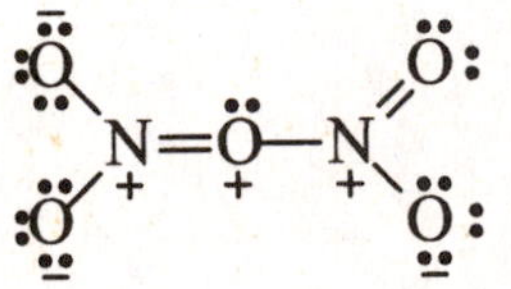   (vi)

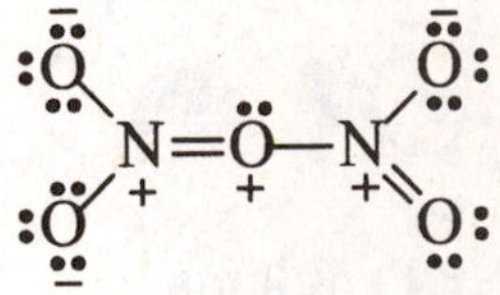

(vii) 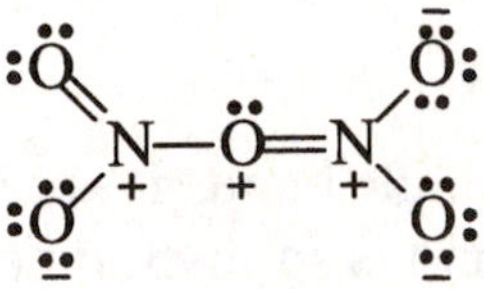   (viii)

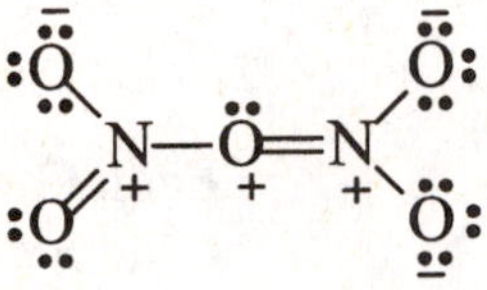

Structures (i) – (iv) make more important contributions to the resonance hybrid because of only –1 and 0 formal charges on the oxygens. A +1 formal charge is unlikely.

5.72

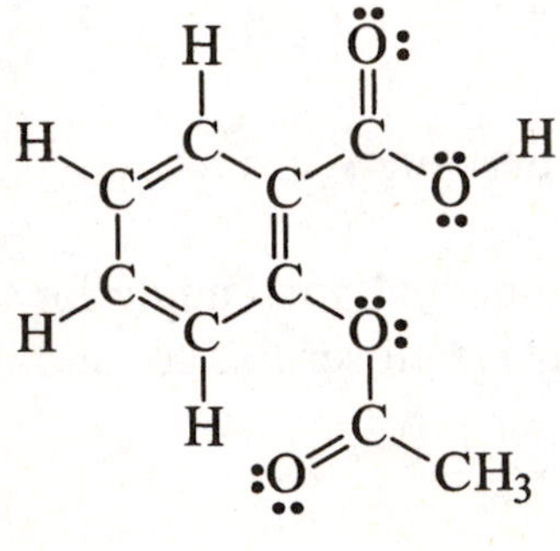

21 σ bonds
 5 π bonds
Each C with a double bond is sp^2 hybridized.
The –CH$_3$ carbon is sp^3 hybridized.

5.74 (a)

	S_2		S_2^{2-}	
σ^*_{3p}	—		—	
π^*_{3p}	↑	↑	↑↓	↑↓
π_{3p}	↑↓	↑↓	↑↓	↑↓
σ_{3p}	↑↓		↑↓	
σ^*_{3s}	↑↓		↑↓	
σ_{3s}	↑↓		↑↓	

(b) S_2 would be paramagnetic with two unpaired electrons in the π^*_{3p} MOs.

(c) Bond order $= \dfrac{\left(\begin{array}{c}\text{number of}\\\text{bonding electrons}\end{array}\right) - \left(\begin{array}{c}\text{number of}\\\text{antibonding electrons}\end{array}\right)}{2}$

S_2 bond order $= \dfrac{8-4}{2} = 2$

(d) S_2^{2-} bond order $= \dfrac{8-6}{2} = 1$

The two added electrons go into the antibonding π^*_{3p} MOs, the bond order drops from 2 to 1, and the bond length in S_2^{2-} should be longer than the bond length in S_2.

5.76 (a)

The left S has 5 electron clouds (4 bonding, 1 lone pair). The geometry about this S is seesaw. The right S has 4 electron clouds (2 bonding, 2 lone pairs). The geometry about this S is bent.

(b)

The left C has 4 electron clouds (4 bonding, 0 lone pairs). The geometry about this C is tetrahedral. The right C has 3 electron clouds (3 bonding, 0 lone pairs). The geometry about this C is trigonal planar. The central two C's have 2 electron clouds (2 bonding, 0 lone pairs). The geometry about these two C's is linear.

5.78

	Number of Bonded Atoms	Number of Lone Pairs	Shape
(a) BF_3	3	0	trigonal planar (120°)
PF_3	3	1	trigonal pyramidal (~107°)

PF_3 has the smaller F–X–F angles.

	Number of Bonded Atoms	Number of Lone Pairs	Shape
(b) PCl_4^+	4	0	tetrahedral (109.5°)
ICl_2^-	2	3	linear (180°)

PCl_4^+ has the smaller Cl–X–Cl angles.

(c) CCl_3^-	3	1	trigonal pyramidal (~107°)
PCl_6^-	6	0	octahedral (90°)

PCl_6^- has the smaller Cl–X–Cl angles.

Multiconcept Problems

5.80 (a)

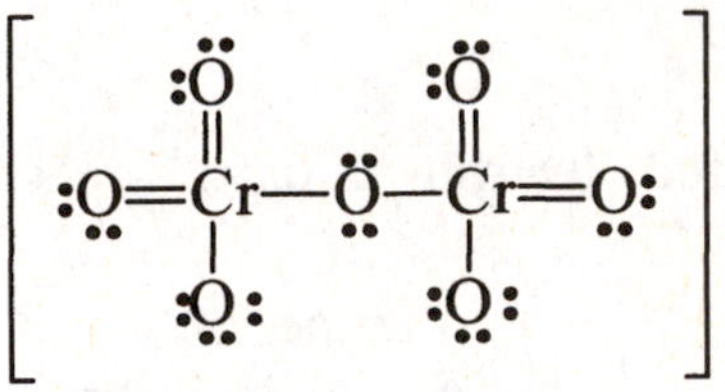

(b) Each Cr atom has 6 pairs of electrons around it. The likely geometry about each Cr atom is tetrahedral because each Cr has 4 charge clouds.

5.82 (a) Each carbon is sp^2 hybridized.
(b) & (c)

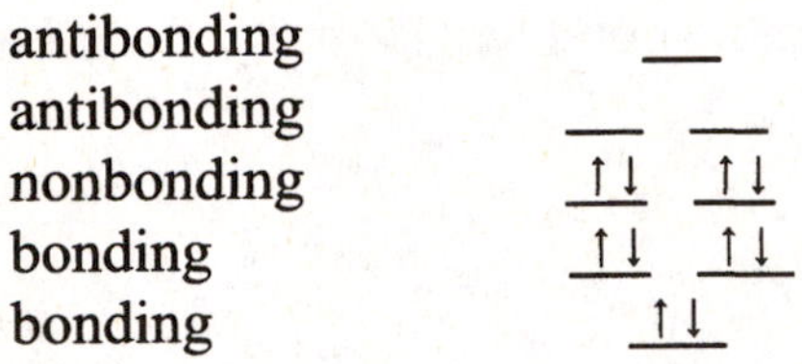

antibonding	—
antibonding	— —
nonbonding	↑↓ ↑↓
bonding	↑↓ ↑↓
bonding	↑↓

(d) The cyclooctatetraene dianion has only paired electrons and is diamagnetic.

6 Chemical Arithmetic: Stoichiometry

6.1 $2 \ NaClO_3 \rightarrow 2 \ NaCl + 3 \ O_2$

6.2   (a) $C_6H_{12}O_6 \rightarrow 2 \ C_2H_6O + 2 \ CO_2$
      (b) $6 \ CO_2 + 6 \ H_2O \rightarrow C_6H_{12}O_6 + 6 \ O_2$
      (c) $4 \ NH_3 + Cl_2 \rightarrow N_2H_4 + 2 \ NH_4Cl$

6.3 $3 \ A_2 + 2 \ B \rightarrow 2 \ BA_3$

6.4   (a) Fe_2O_3: $2(55.85) + 3(16.00) = 159.7$
      (b) H_2SO_4: $2(1.01) + 1(32.07) + 4(16.00) = 98.1$
      (c) $C_6H_8O_7$: $6(12.01) + 8(1.01) + 7(16.00) = 192.1$
      (d) $C_{16}H_{18}N_2O_4S$: $16(12.01) + 18(1.01) + 2(14.01) + 4(16.00) + 1(32.07) = 334.4$

6.5 $C_9H_8O_4$, 180.2;   500 mg = 500 x 10^{-3} g = 0.500 g

$$0.500 \ \text{g} \ \text{x} \ \frac{1 \ \text{mol}}{180.2 \ \text{g}} = 2.77 \ \text{x} \ 10^{-3} \ \text{mol aspirin}$$

$$2.77 \ \text{x} \ 10^{-3} \ \text{mol x} \ \frac{6.022 \ \text{x} \ 10^{23} \ \text{molecules}}{1 \ \text{mol}} = 1.67 \ \text{x} \ 10^{21} \ \text{aspirin molecules}$$

6.6   salicylic acid, $C_7H_6O_3$, 138.1;   acetic anhydride, $C_4H_6O_3$, 102.1
      aspirin, $C_9H_8O_4$, 180.2;   acetic acid, CH_3CO_2H, 60.1

$$\text{(a)} \ \ 4.50 \ \text{g} \ C_7H_6O_3 \ \text{x} \ \frac{1 \ \text{mol} \ C_7H_6O_3}{138.1 \ \text{g} \ C_7H_6O_3} \ \text{x} \ \frac{1 \ \text{mol} \ C_4H_6O_3}{1 \ \text{mol} \ C_7H_6O_3} \ \text{x} \ \frac{102.1 \ \text{g} \ C_4H_6O_3}{1 \ \text{mol} \ C_4H_6O_3} = 3.33 \ \text{g} \ C_4H_6O_3$$

$$\text{(b)} \ \ 4.50 \ \text{g} \ C_7H_6O_3 \ \text{x} \ \frac{1 \ \text{mol} \ C_7H_6O_3}{138.1 \ \text{g} \ C_7H_6O_3} \ \text{x} \ \frac{1 \ \text{mol} \ C_9H_8O_4}{1 \ \text{mol} \ C_7H_6O_3} \ \text{x} \ \frac{180.2 \ \text{g} \ C_9H_8O_4}{1 \ \text{mol} \ C_9H_8O_4} = 5.87 \ \text{g} \ C_9H_8O_4$$

$$\text{(c)} \ \ 4.50 \ \text{g} \ C_7H_6O_3 \ \text{x} \ \frac{1 \ \text{mol} \ C_7H_6O_3}{138.1 \ \text{g} \ C_7H_6O_3} \ \text{x} \ \frac{1 \ \text{mol} \ CH_3CO_2H}{1 \ \text{mol} \ C_7H_6O_3} \ \text{x} \ \frac{60.1 \ \text{g} \ CH_3CO_2H}{1 \ \text{mol} \ CH_3CO_2H} = 1.96 \ \text{g} \ CH_3CO_2H$$

6.7 C_2H_4, 28.1; C_2H_6O, 46.1

$$4.6 \ \text{g} \ C_2H_4 \ \text{x} \ \frac{1 \ \text{mol} \ C_2H_4}{28.1 \ \text{g} \ C_2H_4} \ \text{x} \ \frac{1 \ \text{mol} \ C_2H_6O}{1 \ \text{mol} \ C_2H_4} \ \text{x} \ \frac{46.1 \ \text{g} \ C_2H_6O}{1 \ \text{mol} \ C_2H_6O} = 7.5 \ \text{g} \ C_2H_6O$$
$$\text{(theoretical yield)}$$

$$\text{Percent yield} = \frac{\text{Actual yield}}{\text{Theoretical yield}} \ \text{x} \ 100\% = \frac{4.7 \ \text{g}}{7.5 \ \text{g}} \ \text{x} \ 100\% = 63\%$$

6.8 CH_4, 16.04; CH_2Cl_2, 84.93;   1.85 kg = 1850 g

$$1850 \text{ g } CH_4 \times \frac{1 \text{ mol } CH_4}{16.04 \text{ g } CH_4} \times \frac{1 \text{ mol } CH_2Cl_2}{1 \text{ mol } CH_4} \times \frac{84.93 \text{ g } CH_2Cl_2}{1 \text{ mol } CH_2Cl_2} = 9800 \text{ g } CH_2Cl_2 \text{ (theoretical yield)}$$

Actual yield = (9800 g)(0.431) = 4220 g CH_2Cl_2

6.9 Li_2O, 29.9: 65 kg = 65,000 g; H_2O, 18.0: 80.0 kg = 80,000 g

$$65,000 \text{ g } Li_2O \times \frac{1 \text{ mol } Li_2O}{29.9 \text{ g } Li_2O} = 2.17 \times 10^3 \text{ mol } Li_2O$$

$$80,000 \text{ g } H_2O \times \frac{1 \text{ mol } H_2O}{18.0 \text{ g } H_2O} = 4.44 \times 10^3 \text{ mol } H_2O$$

The reaction stoichiometry between Li_2O and H_2O is one to one. There are twice as many moles of H_2O as there are moles of Li_2O. Therefore, Li_2O is the limiting reactant. $(4.44 \times 10^3 \text{ mol} - 2.17 \times 10^3 \text{ mol}) = 2.27 \times 10^3 \text{ mol } H_2O$ remaining

$$2.27 \times 10^3 \text{ mol } H_2O \times \frac{18.0 \text{ g } H_2O}{1 \text{ mol } H_2O} = 40,860 \text{ g } H_2O = 40.9 \text{ kg} = 41 \text{ kg } H_2O$$

6.10   LiOH, 23.9; CO_2, 44.0

$$500.0 \text{ g LiOH} \times \frac{1 \text{ mol LiOH}}{23.9 \text{ g LiOH}} \times \frac{1 \text{ mol } CO_2}{1 \text{ mol LiOH}} \times \frac{44.0 \text{ g } CO_2}{1 \text{ mol } CO_2} = 921 \text{ g } CO_2$$

6.11   (a) $A + B_2 \rightarrow AB_2$
There is a 1:1 stoichiometry between the two reactants. A is the limiting reactant because there are fewer reactant A's than there are reactant B_2's.
(b) 1.0 mol of AB_2 can be made from 1.0 mol of A and 1.0 mol of B_2.

6.12   For dimethylhydrazine, $C_2H_8N_2$, divide each subscript by 2 to obtain the empirical formula. The empirical formula is CH_4N. $C_2H_8N_2$, 60.1 or 60.1 g/mol

$$\% C = \frac{2 \times 12.0 \text{ g}}{60.1 \text{ g}} \times 100\% = 39.9\%$$

$$\% H = \frac{8 \times 1.01 \text{ g}}{60.1 \text{ g}} \times 100\% = 13.4\%$$

$$\% N = \frac{2 \times 14.0 \text{ g}}{60.1 \text{ g}} \times 100\% = 46.6\%$$

6.13   Assume a 100.0 g sample. From the percent composition data, a 100.0 g sample contains 14.25 g C, 56.93 g O, and 28.83 g Mg.

$$14.25 \text{ g C} \times \frac{1 \text{ mol C}}{12.0 \text{ g C}} = 1.19 \text{ mol C}$$

$$56.93 \text{ g O} \times \frac{1 \text{ mol O}}{16.0 \text{ g O}} = 3.56 \text{ mol O}$$

$$28.83 \text{ g Mg} \times \frac{1 \text{ mol Mg}}{24.3 \text{ g Mg}} = 1.19 \text{ mol Mg}$$

$Mg_{1.19}C_{1.19}O_{3.56}$; divide each subscript by the smallest, 1.19.
$Mg_{1.19/1.19}C_{1.19/1.19}O_{3.56/1.19}$
The empirical formula is $MgCO_3$.

6.14 $C_6H_8O_7$, 192.1 or 192.1 g/mol

$$\% \text{ C} = \frac{6 \times 12.0 \text{ g}}{192.1 \text{ g}} \times 100\% = 37.5\%$$

$$\% \text{ H} = \frac{8 \times 1.01 \text{ g}}{192.1 \text{ g}} \times 100\% = 4.21\%$$

$$\% \text{ O} = \frac{7 \times 16.0 \text{ g}}{192.1 \text{ g}} \times 100\% = 58.3\%$$

6.15 $$1.161 \text{ g H}_2\text{O} \times \frac{1 \text{ mol H}_2\text{O}}{18.0 \text{ g H}_2\text{O}} \times \frac{2 \text{ mol H}}{1 \text{ mol H}_2\text{O}} = 0.129 \text{ mol H}$$

$$2.818 \text{ g CO}_2 \times \frac{1 \text{ mol CO}_2}{44.0 \text{ g CO}_2} \times \frac{1 \text{ mol C}}{1 \text{ mol CO}_2} = 0.0640 \text{ mol C}$$

$$0.129 \text{ mol H} \times \frac{1.01 \text{ g H}}{1 \text{ mol H}} = 0.130 \text{ g H}$$

$$0.0640 \text{ mol C} \times \frac{12.0 \text{ g C}}{1 \text{ mol C}} = 0.768 \text{ g C}$$

1.00 g total − (0.130 g H + 0.768 g C) = 0.102 g O

$$0.102 \text{ g O} \times \frac{1 \text{ mol O}}{16.0 \text{ g O}} = 0.006\,38 \text{ mol O}$$

$C_{0.0640}H_{0.129}O_{0.006\,38}$; divide each subscript by the smallest, 0.006 38.
$C_{0.0640/0.006\,38}H_{0.129/0.006\,38}O_{0.006\,38/0.006\,38}$
$C_{10.03}H_{20.22}O_1$
The empirical formula is $C_{10}H_{20}O$.

6.16   The empirical formula is CH_2O, 30: molecular mass = 150.

$$\frac{\text{molecular mass}}{\text{empirical formula mass}} = \frac{150}{30} = 5 \text{ ; therefore}$$

molecular formula = 5 x empirical formula = $C_{(5 \times 1)}H_{(5 \times 2)}O_{(5 \times 1)} = C_5H_{10}O_5$

6.17   (a) Assume a 100.0 g sample. From the percent composition data, a 100.0 g sample contains 21.86 g H and 78.14 g B.

$$21.86 \text{ g H} \times \frac{1 \text{ mol H}}{1.01 \text{ g H}} = 21.6 \text{ mol H}$$

$$78.14 \text{ g B} \times \frac{1 \text{ mol B}}{10.8 \text{ g B}} = 7.24 \text{ mol B}$$

$B_{7.24} H_{21.6}$; divide each subscript by the smaller, 7.24.

$B_{7.24/7.24} H_{21.6/7.24}$

The empirical formula is BH_3, 13.8.

$27.7 / 13.8 = 2$; molecular formula $= B_{(2 \times 1)}H_{(2 \times 3)} = B_2H_6$

(b) Assume a 100.0 g sample. From the percent composition data, a 100.0 g sample contains 6.71 g H, 40.00 g C, and 53.28 g O.

$$6.71 \text{ g H} \times \frac{1 \text{ mol H}}{1.01 \text{ g H}} = 6.64 \text{ mol H}$$

$$40.00 \text{ g C} \times \frac{1 \text{ mol C}}{12.0 \text{ g C}} = 3.33 \text{ mol C}$$

$$53.28 \text{ g O} \times \frac{1 \text{ mol O}}{16.0 \text{ g O}} = 3.33 \text{ mol O}$$

$C_{3.33} H_{6.64} O_{3.33}$; divide each subscript by the smallest, 3.33.

$C_{3.33/3.33} H_{6.64/3.33} O_{3.33/3.33}$

The empirical formula is CH_2O, 30.0.

$90.08 / 30.0 = 3$; molecular formula $= C_{(3 \times 1)}H_{(3 \times 2)}O_{(3 \times 1)} = C_3H_6O_3$

6.18 (a) 125 mL = 0.125 L; (0.20 mol/L)(0.125 L) = 0.025 mol $NaHCO_3$

(b) 650.0 mL = 0.6500 L; (2.50 mol/L)(0.6500 L) = 1.62 mol H_2SO_4

6.19 (a) NaOH, 40.0; 500.0 mL = 0.5000 L

$$1.25 \frac{\text{mol NaOH}}{\text{L}} \times 0.500 \text{ L} \times \frac{40.0 \text{ g NaOH}}{1 \text{ mol NaOH}} = 25.0 \text{ g NaOH}$$

(b) $C_6H_{12}O_6$, 180.2

$$0.250 \frac{\text{mol } C_6H_{12}O_6}{\text{L}} \times 1.50 \text{ L} \times \frac{180.2 \text{ g } C_6H_{12}O_6}{1 \text{ mol } C_6H_{12}O_6} = 67.6 \text{ g } C_6H_{12}O_6$$

6.20 $C_6H_{12}O_6$, 180.2

$$25.0 \text{ g } C_6H_{12}O_6 \times \frac{1 \text{ mol } C_6H_{12}O_6}{180.2 \text{ g } C_6H_{12}O_6} = 0.1387 \text{ mol } C_6H_{12}O_6$$

$$0.1387 \text{ mol} \times \frac{1 \text{ L}}{0.20 \text{ mol}} = 0.69 \text{ L}; \quad 0.69 \text{ L} = 690 \text{ mL}$$

6.21 $C_{27}H_{46}O$, 386.7; 750 mL = 0.750 L

$$0.005 \frac{\text{mol } C_{27}H_{46}O}{\text{L}} \times 0.750 \text{ L} \times \frac{386.7 \text{ g } C_{27}H_{46}O}{1 \text{ mol } C_{27}H_{46}O} = 1 \text{ g } C_{27}H_{46}O$$

6.22 $M_i \times V_i = M_f \times V_f$; $\quad M_f = \dfrac{M_i \times V_i}{V_f} = \dfrac{3.50 \text{ M} \times 75.0 \text{ mL}}{400.0 \text{ mL}} = 0.656 \text{ M}$

6.23 $M_i \times V_i = M_f \times V_f;$ $V_i = \dfrac{M_f \times V_f}{M_i} = \dfrac{0.500\ M \times 250.0\ mL}{18.0\ M} = 6.94\ mL$

     Dilute 6.94 mL of 18.0 M H_2SO_4 with enough water to make 250.0 mL of solution. The resulting solution will be 0.500 M H_2SO_4.

6.24   50.0 mL = 0.0500 L; (0.100 mol/L)(0.0500 L) = 5.00 x 10^{-3} mol NaOH

     $5.00 \times 10^{-3}\ mol\ NaOH \times \dfrac{1\ mol\ H_2SO_4}{2\ mol\ NaOH} = 2.50 \times 10^{-3}\ mol\ H_2SO_4$

     $volume = 2.50 \times 10^{-3}\ mol \times \dfrac{1\ L}{0.250\ mol} = 0.0100\ L;$ $0.0100\ L = 10.0\ mL\ H_2SO_4$

6.25 $HNO_3(aq) + KOH(aq) \rightarrow KNO_3(aq) + H_2O(l)$
     25.0 mL = 0.0250 L and 68.5 mL = 0.0685 L

     $0.150\ \dfrac{mol\ KOH}{L} \times 0.0250\ L \times \dfrac{1\ mol\ HNO_3}{1\ mol\ KOH} = 3.75 \times 10^{-3}\ mol\ HNO_3$

     $HNO_3\ molarity = \dfrac{3.75 \times 10^{-3}\ mol}{0.0685\ L} = 5.47 \times 10^{-2}\ M$

6.26   From the reaction stoichiometry, moles NaOH = moles CH_3CO_2H
     (0.200 mol/L)(0.0947 L) = 0.018 94 mol NaOH = 0.018 94 mol CH_3CO_2H

     $molarity = \dfrac{0.018\ 94\ mol}{0.0250\ L} = 0.758\ M$

6.27   Because the two volumes are equal (let the volume = y L), and the concentrations are proportional to the number of solute ions.

     $OH^-\ concentration = 1.00\ M \times \dfrac{y\ L}{12\ H^+} \times \dfrac{8\ OH^-}{y\ L} = 0.67\ M$

6.28   Main sources of error in calculating Avogadro's number by spreading oil on a pond are:
     (i)   the assumption that the oil molecules are tiny cubes
     (ii)   the assumption that the oil layer is one molecule thick
     (iii) the assumption of a molecular mass of 900 u for the oil

6.29   area of oil = 2.0 x 10^7 cm^2

     volume of oil = 4.9 cm^3 = area x 4 l = (2.0 x 10^7 cm^2) x 4 l

     $l = \dfrac{4.9\ cm^3}{(2.0 \times 10^7\ cm^2)(4)} = 6.125 \times 10^{-8}\ cm$

     area of oil = 2.0 x 10^7 cm^2 = l^2 x N = (6.125 x 10^{-8} cm)^2 x N

     $N = \dfrac{2.0 \times 10^7\ cm^2}{(6.125 \times 10^{-8}\ cm)^2} = 5.33 \times 10^{21}\ oil\ molecules$

$$\text{moles of oil} = (4.9 \text{ cm}^3) \times (0.95 \text{ g/cm}^3) \times \frac{1 \text{ mol oil}}{900 \text{ g oil}} = 0.0052 \text{ mol oil}$$

$$\text{Avogadro's number} = \frac{5.33 \times 10^{21} \text{ molecules}}{0.0052 \text{ mol}} = 1.0 \times 10^{24} \text{ molecules/mole}$$

Conceptual Problems

6.30   The concentration of a solution is cut in half when the volume is doubled. This is best represented by box (b).

6.32   The molecular formula for cytosine is $C_4H_5N_3O$.

$$\text{mol CO}_2 = 0.001 \text{ mol cyt} \times \frac{4 \text{ C}}{\text{cyt}} \times \frac{1 \text{ CO}_2}{\text{C}} = 0.004 \text{ mol CO}_2$$

$$\text{mol H}_2\text{O} = 0.001 \text{ mol cyt} \times \frac{5 \text{ H}}{\text{cyt}} \times \frac{1 \text{ H}_2\text{O}}{2 \text{ H}} = 0.0025 \text{ mol H}_2\text{O}$$

6.34 $C_{17}H_{18}F_3NO$ $17(12.01) + 18(1.01) + 3(19.00) + 1(14.01) + 1(16.00) = 309.36$

6.36   (a) $A_2 + 3 B_2 \rightarrow 2 AB_3$; B_2 is the limiting reactant because it is completely consumed.
        (b) For 1.0 mol of A_2, 3.0 mol of B_2 are required. Because only 1.0 mol of B_2 is available, B_2 is the limiting reactant.

$$1 \text{ mol B}_2 \times \frac{2 \text{ mol AB}_3}{3 \text{ mol B}_2} = 2/3 \text{ mol AB}_3$$

Section Problem
Balancing Equations (Section 6.1)

6.38   Equation (b) is balanced, (a) is not balanced.

6.40   (a) $Mg + 2 HNO_3 \rightarrow H_2 + Mg(NO_3)_2$
        (b) $CaC_2 + 2 H_2O \rightarrow Ca(OH)_2 + C_2H_2$
        (c) $2 S + 3 O_2 \rightarrow 2 SO_3$
        (d) $UO_2 + 4 HF \rightarrow UF_4 + 2 H_2O$

6.42   (a) $SiCl_4 + 2 H_2O \rightarrow SiO_2 + 4 HCl$
        (b) $P_4O_{10} + 6 H_2O \rightarrow 4 H_3PO_4$
        (c) $CaCN_2 + 3 H_2O \rightarrow CaCO_3 + 2 NH_3$
        (d) $3 NO_2 + H_2O \rightarrow 2 HNO_3 + NO$

Molecular Weights and Stoichiometry (Section 6.3)

6.44   (a) Hg_2Cl_2: $2(200.59) + 2(35.45) = 472.1$
        (b) $C_4H_8O_2$: $4(12.01) + 8(1.01) + 2(16.00) = 88.1$
        (c) CF_2Cl_2: $1(12.01) + 2(19.00) + 2(35.45) = 120.9$

6.46 (a) $C_{33}H_{35}FN_2O_5$: $33(12.01) + 35(1.01) + 1(19.00) + 2(14.01) + 5(16.00) = 558.7$
      (b) $C_{22}H_{27}F_3O_4S$: $22(12.01) + 27(1.01) + 3(19.00) + 4(16.00) + 1(32.06) = 444.6$
      (c) $C_{16}H_{16}ClNO_2S$: $16(12.01) + 16(1.01) + 1(35.45) + 1(14.01) + 2(16.00) + 1(32.06) = 321.8$

6.48 One mole equals the atomic mass or molecular mass in grams.
      (a) Ti, 47.87 g   (b) Br_2, 159.81 g   (c) Hg, 200.59 g   (d) H_2O, 18.02 g

6.50 There are 3 ions (one Mg^{2+} and 2 Cl^-) per formula unit of $MgCl_2$.
      $MgCl_2$, 95.2

$$27.5 \text{ g } MgCl_2 \times \frac{1 \text{ mol } MgCl_2}{95.2 \text{ g } MgCl_2} \times \frac{3 \text{ mol ions}}{1 \text{ mol } MgCl_2} = 0.867 \text{ mol ions}$$

6.52 Molar mass $= \dfrac{3.28 \text{ g}}{0.0275 \text{ mol}} = 119 \text{ g/mol}$; molecular mass $= 119$.

6.54 $FeSO_4$, 151.9; 300 mg = 0.300 g

$$0.300 \text{ g } FeSO_4 \times \frac{1 \text{ mol } FeSO_4}{151.9 \text{ g } FeSO_4} = 1.97 \times 10^{-3} \text{ mol } FeSO_4$$

$$1.97 \times 10^{-3} \text{ mol } FeSO_4 \times \frac{6.022 \times 10^{23} \text{ Fe(II) atoms}}{1 \text{ mol } FeSO_4} = 1.19 \times 10^{21} \text{ Fe(II) atoms}$$

6.56 $C_8H_{10}N_4O_2$, 194.2; 125 mg = 0.125 g

$$0.125 \text{ g caffeine} \times \frac{1 \text{ mol caffeine}}{194.2 \text{ g caffeine}} = 6.44 \times 10^{-4} \text{ mol caffeine}$$

$$0.125 \text{ g caffeine} \times \frac{1 \text{ mol caffeine}}{194.2 \text{ g caffeine}} \times \frac{6.022 \times 10^{23} \text{ molecules}}{1 \text{ mol}} = 3.88 \times 10^{20} \text{ caffeine molecules}$$

6.58 By definition, 6.022×10^{23} particles are 1.000 mole of particles.
      mol Ar = 0.2500 mol and mol "other" = 0.7500 mol

$$\text{mass Ar} = 0.2500 \text{ mol Ar} \times \frac{39.95 \text{ g Ar}}{1 \text{ mol Ar}} = 9.99 \text{ g Ar}$$

      mass "other" = total mass - mass Ar = 25.12 g - 9.99 g = 15.13 g "other"

$$\text{molar mass "other"} = \frac{15.13 \text{ g "other"}}{0.7500 \text{ mol "other"}} = 20.17 \text{ g/mol}$$

      The "other" element is neon (Ne).

6.60 TiO_2, 79.87; $100.0 \text{ kg Ti} \times \dfrac{79.87 \text{ kg } TiO_2}{47.87 \text{ kg Ti}} = 166.8 \text{ kg } TiO_2$

6.62 (a) $2\,Fe_2O_3 + 3\,C \rightarrow 4\,Fe + 3\,CO_2$

(b) Fe_2O_3, 159.7; $525\text{ g }Fe_2O_3 \times \dfrac{1\text{ mol }Fe_2O_3}{159.7\text{ g }Fe_2O_3} \times \dfrac{3\text{ mol }C}{2\text{ mol }Fe_2O_3} = 4.93\text{ mol C}$

(c) $4.93\text{ mol C} \times \dfrac{12.01\text{ g C}}{1\text{ mol C}} = 59.2\text{ g C}$

6.64 (a) $2\,Mg + O_2 \rightarrow 2\,MgO$

(b) Mg, 24.30; O_2, 32.00; MgO, 40.30

$25.0\text{ g Mg} \times \dfrac{1\text{ mol Mg}}{24.30\text{ g Mg}} \times \dfrac{1\text{ mol }O_2}{2\text{ mol Mg}} \times \dfrac{32.00\text{ g }O_2}{1\text{ mol }O_2} = 16.5\text{ g }O_2$

$25.0\text{ g Mg} \times \dfrac{1\text{ mol Mg}}{24.30\text{ g Mg}} \times \dfrac{2\text{ mol MgO}}{2\text{ mol Mg}} \times \dfrac{40.30\text{ g MgO}}{1\text{ mol MgO}} = 41.5\text{ g MgO}$

(c) $25.0\text{ g }O_2 \times \dfrac{1\text{ mol }O_2}{32.00\text{ g }O_2} \times \dfrac{2\text{ mol Mg}}{1\text{ mol }O_2} \times \dfrac{24.30\text{ g Mg}}{1\text{ mol Mg}} = 38.0\text{ g Mg}$

$25.0\text{ g }O_2 \times \dfrac{1\text{ mol }O_2}{32.00\text{ g }O_2} \times \dfrac{2\text{ mol MgO}}{1\text{ mol }O_2} \times \dfrac{40.30\text{ g MgO}}{1\text{ mol MgO}} = 63.0\text{ g MgO}$

6.66 (a) $2\,HgO \rightarrow 2\,Hg + O_2$

(b) HgO, 216.6; Hg, 200.6; O_2, 32.0

$45.5\text{ g HgO} \times \dfrac{1\text{ mol HgO}}{216.6\text{ g HgO}} \times \dfrac{2\text{ mol Hg}}{2\text{ mol HgO}} \times \dfrac{200.6\text{ g Hg}}{1\text{ mol Hg}} = 42.1\text{ g Hg}$

$45.5\text{ g HgO} \times \dfrac{1\text{ mol HgO}}{216.6\text{ g HgO}} \times \dfrac{1\text{ mol }O_2}{2\text{ mol HgO}} \times \dfrac{32.00\text{ g }O_2}{1\text{ mol }O_2} = 3.36\text{ g }O_2$

(c) $33.3\text{ g }O_2 \times \dfrac{1\text{ mol }O_2}{32.00\text{ g }O_2} \times \dfrac{2\text{ mol HgO}}{1\text{ mol }O_2} \times \dfrac{216.6\text{ g HgO}}{1\text{ mol HgO}} = 451\text{ g HgO}$

6.68 $2.00\text{ g Ag} \times \dfrac{1\text{ mol Ag}}{107.9\text{ g Ag}} = 0.0185\text{ mol Ag};\ \ 0.657\text{ g Cl} \times \dfrac{1\text{ mol Cl}}{35.45\text{ g Cl}} = 0.0185\text{ mol Cl}$

$Ag_{0.0185}Cl_{0.0185}$; divide both subscripts by 0.0185. The empirical formula is AgCl.

6.70 N_2H_4, 32.05; I_2, 253.8; HI, 127.9

(a) $36.7\text{ g }N_2H_4 \times \dfrac{1\text{ mol }N_2H_4}{32.05\text{ g }N_2H_4} \times \dfrac{2\text{ mol }I_2}{1\text{ mol }N_2H_4} \times \dfrac{253.8\text{ g }I_2}{1\text{ mol }I_2} = 581\text{ g }I_2$

(b) $115.7\text{ g }N_2H_4 \times \dfrac{1\text{ mol }N_2H_4}{32.05\text{ g }N_2H_4} \times \dfrac{4\text{ mol HI}}{1\text{ mol }N_2H_4} \times \dfrac{127.9\text{ g HI}}{1\text{ mol HI}} = 1847\text{ g HI}$

Limiting Reactants and Reaction Yield (Sections 6.4 and 6.5)

6.72 $3.44 \text{ mol } N_2 \times \dfrac{3 \text{ mol } H_2}{1 \text{ mol } N_2} = 10.3 \text{ mol } H_2$ required.

Because there is only 1.39 mol H_2, H_2 is the limiting reactant.

$1.39 \text{ mol } H_2 \times \dfrac{2 \text{ mol } NH_3}{3 \text{ mol } H_2} \times \dfrac{17.03 \text{ g } NH_3}{1 \text{ mol } NH_3} = 15.8 \text{ g } NH_3$

$1.39 \text{ mol } H_2 \times \dfrac{1 \text{ mol } N_2}{3 \text{ mol } H_2} \times \dfrac{28.01 \text{ g } N_2}{1 \text{ mol } N_2} = 13.0 \text{ g } N_2$ reacted

$3.44 \text{ mol } N_2 \times \dfrac{28.01 \text{ g } N_2}{1 \text{ mol } N_2} = 96.3 \text{ g } N_2$ initially

$(96.3 \text{ g} - 13.0 \text{ g}) = 83.3 \text{ g } N_2$ left over

6.74 C_2H_4, 28.05; Cl_2, 70.91; $C_2H_4Cl_2$, 98.96

$15.4 \text{ g } C_2H_4 \times \dfrac{1 \text{ mol } C_2H_4}{28.05 \text{ g } C_2H_4} = 0.549 \text{ mol } C_2H_4$

$3.74 \text{ g } Cl_2 \times \dfrac{1 \text{ mol } Cl_2}{70.91 \text{ g } Cl_2} = 0.0527 \text{ mol } Cl_2$

Because the reaction stoichiometry between C_2H_4 and Cl_2 is one to one, Cl_2 is the limiting reactant.

$0.0527 \text{ mol } Cl_2 \times \dfrac{1 \text{ mol } C_2H_4Cl_2}{1 \text{ mol } Cl_2} \times \dfrac{98.96 \text{ g } C_2H_4Cl_2}{1 \text{ mol } C_2H_4Cl_2} = 5.22 \text{ g } C_2H_4Cl_2$

6.76 H_2SO_4, 98.08; $NiCO_3$, 118.7; $NiSO_4$, 154.8

(a) $14.5 \text{ g } NiCO_3 \times \dfrac{1 \text{ mol } NiCO_3}{118.7 \text{ g } NiCO_3} \times \dfrac{1 \text{ mol } H_2SO_4}{1 \text{ mol } NiCO_3} \times \dfrac{98.08 \text{ g } H_2SO_4}{1 \text{ mol } H_2SO_4} = 12.0 \text{ g } H_2SO_4$

(b) $14.5 \text{ g } NiCO_3 \times \dfrac{1 \text{ mol } NiCO_3}{118.7 \text{ g } NiCO_3} \times \dfrac{1 \text{ mol } NiSO_4}{1 \text{ mol } NiCO_3} \times \dfrac{154.8 \text{ g } NiSO_4}{1 \text{ mol } NiSO_4} \times 0.789 = 14.9 \text{ g } NiSO_4$

6.78 $CaCO_3$, 100.1; HCl, 36.46

$CaCO_3 + 2 \text{ HCl} \rightarrow CaCl_2 + H_2O + CO_2$

$2.35 \text{ g } CaCO_3 \times \dfrac{1 \text{ mol } CaCO_3}{100.1 \text{ g } CaCO_3} = 0.0235 \text{ mol } CaCO_3$

$2.35 \text{ g } HCl \times \dfrac{1 \text{ mol } HCl}{36.46 \text{ g } HCl} = 0.0645 \text{ mol } HCl$

The reaction stoichiometry is 1 mole of $CaCO_3$ for every 2 moles of HCl. For 0.0235 mol $CaCO_3$, we only need 2(0.0235 mol) = 0.0470 mol HCl. We have 0.0645 mol HCl, therefore $CaCO_3$ is the limiting reactant.

$0.0235 \text{ mol } CaCO_3 \times \dfrac{1 \text{ mol } CO_2}{1 \text{ mol } CaCO_3} \times \dfrac{22.4 \text{ L}}{1 \text{ mol } CO_2} = 0.526 \text{ L } CO_2$

6.80 $CH_3CO_2H + C_5H_{12}O \rightarrow C_7H_{14}O_2 + H_2O$

CH_3CO_2H, 60.05; $C_5H_{12}O$, 88.15; $C_7H_{14}O_2$, 130.19

$$3.58 \text{ g } CH_3CO_2H \text{ x } \frac{1 \text{ mol } CH_3CO_2H}{60.05 \text{ g } CH_3CO_2H} = 0.0596 \text{ mol } CH_3CO_2H$$

$$4.75 \text{ g } C_5H_{12}O \text{ x } \frac{1 \text{ mol } C_5H_{12}O}{88.15 \text{ g } C_5H_{12}O} = 0.0539 \text{ mol } C_5H_{12}O$$

Because the reaction stoichiometry between CH_3CO_2H and $C_5H_{12}O$ is one to one, isopentyl alcohol ($C_5H_{12}O$) is the limiting reactant.

$$0.0539 \text{ mol } C_5H_{12}O \text{ x } \frac{1 \text{ mol } C_7H_{14}O_2}{1 \text{ mol } C_5H_{12}O} \text{ x } \frac{130.19 \text{ g } C_7H_{14}O_2}{1 \text{ mol } C_7H_{14}O_2} = 7.02 \text{ g } C_7H_{14}O_2$$

7.02 g $C_7H_{14}O_2$ is the theoretical yield. Actual yield = (7.02 g)(0.45) = 3.2 g.

6.82 $CH_3CO_2H + C_5H_{12}O \rightarrow C_7H_{14}O_2 + H_2O$

CH_3CO_2H, 60.05; $C_5H_{12}O$, 88.15; $C_7H_{14}O_2$, 130.19

$$1.87 \text{ g } CH_3CO_2H \text{ x } \frac{1 \text{ mol } CH_3CO_2H}{60.05 \text{ g } CH_3CO_2H} = 0.0311 \text{ mol } CH_3CO_2H$$

$$2.31 \text{ g } C_5H_{12}O \text{ x } \frac{1 \text{ mol } C_5H_{12}O}{88.15 \text{ g } C_5H_{12}O} = 0.0262 \text{ mol } C_5H_{12}O$$

Because the reaction stoichiometry between CH_3CO_2H and $C_5H_{12}O$ is one to one, isopentyl alcohol ($C_5H_{12}O$) is the limiting reactant.

$$0.0262 \text{ mol } C_5H_{12}O \text{ x } \frac{1 \text{ mol } C_7H_{14}O_2}{1 \text{ mol } C_5H_{12}O} \text{ x } \frac{130.19 \text{ g } C_7H_{14}O_2}{1 \text{ mol } C_7H_{14}O_2} = 3.41 \text{ g } C_7H_{14}O_2$$

3.41 g $C_7H_{14}O_2$ is the theoretical yield.

$$\% \text{ Yield} = \frac{\text{Actual yield}}{\text{Theoretical yield}} \text{ x } 100\% = \frac{2.96 \text{ g}}{3.41 \text{ g}} \text{ x } 100\% = 86.8\%$$

Formulas and Elemental Analysis (Sections 6.6 and 6.7)

6.84 CH_4N_2O, 60.1

$$\% C = \frac{12.0 \text{ g C}}{60.1 \text{ g}} \text{ x } 100\% = 20.0\%$$

$$\% H = \frac{4 \text{ x } 1.01 \text{ g H}}{60.1 \text{ g}} \text{ x } 100\% = 6.72\%$$

$$\% N = \frac{2 \text{ x } 14.0 \text{ g N}}{60.1 \text{ g}} \text{ x } 100\% = 46.6\%$$

$$\% O = \frac{16.0 \text{ g O}}{60.1 \text{ g}} \text{ x } 100\% = 26.6\%$$

6.86   Assume a 100.0 g sample of liquid. From the percent composition data, a 100.0 g sample of liquid contains 5.57 g H, 28.01 g Cl, and 66.42 g C.

$$66.42 \text{ g C x } \frac{1 \text{ mol C}}{12.01 \text{ g C}} = 5.530 \text{ mol C}$$

$$5.57 \text{ g H x } \frac{1 \text{ mol H}}{1.01 \text{ g H}} = 5.51 \text{ mol H}$$

$$28.01 \text{ g Cl x } \frac{1 \text{ mol Cl}}{35.45 \text{ g Cl}} = 0.7901 \text{ mol Cl}$$

$C_{5.530}H_{5.51}Cl_{0.7901}$; divide each subscript by the smallest, 0.7901.
$C_{5.530/0.7901}H_{5.51/0.7901}Cl_{0.7901/0.7901}$
C_7H_7Cl
The empirical formula is C_7H_7Cl, 126.59.
Because the molecular mass equals the empirical formula mass, the empirical formula is also the molecular formula.

6.88 Assume a 100.0 g sample. From the percent composition data, a 100.0 g sample contains 24.25 g F and 75.75 g Sn.

$$24.25 \text{ g F x } \frac{1 \text{ mol F}}{19.00 \text{ g F}} = 1.276 \text{ mol F}$$

$$75.75 \text{ g Sn x } \frac{1 \text{ mol Sn}}{118.7 \text{ g Sn}} = 0.6382 \text{ mol Sn}$$

$Sn_{0.6382}F_{1.276}$; divide each subscript by the smaller, 0.6382.
$Sn_{0.6382/0.6382}F_{1.276/0.6382}$
The empirical formula is SnF_2.

6.90 Mass of toluene sample = 45.62 mg = 0.045 62 g; mass of CO_2 = 152.5 mg = 0.1525 g; mass of H_2O = 35.67 mg = 0.035 67 g

$$0.1525 \text{ g CO}_2 \text{ x } \frac{1 \text{ mol CO}_2}{44.01 \text{ g CO}_2} \text{ x } \frac{1 \text{ mol C}}{1 \text{ mol CO}_2} = 0.003\ 465 \text{ mol C}$$

$$\text{mass C} = 0.003\ 465 \text{ mol C x } \frac{12.011 \text{ g C}}{1 \text{ mol C}} = 0.041\ 62 \text{ g C}$$

$$0.035\ 67 \text{ g H}_2O \text{ x } \frac{1 \text{ mol H}_2O}{18.02 \text{ g H}_2O} \text{ x } \frac{2 \text{ mol H}}{1 \text{ mol H}_2O} = 0.003\ 959 \text{ mol H}$$

$$\text{mass H} = 0.003\ 959 \text{ mol H x } \frac{1.008 \text{ g H}}{1 \text{ mol H}} = 0.003\ 991 \text{ g H}$$

The (mass C + mass H) = 0.041 62 g + 0.003 991 g = 0.045 61 g. The calculated mass of (C + H) essentially equals the mass of the toluene sample, this means that toluene contains only C and H and no other elements.
$C_{0.003\ 465}H_{0.003\ 959}$; divide each subscript by the smaller, 0.003 465.
$C_{0.003\ 465/0.003\ 465}H_{0.003\ 959/0.003\ 465}$
$CH_{1.14}$; multiply each subscript by 7 to obtain integers.
The empirical formula is C_7H_8.

6.92   Let X equal the molecular mass of cytochrome c.

$$0.0043 = \frac{55.847}{X}; \quad X = \frac{55.847}{0.0043} = 13{,}000$$

6.94   Let X equal the molecular mass of disilane.

$$0.9028 = \frac{2 \times 28.09}{X}; \quad X = \frac{2 \times 28.09}{0.9028} = 62.23$$

$62.23 - 2(\text{Si atomic mass}) = 62.23 - 2(28.09) = 6.05$
6.05 is the total mass of H atoms.

$$6.05 \times \frac{1 \text{ H atom}}{1.01} = 6 \text{ H atoms}; \text{ disilane is } Si_2H_6.$$

6.96   Mass of the sample is 70.042 11
For C_5H_{10}, mass $= 5(12.000\,000) + 10(1.007\,825) = 70.078\,250$
For C_4H_6O, mass $= 4(12.000\,000) + 6(1.007\,825) + 1(15.994\,915) = 70.041\,865$
For $C_3H_6N_2$, mass $= 3(12.000\,000) + 6(1.007\,825) + 2(14.003\,074) = 70.053\,098$
The sample is C_4H_6O.

6.98 $C_{12}Br_{10}$, 943.2; $C_{12}Br_{10}O$, 959.2; 17.33 mg = 0.017 33 g

For $C_{12}Br_{10}$, % C $= \dfrac{12 \times 12.01 \text{ g C}}{943.2 \text{ g}} \times 100\% = 15.28\%$

For $C_{12}Br_{10}O$, % C $= \dfrac{12 \times 12.01 \text{ g C}}{959.2 \text{ g}} \times 100\% = 15.03\%$

Calculate the mass of C in 17.33 mg of CO_2.

$$0.017\,33 \text{ g } CO_2 \times \frac{1 \text{ mol } CO_2}{44.01 \text{ g } CO_2} \times \frac{1 \text{ mol C}}{1 \text{ mol } CO_2} \times \frac{12.01 \text{ g C}}{1 \text{ mol C}} = 0.004\,729 \text{ g} = 4.729 \text{ mg C}$$

Calculate the %C in the 31.472 mg sample.

$$\%C = \frac{4.729 \text{ mg C}}{31.472 \text{ mg}} \times 100\% = 15.03\%$$

Decabrom is $C_{12}Br_{10}O$.

Molarity, Solution Stoichiometry, Dilution, and Titration (Sections 6.8–6.11)

6.100 (a) 35.0 mL = 0.0350 L; $\dfrac{1.200 \text{ mol } HNO_3}{L} \times 0.0350 \text{ L} = 0.0420 \text{ mol } HNO_3$

(b) 175 mL = 0.175 L; $\dfrac{0.67 \text{ mol } C_6H_{12}O_6}{L} \times 0.175 \text{ L} = 0.12 \text{ mol } C_6H_{12}O_6$

6.102 $BaCl_2$, 208.2

$$15.0 \text{ g } BaCl_2 \times \frac{1 \text{ mol } BaCl_2}{208.2 \text{ g } BaCl_2} = 0.0720 \text{ mol } BaCl_2$$

$$0.0720 \text{ mol} \times \frac{1.0 \text{ L}}{0.45 \text{ mol}} = 0.16 \text{ L}; \quad 0.16 \text{ L} = 160 \text{ mL}$$

6.104   NaCl, 58.4;   400 mg = 0.400 g;   100 mL = 0.100 L

$$0.400 \text{ g NaCl} \times \frac{1 \text{ mol NaCl}}{58.4 \text{ g NaCl}} = 0.006\ 85 \text{ mol NaCl}$$

$$\text{molarity} = \frac{0.006\ 85 \text{ mol}}{0.100 \text{ L}} = 0.0685 \text{ M}$$

6.106 $3.045 \text{ g Cu} \times \dfrac{1 \text{ mol Cu}}{63.546 \text{ g Cu}} = 0.047\ 92 \text{ mol Cu};$ 50.0 mL = 0.0500 L

$$Cu(NO_3)_2 \text{ molarity} = \frac{0.047\ 92 \text{ mol}}{0.0500 \text{ L}} = 0.958 \text{ M}$$

6.108 $M_f \times V_f = M_i \times V_i;$ $M_f = \dfrac{M_i \times V_i}{V_f} = \dfrac{12.0 \text{ M} \times 35.7 \text{ mL}}{250.0 \text{ mL}} = 1.71 \text{ M HCl}$

6.110 $2 \text{ HBr(aq)} + K_2CO_3\text{(aq)} \rightarrow 2 \text{ KBr(aq)} + CO_2\text{(g)} + H_2O\text{(l)}$
        K_2CO_3, 138.2;   450 mL = 0.450 L

$$\frac{0.500 \text{ mol HBr}}{L} \times 0.450 \text{ L} = 0.225 \text{ mol HBr}$$

$$0.225 \text{ mol HBr} \times \frac{1 \text{ mol } K_2CO_3}{2 \text{ mol HBr}} \times \frac{138.2 \text{ g } K_2CO_3}{1 \text{ mol } K_2CO_3} = 15.5 \text{ g } K_2CO_3$$

6.112 $H_2C_2O_4$, 90.04

$$3.225 \text{ g } H_2C_2O_4 \times \frac{1 \text{ mol } H_2C_2O_4}{90.04 \text{ g } H_2C_2O_4} \times \frac{2 \text{ mol KMnO}_4}{5 \text{ mol } H_2C_2O_4} = 0.0143 \text{ mol KMnO}_4$$

$$0.0143 \text{ mol} \times \frac{1 \text{ L}}{0.250 \text{ mol}} = 0.0572 \text{ L} = 57.2 \text{ mL}$$

Chapter Problems

6.114   NaCl, 58.4;   KCl, 74.6; $CaCl_2$, 111.0;   500 mL = 0.500 L

$$4.30 \text{ g NaCl} \times \frac{1 \text{ mol NaCl}}{58.4 \text{ g NaCl}} = 0.0736 \text{ mol NaCl}$$

$$0.150 \text{ g KCl} \times \frac{1 \text{ mol KCl}}{74.6 \text{ g KCl}} = 0.002\ 01 \text{ mol KCl}$$

$$0.165 \text{ g } CaCl_2 \times \frac{1 \text{ mol } CaCl_2}{111.0 \text{ g } CaCl_2} = 0.001\ 49 \text{ mol } CaCl_2$$

$$0.0736 \text{ mol} + 0.002\ 01 \text{ mol} + 2(0.001\ 49 \text{ mol}) = 0.0786 \text{ mol } Cl^-$$

$$Na^+ \text{ molarity} = \frac{0.0736 \text{ mol}}{0.500 \text{ L}} = 0.147 \text{ M}$$

$$\text{Ca}^{2+} \text{ molarity} = \frac{0.001\ 49\ \text{mol}}{0.500\ \text{L}} = 0.002\ 98\ \text{M}$$

$$\text{K}^+ \text{ molarity} = \frac{0.002\ 01\ \text{mol}}{0.500\ \text{L}} = 0.004\ 02\ \text{M}$$

$$\text{Cl}^- \text{ molarity} = \frac{0.0786\ \text{mol}}{0.500\ \text{L}} = 0.157\ \text{M}$$

6.116 (a) 1.0×10^{-11} g/mL x $\dfrac{1\ \text{mol Au}}{197.0\ \text{g Au}}$ x $\dfrac{1000\ \text{mL}}{1\ \text{L}} = 5.1 \times 10^{-11}$ mol/L

   (b) 1.0×10^{-11} g/mL x $\dfrac{1000\ \text{mL}}{1\ \text{L}}$ x 1.3×10^{21} L $= 1.3 \times 10^{13}$ g Au

6.118 (a) $C_6H_{12}O_6$, 180.2

$$\% \text{C} = \frac{6 \times 12.01\ \text{g C}}{180.2\ \text{g}} \times 100\% = 39.99\%$$

$$\% \text{H} = \frac{12 \times 1.008\ \text{g H}}{180.2\ \text{g}} \times 100\% = 6.713\%$$

$$\% \text{O} = \frac{6 \times 16.00\ \text{g O}}{180.2\ \text{g}} \times 100\% = 53.27\%$$

   (b) H_2SO_4, 98.08

$$\% \text{H} = \frac{2 \times 1.008\ \text{g H}}{98.08\ \text{g}} \times 100\% = 2.055\%$$

$$\% \text{S} = \frac{32.07\ \text{g S}}{98.08\ \text{g}} \times 100\% = 32.70\%$$

$$\% \text{O} = \frac{4 \times 16.00\ \text{g O}}{98.08\ \text{g}} \times 100\% = 65.25\%$$

   (c) $KMnO_4$, 158.0

$$\% \text{K} = \frac{39.10\ \text{g K}}{158.0\ \text{g}} \times 100\% = 24.75\%$$

$$\% \text{Mn} = \frac{54.94\ \text{g Mn}}{158.0\ \text{g}} \times 100\% = 34.77\%$$

$$\% \text{O} = \frac{4 \times 16.00\ \text{g O}}{158.0\ \text{g}} \times 100\% = 40.51\%$$

   (d) $C_7H_5NO_3S$, 183.2

$$\% \text{C} = \frac{7 \times 12.01\ \text{g C}}{183.2\ \text{g}} \times 100\% = 45.89\%$$

$$\% \text{H} = \frac{5 \times 1.008\ \text{g H}}{183.2\ \text{g}} \times 100\% = 2.751\%$$

$$\% \, N = \frac{14.01 \text{ g N}}{183.2 \text{ g}} \times 100\% = 7.647\%$$

$$\% \, O = \frac{3 \times 16.00 \text{ g O}}{183.2 \text{ g}} \times 100\% = 26.20\%$$

$$\% \, S = \frac{32.07 \text{ g S}}{183.2 \text{ g}} \times 100\% = 17.51\%$$

6.120 WCl_6, 396.6; W_6Cl_{12}, 1528

(a) $6 \, WCl_6 + 8 \, Bi \rightarrow W_6Cl_{12} + 8 \, BiCl_3$

(b) $150.0 \text{ g } WCl_6 \times \dfrac{1 \text{ mol } WCl_6}{396.6 \text{ g } WCl_6} \times \dfrac{8 \text{ mol Bi}}{6 \text{ mol } WCl_6} \times \dfrac{209.0 \text{ g Bi}}{1 \text{ mol Bi}} = 105.4 \text{ g Bi}$

(c) Compute the theoretical yield for W_6Cl_{12} from each reactant to determine the limiting reactant.

$$228 \text{ g } WCl_6 \times \frac{1 \text{ mol } WCl_6}{396.6 \text{ g } WCl_6} \times \frac{1 \text{ mol } W_6Cl_{12}}{6 \text{ mol } WCl_6} \times \frac{1528 \text{ g } W_6Cl_{12}}{1 \text{ mol } W_6Cl_{12}} = 146 \text{ g } W_6Cl_{12}$$

$$175 \text{ g Bi} \times \frac{1 \text{ mol Bi}}{209.0 \text{ g Bi}} \times \frac{1 \text{ mol } W_6Cl_{12}}{8 \text{ mol Bi}} \times \frac{1528 \text{ g } W_6Cl_{12}}{1 \text{ mol } W_6Cl_{12}} = 160 \text{ g } W_6Cl_{12}$$

The smaller theoretical yield (146 g) means that WCl_6 is the limiting reactant and 146 g W_6Cl_{12} are produced.

6.122 Assume a 100.0 g sample of ferrocene. From the percent composition data, a 100.0 g sample contains 5.42 g H, 64.56 g C, and 30.02 g Fe.

$$5.42 \text{ g H} \times \frac{1 \text{ mol H}}{1.01 \text{ g H}} = 5.37 \text{ mol H}$$

$$64.56 \text{ g C} \times \frac{1 \text{ mol C}}{12.01 \text{ g C}} = 5.376 \text{ mol C}$$

$$30.02 \text{ g Fe} \times \frac{1 \text{ mol Fe}}{55.85 \text{ g Fe}} = 0.5375 \text{ mol Fe}$$

$C_{5.376}H_{5.37}Fe_{0.5375}$; divide each subscript by the smallest, 0.5375.
$C_{5.376\,/\,0.5375}H_{5.37\,/\,0.5375}Fe_{0.5375\,/\,0.5375}$
The empirical formula is $C_{10}H_{10}Fe$.

6.124 Na_2SO_4, 142.04; Na_3PO_4, 163.94; Li_2SO_4, 109.95; 100.00 mL = 0.10000 L

$$0.550 \text{ g } Na_2SO_4 \times \frac{1 \text{ mol } Na_2SO_4}{142.04 \text{ g } Na_2SO_4} = 0.003 \, 872 \text{ mol } Na_2SO_4$$

$$1.188 \text{ g } Na_3PO_4 \times \frac{1 \text{ mol } Na_3PO_4}{163.94 \text{ g } Na_3PO_4} = 0.007 \, 247 \text{ mol } Na_3PO_4$$

$$0.223 \text{ g } Li_2SO_4 \times \frac{1 \text{ mol } Li_2SO_4}{109.95 \text{ g } Li_2SO_4} = 0.002 \, 028 \text{ mol } Li_2SO_4$$

$$\text{Na}^+ \text{ molarity} = \frac{(2 \times 0.003\ 872 \text{ mol}) + (3 \times 0.007\ 247 \text{ mol})}{0.100\ 00 \text{ L}} = 0.295 \text{ M}$$

$$\text{Li}^+ \text{ molarity} = \frac{2 \times 0.002\ 028 \text{ mol}}{0.100\ 00 \text{ L}} = 0.0406 \text{ M}$$

$$\text{SO}_4^{2-} \text{ molarity} = \frac{(1 \times 0.003\ 872 \text{ mol}) + (1 \times 0.002\ 028 \text{ mol})}{0.100\ 00 \text{ L}} = 0.0590 \text{ M}$$

$$\text{PO}_4^{3-} \text{ molarity} = \frac{1 \times 0.007\ 247 \text{ mol}}{0.100\ 00 \text{ L}} = 0.0725 \text{ M}$$

6.126 The combustion reaction is: $2 \text{ C}_8\text{H}_{18} + 25 \text{ O}_2 \rightarrow 16 \text{ CO}_2 + 18 \text{ H}_2\text{O}$
C_8H_{18}, 114.23; CO_2, 44.01

$$\text{pounds CO}_2 = 1.00 \text{ gal} \times \frac{3.7854 \text{ L}}{1 \text{ gal}} \times \frac{1000 \text{ mL}}{1 \text{ L}} \times \frac{0.703 \text{ g C}_8\text{H}_{18}}{1 \text{ mL}} \times \frac{1 \text{ mol C}_8\text{H}_{18}}{114.23 \text{ g C}_8\text{H}_{18}} \times$$

$$\frac{16 \text{ mol CO}_2}{2 \text{ mol C}_8\text{H}_{18}} \times \frac{44.01 \text{ g CO}_2}{1 \text{ mol CO}_2} \times \frac{1 \text{ lb}}{453.59 \text{ g}} = 18.1 \text{ lb CO}_2$$

6.128 AgCl, 143.32; CO_2, 44.01; H_2O, 18.02

$$\text{mol Cl in 1.00 g of X} = 1.95 \text{ g AgCl} \times \frac{1 \text{ mol AgCl}}{143.32 \text{ g AgCl}} \times \frac{1 \text{ mol Cl}}{1 \text{ mol AgCl}} = 0.0136 \text{ mol Cl}$$

$$\text{mass Cl} = 0.0136 \text{ mol Cl} \times \frac{35.453 \text{ g Cl}}{1 \text{ mol Cl}} = 0.482 \text{ g Cl}$$

$$\text{mol C in 1.00 g of X} = 0.900 \text{ g CO}_2 \times \frac{1 \text{ mol CO}_2}{44.01 \text{ g CO}_2} \times \frac{1 \text{ mol C}}{1 \text{ mol CO}_2} = 0.0204 \text{ mol C}$$

$$\text{mass C} = 0.0204 \text{ mol C} \times \frac{12.011 \text{ g C}}{1 \text{ mol C}} = 0.245 \text{ g C}$$

$$\text{mol H in 1.00 g of X} = 0.735 \text{ g H}_2\text{O} \times \frac{1 \text{ mol H}_2\text{O}}{18.02 \text{ g H}_2\text{O}} \times \frac{2 \text{ mol H}}{1 \text{ mol H}_2\text{O}} = 0.0816 \text{ mol H}$$

$$\text{mass H} = 0.0816 \text{ mol H} \times \frac{1.008 \text{ g H}}{1 \text{ mol H}} = 0.0823 \text{ g H}$$

mass N = 1.00 g – mass Cl – mass C – mass H = 1.00 – 0.482 g – 0.245 g – 0.0823 g = 0.19 g N

$$\text{mol N in 1.00 g of X} = 0.19 \text{ g N} \times \frac{1 \text{ mol N}}{14.01 \text{ g N}} = 0.014 \text{ mol N}$$

Determine empirical formula.

$\text{C}_{0.0204}\text{H}_{0.0816}\text{N}_{0.014}\text{Cl}_{0.0136}$; divide each subscript by the smallest, 0.0136.

$\text{C}_{0.0204 / 0.0136}\text{H}_{0.0816 / 0.0136}\text{N}_{0.014 / 0.0136}\text{Cl}_{0.0136 / 0.0136}$

$\text{C}_{1.5}\text{H}_6\text{NCl}$, multiply each subscript by 2 to get integers.

The empirical formula is $\text{C}_3\text{H}_{12}\text{N}_2\text{Cl}_2$.

6.130   Let SA stand for salicylic acid.

$$\text{mol C in 1.00 g of SA} = 2.23 \text{ g CO}_2 \times \frac{1 \text{ mol CO}_2}{44.01 \text{ g CO}_2} \times \frac{1 \text{ mol C}}{1 \text{ mol CO}_2} = 0.0507 \text{ mol C}$$

$$\text{mass C} = 0.0507 \text{ mol C} \times \frac{12.011 \text{ g C}}{1 \text{ mol C}} = 0.609 \text{ g C}$$

$$\text{mol H in 1.00 g of SA} = 0.39 \text{ g H}_2\text{O} \times \frac{1 \text{ mol H}_2\text{O}}{18.02 \text{ g H}_2\text{O}} \times \frac{2 \text{ mol H}}{1 \text{ mol H}_2\text{O}} = 0.043 \text{ mol H}$$

$$\text{mass H} = 0.043 \text{ mol H} \times \frac{1.008 \text{ g H}}{1 \text{ mol H}} = 0.043 \text{ g H}$$

mass O = 1.00 g – mass C – mass H = 1.00 – 0.609 g – 0.043 g = 0.35 g O

$$\text{mol O in 1.00 g of} = 0.35 \text{ g N} \times \frac{1 \text{ mol O}}{16.00 \text{ g O}} = 0.022 \text{ mol O}$$

Determine empirical formula.

$C_{0.0507}H_{0.043}O_{0.022}$; divide each subscript by the smallest, 0.022.

$C_{0.0507/0.022}H_{0.043/0.022}O_{0.022/0.022}$

$C_{2.3}H_2O$, multiply each subscript by 3 to get integers.

The empirical formula is $C_7H_6O_3$. The empirical formula mass = 138.12 g/mol.

Because salicylic acid has only one acidic hydrogen, there is a 1 to 1 mol ratio between salicylic acid and NaOH in the acid-base titration.

$$\text{mol SA in 1.00 g SA} = 72.4 \text{ mL} \times \frac{1 \text{ L}}{1000 \text{ mL}} \times \frac{0.100 \text{ mol NaOH}}{1 \text{ L}} \times \frac{1 \text{ mol SA}}{1 \text{ mol NaOH}} =$$
$$0.00724 \text{ mol SA}$$

$$\text{SA molar mass} = \frac{1.00 \text{ g}}{0.00724 \text{ mol}} = 138 \text{ g/mol}$$

Because the empirical formula mass and the molar mass are the same, the empirical formula is the molecular formula for salicylic acid.

6.132   Let X equal the mass of benzoic acid and Y the mass of gallic acid in the 1.00 g mixture.
Therefore, X + Y = 1.00 g.
Because both acids contain only one acidic hydrogen, there is a 1 to 1 mol ratio between each acid and NaOH in the acid-base titration.
In the titration, mol benzoic acid + mol gallic acid = mol NaOH.

$$\text{Therefore, X} \times \frac{1 \text{ mol BA}}{122 \text{ g BA}} + Y \times \frac{1 \text{ mol GA}}{170 \text{ g GA}} = \text{mol NaOH}$$

$$\text{mol NaOH} = 14.7 \text{ mL} \times \frac{1 \text{ L}}{1000 \text{ mL}} \times \frac{0.500 \text{ mol NaOH}}{1 \text{ L}} = 0.00735 \text{ mol NaOH}$$

We have two unknowns, X and Y, and two equations.

X + Y = 1.00 g

$$X \times \frac{1 \text{ mol BA}}{122 \text{ g BA}} + Y \times \frac{1 \text{ mol GA}}{170 \text{ g GA}} = 0.00735 \text{ mol NaOH}$$

Rearrange to get X = 1.00 g – Y and then substitute it into the equation above to solve for Y.

$$(1.00 \text{ g} - Y) \times \frac{1 \text{ mol BA}}{122 \text{ g BA}} + Y \times \frac{1 \text{ mol GA}}{170 \text{ g GA}} = 0.00735 \text{ mol NaOH}$$

$$\frac{1 \text{ mol}}{122} - \frac{Y \text{ mol}}{122 \text{ g}} + \frac{Y \text{ mol}}{170 \text{ g}} = 0.00735 \text{ mol}$$

$$-\frac{Y \text{ mol}}{122 \text{ g}} + \frac{Y \text{ mol}}{170 \text{ g}} = 0.00735 \text{ mol} - \frac{1 \text{ mol}}{122} = -8.47 \times 10^{-4} \text{ mol}$$

$$\frac{(-Y \text{ mol})(170 \text{ g}) + (Y \text{ mol})(122 \text{ g})}{(170 \text{ g})(122 \text{ g})} = -8.47 \times 10^{-4} \text{ mol}$$

$$\frac{-48 \text{ Y mol}}{20740 \text{ g}} = -8.47 \times 10^{-4} \text{ mol}; \quad \frac{48 \text{ Y}}{20740 \text{ g}} = 8.47 \times 10^{-4}$$

$$Y = \frac{(20740 \text{ g})(8.47 \times 10^{-4})}{48} = 0.366 \text{ g}$$

$X = 1.00 \text{ g} - 0.366 \text{ g} = 0.634 \text{ g}$

In the 1.00 g mixture there is 0.63 g of benzoic acid and 0.37 g of gallic acid.

6.134 FeO, 71.85; Fe_2O_3, 159.7

Let X equal the mass of FeO and Y the mass of Fe_2O_3 in the 10.0 g mixture. Therefore, $X + Y = 10.0$ g.

$$\text{mol Fe} = 7.43 \text{ g} \times \frac{1 \text{ mol Fe}}{55.85 \text{ g Fe}} = 0.133 \text{ mol Fe}$$

$$\text{mol FeO} + 2 \times \text{mol Fe}_2O_3 = 0.133 \text{ mol Fe}$$

$$X \times \frac{1 \text{ mol FeO}}{71.85 \text{ g FeO}} + 2 \times \left(Y \times \frac{1 \text{ mol Fe}_2O_3}{159.7 \text{ g Fe}_2O_3} \right) = 0.133 \text{ mol Fe}$$

Rearrange to get $X = 10.0 \text{ g} - Y$ and then substitute it into the equation above to solve for Y.

$$(10.0 \text{ g} - Y) \times \frac{1 \text{ mol FeO}}{71.85 \text{ g FeO}} + 2 \times \left(Y \times \frac{1 \text{ mol Fe}_2O_3}{159.7 \text{ g Fe}_2O_3} \right) = 0.133 \text{ mol Fe}$$

$$\frac{10.0 \text{ mol}}{71.85} - \frac{Y \text{ mol}}{71.85 \text{ g}} + \frac{2 \text{ Y mol}}{159.7 \text{ g}} = 0.133 \text{ mol}$$

$$-\frac{Y \text{ mol}}{71.85 \text{ g}} + \frac{2 \text{ Y mol}}{159.7 \text{ g}} = 0.133 \text{ mol} - \frac{10.0 \text{ mol}}{71.85} = -0.0062 \text{ mol}$$

$$\frac{(-Y \text{ mol})(159.7 \text{ g}) + (2 \text{ Y mol})(71.85 \text{ g})}{(71.85 \text{ g})(159.7 \text{ g})} = -0.0062 \text{ mol}$$

$$\frac{-16.0 \text{ Y mol}}{11474 \text{ g}} = -0.0062 \text{ mol}; \quad \frac{16.0 \text{ Y}}{11474 \text{ g}} = 0.0062$$

$Y = (0.0062)(11474 \text{ g})/16.0 = 4.44 \text{ g} = 4.4 \text{ g Fe}_2O_3$

$X = 10.0 \text{ g} - Y = 10.0 \text{ g} - 4.4 \text{ g} = 5.6 \text{ g FeO}$

6.136 $C_6H_{12}O_6 + 6 O_2 \rightarrow 6 CO_2 + 6 H_2O$; $C_6H_{12}O_6$, 180.16; CO_2, 44.01

$$66.3 \text{ g C}_6H_{12}O_6 \times \frac{1 \text{ mol C}_6H_{12}O_6}{180.16 \text{ g C}_6H_{12}O_6} \times \frac{6 \text{ mol CO}_2}{1 \text{ mol C}_6H_{12}O_6} \times \frac{44.01 \text{ g CO}_2}{1 \text{ mol CO}_2} = 97.2 \text{ g CO}_2$$

$$66.3 \text{ g C}_6\text{H}_{12}\text{O}_6 \text{ x } \frac{1 \text{ mol C}_6\text{H}_{12}\text{O}_6}{180.16 \text{ g C}_6\text{H}_{12}\text{O}_6} \text{ x } \frac{6 \text{ mol CO}_2}{1 \text{ mol C}_6\text{H}_{12}\text{O}_6} \text{ x } \frac{25.4 \text{ L CO}_2}{1 \text{ mol CO}_2} = 56.1 \text{ L CO}_2$$

6.138   Mass of Cu = 2.196 g; mass of S = 2.748 g – 2.196 g = 0.552 g S

(a) $\%\text{Cu} = \dfrac{2.196 \text{ g}}{2.748 \text{ g}} \text{ x } 100\% = 79.91\%$

$\%\text{S} = \dfrac{0.552 \text{ g}}{2.748 \text{ g}} \text{ x } 100\% = 20.1\%$

(b) $2.196 \text{ g Cu x } \dfrac{1 \text{ mol Cu}}{63.55 \text{ g Cu}} = 0.034\,55 \text{ mol Cu}$

$0.552 \text{ g S x } \dfrac{1 \text{ mol S}}{32.07 \text{ g S}} = 0.0172 \text{ mol S}$

$\text{Cu}_{0.03455}\text{S}_{0.0172}$; divide each subscript by the smaller, 0.0172.
$\text{Cu}_{0.03455 / 0.0172}\text{S}_{0.0172 / 0.0172}$
The empirical formula is Cu_2S.

(c) Cu_2S, 159.16

$$\frac{5.6 \text{ g Cu}_2\text{S}}{1 \text{ cm}^3} \text{ x } \frac{1 \text{ mol Cu}_2\text{S}}{159.16 \text{ g Cu}_2\text{S}} \text{ x } \frac{2 \text{ mol Cu}^+ \text{ ions}}{1 \text{ mol Cu}_2\text{S}} \text{ x } \frac{6.022 \text{ x } 10^{23} \text{ Cu}^+ \text{ ions}}{1 \text{ mol Cu}^+ \text{ ions}}$$

$= 4.2 \text{ x } 10^{22} \text{ Cu}^+ \text{ ions/cm}^3$

6.140 NH_4NO_3, 80.04; $(\text{NH}_4)_2\text{HPO}_4$, 132.06
Assume you have a 100.0 g sample of the mixture.
Let X = grams of NH_4NO_3 and $(100.0 - \text{X})$ = grams of $(\text{NH}_4)_2\text{HPO}_4$.
Both compounds contain 2 nitrogen atoms per formula unit.
Because the mass % N in the sample is 30.43%, the 100.0 g sample contains 30.43 g N.

$$\text{mol NH}_4\text{NO}_3 = (\text{X}) \text{ x } \frac{1 \text{ mol NH}_4\text{NO}_3}{80.04 \text{ g}}$$

$$\text{mol } (\text{NH}_4)_2\text{HPO}_4 = (100.0 - \text{X}) \text{ x } \frac{1 \text{ mol } (\text{NH}_4)_2\text{HPO}_4}{132.06 \text{ g}}$$

$$\text{mass N} = \left(\left((\text{X}) \text{ x } \frac{1 \text{ mol NH}_4\text{NO}_3}{80.04 \text{ g}} \right) + \left((100.0 - \text{X}) \text{ x } \frac{1 \text{ mol } (\text{NH}_4)_2\text{HPO}_4}{132.06 \text{ g}} \right) \right) \text{ x }$$

$$\left(\frac{2 \text{ mol N}}{1 \text{ mol ammonium cmpds}} \right) \text{ x } \left(\frac{14.0067 \text{ g N}}{1 \text{ mol N}} \right) = 30.43 \text{ g}$$

Solve for X.

$$\left(\frac{\text{X}}{80.04} + \frac{100.0 - \text{X}}{132.06} \right) (2)(14.0067) = 30.43$$

$$\left(\frac{\text{X}}{80.04} + \frac{100.0 - \text{X}}{132.06} \right) = 1.08627$$

$$\frac{(132.06)(\text{X}) + (100.0 - \text{X})(80.04)}{(80.04)(132.06)} = 1.08627$$

$(132.06)(X) + (100.0 - X)(80.04) = (1.08627)(80.04)(132.06)$

$132.06X + 8004 - 80.04X = 11481.96$

$132.06X - 80.04X = 11481.96 - 8004$

$52.02X = 3477.96$

$X = \dfrac{3477.96}{52.02} = 66.86 \text{ g } NH_4NO_3$

$(100.0 - X) = (100.0 - 66.86) = 33.14 \text{ g } (NH_4)_2HPO_4$

$\dfrac{mass_{NH_4NO_3}}{mass_{(NH_4)_2HPO_4}} = \dfrac{66.86 \text{ g}}{33.14 \text{ g}} = 2.018$

The mass ratio of NH_4NO_3 to $(NH_4)_2HPO_4$ in the mixture is 2 to 1.

Multiconcept Problems

6.142 (a) 56.0 mL = 0.0560 L

$mol \ X_2 = (0.0560 \text{ L } X_2)\left(\dfrac{1 \text{ mol}}{22.41 \text{ L}}\right) = 0.00250 \text{ mol } X_2$

$mass \ X_2 = 1.12 \text{ g } MX_2 - 0.720 \text{ g } MX = 0.40 \text{ g } X_2$

$molar \ mass \ X_2 = \dfrac{0.40 \text{ g}}{0.00250 \text{ mol}} = 160 \text{ g/mol}$

atomic mass of X = 160/2 = 80;   X is Br.

(b) $mol \ MX = 0.00250 \text{ mol } X_2 \ \times \ \dfrac{2 \text{ mol } MX}{1 \text{ mol } X_2} = 0.00500 \text{ mol } MX$

$mass \ of \ X \ in \ MX = 0.00500 \text{ mol } MX \times \dfrac{1 \text{ mol } X}{1 \text{ mol } MX} \times \dfrac{80 \text{ g } X}{1 \text{ mol } X} = 0.40 \text{ g } X$

mass of M in MX = 0.720 g MX – 0.40 g X = 0.32 g M

$molar \ mass \ M = \dfrac{0.32 \text{ g}}{0.00500 \text{ mol}} = 64 \text{ g/mol}$

atomic mass of X = 64;   M is Cu.

6.144 AgCl, 143.32

(a) $mass \ Cl \ in \ AgCl = 1.126 \text{ g } AgCl \ \times \ \dfrac{35.453 \text{ g } Cl}{143.32 \text{ g } AgCl} = 0.279 \text{ g } Cl$

$\%Cl \ in \ alkaline \ earth \ chloride = \dfrac{0.279 \text{ g } Cl}{0.436 \text{ g}} \times 100\% = 64.0\% \ Cl$

(b) Because M is an alkaline earth metal, M is a 2+ cation.

For MCl_2, mass of M = 0.436 g – 0.279 g = 0.157 g M

$mol \ M = 0.279 \text{ g } Cl \times \dfrac{1 \text{ mol } Cl}{35.453 \text{ g } Cl} \times \dfrac{1 \text{ mol } M}{2 \text{ mol } Cl} = 0.003 \ 93 \text{ mol } M$

$molar \ mass \ for \ M = \dfrac{0.157 \text{ g}}{0.003 \ 93 \text{ mol}} = 39.9 \text{ g/mol}; \quad M = Ca$

(c) $Ca(s) + Cl_2(g) \rightarrow CaCl_2(s)$

$CaCl_2(aq) + 2 AgNO_3(aq) \rightarrow 2 AgCl(s) + Ca(NO_3)_2(aq)$

(d) $1.005 \text{ g Ca} \times \dfrac{1 \text{ mol Ca}}{40.078 \text{ g Ca}} = 0.0251 \text{ mol Ca}$

$1.91 \times 10^{22} \text{ Cl}_2 \text{ molecules} \times \dfrac{1 \text{ mol Cl}_2}{6.022 \times 10^{23} \text{ Cl}_2 \text{ molecules}} = 0.0317 \text{ mol Cl}_2$

Because the stoichiometry between Ca and Cl_2 is one to one, the Cl_2 is in excess.

Mass Cl_2 unreacted $= (0.0317 - 0.0251) \text{ mol Cl}_2 \times \dfrac{70.91 \text{ g Cl}_2}{1 \text{ mol Cl}_2} = 0.47 \text{ g Cl}_2$ unreacted

6.146 (a) (i) $M_2O_3(s) + 3 C(s) + 3 Cl_2(g) \rightarrow 2 MCl_3(l) + 3 CO(g)$

(ii) $2 MCl_3(l) + 3 H_2(g) \rightarrow 2 M(s) + 6 HCl(g)$

(b) $HCl(aq) + NaOH(aq) \rightarrow H_2O(l) + NaCl(aq)$

$144.2 \text{ mL} = 0.1442 \text{ L}$

mol NaOH $= (0.511 \text{ mol/L})(0.1442 \text{ L}) = 0.07369 \text{ mol NaOH}$

mol HCl $= 0.07369 \text{ mol NaOH} \times \dfrac{1 \text{ mol HCl}}{1 \text{ mol NaOH}} = 0.07369 \text{ mol HCl}$

mol M $= 0.07369 \text{ mol HCl} \times \dfrac{2 \text{ mol M}}{6 \text{ mol HCl}} = 0.02456 \text{ mol M}$

mol $M_2O_3 = 0.02456 \text{ mol M} \times \dfrac{2 \text{ mol MCl}_3}{2 \text{ mol M}} \times \dfrac{1 \text{ mol M}_2O_3}{2 \text{ mol MCl}_3} = 0.01228 \text{ mol M}_2O_3$

molar mass $M_2O_3 = \dfrac{0.855 \text{ g}}{0.01228 \text{ mol}} = 69.6 \text{ g/mol}$; molecular mass $M_2O_3 = 69.6$

atomic mass of M $= \dfrac{69.6 - (3 \times 16.0)}{2} = 10.8$; M = B

(c) mass of M $= 0.02456 \text{ mol M} \times \dfrac{10.81 \text{ g M}}{1 \text{ mol M}} = 0.265 \text{ g M}$

Reactions in Aqueous Solution

7.1 $FeBr_3$ contains 3 Br^- ions. The molar concentration of Br^- ions = 3 x 0.225 M = 0.675 M.

7.2 A_2Y is the strongest electrolyte because it is completely dissociated into ions.
A_2X is the weakest electrolyte because it is the least dissociated of the three substances.

7.3 (a) precipitation (b) redox (c) acid-base neutralization

7.4 (a) Ionic equation:
$2 Ag^+(aq) + 2 NO_3^-(aq) + 2 Na^+(aq) + CrO_4^{2-}(aq) \rightarrow Ag_2CrO_4(s) + 2 Na^+(aq) + 2 NO_3^-(aq)$
Delete spectator ions from the ionic equation to get the net ionic equation.
Net ionic equation: $2 Ag^+(aq) + CrO_4^{2-}(aq) \rightarrow Ag_2CrO_4(s)$
(b) Ionic equation:
$2 H^+(aq) + SO_4^{2-}(aq) + MgCO_3(s) \rightarrow H_2O(l) + CO_2(g) + Mg^{2+}(aq) + SO_4^{2-}(aq)$
Delete spectator ions from the ionic equation to get the net ionic equation.
Net ionic equation: $2 H^+(aq) + MgCO_3(s) \rightarrow H_2O(l) + CO_2(g) + Mg^{2+}(aq)$
(c) Ionic equation:
$Hg^{2+}(aq) + 2 NO_3^-(aq) + 2 NH_4^+(aq) + 2 I^-(aq) \rightarrow HgI_2(s) + 2 NH_4^+(aq) + 2 NO_3^-(aq)$
Delete spectator ions from the ionic equation to get the net ionic equation.
Net ionic equation: $Hg^{2+}(aq) + 2 I^-(aq) \rightarrow HgI_2(s)$

7.5 (a) $CdCO_3$, insoluble (b) MgO, insoluble (c) Na_2S, soluble
(d) $PbSO_4$, insoluble (e) $(NH_4)_3PO_4$, soluble (f) $HgCl_2$, soluble

7.6 (a) Ionic equation:
$Ni^{2+}(aq) + 2 Cl^-(aq) + 2 NH_4^+(aq) + S^{2-}(aq) \rightarrow NiS(s) + 2 NH_4^+(aq) + 2 Cl^-(aq)$
Delete spectator ions from the ionic equation to get the net ionic equation.
Net ionic equation: $Ni^{2+}(aq) + S^{2-}(aq) \rightarrow NiS(s)$
(b) Ionic equation:
$2 Na^+(aq) + CrO_4^{2-}(aq) + Pb^{2+}(aq) + 2 NO_3^-(aq) \rightarrow PbCrO_4(s) + 2 Na^+(aq) + 2 NO_3^-(aq)$
Delete spectator ions from the ionic equation to get the net ionic equation.
Net ionic equation: $Pb^{2+}(aq) + CrO_4^{2-}(aq) \rightarrow PbCrO_4(s)$
(c) Ionic equation:
$2 Ag^+(aq) + 2 ClO_4^-(aq) + Ca^{2+}(aq) + 2 Br^-(aq) \rightarrow 2 AgBr(s) + Ca^{2+}(aq) + 2 ClO_4^-(aq)$
Delete spectator ions from the ionic equation and reduce coefficients to get the net ionic equation.
Net ionic equation: $Ag^+(aq) + Br^-(aq) \rightarrow AgBr(s)$
(d) Ionic equation:
$Zn^{2+}(aq) + 2 Cl^-(aq) + 2 K^+(aq) + CO_3^{2-}(aq) \rightarrow ZnCO_3(s) + 2 K^+(aq) + 2 Cl^-(aq)$
Delete spectator ions from the ionic equation to get the net ionic equation.
Net ionic equation: $Zn^{2+}(aq) + CO_3^{2-}(aq) \rightarrow ZnCO_3(s)$

7.7 $3 CaCl_2(aq) + 2 Na_3PO_4(aq) \rightarrow Ca_3(PO_4)_2(s) + 6 NaCl(aq)$
Ionic equation:
$3 Ca^{2+}(aq) + 6 Cl^-(aq) + 6 Na^+(aq) + 2 PO_4^{3-}(aq) \rightarrow Ca_3(PO_4)_2(s) + 6 Na^+(aq) + 6 Cl^-(aq)$
Delete spectator ions from the ionic equation to get the net ionic equation.
Net ionic equation: $3 Ca^{2+}(aq) + 2 PO_4^{3-}(aq) \rightarrow Ca_3(PO_4)_2(s)$

7.8   A precipitate results from the reaction. The precipitate contains cations and anions in a
3:2 ratio. The precipitate is either $Mg_3(PO_4)_2$ or $Zn_3(PO_4)_2$.

7.9   (a) HIO_4, periodic acid   (b) $HBrO_2$, bromous acid   (c) H_2CrO_4, chromic acid

7.10   (a) H_3PO_3 (b) H_2Se

7.11   (a) Ionic equation:
$2 Cs^+(aq) + 2 OH^-(aq) + 2 H^+(aq) + SO_4^{2-}(aq) \rightarrow 2 Cs^+(aq) + SO_4^{2-}(aq) + 2 H_2O(l)$
Delete spectator ions from the ionic equation and reduce coefficients to get the net ionic
equation.
Net ionic equation: $H^+(aq) + OH^-(aq) \rightarrow H_2O(l)$
(b) Ionic equation:
$Ca^{2+}(aq) + 2 OH^-(aq) + 2 CH_3CO_2H(aq) \rightarrow Ca^{2+}(aq) + 2 CH_3CO_2^-(aq) + 2 H_2O(l)$
Delete spectator ions from the ionic equation and reduce coefficients to get the net ionic
equation.
Net ionic equation: $CH_3CO_2H(aq) + OH^-(aq) \rightarrow CH_3CO_2^-(aq) + H_2O(l)$

7.12   HY is the strongest acid because it is completely dissociated.
HX is the weakest acid because it is the least dissociated.

7.13   (a) $SnCl_4$:   Cl -1, Sn $+4$ (b) CrO_3:   O -2, Cr $+6$
        (c) $VOCl_3$:   O -2, Cl -1, V $+5$ (d) V_2O_3:   O -2, V $+3$
        (e) HNO_3:   O -2, H $+1$, N $+5$ (f) $FeSO_4$:   O -2, S $+6$, Fe $+2$

7.14 $2 Cu^{2+}(aq) + 4 I^-(aq) \rightarrow 2 CuI(s) + I_2(aq)$
oxidation numbers: Cu^{2+} $+2$; I^- -1;   CuI: Cu $+1$, I -1; I_2: 0
oxidizing agent (oxidation number decreases), Cu^{2+}
reducing agent (oxidation number increases), I^-

7.15   (a) $SnO_2(s) + 2 C(s) \rightarrow Sn(s) + 2 CO(g)$
C is oxidized (its oxidation number increases from 0 to +2). C is the reducing agent.
The Sn in SnO_2 is reduced (its oxidation number decreases from +4 to 0). SnO_2 is the
oxidizing agent.
(b) $Sn^{2+}(aq) + 2 Fe^{3+}(aq) \rightarrow Sn^{4+}(aq) + 2 Fe^{2+}(aq)$
Sn^{2+} is oxidized (its oxidation number increases from +2 to +4). Sn^{2+} is the reducing agent.
Fe^{3+} is reduced (its oxidation number decreases from +3 to +2). Fe^{3+} is the oxidizing agent.

(c) $4 NH_3(g) + 5 O_2(g) \rightarrow 4 NO(g) + 6 H_2O(l)$
The N in NH_3 is oxidized (its oxidation number increases from -3 to $+2$). NH_3 is the reducing agent.
Each O in O_2 is reduced (its oxidation number decreases from 0 to -2). O_2 is the oxidizing agent.

7.16   (a) Pt is below H in the activity series; therefore NO REACTION.
(b) Mg is below Ca in the activity series; therefore NO REACTION.

7.17   Because B will reduce A^+, B is above A in the activity series. Because B will not reduce C^+, C is above B in the activity series. Therefore C must be above A in the activity series and C will reduce A^+.

7.18   "Any element higher in the activity series will react with the ion of any element lower in the activity series."
$A + D^+ \rightarrow A^+ + D$; therefore A is higher than D.
$B^+ + D \rightarrow B + D^+$; therefore D is higher than B.
$C^+ + D \rightarrow C + D^+$; therefore D is higher than C.
$B + C^+ \rightarrow B^+ + C$; therefore B is higher than C.
The net result is A > D > B > C.

7.19   (a) $MnO_4^-(aq) \rightarrow MnO_2(s)$ (reduction)
    $IO_3^-(aq) \rightarrow IO_4^-(aq)$ (oxidation)
(b) $NO_3^-(aq) \rightarrow NO_2(g)$ (reduction)
    $SO_2(aq) \rightarrow SO_4^{2-}(aq)$ (oxidation)

7.20 $NO_3^-(aq) + Cu(s) \rightarrow NO(g) + Cu^{2+}(aq)$
$[Cu(s) \rightarrow Cu^{2+}(aq) + 2 e^-] \times 3$ (oxidation half reaction)

$NO_3^-(aq) \rightarrow NO(g)$
$NO_3^-(aq) \rightarrow NO(g) + 2 H_2O(l)$
$4 H^+(aq) + NO_3^-(aq) \rightarrow NO(g) + 2 H_2O(l)$
$[3 e^- + 4 H^+(aq) + NO_3^-(aq) \rightarrow NO(g) + 2 H_2O(l)] \times 2$ (reduction half reaction)

Combine the two half reactions.
$2 NO_3^-(aq) + 8 H^+(aq) + 3 Cu(s) \rightarrow 3 Cu^{2+}(aq) + 2 NO(g) + 4 H_2O(l)$

7.21 $Fe(OH)_2(s) + O_2(g) \rightarrow Fe(OH)_3(s)$
$[Fe(OH)_2(s) + OH^-(aq) \rightarrow Fe(OH)_3(s) + e^-] \times 4$ (oxidation half reaction)

$O_2(g) \rightarrow 2 H_2O(l)$
$4 H^+(aq) + O_2(g) \rightarrow 2 H_2O(l)$
$4 e^- + 4 H^+(aq) + O_2(g) \rightarrow 2 H_2O(l)$
$4 e^- + 4 H^+(aq) + 4 OH^-(aq) + O_2(g) \rightarrow 2 H_2O(l) + 4 OH^-(aq)$
$4 e^- + 4 H_2O(l) + O_2(g) \rightarrow 2 H_2O(l) + 4 OH^-(aq)$
$4 e^- + 2 H_2O(l) + O_2(g) \rightarrow 4 OH^-(aq)$ (reduction half reaction)

Combine the two half reactions.

$4\ Fe(OH)_2(s) + 4\ OH^-(aq) + 2\ H_2O(l) + O_2(g) \rightarrow 4\ Fe(OH)_3(s) + 4\ OH^-(aq)$

$4\ Fe(OH)_2(s) + 2\ H_2O(l) + O_2(g) \rightarrow 4\ Fe(OH)_3(s)$

7.22   31.50 mL = 0.031 50 L; 10.00 mL = 0.010 00 L

$$0.031\ 50\ L \times \frac{0.105\ mol\ BrO_3^-}{1\ L} \times \frac{6\ mol\ Fe^{2+}}{1\ mol\ BrO_3^-} = 1.98 \times 10^{-2}\ mol\ Fe^{2+}$$

$$molarity = \frac{1.98 \times 10^{-2}\ mol\ Fe^{2+}}{0.010\ 00\ L} = 1.98\ M\ Fe^{2+}\ solution$$

7.23 $Pb(s) + HSO_4^-(aq) \rightarrow PbSO_4(s)$

$Pb(s) + HSO_4^-(aq) \rightarrow PbSO_4(s) + H^+(aq)$

$Pb(s) + HSO_4^-(aq) \rightarrow PbSO_4(s) + H^+(aq) + 2\ e^-$ (oxidation half reaction)

$PbO_2(s) + HSO_4^-(aq) \rightarrow PbSO_4(s)$

$PbO_2(s) + HSO_4^-(aq) \rightarrow PbSO_4(s) + 2\ H_2O(l)$

$PbO_2(s) + HSO_4^-(aq) + 3\ H^+(aq) \rightarrow PbSO_4(s) + 2\ H_2O(l)$

$PbO_2(s) + HSO_4^-(aq) + 3\ H^+(aq) + 2\ e^- \rightarrow PbSO_4(s) + 2\ H_2O(l)$

(reduction half reaction)

Combine the two half reactions.

$Pb(s) + PbO_2(s) + 2\ HSO_4^-(aq) + 3\ H^+(aq) \rightarrow 2\ PbSO_4(s) + 2\ H_2O(l) + H^+(aq)$

$Pb(s) + PbO_2(s) + 2\ HSO_4^-(aq) + 2\ H^+(aq) \rightarrow 2\ PbSO_4(s) + 2\ H_2O(l)$

7.24   For a green process look for a solvent that is safe, non-toxic, non-polluting, and renewable. H_2O would be an excellent green solvent.

Conceptual Problems

7.26   In the precipitate there are two cations (blue) for each anion (red). Looking at the ions in the list, the anion must have a –2 charge and the cation a +1 charge for charge neutrality of the precipitate. The cation must be Ag^+ because all Na^+ salts are soluble. Ag_2CrO_4 and Ag_2CO_3 are insoluble and consistent with the observed result.

7.28   The concentration in the buret is three times that in the flask. The NaOCl concentration is 0.040 M. Because the I^- concentration in the buret is three times the OCl^- concentration in the flask and the reaction requires 2 I^- ions per OCl^- ion, 2/3 or 67% of the I^- solution from the buret must be added to the flask to react with all of the OCl^-.

7.30   (a) $Sr^+ + At \rightarrow Sr + At^+$ No reaction.

(b) $Si + At^+ \rightarrow Si^+ + At$ Reaction would occur.

(c) $Sr + Si^+ \rightarrow Sr^+ + Si$ Reaction would occur.

Section Problems
Aqueous Reactions, Net Ionic Equations, and Electrolytes (Sections 7.1–7.3)

7.32   (a) precipitation   (b) redox   (c) acid-base neutralization

7.34   (a) Ionic equation:
$Hg^{2+}(aq) + 2\ NO_3^-(aq) + 2\ Na^+(aq) + 2\ I^-(aq) \rightarrow 2\ Na^+(aq) + 2\ NO_3^-(aq) + HgI_2(s)$
Delete spectator ions from the ionic equation to get the net ionic equation.
Net ionic equation: $Hg^{2+}(aq) + 2\ I^-(aq) \rightarrow HgI_2(s)$

(b) $2\ HgO(s) \xrightarrow{\text{Heat}} 2\ Hg(l) + O_2(g)$
(c) Ionic equation:
$H_3PO_4(aq) + 3\ K^+(aq) + 3\ OH^-(aq) \rightarrow 3\ K^+(aq) + PO_4^{3-}(aq) + 3\ H_2O(l)$
Delete spectator ions from the ionic equation to get the net ionic equation.
Net ionic equation: $H_3PO_4(aq) + 3\ OH^-(aq) \rightarrow PO_4^{3-}(aq) + 3\ H_2O(l)$

7.36   (a) strong electrolyte, bright (b) nonelectrolyte, dark   (c) weak electrolyte, dim

7.38 $Ba(OH)_2$ is soluble in aqueous solution, dissociates into $Ba^{2+}(aq)$ and $2\ OH^-(aq)$, and conducts electricity. In aqueous solution, H_2SO_4 dissociates into $H^+(aq)$ and $HSO_4^-(aq)$. H_2SO_4 solutions conduct electricity. When equal molar solutions of $Ba(OH)_2$ and H_2SO_4 are mixed, the insoluble $BaSO_4$ is formed along with two H_2O. In water, $BaSO_4$ does not produce any appreciable amount of ions and the mixture does not conduct electricity.

7.40   (a) HBr, strong electrolyte   (b) HF, weak electrolyte
(c) $NaClO_4$, strong electrolyte   (d) $(NH_4)_2CO_3$, strong electrolyte
(e) NH_3, weak electrolyte   (f) C_2H_5OH, nonelectrolyte

7.42   (a) K_2CO_3 contains 3 ions (2 K^+ and 1 CO_3^{2-}).
The molar concentration of ions = 3 x 0.750 M = 2.25 M.
(b) $AlCl_3$ contains 4 ions (1 Al^{3+} and 3 Cl^-).
The molar concentration of ions = 4 x 0.355 M = 1.42 M.

Precipitation Reactions and Solubility Guidelines (Section 7.4)

7.44   (a) Ag_2O, insoluble   (b) $Ba(NO_3)_2$, soluble
(c) $SnCO_3$, insoluble   (d) Fe_2O_3, insoluble

7.46   (a) No precipitate will form.   (b) $FeCl_2(aq) + 2\ KOH(aq) \rightarrow Fe(OH)_2(s) + 2\ KCl(aq)$
(c) No precipitate will form.   (d) No precipitate will form.

7.48   (a) 0.10 M $LiNO_3$ will not form a precipitate.
(b) $BaSO_4(s)$ will precipitate.   (c) $AgCl(s)$ will precipitate.

7.50   (a) $Pb(NO_3)_2(aq) + Na_2SO_4(aq) \rightarrow PbSO_4(s) + 2\ NaNO_3(aq)$
(b) $3\ MgCl_2(aq) + 2\ K_3PO_4(aq) \rightarrow Mg_3(PO_4)_2(s) + 6\ KCl(aq)$
(c) $ZnSO_4(aq) + Na_2CrO_4(aq) \rightarrow ZnCrO_4(s) + Na_2SO_4(aq)$

7.52 $Ag^+(aq) + NO_3^-(aq) + H^+(aq) + Cl^-(aq) \rightarrow AgCl(s) + H^+(aq) + NO_3^-(aq)$

$\text{mol Cl}^- = 30.0 \text{ mL} \times \dfrac{1 \text{ L}}{1000 \text{ mL}} \times 0.150 \text{ mol/L} = 0.00450 \text{ mol}$

$\text{mol Ag}^+ = 25.0 \text{ mL} \times \dfrac{1 \text{ L}}{1000 \text{ mL}} \times 0.200 \text{ mol/L} = 0.00500 \text{ mol}$

Because the mole ratio between Ag^+ and Cl^- is 1 to 1, Cl^- is the limiting reactant because it is the smaller mol quantity. 0.00450 mol of AgCl is produced.
AgCl, 143.32

$\text{mass AgCl} = 0.00450 \text{ mol AgCl} \times \dfrac{143.32 \text{ g AgCl}}{1 \text{ mol AgCl}} = 0.645 \text{ g AgCl}$

7.54   Add HCl(aq); it will selectively precipitate AgCl(s).

7.56 Ag^+ is eliminated because it would have precipitated as AgCl(s); Ba^{2+} is eliminated because it would have precipitated as $BaSO_4$(s). The solution might contain Cs^+ and/or NH_4^+. Neither of these will precipitate with OH^-, SO_4^{2-}, or Cl^-.

Acids, Bases, and Neutralization Reactions (Section 7.5)

7.58   Add the solution to an active metal, such as magnesium. Bubbles of H_2 gas indicate the presence of an acid.

7.60   (a) $2 H^+(aq) + 2 ClO_4^-(aq) + Ca^{2+}(aq) + 2 OH^-(aq) \rightarrow Ca^{2+}(aq) + 2 ClO_4^-(aq) + 2 H_2O(l)$
        (b) $CH_3CO_2H(aq) + Na^+(aq) + OH^-(aq) \rightarrow CH_3CO_2^-(aq) + Na^+(aq) + H_2O(l)$

7.62   (a) $LiOH(aq) + HI(aq) \rightarrow LiI(aq) + H_2O(l)$
        Ionic equation: $Li^+(aq) + OH^-(aq) + H^+(aq) + I^-(aq) \rightarrow Li^+(aq) + I^-(aq) + H_2O(l)$
        Delete spectator ions from the ionic equation to get the net ionic equation.
        Net ionic equation: $H^+(aq) + OH^-(aq) \rightarrow H_2O(l)$
        (b) $2 HBr(aq) + Ca(OH)_2(aq) \rightarrow CaBr_2(aq) + 2 H_2O(l)$
        Ionic equation:
        $2 H^+(aq) + 2 Br^-(aq) + Ca^{2+}(aq) + 2 OH^-(aq) \rightarrow Ca^{2+}(aq) + 2 Br^-(aq) + 2 H_2O(l)$
        Delete spectator ions from the ionic equation to get the net ionic equation.
        Net ionic equation: $H^+(aq) + OH^-(aq) \rightarrow H_2O(l)$

7.64   (a) $HBr(aq) + KOH(aq) \rightarrow KBr(aq) + H_2O(l)$
        50.0 mL = 0.0500 L and 30.0 mL = 0.0300 L
        (0.100 mol/L HBr)(0.0500 L) = 0.005 00 mol HBr
        (0.200 mol/L KOH)(0.0300 L) = 0.006 00 mol KOH
        HBr and KOH react in a one to one mole ratio. The resulting solution is basic because there is an excess of KOH.
        (b) $2 HCl(aq) + Ba(OH)_2(aq) \rightarrow BaCl_2(aq) + 2 H_2O(l))$
        100.0 mL = 0.1000 L and 75 mL = 0.0750 L
        (0.0750 mol/L HCl)(0.1000 L) = 0.007 50 mol HCl
        (0.100 mol/L Ba(OH)$_2$)(0.0750 L) = 0.007 50 mol Ba(OH)$_2$

$$0.007\ 50\ mol\ HCl \times \frac{1\ mol\ Ba(OH)_2}{2\ mol\ HCl} = 0.003\ 75\ mol\ Ba(OH)_2\ needed$$

The resulting solution is basic because there is an excess of $Ba(OH)_2$.

7.66   (a) $LiOH(aq) + HBr(aq) \rightarrow LiBr(aq) + H_2O(l)$
25.0 mL = 0.0250 L and 75.0 mL = 0.0750 L
(0.240 mol/L LiOH)(0.0250 L) = 0.006 00 mol LiOH
(0.200 mol/L HBr)(0.0750 L) = 0.0150 mol HBr
HBr and LiOH react in a one to one mole ratio.
mol HBr left over = 0.0150 mol HBr - 0.006 00 mol LiOH = 0.0090 mol HBr
$KOH(aq) + HBr(aq) \rightarrow KBr(aq) + H_2O(l)$
HBr and KOH react in a one to one mole ratio.
0.0090 mol KOH will neutralize the solution.

$$volume = \frac{mol}{M} = \frac{0.0090\ mol\ KOH}{1.00\ mol/L} = 0.0090\ L = 9.0\ mL\ KOH$$

(b) $HCl(aq) + NaOH(aq) \rightarrow NaCl(aq) + H_2O(l)$
45.0 mL = 0.0450 L and 10.0 mL = 0.0100 L
(0.300 mol/L HCl)(0.0450 L) = 0.0135 mol HCl
(0.250 mol/L NaOH)(0.0100 L) = 0.002 50 mol NaOH
HCl and NaOH react in a one to one mole ratio.
mol HCl left over = 0.0135 mol HCl - 0.002 50 mol NaOH = 0.0110 mol HCl
$KOH(aq) + HCl(aq) \rightarrow KCl(aq) + H_2O(l)$
HCl and KOH react in a one to one mole ratio.
0.0110 mol KOH will neutralize the solution.

$$volume = \frac{mol}{M} = \frac{0.0110\ mol\ KOH}{1.00\ mol/L} = 0.0110\ L = 11.0\ mL\ KOH$$

Redox Reactions and Oxidation Numbers (Sections 7.6–7.8)

7.68   The best reducing agents are at the bottom left of the periodic table. The best oxidizing agents are at the top right of the periodic table (excluding the noble gases).

7.70   (a) An oxidizing agent gains electrons.
(b) A reducing agent loses electrons.
(c) A substance undergoing oxidation loses electrons.
(d) A substance undergoing reduction gains electrons.

7.72   (a) NO_2 O –2, N +4   (b) SO_3 O –2, S +6
(c) $COCl_2$ O –2, Cl –1, C +4   (d) CH_2Cl_2 Cl –1, H +1, C 0
(e) $KClO_3$ O –2, K +1, Cl +5   (f) HNO_3 O –2, H +1, N +5

7.74   (a) ClO_3^- O –2, Cl +5   (b) SO_3^{2-} O –2, S +4
(c) $C_2O_4^{2-}$ O –2, C +3   (d) NO_2^- O –2, N +3
(e) BrO^- O –2, Br +1   (f) AsO_4^{3-} O –2, As +5

7.76   (a) $Ca(s) + Sn^{2+}(aq) \rightarrow Ca^{2+}(aq) + Sn(s)$
Ca(s) is oxidized (oxidation number increases from 0 to +2).
$Sn^{2+}(aq)$ is reduced (oxidation number decreases from +2 to 0).
(b) $ICl(s) + H_2O(l) \rightarrow HCl(aq) + HOI(aq)$
No oxidation numbers change. The reaction is not a redox reaction.

7.78   (a) Zn is below Na^+; therefore no reaction.
(b) Pt is below H^+; therefore no reaction.
(c) Au is below Ag^+; therefore no reaction.
(d) Ag is above Au^{3+}; the reaction is $Au^{3+}(aq) + 3\ Ag(s) \rightarrow 3\ Ag^+(aq) + Au(s)$.

7.80   (a) "Any element higher in the activity series will react with the ion of any element lower in the activity series."
$A + B^+ \rightarrow A^+ + B$; therefore A is higher than B.
$C^+ + D \rightarrow$ no reaction; therefore C is higher than D.
$B + D^+ \rightarrow B^+ + D$; therefore B is higher than D.
$B + C^+ \rightarrow B^+ + C$; therefore B is higher than C.
The net result is A > B > C > D.
(b) (1) C is below A^+; therefore no reaction.
      (2) D is below A^+; therefore no reaction.

Balancing Redox Reactions (Section 7.9)

7.82   (a) N oxidation number decreases from +5 to +2; reduction.
(b) Zn oxidation number increases from 0 to +2; oxidation.
(c) Ti oxidation number increases from +3 to +4; oxidation.
(d) Sn oxidation number decreases from +4 to +2; reduction.

7.84   (a) $NO_3^-(aq) \rightarrow NO(g)$
            $NO_3^-(aq) \rightarrow NO(g) + 2\ H_2O(l)$
            $4\ H^+(aq) + NO_3^-(aq) \rightarrow NO(g) + 2\ H_2O(l)$
            $3\ e^- + 4\ H^+(aq) + NO_3^-(aq) \rightarrow NO(g) + 2\ H_2O(l)$
    (b) $Zn(s) \rightarrow Zn^{2+}(aq) + 2\ e^-$
    (c) $Ti^{3+}(aq) \rightarrow TiO_2(s)$
            $Ti^{3+}(aq) + 2\ H_2O(l) \rightarrow TiO_2(s)$
            $Ti^{3+}(aq) + 2\ H_2O(l) \rightarrow TiO_2(s) + 4\ H^+(aq)$
            $Ti^{3+}(aq) + 2\ H_2O(l) \rightarrow TiO_2(s) + 4\ H^+(aq) + e^-$
    (d) $Sn^{4+}(aq) + 2\ e^- \rightarrow Sn^{2+}(aq)$

7.86   (a) $Te(s) + NO_3^-(aq) \rightarrow TeO_2(s) + NO(g)$
oxidation: $Te(s) \rightarrow TeO_2(s)$
reduction: $NO_3^-(aq) \rightarrow NO(g)$
(b) $H_2O_2(aq) + Fe^{2+}(aq) \rightarrow Fe^{3+}(aq) + H_2O(l)$
oxidation: $Fe^{2+}(aq) \rightarrow Fe^{3+}(aq)$
reduction: $H_2O_2(aq) \rightarrow H_2O(l)$

7.88   (a) $Cr_2O_7^{2-}(aq) \rightarrow Cr^{3+}(aq)$

              $Cr_2O_7^{2-}(aq) \rightarrow 2\ Cr^{3+}(aq)$

              $Cr_2O_7^{2-}(aq) \rightarrow 2\ Cr^{3+}(aq) + 7\ H_2O(l)$

              $14\ H^+(aq) + Cr_2O_7^{2-}(aq) \rightarrow 2\ Cr^{3+}(aq) + 7\ H_2O(l)$

              $14\ H^+(aq) + Cr_2O_7^{2-}(aq) + 6\ e^- \rightarrow 2\ Cr^{3+}(aq) + 7\ H_2O(l)$

  (b) $CrO_4^{2-}(aq) \rightarrow Cr(OH)_4^-(aq)$

              $4\ H^+(aq) + CrO_4^{2-}(aq) \rightarrow Cr(OH)_4^-(aq)$

              $4\ H^+(aq) + 4\ OH^-(aq) + CrO_4^{2-}(aq) \rightarrow Cr(OH)_4^-(aq) + 4\ OH^-(aq)$

              $4\ H_2O(l) + CrO_4^{2-}(aq) \rightarrow Cr(OH)_4^-(aq) + 4\ OH^-(aq)$

              $4\ H_2O(l) + CrO_4^{2-}(aq) + 3\ e^- \rightarrow Cr(OH)_4^-(aq) + 4\ OH^-(aq)$

  (c) $Bi^{3+}(aq) \rightarrow BiO_3^-(aq)$

              $Bi^{3+}(aq) + 3\ H_2O(l) \rightarrow BiO_3^-(aq)$

              $Bi^{3+}(aq) + 3\ H_2O(l) \rightarrow BiO_3^-(aq) + 6\ H^+(aq)$

              $Bi^{3+}(aq) + 3\ H_2O(l) + 6\ OH^-(aq) \rightarrow BiO_3^-(aq) + 6\ H^+(aq) + 6\ OH^-(aq)$

              $Bi^{3+}(aq) + 3\ H_2O(l) + 6\ OH^-(aq) \rightarrow BiO_3^-(aq) + 6\ H_2O(l)$

              $Bi^{3+}(aq) + 6\ OH^-(aq) \rightarrow BiO_3^-(aq) + 3\ H_2O(l)$

              $Bi^{3+}(aq) + 6\ OH^-(aq) \rightarrow BiO_3^-(aq) + 3\ H_2O(l) + 2\ e^-$

  (d) $ClO^-(aq) \rightarrow Cl^-(aq)$

              $ClO^-(aq) \rightarrow Cl^-(aq) + H_2O(l)$

              $2\ H^+(aq) + ClO^-(aq) \rightarrow Cl^-(aq) + H_2O(l)$

              $2\ H^+(aq) + 2\ OH^-(aq) + ClO^-(aq) \rightarrow Cl^-(aq) + H_2O(l) + 2\ OH^-(aq)$

              $2\ H_2O(l) + ClO^-(aq) \rightarrow Cl^-(aq) + H_2O(l) + 2\ OH^-(aq)$

              $H_2O(l) + ClO^-(aq) \rightarrow Cl^-(aq) + 2\ OH^-(aq)$

              $H_2O(l) + ClO^-(aq) + 2\ e^- \rightarrow Cl^-(aq) + 2\ OH^-(aq)$

7.90   (a) $MnO_4^-(aq) \rightarrow MnO_2(s)$

              $MnO_4^-(aq) \rightarrow MnO_2(s) + 2\ H_2O(l)$

              $4\ H^+(aq) + MnO_4^-(aq) \rightarrow MnO_2(s) + 2\ H_2O(l)$

              $[4\ H^+(aq) + MnO_4^-(aq) + 3\ e^- \rightarrow MnO_2(s) + 2\ H_2O(l)] \times 2$ (reduction half reaction)

              $IO_3^-(aq) \rightarrow IO_4^-(aq)$

              $H_2O(l) + IO_3^-(aq) \rightarrow IO_4^-(aq)$

              $H_2O(l) + IO_3^-(aq) \rightarrow IO_4^-(aq) + 2\ H^+(aq)$

              $[H_2O(l) + IO_3^-(aq) \rightarrow IO_4^-(aq) + 2\ H^+(aq) + 2\ e^-] \times 3$ (oxidation half reaction)

              Combine the two half reactions.

              $8\ H^+(aq) + 3\ H_2O(l) + 2\ MnO_4^-(aq) + 3\ IO_3^-(aq) \rightarrow$

                                $6\ H^+(aq) + 4\ H_2O(l) + 2\ MnO_2(s) + 3\ IO_4^-(aq)$

              $2\ H^+(aq) + 2\ MnO_4^-(aq) + 3\ IO_3^-(aq) \rightarrow 2\ MnO_2(s) + 3\ IO_4^-(aq) + H_2O(l)$

              $2\ H^+(aq) + 2\ OH^-(aq) + 2\ MnO_4^-(aq) + 3\ IO_3^-(aq) \rightarrow$

                                $2\ MnO_2(s) + 3\ IO_4^-(aq) + H_2O(l) + 2\ OH^-(aq)$

              $2\ H_2O(l) + 2\ MnO_4^-(aq) + 3\ IO_3^-(aq) \rightarrow$

                                $2\ MnO_2(s) + 3\ IO_4^-(aq) + H_2O(l) + 2\ OH^-(aq)$

              $H_2O(l) + 2\ MnO_4^-(aq) + 3\ IO_3^-(aq) \rightarrow 2\ MnO_2(s) + 3\ IO_4^-(aq) + 2\ OH^-(aq)$

(b) $Cu(OH)_2(s) \rightarrow Cu(s)$

$Cu(OH)_2(s) \rightarrow Cu(s) + 2\ H_2O(l)$

$2\ H^+(aq) + Cu(OH)_2(s) \rightarrow Cu(s) + 2\ H_2O(l)$

$[2\ H^+(aq) + Cu(OH)_2(s) + 2\ e^- \rightarrow Cu(s) + 2\ H_2O(l)]\ x\ 2$ (reduction half reaction)

$N_2H_4(aq) \rightarrow N_2(g)$

$N_2H_4(aq) \rightarrow N_2(g) + 4\ H^+(aq)$

$N_2H_4(aq) \rightarrow N_2(g) + 4\ H^+(aq) + 4\ e^-$ (oxidation half reaction)

Combine the two half reactions.

$4\ H^+(aq) + 2\ Cu(OH)_2(s) + N_2H_4(aq) \rightarrow 2\ Cu(s) + 4\ H_2O(l) + N_2(g) + 4\ H^+(aq)$

$2\ Cu(OH)_2(s) + N_2H_4(aq) \rightarrow 2\ Cu(s) + 4\ H_2O(l) + N_2(g)$

(c) $Fe(OH)_2(s) \rightarrow Fe(OH)_3(s)$

$Fe(OH)_2(s) + H_2O(l) \rightarrow Fe(OH)_3(s)$

$Fe(OH)_2(s) + H_2O(l) \rightarrow Fe(OH)_3(s) + H^+(aq)$

$[Fe(OH)_2(s) + H_2O(l) \rightarrow Fe(OH)_3(s) + H^+(aq) + e^-]\ x\ 3$ (oxidation half reaction)

$CrO_4^{2-}(aq) \rightarrow Cr(OH)_4^-(aq)$

$4\ H^+(aq) + CrO_4^{2-}(aq) \rightarrow Cr(OH)_4^-(aq)$

$4\ H^+(aq) + CrO_4^{2-}(aq) + 3\ e^- \rightarrow Cr(OH)_4^-(aq)$ (reduction half reaction)

Combine the two half reactions.

$3\ Fe(OH)_2(s) + 3\ H_2O(l) + 4\ H^+(aq) + CrO_4^{2-}(aq) \rightarrow$
$$3\ Fe(OH)_3(s) + 3\ H^+(aq) + Cr(OH)_4^-(aq)$$

$3\ Fe(OH)_2(s) + 3\ H_2O(l) + H^+(aq) + CrO_4^{2-}(aq) \rightarrow 3\ Fe(OH)_3(s) + Cr(OH)_4^-(aq)$

$3\ Fe(OH)_2(s) + 3\ H_2O(l) + H^+(aq) + OH^-(aq) + CrO_4^{2-}(aq) \rightarrow$
$$3\ Fe(OH)_3(s) + Cr(OH)_4^-(aq) + OH^-(aq)$$

$3\ Fe(OH)_2(s) + 4\ H_2O(l) + CrO_4^{2-}(aq) \rightarrow 3\ Fe(OH)_3(s) + Cr(OH)_4^-(aq) + OH^-(aq)$

(d) $ClO_4^-(aq) \rightarrow ClO_2^-(aq)$

$ClO_4^-(aq) \rightarrow ClO_2^-(aq) + 2\ H_2O(l)$

$4\ H^+(aq) + ClO_4^-(aq) \rightarrow ClO_2^-(aq) + 2\ H_2O(l)$

$4\ H^+(aq) + ClO_4^-(aq) + 4\ e^- \rightarrow ClO_2^-(aq) + 2\ H_2O(l)$ (reduction half reaction)

$H_2O_2(aq) \rightarrow O_2(g)$

$H_2O_2(aq) \rightarrow O_2(g) + 2\ H^+(aq)$

$[H_2O_2(aq) \rightarrow O_2(g) + 2\ H^+(aq) + 2\ e^-]\ x\ 2$ (oxidation half reaction)

Combine the two half reactions.

$4\ H^+(aq) + ClO_4^-(aq) + 2\ H_2O_2(aq) \rightarrow ClO_2^-(aq) + 2\ H_2O(l) + 2\ O_2(g) + 4\ H^+(aq)$

$ClO_4^-(aq) + 2\ H_2O_2(aq) \rightarrow ClO_2^-(aq) + 2\ H_2O(l) + 2\ O_2(g)$

7.92   (a) $Zn(s) \rightarrow Zn^{2+}(aq)$

$Zn(s) \rightarrow Zn^{2+}(aq) + 2\ e^-$ (oxidation half reaction)

$VO^{2+}(aq) \rightarrow V^{3+}(aq)$

$VO^{2+}(aq) \rightarrow V^{3+}(aq) + H_2O(l)$

$2 H^+(aq) + VO^{2+}(aq) \rightarrow V^{3+}(aq) + H_2O(l)$

$[2 H^+(aq) + VO^{2+}(aq) + e^- \rightarrow V^{3+}(aq) + H_2O(l)]$ x 2   (reduction half reaction)

Combine the two half reactions.

$Zn(s) + 2 VO^{2+}(aq) + 4 H^+(aq) \rightarrow Zn^{2+}(aq) + 2 V^{3+}(aq) + 2 H_2O(l)$

(b) $Ag(s) \rightarrow Ag^+(aq)$

$Ag(s) \rightarrow Ag^+(aq) + e^-$ (oxidation half reaction)

$NO_3^-(aq) \rightarrow NO_2(g)$

$NO_3^-(aq) \rightarrow NO_2(g) + H_2O(l)$

$2 H^+(aq) + NO_3^-(aq) \rightarrow NO_2(g) + H_2O(l)$

$2 H^+(aq) + NO_3^-(aq) + e^- \rightarrow NO_2(g) + H_2O(l)$ (reduction half reaction)

Combine the two half reactions.

$2 H^+(aq) + Ag(s) + NO_3^-(aq) \rightarrow Ag^+(aq) + NO_2(g) + H_2O(l)$

(c) $Mg(s) \rightarrow Mg^{2+}(aq)$

$[Mg(s) \rightarrow Mg^{2+}(aq) + 2 e^-]$ x 3   (oxidation half reaction)

$VO_4^{3-}(aq) \rightarrow V^{2+}(aq)$

$VO_4^{3-}(aq) \rightarrow V^{2+}(aq) + 4 H_2O(l)$

$8 H^+(aq) + VO_4^{3-}(aq) \rightarrow V^{2+}(aq) + 4 H_2O(l)$

$[8 H^+(aq) + VO_4^{3-}(aq) + 3 e^- \rightarrow V^{2+}(aq) + 4 H_2O(l)]$ x 2   (reduction half reaction)

Combine the two half reactions.

$3 Mg(s) + 16 H^+(aq) + 2 VO_4^{3-}(aq) \rightarrow 3 Mg^{2+}(aq) + 2 V^{2+}(aq) + 8 H_2O(l)$

(d) $I^-(aq) \rightarrow I_3^-(aq)$

$3 I^-(aq) \rightarrow I_3^-(aq)$

$[3 I^-(aq) \rightarrow I_3^-(aq) + 2 e^-]$ x 8   (oxidation half reaction)

$IO_3^-(aq) \rightarrow I_3^-(aq)$

$3 IO_3^-(aq) \rightarrow I_3^-(aq)$

$3 IO_3^-(aq) \rightarrow I_3^-(aq) + 9 H_2O(l)$

$18 H^+(aq) + 3 IO_3^-(aq) \rightarrow I_3^-(aq) + 9 H_2O(l)$

$18 H^+(aq) + 3 IO_3^-(aq) + 16 e^- \rightarrow I_3^-(aq) + 9 H_2O(l)$ (reduction half reaction)

Combine the two half reactions.

$18 H^+(aq) + 3 IO_3^-(aq) + 24 I^-(aq) \rightarrow 9 I_3^-(aq) + 9 H_2O(l)$

Divide each coefficient by 3.

$6 H^+(aq) + IO_3^-(aq) + 8 I^-(aq) \rightarrow 3 I_3^-(aq) + 3 H_2O(l)$

Redox Titrations (Section 7.10)

7.94 $I_2(aq) + 2 S_2O_3^{2-}(aq) \rightarrow S_4O_6^{2-}(aq) + 2 I^-(aq);$ $\quad$ 35.20 mL = 0.032 50 L

$$0.035\ 20\ L \times \frac{0.150\ mol\ S_2O_3^{2-}}{L} \times \frac{1\ mol\ I_2}{2\ mol\ S_2O_3^{2-}} \times \frac{253.8\ g\ I_2}{1\ mol\ I_2} = 0.670\ g\ I_2$$

7.96 $\quad$ 46.99 mL = 0.046 99 L; 50.00 mL = 0.050 00 L

$(0.2004\ mol/L\ Cr_2O_7^{2-})(0.046\ 99\ L) = 0.009\ 417\ mol\ Cr_2O_7^{2-}$

$$0.009\ 417\ mol\ Cr_2O_7^{2-} \times \frac{6\ mol\ Fe^{2+}}{1\ mol\ Cr_2O_7^{2-}} = 0.056\ 50\ mol\ Fe^{2+}$$

$$[Fe^{2+}] = \frac{0.056\ 50\ mol}{0.050\ 00\ L} = 1.130\ M$$

7.98 $\quad$ $3\ H_3AsO_3(aq) + BrO_3^-(aq) \rightarrow Br^-(aq) + 3\ H_3AsO_4(aq)$

22.35 mL = 0.022 35 L and 50.00 mL = 0.050 00 L

$$0.022\ 35\ L \times \frac{0.100\ mol\ BrO_3^-}{L} \times \frac{3\ mol\ H_3AsO_3}{1\ mol\ BrO_3^-} = 6.70 \times 10^{-3}\ mol\ H_3AsO_3$$

$$molarity = \frac{6.70 \times 10^{-3}\ mol}{0.050\ 00\ L} = 0.134\ M\ As(III)$$

7.100 $\quad$ $2\ Fe^{3+}(aq) + Sn^{2+}(aq) \rightarrow 2\ Fe^{2+}(aq) + Sn^{4+}(aq);$ $\quad$ 13.28 mL = 0.013 28 L

$$0.013\ 28\ L \times \frac{0.1015\ mol\ Sn^{2+}}{L} \times \frac{2\ mol\ Fe^{3+}}{1\ mol\ Sn^{2+}} \times \frac{55.845\ g\ Fe^{3+}}{1\ mol\ Fe^{3+}} = 0.1506\ g\ Fe^{3+}$$

$$mass\ \%\ Fe = \frac{0.1506\ g}{0.1875\ g} \times 100\% = 80.32\%$$

7.102 $\quad$ $C_2H_5OH(aq) + 2\ Cr_2O_7^{2-}(aq) + 16\ H^+(aq) \rightarrow 2\ CO_2(g) + 4\ Cr^{3+}(aq) + 11\ H_2O(l)$

C_2H_5OH, 46.07; 8.76 mL = 0.008 76 L

$$0.008\ 76\ L \times \frac{0.049\ 88\ mol\ Cr_2O_7^{2-}}{L} \times \frac{1\ mol\ C_2H_5OH}{2\ mol\ Cr_2O_7^{2-}} \times \frac{46.07\ g\ C_2H_5OH}{1\ mol\ C_2H_5OH}$$

$$= 0.010\ 07\ g\ C_2H_5OH$$

$$mass\ \%\ C_2H_5OH = \frac{0.010\ 07\ g}{10.002\ g} \times 100\% = 0.101\%$$

Chapter Problems

7.104 $\quad$ (a) $\quad$ $[Fe(CN)_6]^{3-}(aq) \rightarrow [Fe(CN)_6]^{4-}(aq)$

$\qquad\qquad$ $([Fe(CN)_6]^{3-}(aq) + e^- \rightarrow [Fe(CN)_6]^{4-}(aq)) \times 4$ $\qquad$ (reduction half reaction)

$N_2H_4(aq) \rightarrow N_2(g)$

$N_2H_4(aq) \rightarrow N_2(g) + 4 H^+(aq)$

$N_2H_4(aq) \rightarrow N_2(g) + 4 H^+(aq) + 4 e^-$

$N_2H_4(aq) + 4 OH^-(aq) \rightarrow N_2(g) + 4 H^+(aq) + 4 OH^-(aq) + 4 e^-$

$N_2H_4(aq) + 4 OH^-(aq) \rightarrow N_2(g) + 4 H_2O(l) + 4 e^-$ (oxidation half reaction)

Combine the two half reactions.

$4 [Fe(CN)_6]^{3-}(aq) + N_2H_4(aq) + 4 OH^-(aq) \rightarrow$
$$4 [Fe(CN)_6]^{4-}(aq) + N_2(g) + 4 H_2O(l)$$

(b) $Cl_2(g) \rightarrow Cl^-(aq)$

$Cl_2(g) \rightarrow 2 Cl^-(aq)$

$Cl_2(g) + 2 e^- \rightarrow 2 Cl^-(aq)$ (reduction half reaction)

$SeO_3^{2-}(aq) \rightarrow SeO_4^{2-}(aq)$

$SeO_3^{2-}(aq) + H_2O(l) \rightarrow SeO_4^{2-}(aq)$

$SeO_3^{2-}(aq) + H_2O(l) \rightarrow SeO_4^{2-}(aq) + 2 H^+(aq)$

$SeO_3^{2-}(aq) + H_2O(l) \rightarrow SeO_4^{2-}(aq) + 2 H^+(aq) + 2 e^-$

$SeO_3^{2-}(aq) + H_2O(l) + 2 OH^-(aq) \rightarrow SeO_4^{2-}(aq) + 2 H^+(aq) + 2 OH^-(aq) + 2 e^-$

$SeO_3^{2-}(aq) + H_2O(l) + 2 OH^-(aq) \rightarrow SeO_4^{2-}(aq) + 2 H_2O(l) + 2 e^-$

$SeO_3^{2-}(aq) + 2 OH^-(aq) \rightarrow SeO_4^{2-}(aq) + H_2O(l) + 2 e^-$ (oxidation half reaction)

Combine the two half reactions.

$SeO_3^{2-}(aq) + Cl_2(g) + 2 OH^-(aq) \rightarrow SeO_4^{2-}(aq) + 2 Cl^-(aq) + H_2O(l)$

(c) $Co^{2+}(aq) \rightarrow Co(OH)_3(s)$

$Co^{2+}(aq) + 3 H_2O(l) \rightarrow Co(OH)_3(s)$

$Co^{2+}(aq) + 3 H_2O(l) \rightarrow Co(OH)_3(s) + 3 H^+(aq)$

$[Co^{2+}(aq) + 3 H_2O(l) \rightarrow Co(OH)_3(s) + 3 H^+(aq) + e^-] \times 2$
$$\text{(oxidation half reaction)}$$

$HO_2^-(aq) \rightarrow H_2O(l)$

$HO_2^-(aq) \rightarrow 2 H_2O(l)$

$3 H^+(aq) + HO_2^-(aq) \rightarrow 2 H_2O(l)$

$3 H^+(aq) + HO_2^-(aq) + 2 e^- \rightarrow 2 H_2O(l)$ (reduction half reaction)

Combine the two half reactions.

$2 Co^{2+}(aq) + 6 H_2O(l) + 3 H^+(aq) + HO_2^-(aq) \rightarrow$
$$2 Co(OH)_3(s) + 6 H^+(aq) + 2 H_2O(l)$$

$2 Co^{2+}(aq) + 4 H_2O(l) + HO_2^-(aq) \rightarrow 2 Co(OH)_3(s) + 3 H^+(aq)$

$2 Co^{2+}(aq) + 4 H_2O(l) + HO_2^-(aq) + 3 OH^-(aq) \rightarrow$
$$2 Co(OH)_3(s) + 3 H^+(aq) + 3 OH^-(aq)$$

$2 Co^{2+}(aq) + 4 H_2O(l) + HO_2^-(aq) + 3 OH^-(aq) \rightarrow$
$$2 Co(OH)_3(s) + 3 H_2O(l)$$

$2 Co^{2+}(aq) + H_2O(l) + HO_2^-(aq) + 3 OH^-(aq) \rightarrow 2 Co(OH)_3(s)$

7.106 (a) C_2H_6 H +1,   C –3
       (b) $Na_2B_4O_7$ O –2,   Na +1,   B +3
       (c) Mg_2SiO_4 O –2,   Mg +2,   Si +4

7.108 (a) "Any element higher in the activity series will react with the ion of any element lower in the activity series."

     $C + B^+ \rightarrow C^+ + B$; therefore C is higher than B.

     $A^+ + D \rightarrow$ no reaction; therefore A is higher than D.

     $C^+ + A \rightarrow$ no reaction; therefore C is higher than A.

     $D + B^+ \rightarrow D^+ + B$; therefore D is higher than B.

     The net result is C > A > D > B.

     (b) (1) The reaction, $A^+ + C \rightarrow A + C^+$, will occur because C is above A in the activity series.

       (2) The reaction, $A^+ + B \rightarrow A + B^+$, will not occur because B is below A in the activity series.

7.110   10.49 mL = 0.010 49 L

     (0.100 mol/L $S_2O_3^{2-}$)(0.010 49 L) = 0.001 049 mol $S_2O_3^{2-}$

$$0.001\ 049\ \text{mol}\ S_2O_3^{2-} \times \frac{1\ \text{mol}\ I_3^-}{2\ \text{mol}\ S_2O_3^{2-}} \times \frac{2\ \text{mol}\ Cu^{2+}}{1\ \text{mol}\ I_3^-} = 0.001\ 049\ \text{mol}\ Cu^{2-}$$

$$0.001\ 049\ \text{mol}\ Cu^{2-} \times \frac{63.55\ \text{g}\ Cu}{1\ \text{mol}\ Cu^{2+}} = 0.066\ 66\ \text{g}\ Cu$$

$$\text{mass \% Cu} = \frac{0.066\ 66\ \text{g}\ Cu}{14.98\ \text{g}\ \text{sample}} \times 100\% = 0.4450\ \%\ Cu$$

7.112 $MgF_2(s) \rightleftharpoons Mg^{2+}(aq) + 2\ F^-(aq)$

              x   2x

     $[Mg^{2+}] = x = 2.6 \times 10^{-4}$ M and $[F^-] = 2x = 2(2.6 \times 10^{-4}$ M$) = 5.2 \times 10^{-4}$ M in a saturated solution.

     $K_{sp} = [Mg^{2+}][F^-]^2 = (2.6 \times 10^{-4}$ M$)(5.2 \times 10^{-4}$ M$)^2 = 7.0 \times 10^{-11}$

7.114 (a) Add HCl to precipitate Hg_2Cl_2. $Hg_2^{2+}(aq) + 2Cl^-(aq) \rightarrow Hg_2Cl_2(s)$
       (b) Add H_2SO_4 to precipitate $PbSO_4$. $Pb^{2+}(aq) + SO_4^{2-}(aq) \rightarrow PbSO_4(s)$
       (c) Add Na_2CO_3 to precipitate $CaCO_3$. $Ca^{2+}(aq) + CO_3^{2-}(aq) \rightarrow CaCO_3(s)$
       (d) Add Na_2SO_4 to precipitate $BaSO_4$. $Ba^{2+}(aq) + SO_4^{2-}(aq) \rightarrow BaSO_4(s)$

7.116 All four reactions are redox reactions.

     (a) $Mn(OH)_2(s) \rightarrow Mn(OH)_3(s)$

          $Mn(OH)_2(s) + OH^-(aq) \rightarrow Mn(OH)_3(s)$

          $[Mn(OH)_2(s) + OH^-(aq) \rightarrow Mn(OH)_3(s) + e^-] \times 2$ (oxidation half reaction)

          $H_2O_2(aq) \rightarrow 2\ H_2O(l)$

          $2\ H^+(aq) + H_2O_2(aq) \rightarrow 2\ H_2O(l)$

          $2\ e^- + 2\ H^+(aq) + H_2O_2(aq) \rightarrow 2\ H_2O(l)$

$2 e^- + 2 OH^-(aq) + 2 H^+(aq) + H_2O_2(aq) \rightarrow 2 H_2O(l) + 2 OH^-(aq)$
$2 e^- + 2 H_2O(l) + H_2O_2(aq) \rightarrow 2 H_2O(l) + 2 OH^-(aq)$
$2 e^- + H_2O_2(aq) \rightarrow 2 OH^-(aq)$ (reduction half reaction)

Combine the two half reactions.
$2 Mn(OH)_2(s) + 2 OH^-(aq) + H_2O_2(aq) \rightarrow 2 Mn(OH)_3(s) + 2 OH^-(aq)$
$2 Mn(OH)_2(s) + H_2O_2(aq) \rightarrow 2 Mn(OH)_3(s)$

(b) $[MnO_4^{2-}(aq) \rightarrow MnO_4^-(aq) + e^-] \times 2$ (oxidation half reaction)

$MnO_4^{2-}(aq) \rightarrow MnO_2(s)$
$MnO_4^{2-}(aq) \rightarrow MnO_2(s) + 2 H_2O(l)$
$4 H^+(aq) + MnO_4^{2-}(aq) \rightarrow MnO_2(s) + 2 H_2O(l)$
$2 e^- + 4 H^+(aq) + MnO_4^{2-}(aq) \rightarrow MnO_2(s) + 2 H_2O(l)$ (reduction half reaction)

Combine the two half reactions.
$4 H^+(aq) + 3 MnO_4^{2-}(aq) \rightarrow MnO_2(s) + 2 MnO_4^-(aq) + 2 H_2O(l)$

(c) $I^-(aq) \rightarrow I_3^-(aq)$
$3 I^-(aq) \rightarrow I_3^-(aq)$
$[3 I^-(aq) \rightarrow I_3^-(aq) + 2 e^-] \times 8$ (oxidation half reaction)

$IO_3^-(aq) \rightarrow I_3^-(aq)$
$3 IO_3^-(aq) \rightarrow I_3^-(aq)$
$3 IO_3^-(aq) \rightarrow I_3^-(aq) + 9 H_2O(l)$
$18 H^+(aq) + 3 IO_3^-(aq) \rightarrow I_3^-(aq) + 9 H_2O(l)$
$16 e^- + 18 H^+(aq) + 3 IO_3^-(aq) \rightarrow I_3^-(aq) + 9 H_2O(l)$ (reduction half reaction)

Combine the two half reactions.
$24 I^-(aq) + 3 IO_3^-(aq) + 18 H^+(aq) \rightarrow 9 I_3^-(aq) + 9 H_2O(l)$
Divide all coefficients by 3.
$8 I^-(aq) + IO_3^-(aq) + 6 H^+(aq) \rightarrow 3 I_3^-(aq) + 3 H_2O(l)$

(d) $P(s) \rightarrow HPO_3^{2-}(aq)$
$3 H_2O(l) + P(s) \rightarrow HPO_3^{2-}(aq)$
$3 H_2O(l) + P(s) \rightarrow HPO_3^{2-}(aq) + 5 H^+(aq)$
$[3 H_2O(l) + P(s) \rightarrow HPO_3^{2-}(aq) + 5 H^+(aq) + 3 e^-] \times 2$
(oxidation half reaction)
$PO_4^{3-}(aq) \rightarrow HPO_3^{2-}(aq)$
$PO_4^{3-}(aq) \rightarrow HPO_3^{2-}(aq) + H_2O(l)$
$3 H^+(aq) + PO_4^{3-}(aq) \rightarrow HPO_3^{2-}(aq) + H_2O(l)$
$[2 e^- + 3 H^+(aq) + PO_4^{3-}(aq) \rightarrow HPO_3^{2-}(aq) + H_2O(l)] \times 3$
(reduction half reaction)
Combine the two half reactions and add OH^-.
$6 H_2O(l) + 2 P(s) + 9 H^+(aq) + 3 PO_4^{3-}(aq) \rightarrow$
$5 HPO_3^{2-}(aq) + 10 H^+(aq) + 3 H_2O(l)$

$$3 H_2O(l) + 2 P(s) + 3 PO_4^{3-}(aq) \rightarrow 5 HPO_3^{2-}(aq) + H^+(aq)$$

$$3 H_2O(l) + 2 P(s) + 3 PO_4^{3-}(aq) + OH^-(aq) \rightarrow$$
$$5 HPO_3^{2-}(aq) + H^+(aq) + OH^-(aq)$$

$$3 H_2O(l) + 2 P(s) + 3 PO_4^{3-}(aq) + OH^-(aq) \rightarrow 5 HPO_3^{2-}(aq) + H_2O(l)$$

$$2 H_2O(l) + 2 P(s) + 3 PO_4^{3-}(aq) + OH^-(aq) \rightarrow 5 HPO_3^{2-}(aq)$$

7.118 (a) $S_4O_6^{2-}(aq) \rightarrow H_2S(aq)$

$S_4O_6^{2-}(aq) \rightarrow 4 H_2S(aq)$

$S_4O_6^{2-}(aq) \rightarrow 4 H_2S(aq) + 6 H_2O(l)$

$20 H^+(aq) + S_4O_6^{2-}(aq) \rightarrow 4 H_2S(aq) + 6 H_2O(l)$

$18 e^- + 20 H^+(aq) + S_4O_6^{2-}(aq) \rightarrow 4 H_2S(aq) + 6 H_2O(l)$ (reduction half reaction)

$Al(s) \rightarrow Al^{3+}(aq)$

$[Al(s) \rightarrow Al^{3+}(aq) + 3 e^-] \times 6$ (oxidation half reaction)

Combine the two half reactions.

$20 H^+(aq) + S_4O_6^{2-}(aq) + 6 Al(s) \rightarrow 4 H_2S(aq) + 6 Al^{3+}(aq) + 6 H_2O(l)$

(b) $S_2O_3^{2-}(aq) \rightarrow S_4O_6^{2-}(aq)$

$2 S_2O_3^{2-}(aq) \rightarrow S_4O_6^{2-}(aq)$

$[2 S_2O_3^{2-}(aq) \rightarrow S_4O_6^{2-}(aq) + 2 e^-] \times 3$ (oxidation half reaction)

$Cr_2O_7^{2-}(aq) \rightarrow Cr^{3+}(aq)$

$Cr_2O_7^{2-}(aq) \rightarrow 2 Cr^{3+}(aq)$

$Cr_2O_7^{2-}(aq) \rightarrow 2 Cr^{3+}(aq) + 7 H_2O(l)$

$14 H^+(aq) + Cr_2O_7^{2-}(aq) \rightarrow 2 Cr^{3+}(aq) + 7 H_2O(l)$

$6 e^- + 14 H^+(aq) + Cr_2O_7^{2-}(aq) \rightarrow 2 Cr^{3+}(aq) + 7 H_2O(l)$ (reduction half reaction)

Combine the two half reactions.

$14 H^+(aq) + 6 S_2O_3^{2-}(aq) + Cr_2O_7^{2-}(aq) \rightarrow 3 S_4O_6^{2-}(aq) + 2 Cr^{3+}(aq) + 7 H_2O(l)$

(c) $ClO_3^-(aq) \rightarrow Cl^-(aq)$

$ClO_3^-(aq) \rightarrow Cl^-(aq) + 3 H_2O(l)$

$6 H^+(aq) + ClO_3^-(aq) \rightarrow Cl^-(aq) + 3 H_2O(l)$

$[6 e^- + 6 H^+(aq) + ClO_3^-(aq) \rightarrow Cl^-(aq) + 3 H_2O(l)] \times 14$ (reduction half reaction)

$As_2S_3(s) \rightarrow H_2AsO_4^-(aq) + HSO_4^-(aq)$

$As_2S_3(s) \rightarrow 2 H_2AsO_4^-(aq) + 3 HSO_4^-(aq)$

$20 H_2O(l) + As_2S_3(s) \rightarrow 2 H_2AsO_4^-(aq) + 3 HSO_4^-(aq)$

$20 H_2O(l) + As_2S_3(s) \rightarrow 2 H_2AsO_4^-(aq) + 3 HSO_4^-(aq) + 33 H^+(aq)$

$[20 H_2O(l) + As_2S_3(s) \rightarrow 2 H_2AsO_4^-(aq) + 3 HSO_4^-(aq) + 33 H^+(aq) + 28 e^-] \times 3$
(oxidation half reaction)

Combine the two half reactions.

$84 H^+(aq) + 60 H_2O(l) + 14 ClO_3^-(aq) + 3 As_2S_3(s) \rightarrow$
$14 Cl^-(aq) + 6 H_2AsO_4^-(aq) + 9 HSO_4^-(aq) + 42 H_2O(l) + 99 H^+(aq)$

$18 H_2O(l) + 14 ClO_3^-(aq) + 3 As_2S_3(s) \rightarrow$
$14 Cl^-(aq) + 6 H_2AsO_4^-(aq) + 9 HSO_4^-(aq) + 15 H^+(aq)$

(d) $IO_3^-(aq) \rightarrow I^-(aq)$

$IO_3^-(aq) \rightarrow I^-(aq) + 3 H_2O(l)$

$6 H^+(aq) + IO_3^-(aq) \rightarrow I^-(aq) + 3 H_2O(l)$

$[6 e^- + 6 H^+(aq) + IO_3^-(aq) \rightarrow I^-(aq) + 3 H_2O(l)] \times 7$ (reduction half reaction)

$Re(s) \rightarrow ReO_4^-(aq)$

$4 H_2O(l) + Re(s) \rightarrow ReO_4^-(aq)$

$4 H_2O(l) + Re(s) \rightarrow ReO_4^-(aq) + 8 H^+(aq)$

$[4 H_2O(l) + Re(s) \rightarrow ReO_4^-(aq) + 8 H^+(aq) + 7 e^-] \times 6$ (oxidation half reaction)

Combine the two half reactions.

$42 H^+(aq) + 24 H_2O(l) + 7 IO_3^-(aq) + 6 Re(s) \rightarrow$

$\qquad\qquad 7 I^-(aq) + 6 ReO_4^-(aq) + 21 H_2O(l) + 48 H^+(aq)$

$3 H_2O(l) + 7 IO_3^-(aq) + 6 Re(s) \rightarrow 7 I^-(aq) + 6 ReO_4^-(aq) + 6 H^+(aq)$

(e) $HSO_4^-(aq) + Pb_3O_4(s) \rightarrow PbSO_4(s)$

$3 HSO_4^-(aq) + Pb_3O_4(s) \rightarrow 3 PbSO_4(s)$

$3 HSO_4^-(aq) + Pb_3O_4(s) \rightarrow 3 PbSO_4(s) + 4 H_2O(l)$

$5 H^+(aq) + 3 HSO_4^-(aq) + Pb_3O_4(s) \rightarrow 3 PbSO_4(s) + 4 H_2O(l)$

$[2 e^- + 5 H^+(aq) + 3 HSO_4^-(aq) + Pb_3O_4(s) \rightarrow 3 PbSO_4(s) + 4 H_2O(l)] \times 10$

$\qquad\qquad$(reduction half reaction)

$As_4(s) \rightarrow H_2AsO_4^-(aq)$

$As_4(s) \rightarrow 4 H_2AsO_4^-(aq)$

$16 H_2O(l) + As_4(s) \rightarrow 4 H_2AsO_4^-(aq)$

$16 H_2O(l) + As_4(s) \rightarrow 4 H_2AsO_4^-(aq) + 24 H^+(aq)$

$16 H_2O(l) + As_4(s) \rightarrow 4 H_2AsO_4^-(aq) + 24 H^+(aq) + 20 e^-$ (oxidation half reaction)

Combine the two half reactions.

$26 H^+(aq) + 30 HSO_4^-(aq) + As_4(s) + 10 Pb_3O_4(s) \rightarrow$

$\qquad\qquad 4 H_2AsO_4^-(aq) + 30 PbSO_4(s) + 24 H_2O(l)$

(f) $HNO_2(aq) \rightarrow NO_3^-(aq)$

$H_2O(l) + HNO_2(aq) \rightarrow NO_3^-(aq)$

$H_2O(l) + HNO_2(aq) \rightarrow NO_3^-(aq) + 3 H^+(aq)$

$H_2O(l) + HNO_2(aq) \rightarrow NO_3^-(aq) + 3 H^+(aq) + 2 e^-$ (oxidation half reaction)

$HNO_2(aq) \rightarrow NO(g)$

$HNO_2(aq) \rightarrow NO(g) + H_2O(l)$

$H^+(aq) + HNO_2(aq) \rightarrow NO(g) + H_2O(l)$

$[1 e^- + H^+(aq) + HNO_2(aq) \rightarrow NO(g) + H_2O(l)] \times 2$ (reduction half reaction)

Combine the two half reactions.

$3 HNO_2(aq) \rightarrow NO_3^-(aq) + 2 NO(g) + H_2O(l) + H^+(aq)$

7.120 CuO, 79.55; Cu$_2$O, 143.09

Let X equal the mass of CuO and Y the mass of Cu$_2$O in the 10.50 g mixture. Therefore, X + Y = 10.50 g.

$$\text{mol Cu} = 8.66 \text{ g} \times \frac{1 \text{ mol Cu}}{63.546 \text{ g Cu}} = 0.1363 \text{ mol Cu}$$

$$\text{mol CuO} + 2 \times \text{mol Cu}_2\text{O} = 0.1363 \text{ mol Cu}$$

$$X \times \frac{1 \text{ mol CuO}}{79.55 \text{ g CuO}} + 2 \times \left(Y \times \frac{1 \text{ mol Cu}_2\text{O}}{143.09 \text{ g Cu}_2\text{O}} \right) = 0.1363 \text{ mol Cu}$$

Rearrange to get X = 10.50 g – Y and then substitute it into the equation above to solve for Y.

$$(10.50 \text{ g} - Y) \times \frac{1 \text{ mol CuO}}{79.55 \text{ g CuO}} + 2 \times \left(Y \times \frac{1 \text{ mol Cu}_2\text{O}}{143.09 \text{ g Cu}_2\text{O}} \right) = 0.1363 \text{ mol Cu}$$

$$\frac{10.50 \text{ mol}}{79.55} - \frac{Y \text{ mol}}{79.55 \text{ g}} + \frac{2 \text{ Y mol}}{143.09 \text{ g}} = 0.1363 \text{ mol}$$

$$-\frac{Y \text{ mol}}{79.55 \text{ g}} + \frac{2 \text{ Y mol}}{143.09 \text{ g}} = 0.1363 \text{ mol} - \frac{10.50 \text{ mol}}{79.55} = 0.0043 \text{ mol}$$

$$\frac{(- \text{ Y mol})(143.09 \text{ g}) + (2 \text{ Y mol})(79.55 \text{ g})}{(79.55 \text{ g})(143.09 \text{ g})} = 0.0043 \text{ mol}$$

$$\frac{16.01 \text{ Y mol}}{11383 \text{ g}} = 0.0043 \text{ mol}; \quad \frac{16.01 \text{ Y}}{11383 \text{ g}} = 0.0043$$

Y = (0.0043)(11383 g)/16.01 = 3.06 g Cu$_2$O

X = 10.50 g – Y = 10.50 g – 3.06 g = 7.44 g CuO

7.122 Mass O in M$_2$O$_3$ = 1.890 g M$_2$O$_3$ - 1.000 g M = 0.890 g O

$$0.890 \text{ g O} \times \frac{1 \text{ mol O}}{16.00 \text{ g O}} = 0.0556 \text{ mol O}$$

$$0.0556 \text{ mol O} \times \frac{2 \text{ mol M}}{3 \text{ mol O}} = 0.0371 \text{ mol M}$$

$$\text{molar mass M} = \frac{1.000 \text{ g M}}{0.0371 \text{ mol M}} = 26.95 \text{ g/mol; M is Al}$$

7.124 (a) 2 H$^+$ + 2 Cl$^-$ + H$_2$O$_2$ → Cl$_2$ + 2 H$_2$O, H$_2$O$_2$ reduced
       (b) 6 H$^+$ + 2 MnO$_4^-$ + 5 H$_2$O$_2$ → 2 Mn^{2+} + 5 O$_2$ + 8 H$_2$O, H$_2$O$_2$ oxidized
       (c) H$_2$O$_2$ + Cl$_2$ → 2 H$^+$ + 2 Cl$^-$ + O$_2$, H$_2$O$_2$ oxidized

7.126 48.39 mL = 0.048 39 L

(0.1116 mol/L MnO$_4^-$)(0.048 39 L) = 5.400 $\times$ 10^{-3} mol MnO$_4^-$

$$5.400 \times 10^{-3} \text{ mol MnO}_4^- \times \frac{5 \text{ mol Fe}^{2+}}{1 \text{ mol MnO}_4^-} \times \frac{55.84 \text{ g Fe}^{2+}}{1 \text{ mol Fe}^{2+}} = 1.508 \text{ g Fe}^{2+}$$

$$\text{mass \% Fe} = \frac{1.508 \text{ g Fe}^{2+}}{2.368 \text{ g}} \times 100\% = 63.68\%$$

Multiconcept Problems

7.128   NaOH, 40.00; Ba(OH)$_2$, 171.34
Let X equal the mass of NaOH and Y the mass of Ba(OH)$_2$ in the 10.0 g mixture.
Therefore, X + Y = 10.0 g.

$$\text{mol HCl} = 108.9 \text{ mL} \times \frac{1 \text{ L}}{1000 \text{ mL}} \times \frac{1.50 \text{ mol HCl}}{1 \text{ L}} = 0.163 \text{ mol HCl}$$

$$\text{mol NaOH} + 2 \times \text{mol Ba(OH)}_2 = 0.163 \text{ mol HCl}$$

$$X \times \frac{1 \text{ mol NaOH}}{40.00 \text{ g NaOH}} + 2 \times \left(Y \times \frac{1 \text{ mol Ba(OH)}_2}{171.34 \text{ g Ba(OH)}_2} \right) = 0.163 \text{ mol HCl}$$

Rearrange to get X = 10.0 g – Y and then substitute it into the equation above to solve for Y.

$$(10.0 \text{ g} - Y) \times \frac{1 \text{ mol NaOH}}{40.00 \text{ g NaOH}} + 2 \times \left(Y \times \frac{1 \text{ mol Ba(OH)}_2}{171.34 \text{ g Ba(OH)}_2} \right) = 0.163 \text{ mol HCl}$$

$$\frac{10.00 \text{ mol}}{40.00} - \frac{Y \text{ mol}}{40.00 \text{ g}} + \frac{2 \text{ Y mol}}{171.34 \text{ g}} = 0.163 \text{ mol}$$

$$-\frac{Y \text{ mol}}{40.00 \text{ g}} + \frac{2 \text{ Y mol}}{171.34 \text{ g}} = 0.163 \text{ mol} - \frac{10.00 \text{ mol}}{40.00} = -0.087 \text{ mol}$$

$$\frac{(-Y \text{ mol})(171.34 \text{ g}) + (2 \text{ Y mol})(40.00 \text{ g})}{(40.00 \text{ g})(171.34 \text{ g})} = -0.087 \text{ mol}$$

$$\frac{-91.34 \text{ Y mol}}{6853.6 \text{ g}} = -0.087 \text{ mol}; \quad \frac{91.34 \text{ Y}}{6853.6 \text{ g}} = 0.087$$

Y = (0.087)(6853.6 g)/91.34 = 6.5 g Ba(OH)$_2$
X = 10.0 g – Y = 10.0 g – 6.5 g = 3.5 g NaOH

7.130   KNO$_3$, 101.10; BaCl$_2$, 208.24; NaCl, 58.44; BaSO$_4$, 233.40; AgCl, 143.32
(a) The two precipitates are BaSO$_4$(s) and AgCl(s).
(b) H$_2$SO$_4$ only reacts with BaCl$_2$.
H$_2$SO$_4$(aq) + BaCl$_2$(aq) → BaSO$_4$(s) + 2 HCl(aq)
Calculate the number of moles of BaCl$_2$ in 100.0 g of the mixture.

$$\text{mol BaCl}_2 = 67.3 \text{ g BaSO}_4 \times \frac{1 \text{ mol BaSO}_4}{233.40 \text{ g BaSO}_4} \times \frac{1 \text{ mol BaCl}_2}{1 \text{ mol BaSO}_4} = 0.288 \text{ mol BaCl}_2$$

Calculate mass and moles of BaCl$_2$ in 250.0 g sample.

$$\text{mass BaCl}_2 = 0.288 \text{ mol BaCl}_2 \times \frac{208.24 \text{ g BaCl}_2}{1 \text{ mol BaCl}_2} \times \frac{250.0 \text{ g}}{100.0 \text{ g}} = 150. \text{ g BaCl}_2$$

$$\text{mol BaCl}_2 = 150. \text{ g BaCl}_2 \times \frac{1 \text{ mol BaCl}_2}{208.24 \text{ g BaCl}_2} = 0.720 \text{ mol BaCl}_2$$

AgNO$_3$ reacts with both NaCl and BaCl$_2$ in the remaining 150.0 g of the mixture.
3 AgNO$_3$(aq) + NaCl(aq) + BaCl$_2$(aq) → 3 AgCl(s) + NaNO$_3$(aq) + Ba(NO$_3$)$_2$(aq)
Calculate the moles of AgCl that would have been produced from the 250.0 g mixture.

$$\text{mol AgCl} = 197.6 \text{ g AgCl} \times \frac{1 \text{ mol AgCl}}{143.32 \text{ g AgCl}} \times \frac{250.0 \text{ g}}{150.0 \text{ g}} = 2.30 \text{ mol AgCl}$$

mol AgCl = 2 x (mol BaCl$_2$) + mol NaCl
Calculate the moles and mass of NaCl in the 250.0 g mixture.
2.30 mol AgCl = 2 x 0.720 mol BaCl$_2$ + mol NaCl
mol NaCl = 2.30 mol − 2(0.720 mol) = 0.86 mol NaCl

$$\text{mass NaCl} = 0.86 \text{ mol NaCl x } \frac{58.44 \text{ g NaCl}}{1 \text{ mol NaCl}} = 50. \text{ g NaCl}$$

Calculate the mass of KNO$_3$ in the 250.0 g mixture.
total mass = mass BaCl$_2$ + mass NaCl + mass KNO$_3$
250.0 g = 150. g BaCl$_2$ + 50. g NaCl + mass KNO$_3$
mass KNO$_3$ = 250.0 g − 150. g BaCl$_2$ − 50. g NaCl = 50. g KNO$_3$

7.132 (a) Cr^{2+}(aq) + Cr$_2$O$_7^{2-}$(aq) → Cr^{3+}(aq)
[Cr^{2+}(aq) → Cr^{3+}(aq) + e$^-$] x 6   (oxidation half reaction)

Cr$_2$O$_7^{2-}$(aq) → Cr^{3+}(aq)
Cr$_2$O$_7^{2-}$(aq) → 2 Cr^{3+}(aq)
Cr$_2$O$_7^{2-}$(aq) → 2 Cr^{3+}(aq) + 7 H$_2$O(l)
14 H$^+$(aq) + Cr$_2$O$_7^{2-}$(aq) → 2 Cr^{3+}(aq) + 7 H$_2$O(l)
6 e$^-$ + 14 H$^+$(aq) + Cr$_2$O$_7^{2-}$(aq) → 2 Cr^{3+}(aq) + 7 H$_2$O(l)   (reduction half reaction)

Combine the two half reactions.
14 H$^+$(aq) + Cr$_2$O$_7^{2-}$(aq) + 6 Cr^{2+}(aq) → 8 Cr^{3+}(aq) + 7 H$_2$O(l)

(b) total volume = 100.0 ml + 20.0 mL = 120.0 mL = 0.1200 L
Initial moles:

$$0.120 \frac{\text{mol Cr(NO}_3)_2}{1 \text{ L}} \text{ x } 0.1000 \text{ L} = 0.0120 \text{ mol Cr(NO}_3)_2$$

$$0.500 \frac{\text{mol HNO}_3}{1 \text{ L}} \text{ x } 0.1000 \text{ L} = 0.0500 \text{ mol HNO}_3$$

$$0.250 \frac{\text{mol K}_2\text{Cr}_2\text{O}_7}{1 \text{ L}} \text{ x } 0.0200 \text{ L} = 0.005 \ 00 \text{ mol K}_2\text{Cr}_2\text{O}_7$$

Check for the limiting reactant. 0.0120 mol of Cr^{2+} requires (0.0120)/6 = 0.00200 mol
Cr$_2$O$_7^{2-}$ and (14/6)(0.0120) = 0.0280 mol H$^+$. Both are in excess of the required amounts,
so Cr^{2+} is the limiting reactant.

	14 H$^+$(aq) +	Cr$_2$O$_7^{2-}$(aq) +	6 Cr^{2+}(aq) →	8 Cr^{3+}(aq) + 7 H$_2$O(l)
Initial moles	0.0500	0.00500	0.0120	0
Change	−14x	−x	−6x	+8x
Because Cr^{2+} is the limiting reactant, 6x = 0.0120 and x = 0.00200				
Final moles	0.0220	0.00300	0	0.0160

$$\text{mol K}^+ = 0.00500 \text{ mol K}_2\text{Cr}_2\text{O}_7 \text{ x } \frac{2 \text{ mol K}^+}{1 \text{ mol K}_2\text{Cr}_2\text{O}_7} = 0.0100 \text{ mol K}^+$$

$$\text{mol NO}_3^- = 0.0120 \text{ mol Cr(NO}_3)_2 \times \frac{2 \text{ mol NO}_3^-}{1 \text{ mol Cr(NO}_3)_2}$$

$$+ \; 0.0500 \text{ mol HNO}_3 \times \frac{1 \text{ mol NO}_3^-}{1 \text{ mol HNO}_3} = 0.0740 \text{ mol NO}_3^-$$

mol H^+ = 0.0220 mol;   mol $Cr_2O_7^{2-}$ = 0.00300 mol;   mol Cr^{3+} = 0.01600 mol

Check for charge neutrality.
Total moles of +charge = 0.0100 + 0.0220 + 3 x (0.01600) = 0.0800 mol +charge
Total moles of –charge = 0.0740 + 2 x (0.00300) = 0.0800 mol –charge
The charges balance and there is electrical neutrality in the solution after the reaction.

$$K^+ \text{ molarity} = \frac{0.0100 \text{ mol K}^+}{0.1200 \text{ L}} = 0.0833 \text{ M}$$

$$\text{NO}_3^- \text{ molarity} = \frac{0.0740 \text{ mol NO}_3^-}{0.1200 \text{ L}} = 0.617 \text{ M}$$

$$H^+ \text{ molarity} = \frac{0.0220 \text{ mol H}^+}{0.1200 \text{ L}} = 0.183 \text{ M}$$

$$Cr_2O_7^{2-} \text{ molarity} = \frac{0.00300 \text{ mol Cr}_2O_7^{2-}}{0.1200 \text{ L}} = 0.0250 \text{ M}$$

$$Cr^{3+} \text{ molarity} = \frac{0.0160 \text{ mol Cr}^{3+}}{0.1200 \text{ L}} = 0.133 \text{ M}$$

7.134 (a) (1) $Cu(s) \rightarrow Cu^{2+}(aq)$
        $[Cu(s) \rightarrow Cu^{2+}(aq) + 2 \text{ e}^-] \times 3$ (oxidation half reaction)

        $NO_3^-(aq) \rightarrow NO(g)$
        $NO_3^-(aq) \rightarrow NO(g) + 2 H_2O(l)$
        $4 H^+(aq) + NO_3^-(aq) \rightarrow NO(g) + 2 H_2O(l)$
        $[3 \text{ e}^- + 4 H^+(aq) + NO_3^-(aq) \rightarrow NO(g) + 2 H_2O(l)] \times 2$ (reduction half reaction)

        Combine the two half reactions.
        $3 Cu(s) + 8 H^+(aq) + 2 NO_3^-(aq) \rightarrow 3 Cu^{2+}(aq) + 2 NO(g) + 4 H_2O(l)$

    (2) $Cu^{2+}(aq) + SCN^-(aq) \rightarrow CuSCN(s)$
        $[\text{e}^- + Cu^{2+}(aq) + SCN^-(aq) \rightarrow CuSCN(s)] \times 2$ (reduction half reaction)

        $HSO_3^-(aq) \rightarrow HSO_4^-(aq)$
        $H_2O(l) + HSO_3^-(aq) \rightarrow HSO_4^-(aq)$
        $H_2O(l) + HSO_3^-(aq) \rightarrow HSO_4^-(aq) + 2 H^+(aq)$
        $H_2O(l) + HSO_3^-(aq) \rightarrow HSO_4^-(aq) + 2 H^+(aq) + 2 \text{ e}^-$
                                        (oxidation half reaction)

111

Combine the two half reactions.
$$2\ Cu^{2+}(aq)\ +\ 2\ SCN^-(aq)\ +\ H_2O(l)\ +\ HSO_3^-(aq)\ \rightarrow$$
$$2\ CuSCN(s)\ +\ HSO_4^-(aq)\ +\ 2\ H^+(aq)$$

(3) $Cu^+(aq)\ \rightarrow\ Cu^{2+}(aq)$
$[Cu^+(aq)\ \rightarrow\ Cu^{2+}(aq)\ +\ e^-]\ x\ 10$ (oxidation half reaction)

$IO_3^-(aq)\ \rightarrow\ I_2(aq)$
$2\ IO_3^-(aq)\ \rightarrow\ I_2(aq)$
$2\ IO_3^-(aq)\ \rightarrow\ I_2(aq)\ +\ 6\ H_2O(l)$
$12\ H^+(aq)\ +\ 2\ IO_3^-(aq)\ \rightarrow\ I_2(aq)\ +\ 6\ H_2O(l)$
$10\ e^-\ +\ 12\ H^+(aq)\ +\ 2\ IO_3^-(aq)\ \rightarrow\ I_2(aq)\ +\ 6\ H_2O(l)$ (reduction half reaction)

Combine the two half reactions.
$$10\ Cu^+(aq)\ +\ 12\ H^+(aq)\ +\ 2\ IO_3^-(aq)\ \rightarrow\ 10\ Cu^{2+}(aq)\ +\ I_2(aq)\ +\ 6\ H_2O(l)$$

(4) $I_2(aq)\ \rightarrow\ I^-(aq)$
$I_2(aq)\ \rightarrow\ 2\ I^-(aq)$
$2\ e^-\ +\ I_2(aq)\ \rightarrow\ 2\ I^-(aq)$ (reduction half reaction)

$S_2O_3^{2-}(aq)\ \rightarrow\ S_4O_6^{2-}(aq)$
$2\ S_2O_3^{2-}(aq)\ \rightarrow\ S_4O_6^{2-}(aq)$
$2\ S_2O_3^{2-}(aq)\ \rightarrow\ S_4O_6^{2-}(aq)\ +\ 2\ e^-$ (oxidation half reaction)

Combine the two half reactions.
$$I_2(aq)\ +\ 2\ S_2O_3^{2-}(aq)\ \rightarrow\ 2\ I^-(aq)\ +\ S_4O_6^{2-}(aq)$$

(5) $2\ ZnNH_4PO_4\ \rightarrow\ Zn_2P_2O_7\ +\ H_2O\ +\ 2\ NH_3$

(b) $10.82\ mL = 0.01082\ L$
mol $S_2O_3^{2-}$ = $(0.1220\ mol/L)(0.01082\ L) = 0.00132$ mol $S_2O_3^{2-}$

mol I_2 = 0.00132 mol $S_2O_3^{2-}$ x $\dfrac{1\ mol\ I_2}{2\ mol\ S_2O_3^{2-}}$ = $6.60\ x\ 10^{-4}$ mol I_2

mol Cu^+ = $6.60\ x\ 10^{-4}$ mol I_2 x $\dfrac{10\ mol\ Cu^+}{1\ mol\ I_2}$ = $6.60\ x\ 10^{-3}$ mol Cu^+ (Cu)

g Cu = $(6.60\ x\ 10^{-3}\ mol)(63.546\ g/mol) = 0.419$ g Cu

mass % Cu in brass = $\dfrac{0.419\ g\ Cu}{0.544\ g\ brass}$ x 100% = 77.1% Cu

(c) $Zn_2P_2O_7$, 304.72

mass % Zn in $Zn_2P_2O_7$ = $\dfrac{2\ x\ 65.39\ g}{304.72\ g}$ x 100% = 42.92%

mass of Zn in $Zn_2P_2O_7$ = $(0.4292)(0.246\ g) = 0.106$ g Zn

mass % Zn in brass = $\dfrac{0.106\ g\ Zn}{0.544\ g\ brass}$ x 100% = 19.5% Zn

7.136 (a) $H_3MO_3(aq) \rightarrow H_3MO_4(aq)$
$H_3MO_3(aq) + H_2O(l) \rightarrow H_3MO_4(aq)$
$H_3MO_3(aq) + H_2O(l) \rightarrow H_3MO_4(aq) + 2 H^+(aq)$
$[H_3MO_3(aq) + H_2O(l) \rightarrow H_3MO_4(aq) + 2 H^+(aq) + 2 e^-] \times 5$ (oxidation half reaction)

$MnO_4^-(aq) \rightarrow Mn^{2+}(aq)$
$MnO_4^-(aq) \rightarrow Mn^{2+}(aq) + 4 H_2O(l)$
$MnO_4^-(aq) + 8 H^+(aq) \rightarrow Mn^{2+}(aq) + 4 H_2O(l)$
$[MnO_4^-(aq) + 8 H^+(aq) + 5 e^- \rightarrow Mn^{2+}(aq) + 4 H_2O(l)] \times 2$ (reduction half reaction)

Combine the two half reactions.
$5 H_3MO_3(aq) + 5 H_2O(l) + 2 MnO_4^-(aq) + 16 H^+(aq) \rightarrow$
$$5 H_3MO_4(aq) + 10 H^+(aq) + 2 Mn^{2+}(aq) + 8 H_2O(l)$$
$5 H_3MO_3(aq) + 2 MnO_4^-(aq) + 6 H^+(aq) \rightarrow 5 H_3MO_4(aq) + 2 Mn^{2+}(aq) + 3 H_2O(l)$

(b) 10.7 mL = 0.0107 L
mol MnO_4^- = (0.0107 L)(0.100 mol/L) = 1.07 x 10^{-3} mol MnO_4^-

mol H_3MO_3 = 1.07 x 10^{-3} mol MnO_4^- x $\dfrac{5 \text{ mol } H_3MO_3}{2 \text{ mol } MnO_4^-}$ = 2.67 x 10^{-3} mol H_3MO_3

mol M_2O_3 = 2.67 x 10^{-3} mol H_3MO_3 x $\dfrac{1 \text{ mol } M_2O_3}{2 \text{ mol } H_3MO_3}$ = 1.34 x 10^{-3} mol M_2O_3

mol M in M_2O_3 = 1.34 x 10^{-3} mol M_2O_3 x $\dfrac{2 \text{ mol M}}{1 \text{ mol } M_2O_3}$ = 2.68 x 10^{-3} mol M

(c) M molar mass = $\dfrac{0.200 \text{ g}}{2.68 \times 10^{-3} \text{ mol}}$ = 74.6 g/mol; M atomic mass = 74.6

M is As.

114

8

Thermochemistry: Chemical Energy

8.1   (a) and (b) are state functions; (c) is not.

8.2 $\Delta V = (4.3\ L - 8.6\ L) = -4.3\ L$

$w = -P\Delta V = -(44\ atm)(-4.3\ L) = +189.2\ L \cdot atm$

$w = (189.2\ L \cdot atm)(101\ \dfrac{J}{L \cdot atm}) = +1.9 \times 10^4\ J$

The positive sign for the work indicates that the surroundings do work on the system. Energy flows into the system.

8.3 $w = -P\Delta V = -(2.5\ atm)(3\ L - 2\ L) = -2.5\ L \cdot atm$

$w = (-2.5\ L \cdot atm)\left(101\ \dfrac{J}{L \cdot atm}\right) = -252.5\ J = -250\ J = -0.25\ kJ$

The negative sign indicates that the expanding system loses work energy and does work on the surroundings.

8.4   (a) $w = -P\Delta V$ is positive and $P\Delta V$ is negative for this reaction because the system volume is decreased at constant pressure.

(b) $P\Delta V$ is small compared to ΔE.

$\Delta H = \Delta E + P\Delta V$; ΔH is negative. Its value is slightly more negative than ΔE.

8.5 $q = (\text{specific heat}) \times m \times \Delta T = (4.18\ \dfrac{J}{g \cdot {}^{\circ}C})(350\ g)(3\ ^{\circ}C - 25\ ^{\circ}C) = -3.2 \times 10^4\ J$

$q = -3.2 \times 10^4\ J \times \dfrac{1\ kJ}{1000\ J} = -32\ kJ$

8.6 $q = (\text{specific heat}) \times m \times \Delta T$

$\text{specific heat} = \dfrac{q}{m \times \Delta T} = \dfrac{97.2\ J}{(75.0\ g)(10.0\ ^{\circ}C)} = 0.130\ J/(g \cdot {}^{\circ}C)$

8.7 $25.0\ mL = 0.0250\ L$ and $50.0\ mL = 0.0500\ L$

$mol\ H_2SO_4 = (1.00\ mol/L)(0.0250\ L) = 0.0250\ mol\ H_2SO_4$

$mol\ NaOH = (1.00\ mol/L)(0.0500\ L) = 0.0500\ mol\ NaOH$

NaOH and H_2SO_4 are present in a 2:1 mol ratio. This matches the stoichiometric ratio in the balanced equation.

$q = (\text{specific heat}) \times m \times \Delta T$

$m = (25.0\ mL + 50.0\ mL)(1.00\ g/mL) = 75.0\ g$

$q = (4.18\ \dfrac{J}{g \cdot {}^{\circ}C})(75.0\ g)(33.9\ ^{\circ}C - 25.0\ ^{\circ}C) = 2790\ J$

mol H_2SO_4 = 0.0250 L x 1.00 $\dfrac{mol}{L}$ H_2SO_4 = 0.0250 mol H_2SO_4

Heat evolved per mole of H_2SO_4 = $\dfrac{2.79 \times 10^3 \text{ J}}{0.0250 \text{ mol } H_2SO_4}$ = 1.1 x 10^5 J/mol H_2SO_4

Because the reaction evolves heat, the sign for ΔH is negative.

ΔH = –1.1 x 10^5 J x $\dfrac{1 \text{ kJ}}{1000 \text{ J}}$ = –1.1 x 10^2 kJ

8.8 $\Delta H°$ = – 484 $\dfrac{kJ}{2 \text{ mol } H_2}$

$P\Delta V$ = (1.00 atm)(–5.6 L) = –5.6 L· atm

$P\Delta V$ = (–5.6 L· atm)(101 $\dfrac{J}{L \cdot atm}$) = –565.6 J = –570 J = –0.57 kJ

w = –$P\Delta V$ = 570 J = 0.57 kJ

ΔH = $\dfrac{-121 \text{ kJ}}{0.50 \text{ mol } H_2}$

ΔE = ΔH – $P\Delta V$ = –121 kJ – (–0.57 kJ) = –120.43 kJ = –120 kJ

8.9 ΔV = 448 L and assume P = 1.00 atm

w = –$P\Delta V$ = –(1.00 atm)(448 L) = – 448 L· atm

w = –(448 L· atm)(101 $\dfrac{J}{L \cdot atm}$) = – 4.52 x 10^4 J

w = – 4.52 x 10^4 J x $\dfrac{1 \text{ kJ}}{1000 \text{ J}}$ = – 45.2 kJ

8.10   (a) C_3H_8, 44.10; $\Delta H°$ = –2220 kJ/mol C_3H_8

15.5 g x $\dfrac{1 \text{ mol } C_3H_8}{44.10 \text{ g } C_3H_8}$ x $\dfrac{-2220 \text{ kJ}}{1 \text{ mol } C_3H_8}$ = –780. kJ

780. kJ of heat is evolved.

(b) $Ba(OH)_2 \cdot 8 H_2O$, 315.5; $\Delta H°$ = +80.3 kJ/mol $Ba(OH)_2 \cdot 8 H_2O$

4.88 g x $\dfrac{1 \text{ mol } Ba(OH)_2 \cdot 8 H_2O}{315.5 \text{ g } Ba(OH)_2 \cdot 8 H_2O}$ x $\dfrac{80.3 \text{ kJ}}{1 \text{ mol } Ba(OH)_2 \cdot 8 H_2O}$ = +1.24 kJ

1.24 kJ of heat is absorbed.

8.11 CH_3NO_2, 61.04

q = 100.0 g CH_3NO_2 x $\dfrac{1 \text{ mol } CH_3NO_2}{61.04 \text{ g } CH_3NO_2}$ x $\dfrac{2441.6 \text{ kJ}}{4 \text{ mol } CH_3NO_2}$ = 1.000 x 10^3 kJ

8.12 $CH_4(g)$ + $Cl_2(g)$ → $CH_3Cl(g)$ + $HCl(g)$ $\Delta H°_1$ = –98.3 kJ
         $\underline{CH_3Cl(g) + Cl_2(g) → CH_2Cl_2(g) + HCl(g)}$ $\Delta H°_2$ = –104 kJ
   Sum $CH_4(g)$ + 2 $Cl_2(g)$ → $CH_2Cl_2(g)$ + 2 HCl(g)
         $\Delta H°$ = $\Delta H°_1$ + $\Delta H°_2$ = –202 kJ

8.13 (a) $A + 2B \rightarrow D$; $\Delta H° = -100 \text{ kJ} + (-50 \text{ kJ}) = -150 \text{ kJ}$

(b) The red arrow corresponds to step 1: $A + B \rightarrow C$
The green arrow corresponds to step 2: $C + B \rightarrow D$
The blue arrow corresponds to the overall reaction.

(c) The top energy level represents $A + 2B$.
The middle energy level represents $C + B$.
The bottom energy level represents D.

8.14

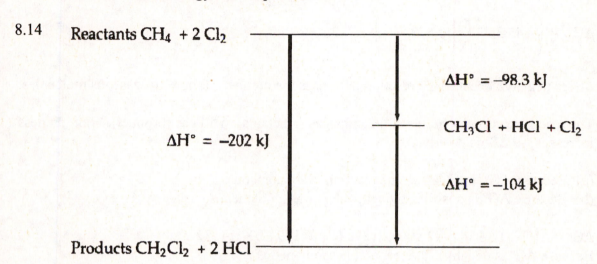

Reactants $CH_4 + 2\,Cl_2$

$\Delta H° = -98.3 \text{ kJ}$

$CH_3Cl + HCl + Cl_2$

$\Delta H° = -202 \text{ kJ}$

$\Delta H° = -104 \text{ kJ}$

Products $CH_2Cl_2 + 2\,HCl$

8.15 $4\,NH_3(g) + 5\,O_2(g) \rightarrow 4\,NO(g) + 6\,H_2O(g)$
$\Delta H°_{rxn} = [4\,\Delta H°_f(NO) + 6\,\Delta H°_f(H_2O)] - [4\,\Delta H°_f(NH_3)]$
$\Delta H°_{rxn} = [(4 \text{ mol})(91.3 \text{ kJ/mol}) + (6 \text{ mol})(-241.8 \text{ kJ/mol})] - [(4 \text{ mol})(-46.1 \text{ kJ/mol})]$
$\Delta H°_{rxn} = -901.2 \text{ kJ}$

8.16 $6\,CO_2(g) + 6\,H_2O(l) \rightarrow C_6H_{12}O_6(s) + 6\,O_2(g)$
$\Delta H°_{rxn} = \Delta H°_f(C_6H_{12}O_6) - [6\,\Delta H°_f(CO_2) + 6\,\Delta H°_f(H_2O(l))]$
$\Delta H°_{rxn} = [(1 \text{ mol})(-1273.3 \text{ kJ/mol})] - [(6 \text{ mol})(-393.5 \text{ kJ/mol}) + (6 \text{ mol})(-285.8 \text{ kJ/mol})]$
$\Delta H°_{rxn} = +2802.5 \text{ kJ} = +2803 \text{ kJ}$

8.17 $H_2C=CH_2(g) + H_2O(g) \rightarrow C_2H_5OH(g)$
$\Delta H°_{rxn} = D \text{ (Reactant bonds)} - D \text{ (Product bonds)}$
$\Delta H°_{rxn} = (D_{C=C} + 4\,D_{C-H} + 2\,D_{O-H}) - (D_{C-C} + D_{C-O} + 5\,D_{C-H} + D_{O-H})$
$\Delta H°_{rxn} = [(1 \text{ mol})(728 \text{ kJ/mol}) + (4 \text{ mol})(410 \text{ kJ/mol}) + (2 \text{ mol})(460 \text{ kJ/mol})]$
$- [(1 \text{ mol})(350 \text{ kJ/mol}) + (1 \text{ mol})(350 \text{ kJ/mol}) + (5 \text{ mol})(410 \text{ kJ/mol}) + (1 \text{ mol})(460 \text{ kJ/mol})]$
$\Delta H°_{rxn} = +78 \text{ kJ}$

8.18 $2\,NH_3(g) + Cl_2(g) \rightarrow N_2H_4(g) + 2\,HCl(g)$
$\Delta H°_{rxn} = D \text{ (Reactant bonds)} - D \text{ (Product bonds)}$
$\Delta H°_{rxn} = (6\,D_{N-H} + D_{Cl-Cl}) - (D_{N-N} + 4\,D_{N-H} + 2\,D_{H-Cl})$
$\Delta H°_{rxn} = [(6 \text{ mol})(390 \text{ kJ/mol}) + (1 \text{ mol})(243 \text{ kJ/mol})]$
$- [(1 \text{ mol})(240 \text{ kJ/mol}) + (4 \text{ mol})(390 \text{ kJ/mol}) + (2 \text{ mol})(432 \text{ kJ/mol})] = -81 \text{ kJ}$

8.19 $C_4H_{10}(l) + \dfrac{13}{2} O_2(g) \rightarrow 4 CO_2(g) + 5 H_2O(g)$

$\Delta H°_{rxn} = [4 \Delta H°_f (CO_2) + 5 \Delta H°_f (H_2O)] - \Delta H°_f (C_4H_{10})$
$\Delta H°_{rxn} = [(4 \text{ mol})(-393.5 \text{ kJ/mol}) + (5 \text{ mol})(-241.8 \text{ kJ/mol})] - [(1 \text{ mol})(-147.5 \text{ kJ/mol})]$
$\Delta H°_{rxn} = -2635.5 \text{ kJ}; \quad \Delta H°_C = -2635.5 \text{ kJ/mol}$

$C_4H_{10}, 58.12; \quad \Delta H°_C = \left(-2635.5 \dfrac{\text{kJ}}{\text{mol}}\right)\left(\dfrac{1 \text{ mol}}{58.12 \text{ g}}\right) = -45.35 \text{ kJ/g}$

$\Delta H°_C = \left(-45.35 \dfrac{\text{kJ}}{\text{g}}\right)\left(0.579 \dfrac{\text{g}}{\text{mL}}\right) = -26.3 \text{ kJ/mL}$

8.20 $\Delta S°$ is negative because the reaction decreases the number of moles of gaseous molecules.

8.21 The reaction proceeds from a solid and a gas (reactants) to all gas (product). Randomness increases, so $\Delta S°$ is positive.

8.22 (a) Because $\Delta G°$ is negative, the reaction is spontaneous.
(b) Because $\Delta G°$ is positive, the reaction is nonspontaneous.

8.23 $\Delta G° = \Delta H° - T\Delta S° = (-92.2 \text{ kJ}) - (298 \text{ K})(-0.199 \text{ kJ/K}) = -32.9 \text{ kJ}$
Because $\Delta G°$ is negative, the reaction is spontaneous.
Set $\Delta G° = 0$ and solve for T.

$\Delta G° = 0 = \Delta H° - T\Delta S°; \quad T = \dfrac{\Delta H°}{\Delta S°} = \dfrac{-92.2 \text{ kJ}}{-0.199 \text{ kJ/K}} = 463 \text{ K} = 190 \text{ °C}$

8.24 (a) $2 A_2 + B_2 \rightarrow 2 A_2B$
(b) Because the reaction is exothermic, ΔH is negative. There are more reactant molecules than product molecules. The randomness of the system decreases on going from reactant to product, therefore ΔS is negative.
(c) Because $\Delta G = \Delta H - T\Delta S$, a reaction with both ΔH and ΔS negative is favored at low temperatures where the negative ΔH term is larger than the positive $- T\Delta S$, and ΔG is negative.

8.25 $C_2H_6O + 3 O_2 \rightarrow 2 CO_2 + 3 H_2O$
$2 C_{19}H_{38}O_2 + 55 O_2 \rightarrow 38 CO_2 + 38 H_2O$

8.26 Because the standard heat of formation of $CO_2(g)$ (–393.5 kJ/mol) is more negative than that of $H_2O(g)$ (–241.8 kJ/mol), formation of CO_2 releases more heat than formation of H_2O. According to the balanced equations (Problem 8.25), combustion of ethanol yields a 2:3 ratio of CO_2 to H_2O, whereas combustion of biodiesel yields a 1:1 ratio. Thus biodiesel has a more favorable (more negative) combustion enthalpy per gram than ethanol.

Conceptual Problems

8.28 (a) $w = -P\Delta V$, $\Delta V > 0$; therefore $w < 0$ and the system is doing work on the surroundings.
(b) Since the temperature has increased there has been an enthalpy change. The system evolved heat, the reaction is exothermic, and $\Delta H < 0$.

8.30

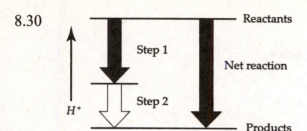

8.32 $\Delta H = \Delta E + P\Delta V$

$\Delta H - \Delta E = P\Delta V$

$\Delta V = \dfrac{\Delta H - \Delta E}{P} = \dfrac{[-35.0 \text{ kJ} - (-34.8 \text{ kJ})]}{1 \text{ atm}} \times \dfrac{1 \text{ L} \cdot \text{atm}}{101 \times 10^{-3} \text{ kJ}} = -2 \text{ L}$

$\Delta V = -2 \text{ L} = V_{\text{final}} - V_{\text{initial}} = V_{\text{final}} - 5 \text{ L}; \qquad V_{\text{final}} = -2 \text{ L} - (-5 \text{ L}) = 3 \text{ L}$

The volume decreases from 5 L to 3 L.

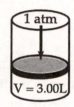

8.34   The change is the spontaneous conversion of a liquid to a gas. ΔG is negative because the change is spontaneous. The conversion of a liquid to a gas is endothermic, therefore ΔH is positive. ΔS is positive because randomness increases when a liquid is converted to a gas.

Section Problems
Heat, Work, and Energy (Sections 8.1–8.3)

8.36   Heat is the energy transferred from one object to another as the result of a temperature difference between them. Temperature is a measure of the kinetic energy of molecular motion.
Energy is the capacity to do work or supply heat. Work is defined as the distance moved times the force that opposes the motion (w = d x F).
Kinetic energy is the energy of motion. Potential energy is stored energy.

8.38   Car: $E_K = \frac{1}{2}(1400 \text{ kg})\left(\dfrac{115 \times 10^3 \text{ m}}{3600 \text{ s}}\right)^2 = 7.1 \times 10^5 \text{ J}$

Truck: $E_K = \frac{1}{2}(12{,}000 \text{ kg})\left(\dfrac{38 \times 10^3 \text{ m}}{3600 \text{ s}}\right)^2 = 6.7 \times 10^5 \text{ J}$

The car has more kinetic energy.

8.40 $w = -P\Delta V = -(3.6 \text{ atm})(3.4 \text{ L} - 3.2 \text{ L}) = -0.72 \text{ L} \cdot \text{atm}$

$w = (-0.72 \text{ L} \cdot \text{atm})\left(\dfrac{101 \text{ J}}{1 \text{ L} \cdot \text{atm}}\right) = -72.7 \text{ J} = -70 \text{ J};$ The energy change is negative.

8.42 (a) 100 W = 100 J/s

$$50 \text{ Cal} \times \frac{1000 \text{ cal}}{1 \text{ Cal}} \times \frac{4.184 \text{ J}}{1 \text{ cal}} \times \frac{1 \text{ s}}{100 \text{ J}} \times \frac{1 \text{ min}}{60 \text{ s}} = 35 \text{ min}$$

(b) 23W = 23 J/s

$$50 \text{ Cal} \times \frac{1000 \text{ cal}}{1 \text{ Cal}} \times \frac{4.184 \text{ J}}{1 \text{ cal}} \times \frac{1 \text{ s}}{23 \text{ J}} \times \frac{1 \text{ min}}{60 \text{ s}} = 150 \text{ min}$$

8.44 (a) $w = -P\Delta V = -(0.975 \text{ atm})(18.0 \text{ L} - 12.0 \text{ L}) = -5.85 \text{ L} \cdot \text{atm}$

$$w = (-5.85 \text{ L} \cdot \text{atm})\left(\frac{101.325 \text{ J}}{1 \text{ L} \cdot \text{atm}}\right) = -593 \text{ J}$$

(b) $\Delta V = 6.0 \text{ L} = 6000 \text{ mL} = 6000 \text{ cm}^3$; $r = \dfrac{17.0 \text{ cm}}{2} = 8.50 \text{ cm}$

$$V = \pi r^2 h; \quad h = \frac{V}{\pi r^2} = \frac{6000 \text{ cm}^3}{\pi(8.50 \text{ cm})^2} = 26.4 \text{ cm}$$

Calorimetry and Heat Capacity (Section 8.5)

8.46 Heat capacity is the amount of heat required to raise the temperature of a substance a given amount. Specific heat is the amount of heat necessary to raise the temperature of exactly 1 g of a substance by exactly 1 °C.

8.48 Na, 22.99

$$\text{specific heat} = 28.2 \frac{\text{J}}{\text{mol} \cdot {}^\circ\text{C}} \times \frac{1 \text{ mol}}{22.99 \text{ g}} = 1.23 \text{ J/(g} \cdot {}^\circ\text{C)}$$

8.50 Mass of solution = 50.0 g + 1.045 g = 51.0 g
$q = $ (specific heat) x m x ΔT

$$q = \left(4.18 \frac{\text{J}}{\text{g} \cdot {}^\circ\text{C}}\right)(51.0 \text{ g})(32.3 \,{}^\circ\text{C} - 25.0 \,{}^\circ\text{C}) = 1.56 \times 10^3 \text{ J} = 1.56 \text{ kJ}$$

CaO, 56.08; $\text{mol CaO} = 1.045 \text{ g} \times \dfrac{1 \text{ mol}}{56.08 \text{ g}} = 0.018\ 63 \text{ mol}$

$$\text{Heat evolved per mole of CaO} = \frac{1.56 \text{ kJ}}{0.018\ 63 \text{ mol}} = 83.7 \text{ kJ/mol CaO}$$

Because the reaction evolves heat, the sign for ΔH is negative. ΔH = −83.7 kJ

8.52 NaOH, 40.00; HCl, 36.46

$$8.00 \text{ g NaOH} \times \frac{1 \text{ mol NaOH}}{40.00 \text{ g NaOH}} = 0.200 \text{ mol NaOH}$$

$$8.00 \text{ g HCl} \times \frac{1 \text{ mol HCl}}{36.46 \text{ g HCl}} = 0.219 \text{ mol HCl}$$

Because the reaction stoichiometry between NaOH and HCl is 1:1, the NaOH is the limiting reactant.

$q_P = -q_{soln} = -(\text{specific heat}) \times m \times \Delta T$

$q_P = -\left(4.18\ \dfrac{J}{g \cdot {}^{\circ}C}\right)(316\ g)(33.5\ {}^{\circ}C - 25.0\ {}^{\circ}C) = -1.12 \times 10^4\ J = -11.2\ kJ$

$\Delta H = q_p/n = (-11.2\ kJ)/(0.200\ mol) = -56\ kJ/mol$

When 10.00 g of HCl in 248.0 g of water is added, the same temperature increase is observed because the mass of NaOH is the same and it is still the limiting reactant. The mass of the solution is also the same.

Energy and Enthalpy (Sections 8.4, 8.6–8.7)

8.54　$\Delta E = q_v$ is the heat change associated with a reaction at constant volume. Since $\Delta V = 0$, no PV work is done.

$\Delta H = q_p$ is the heat change associated with a reaction at constant pressure. Since $\Delta V \neq 0$, PV work can also be done.

8.56　$\Delta H = \Delta E + P\Delta V$; ΔH and ΔE are nearly equal when there are no gases involved in a chemical reaction, or, if gases are involved, $\Delta V = 0$ (that is, there are the same number of reactant and product gas molecules).

8.58　$P\Delta V = -7.6\ J$　　(from Problem 8.41)

$\Delta H = \Delta E + P\Delta V$

$\Delta E = \Delta H - P\Delta V = -0.31\ kJ - (-7.6 \times 10^{-3}\ kJ) = -0.30\ kJ$

8.60　$\Delta H = \Delta E + P\Delta V = 44.0\ kJ + (1.00\ atm)(14.0\ L)\left(\dfrac{101.325\ J}{1\ L \cdot atm} \times \dfrac{1\ kJ}{1000\ J}\right) = 45.4\ kJ$

8.62　$\Delta H = -1256.2\ kJ/mol\ C_2H_2$;　C_2H_2, 26.04

$w = -P\Delta V = -(1.00\ atm)(-2.80\ L) = 2.80\ L \cdot atm$

$w = (2.80\ L \cdot atm)\left(\dfrac{101\ J}{1\ L \cdot atm}\right) = 283\ J = 0.283\ kJ$

$6.50\ g \times \dfrac{1\ mol\ C_2H_2}{26.04\ g\ C_2H_2} = 0.250\ mol\ C_2H_2$

$q = (-1256.2\ kJ/mol)(0.250\ mol) = -314\ kJ$

$\Delta E = \Delta H - P\Delta V = -314\ kJ - (-0.283\ kJ) = -314\ kJ$

8.64　$C_4H_{10}O$, 74.12;　mass of $C_4H_{10}O = (0.7138\ g/mL)(100\ mL) = 71.38\ g$

$mol\ C_4H_{10}O = 71.38\ g \times \dfrac{1\ mol}{74.12\ g} = 0.9630\ mol$

$q = n \times \Delta H_{vap} = 0.9630\ mol \times 26.5\ kJ/mol = 25.5\ kJ$

8.66　Al, 26.98

$mol\ Al = 5.00\ g \times \dfrac{1\ mol}{26.98\ g} = 0.1853\ mol$

$$q = n \times \Delta H^\circ = 0.1853 \text{ mol Al} \times \frac{-1408.4 \text{ kJ}}{2 \text{ mol Al}} = -131 \text{ kJ}; \quad 131 \text{ kJ is released.}$$

8.68 Fe_2O_3, 159.7; $\quad$ mol $Fe_2O_3 = 2.50 \text{ g} \times \dfrac{1 \text{ mol}}{159.7 \text{ g}} = 0.015 \; 65 \text{ mol}$

$$q = n \times \Delta H^\circ = 0.015\;65 \text{ mol } Fe_2O_3 \times \frac{-24.8 \text{ kJ}}{1 \text{ mol } Fe_2O_3} = -0.388 \text{ kJ}; \quad 0.388 \text{ kJ is evolved.}$$

Because ΔH is negative, the reaction is exothermic.

Hess's Law and Heats of Formation (Sections 8.8 and 8.9)

8.70 The standard state of an element is its most stable form at 1 atm and the specified temperature, usually 25 °C.

8.72 Hess's Law – the overall enthalpy change for a reaction is equal to the sum of the enthalpy changes for the individual steps in the reaction.
Hess's Law works because of the law of conservation of energy.

8.74 (a) Cl_2, gas $\qquad$ (b) Hg, liquid $\qquad$ (c) CO_2, gas $\qquad$ (d) Ga, solid

8.76 (a) $2 \text{ Fe(s)} + 3/2 \text{ O}_2(g) \rightarrow Fe_2O_3(s)$
(b) $12 \text{ C(s)} + 11 \text{ H}_2(g) + 11/2 \text{ O}_2(g) \rightarrow C_{12}H_{22}O_{11}(s)$
(c) $U(s) + 3 F_2(g) \rightarrow UF_6(s)$

8.78 $\qquad$ $S(s) + O_2(g) \rightarrow SO_2(g)$ $\qquad\qquad$ $\Delta H^\circ_1 = -296.8 \text{ kJ}$
$\qquad\qquad$ $\underline{SO_2 + \frac{1}{2} O_2(g) \rightarrow SO_3(g)}$ $\qquad\qquad$ $\Delta H^\circ_2 = -98.9 \text{ kJ}$
Sum $\quad S(s) + 3/2 O_2(g) \rightarrow SO_3(g)$ $\qquad\quad$ $\Delta H^\circ_3 = \Delta H^\circ_1 + \Delta H^\circ_2$
$\quad \Delta H^\circ_f = \Delta H^\circ_3 = -296.8 \text{ kJ} + (-98.9 \text{ kJ}) = -395.7 \text{ kJ/mol}$

8.80 $\qquad$ $SO_3(g) + H_2O(l) \rightarrow H_2SO_4(aq)$ $\qquad\qquad$ $\Delta H^\circ_1 = -227.8 \text{ kJ}$
$\qquad\qquad$ $H_2(g) + \frac{1}{2} O_2(g) \rightarrow H_2O(l)$ $\qquad\qquad$ $\Delta H^\circ_2 = \Delta H^\circ_f = -285.8 \text{ kJ}$
$\qquad\qquad$ $\underline{S(s) + 3/2 O_2(g) \rightarrow SO_3(g) \qquad\qquad}$ $\Delta H^\circ_3 = \Delta H^\circ_f = -395.7$
Sum $\quad S(s) + H_2(g) + 2 O_2(g) \rightarrow H_2SO_4(aq)$ $\qquad$ $\Delta H^\circ_f (H_2SO_4) = ?$
$\quad \Delta H^\circ_f (H_2SO_4) = \Delta H^\circ_1 + \Delta H^\circ_2 + \Delta H^\circ_3 = -909.3 \text{ kJ}$

8.82 $C_8H_8(l) + 10 O_2(g) \rightarrow 8 CO_2(g) + 4 H_2O(l)$
$\Delta H^\circ_{rxn} = \Delta H^\circ_c = -4395 \text{ kJ}$
$\Delta H^\circ_{rxn} = [8 \, \Delta H^\circ_f(CO_2) + 4 \, \Delta H^\circ_f(H_2O)] - \Delta H^\circ_f(C_8H_8)$
$-4395 \text{ kJ} = [(8 \text{ mol})(-393.5 \text{ kJ/mol}) + (4 \text{ mol})(-285.8 \text{ kJ/mol})] - [(1 \text{ mol})(\Delta H^\circ_f(C_8H_8))]$
Solve for $\Delta H^\circ_f(C_8H_8)$
$-4395 \text{ kJ} = -4291.2 \text{ kJ} - (1 \text{ mol})(\Delta H^\circ_f(C_8H_8))$
$-103.8 \text{ kJ} = -(1 \text{ mol})(\Delta H^\circ_f(C_8H_8))$
$\Delta H^\circ_f(C_8H_8) = \dfrac{-103.8 \text{ kJ}}{-1 \text{ mol}} = +103.8 \text{ kJ/mol} = +104 \text{ kJ/mol}$

8.84 $\Delta H^\circ_{rxn} = \Delta H^\circ_f(\text{MTBE}) - [\Delta H^\circ_f(\text{2-Methylpropene}) + \Delta H^\circ_f(\text{CH}_3\text{OH})]$
      $-57.5 \text{ kJ} = -313.6 \text{ kJ} - [(1 \text{ mol})(\Delta H^\circ_f(\text{2-Methylpropene})) + (-239.2 \text{ kJ})]$
      Solve for $\Delta H^\circ_f(\text{2-Methylpropene})$.
      $-16.9 \text{ kJ} = (1 \text{ mol})(\Delta H^\circ_f(\text{2-Methylpropene}))$
      $\Delta H^\circ_f(\text{2-Methylpropene}) = -16.9 \text{ kJ/mol}$

8.86 $CaCO_3(s) \rightarrow CaO(s) + CO_2(g)$
      $\Delta H^\circ_{rxn} = [\Delta H^\circ_f(\text{CaO}) + \Delta H^\circ_f(\text{CO}_2)] - \Delta H^\circ_f(\text{CaCO}_3)$
      $\Delta H^\circ_{rxn} = [(1 \text{ mol})(-634.9 \text{ kJ/mol}) + (1 \text{ mol})(-393.5 \text{ kJ/mol})] - [(1 \text{ mol})(-1207.6 \text{ kJ/mol})]$
      $\Delta H^\circ_{rxn} = +179.2 \text{ kJ}$

Bond Dissociation Energies (Section 8.10)

8.88 $H_2C=CH_2(g) + H_2(g) \rightarrow CH_3CH_3(g)$
      $\Delta H^\circ_{rxn} = D(\text{Reactant bonds}) - D(\text{Product bonds})$
      $\Delta H^\circ_{rxn} = (D_{C=C} + 4 D_{C-H} + D_{H-H}) - (6 D_{C-H} + D_{C-C})$
      $\Delta H^\circ_{rxn} = [(1 \text{ mol})(728 \text{ kJ/mol}) + (4 \text{ mol})(410 \text{ kJ/mol}) + (1 \text{ mol})(436 \text{ kJ/mol})]$
      $\quad - [(6 \text{ mol})(410 \text{ kJ/mol}) + (1 \text{ mol})(350 \text{ kJ/mol})] = -6 \text{ kJ}$

8.90 $C_4H_{10} + 13/2 \, O_2 \rightarrow 4 \, CO_2 + 5 \, H_2O$
      $\Delta H^\circ_{rxn} = D(\text{Reactant bonds}) - D(\text{Product bonds})$
      $\Delta H^\circ_{rxn} = (3 D_{C-C} + 10 D_{C-H} + 13/2 D_{O=O}) - (8 D_{C=O} + 10 D_{O-H})$
      $\Delta H^\circ_{rxn} = [(3 \text{ mol})(350 \text{ kJ/mol}) + (10 \text{ mol})(410 \text{ kJ/mol}) + (13/2 \text{ mol})(498 \text{ kJ/mol})]$
      $\quad - [(8 \text{ mol})(804 \text{ kJ/mol}) + (10 \text{ mol})(460 \text{ kJ/mol})] = -2645 \text{ kJ}$

Free Energy and Entropy (Sections 8.12 and 8.13)

8.92 Entropy is a measure of molecular randomness.

8.94 A reaction can be spontaneous yet endothermic if ΔS is positive (more randomness) and the $T\Delta S$ term is larger than ΔH.

8.96 (a) positive (more randomness)   (b) negative (less randomness)

8.98 (a) zero   (equilibrium)   (b) zero   (equilibrium)
      (c) negative (spontaneous)

8.100 ΔS is positive. The reaction increases the total number of molecules.

8.102 $\Delta G = \Delta H - T\Delta S$
       (a) $\Delta G = -48 \text{ kJ} - (400 \text{ K})(135 \times 10^{-3} \text{ kJ/K}) = -102 \text{ kJ}$
       $\Delta G < 0$, spontaneous; $\Delta H < 0$, exothermic.
       (b) $\Delta G = -48 \text{ kJ} - (400 \text{ K})(-135 \times 10^{-3} \text{ kJ/K}) = +6 \text{ kJ}$
       $\Delta G > 0$, nonspontaneous; $\Delta H < 0$, exothermic.
       (c) $\Delta G = +48 \text{ kJ} - (400 \text{ K})(135 \times 10^{-3} \text{ kJ/K}) = -6 \text{ kJ}$
       $\Delta G < 0$, spontaneous; $\Delta H > 0$, endothermic.

(d) $\Delta G = +48 \text{ kJ} - (400 \text{ K})(-135 \times 10^{-3} \text{ kJ/K}) = +102 \text{ kJ}$

$\Delta G > 0$, nonspontaneous; $\Delta H > 0$, endothermic.

8.104 $\Delta G = \Delta H - T\Delta S$;   Set $\Delta G = 0$ and solve for T (the crossover temperature).

$T = \dfrac{\Delta H}{\Delta S} = \dfrac{-33 \text{ kJ}}{-0.058 \text{ kJ/K}} = 570 \text{ K}$

8.106 (a) $\Delta H < 0$ and $\Delta S > 0$; reaction is spontaneous at all temperatures.

(b) $\Delta H < 0$ and $\Delta S < 0$; reaction has a crossover temperature.

(c) $\Delta H > 0$ and $\Delta S > 0$; reaction has a crossover temperature.

(d) $\Delta H > 0$ and $\Delta S < 0$; reaction is nonspontaneous at all temperatures.

8.108 $T = -114.1 \text{ °C} = 273.15 + (-114.1) = 159.0 \text{ K}$

$\Delta G_{fus} = \Delta H_{fus} - T\Delta S_{fus}$; $\Delta G = 0$ at the melting point temperature.

Set $\Delta G = 0$ and solve for ΔS_{fus}.

$\Delta G = 0 = \Delta H_{fus} - T\Delta S_{fus}$

$\Delta S_{fus} = \dfrac{\Delta H_{fus}}{T} = \dfrac{5.02 \text{ kJ/mol}}{159.0 \text{ K}} = 0.0316 \text{ kJ/(K·mol)} = 31.6 \text{ J/(K·mol)}$

Chapter Problems

8.110 $Mg(s) + 2 HCl(aq) \rightarrow MgCl_2(aq) + H_2(g)$

$\text{mol Mg} = 1.50 \text{ g} \times \dfrac{1 \text{ mol}}{24.3 \text{ g}} = 0.0617 \text{ mol Mg}$

$\text{mol HCl} = 0.200 \text{ L} \times 6.00 \dfrac{\text{mol}}{\text{L}} = 1.20 \text{ mol HCl}$

There is an excess of HCl. Mg is the limiting reactant.

$q = \left(4.18 \dfrac{\text{J}}{\text{g·°C}}\right)(200 \text{ g})(42.9 \text{ °C} - 25.0 \text{ °C}) + \left(776 \dfrac{\text{J}}{\text{°C}}\right)(42.9 \text{ °C} - 25.0 \text{ °C}) = 2.89 \times 10^4 \text{ J}$

$q = 2.89 \times 10^4 \text{ J} \times \dfrac{1 \text{ kJ}}{1000 \text{ J}} = 28.9 \text{ kJ}$

Heat evolved per mole of Mg $= \dfrac{28.9 \text{ kJ}}{0.0617 \text{ mol}} = 468 \text{ kJ/mol}$

Because the reaction evolves heat, the sign for ΔH is negative. $\Delta H = -468 \text{ kJ}$

8.112 $2 NO(g) + O_2(g) \rightarrow 2 NO_2(g)$ $\Delta H°_1 = 2(-58.1 \text{ kJ})$

          $\underline{2 NO_2(g) \rightarrow N_2O_4(g)}$ $\Delta H°_2 = -55.3 \text{ kJ}$

Sum $2 NO(g) + O_2(g) \rightarrow N_2O_4(g)$

          $\Delta H° = \Delta H°_1 + \Delta H°_2 = -171.5 \text{ kJ}$

8.114 $\Delta G_{fus} = \Delta H_{fus} - T\Delta S_{fus}$; at the melting point $\Delta G = 0$. Set $\Delta G = 0$ and solve for T (the melting point).

$$\Delta G = 0 = \Delta H_{fus} - T\Delta S_{fus}; \qquad T = \frac{\Delta H_{fus}}{\Delta S_{fus}} = \frac{9.95 \text{ kJ}}{0.0357 \text{ kJ/K}} = 279 \text{ K}$$

8.116 $\Delta H°_{rxn} = D$ (Reactant bonds) $- D$ (Product bonds)
(a) $2 \text{ CH}_4(g) \rightarrow \text{C}_2\text{H}_6(g) + \text{H}_2(g)$
$\Delta H°_{rxn} = (8 \, D_{C-H}) - (D_{C-C} + 6 \, D_{C-H} + D_{H-H})$
$\Delta H°_{rxn} = [(8 \text{ mol})(410 \text{ kJ/mol})] - [(1 \text{ mol})(350 \text{ kJ/mol}) + (6 \text{ mol})(410 \text{ kJ/mol})$
$+ (1 \text{ mol})(436 \text{ kJ/mol})] = +34 \text{ kJ}$
(b) $\text{C}_2\text{H}_6(g) + \text{F}_2(g) \rightarrow \text{C}_2\text{H}_5\text{F}(g) + \text{HF}(g)$
$\Delta H°_{rxn} = (6 \, D_{C-H} + D_{C-C} + D_{F-F}) - (5 \, D_{C-H} + D_{C-C} + D_{C-F} + D_{H-F})$
$\Delta H°_{rxn} = [(6 \text{ mol})(410 \text{ kJ/mol}) + (1 \text{ mol})(350 \text{ kJ/mol}) + (1 \text{ mol})(159 \text{ kJ/mol})]$
$\qquad - [(5 \text{ mol})(410 \text{ kJ/mol}) + (1 \text{ mol})(350 \text{ kJ/mol}) + (1 \text{ mol})(450 \text{ kJ/mol})$
$\qquad\qquad + (1 \text{ mol})(570 \text{ kJ/mol})] = -451 \text{ kJ}$
(c) $\text{N}_2(g) + 3 \text{ H}_2(g) \rightarrow 2 \text{ NH}_3(g)$
The bond dissociation energy for N_2 is 945 kJ/mol.
$\Delta H°_{rxn} = (D_{N_2} + 3 \, D_{H-H}) - (6 \, D_{N-H})$
$\Delta H°_{rxn} = [(1 \text{ mol})(945 \text{ kJ/mol}) + (3 \text{ mol})(436 \text{ kJ/mol})] - [(6 \text{ mol})(390 \text{ kJ/mol})] = -87 \text{ kJ}$

8.118 (a) $2 \text{ C}_8\text{H}_{18}(l) + 25 \text{ O}_2(g) \rightarrow 16 \text{ CO}_2(g) + 18 \text{ H}_2\text{O}(l)$
(b) $\text{C}_8\text{H}_{18}(l) + 25/2 \text{ O}_2(g) \rightarrow 8 \text{ CO}_2(g) + 9 \text{ H}_2\text{O}(l)$
$\Delta H°_{rxn} = \Delta H°_c = -5461 \text{ kJ}$
$\Delta H°_{rxn} = [8 \, \Delta H°_f(\text{CO}_2) + 9 \, \Delta H°_f(\text{H}_2\text{O})] - \Delta H°_f(\text{C}_8\text{H}_{18})$
$-5461 \text{ kJ} = [(8 \text{ mol})(-393.5 \text{ kJ/mol}) + (9 \text{ mol})(-285.8 \text{ kJ/mol})] - [(1 \text{ mol})(\Delta H°_f(\text{C}_8\text{H}_{18}))]$
Solve for $\Delta H°_f(\text{C}_8\text{H}_{18})$.
$-5461 \text{ kJ} = -5720.2 \text{ kJ} - [(1 \text{ mol})(\Delta H°_f(\text{C}_8\text{H}_{18}))]$
$259 \text{ kJ} = -(1 \text{ mol})(\Delta H°_f(\text{C}_8\text{H}_{18}))$
$\Delta H°_f(\text{C}_8\text{H}_{18}) = -259 \text{ kJ/mol}$

8.120 (a) $\Delta S_{total} = \Delta S_{system} + \Delta S_{surr}$ and $\Delta S_{surr} = -\Delta H/T$
$\Delta S_{total} = \Delta S_{system} + (-\Delta H/T) = \Delta S_{system} - \Delta H/T$
$\Delta S_{system} = \Delta S_{total} + \Delta H/T$
$\Delta G = \Delta H - T\Delta S$ (substitute ΔS_{system} for ΔS in this equation)
$\Delta G = \Delta H - T(\Delta S_{total} + \Delta H/T) = -T\Delta S_{total}$
$\Delta G = -T\Delta S_{total}$ For a spontaneous reaction, if $\Delta S_{total} > 0$ then $\Delta G < 0$.
(b) $\Delta G° = \Delta H° - T\Delta S°$

$\Delta H° = \Delta G° + T\Delta S°$
$$\Delta S_{surr} = -\frac{\Delta H°}{T} = -\frac{[\Delta G° + T\Delta S°]}{T} = -\frac{[2879 \times 10^3 \text{ J/mol} + (298 \text{ K})(-262 \text{ J/(K}\cdot\text{mol}))]}{298 \text{ K}}$$

$\Delta S_{surr} = -9399 \text{ J/(K}\cdot\text{mol})$

8.122 $2 CH_4(g) + 4 O_2(g) \rightarrow 2 CO_2(g) + 4 H_2O(l)$ $\Delta H°_1 = 2(-890.3 \text{ kJ})$

               $C_2H_6(g) \rightarrow C_2H_4(g) + H_2(g)$ $\Delta H°_2 = +136.3 \text{ kJ}$

               $2 CO_2(g) + 3 H_2O(l) \rightarrow C_2H_6(g) + 7/2 O_2(g)$ $\Delta H°_3 = \dfrac{3120.8 \text{ kJ}}{2}$

               $\underline{H_2O(l) \rightarrow H_2(g) + 1/2 O_2(g)}$ $\underline{\Delta H°_4 = +285.8 \text{ kJ}}$

    Sum $2 CH_4(g) \rightarrow C_2H_4(g) + 2 H_2(g)$ $\Delta H° = +201.9 \text{ kJ}$

8.124 $q_{ice\ tea} = -q_{ice}$

       $q_{ice\ tea} = (4.18 \dfrac{J}{g \cdot °C})(400.0 \text{ g})(10.0\ °C - 80.0\ °C) = -1.17 \times 10^5 \text{ J}$

       $H_2O, 18.02$

       $q_{ice} = 1.17 \times 10^5 \text{ J} = (6.01 \text{ kJ/mol})\left(\dfrac{1000 \text{ J}}{1 \text{ kJ}}\right)\left(m_{ice} \times \dfrac{1 \text{ mol } H_2O}{18.02 \text{ g } H_2O}\right)$

                                     $+ \left(4.18 \dfrac{J}{(g \cdot °C)}\right)(m_{ice})(10.0\ °C - 0.0\ °C)$

    Solve for the mass of ice, m_{ice}.

    $1.17 \times 10^5 \text{ J} = (3.34 \times 10^2 \text{ J/g})(m_{ice}) + (41.8 \text{ J/g})(m_{ice}) = (3.76 \times 10^2 \text{ J/g})(m_{ice})$

    $m_{ice} = \dfrac{1.17 \times 10^5 \text{ J}}{3.76 \times 10^2 \text{ J/g}} = 311 \text{ g of ice}$

8.126 $CsOH(aq) + HCl(aq) \rightarrow CsCl(aq) + H_2O(l)$

       $\text{mol CsOH} = 0.100 \text{ L} \times \dfrac{0.200 \text{ mol CsOH}}{1.00 \text{ L}} = 0.0200 \text{ mol CsOH}$

       $\text{mol HCl} = 0.050 \text{ L} \times \dfrac{0.400 \text{ mol HCl}}{1.00 \text{ L}} = 0.0200 \text{ mol HCl}$

    The reactants were mixed in equal mole amounts.

    Total volume = 150 mL and has a mass of 150 g.

       $q_{solution} = \left(4.2 \dfrac{J}{(g \cdot °C)}\right)(150 \text{ g})(24.28\ °C - 22.50\ °C) = 1121 \text{ J}$

       $q_{reaction} = -q_{solution} = -1121 \text{ J}$

       $\Delta H = \dfrac{q_{reaction}}{\text{mol CsOH}} = \dfrac{-1121 \text{ J}}{0.0200 \text{ mol CsOH}} \times \dfrac{1 \text{ kJ}}{1000 \text{ J}} = -56 \text{ kJ/mol CsOH}$

8.128 ΔH

       $4 CO(g) + 2 O_2(g) \rightarrow 4 CO_2(g)$ $2(-566.0 \text{ kJ})$

       $2 NO_2(g) \rightarrow 2 NO(g) + O_2(g)$ $+116.2 \text{ kJ}$

       $\underline{2 NO(g) \rightarrow O_2(g) + N_2(g)}$ $\underline{2(-91.3 \text{ kJ})}$

       $4 CO(g) + 2 NO_2(g) \rightarrow 4 CO_2(g) + N_2(g)$ -1198.4 kJ

Multiconcept Problems

8.130 (a)

(b) $C(g) + \frac{1}{2} O_2(g) + Cl_2(g) \rightarrow COCl_2(g)$

$\Delta H°_f = \Delta H°_f(C(g)) + (\frac{1}{2} D_{O=O} + D_{Cl-Cl}) - (D_{C=O} + 2 D_{C-Cl})$

$\Delta H°_f = (716.7 \text{ kJ}) + [(\frac{1}{2} \text{ mol})(498 \text{ kJ/mol}) + (1 \text{ mol})(243 \text{ kJ/mol})]$

$- [(1 \text{ mol})(732 \text{ kJ/mol}) + (2 \text{ mol})(330 \text{ kJ/mol})]$

$\Delta H°_{rxn} = -183$ kJ per mol $COCl_2$;   From Appendix B, $\Delta H°_f(COCl_2) = -219.1$ kJ/mol

The calculation of $\Delta H°_f$ from bond energies is only an estimate because the bond energies are average values derived from many different compounds.

8.132 (a) $2 K(s) + 2 H_2O(l) \rightarrow 2 KOH(aq) + H_2(g)$

(b) $\Delta H°_{rxn} = [2 \Delta H°_f(KOH)] - [2 \Delta H°_f(H_2O)]$

$\Delta H°_{rxn} = [(2 \text{ mol})(-482.4 \text{ kJ/mol})] - [(2 \text{ mol})(-285.8 \text{ kJ/mol})] = -393.2$ kJ

(c) The reaction produces 393.2 kJ/ 2 mol K = 196.6 kJ/ mol K. Assume that the mass of the water does not change and that the specific heat = 4.18 J/(g·°C) for the solution that is produced.

$$q = 7.55 \text{ g K} \times \frac{1 \text{ mol K}}{39.10 \text{ g K}} \times \frac{196.6 \text{ kJ}}{1 \text{ mol K}} \times \frac{1000 \text{ J}}{1 \text{ kJ}} = 3.80 \times 10^4 \text{ J}$$

q = (specific heat) x m x ΔT

$$\Delta T = \frac{q}{(\text{specific heat}) \times m} = \frac{3.80 \times 10^4 \text{ J}}{[4.18 \text{ J/(g·°C)}](400.0 \text{ g})} = 22.7 °C$$

$\Delta T = T_{final} - T_{initial}$

$T_{final} = \Delta T + T_{initial} = 22.7 °C + 25.0 °C = 47.7 °C$

(d) $7.55 \text{ g K} \times \frac{1 \text{ mol K}}{39.10 \text{ g K}} \times \frac{2 \text{ mol KOH}}{2 \text{ mol K}} = 0.193$ mol KOH

Assume that the mass of the solution does not change during the reaction and that the solution has a density of 1.00 g/mL.

solution volume $= 400.0 \text{ g} \times \frac{1.00 \text{ mL}}{1 \text{ g}} \times \frac{1 \text{ L}}{1000 \text{ mL}} = 0.400$ L

molarity $= \frac{0.193 \text{ mol KOH}}{0.400 \text{ L}} = 0.483$ M

$2 KOH(aq) + H_2SO_4(aq) \rightarrow K_2SO_4(aq) + 2 H_2O(l)$

$0.193 \text{ mol KOH} \times \frac{1 \text{ mol H}_2SO_4}{2 \text{ mol KOH}} \times \frac{1000 \text{ mL}}{0.554 \text{ mol H}_2SO_4} = 174$ mL of 0.554 M H_2SO_4

8.134   Assume 100.0 g of Y.

$$\text{mol F} = 61.7 \text{ g F} \times \frac{1 \text{ mol F}}{19.00 \text{ g F}} = 3.25 \text{ mol F}$$

$$\text{mol Cl} = 38.3 \text{ g Cl} \times \frac{1 \text{ mol Cl}}{35.45 \text{ g Cl}} = 1.08 \text{ mol Cl}$$

$Cl_{1.08}F_{3.25}$; divide each subscript by the smaller of the two, 1.08.

$Cl_{1.08 / 1.08}F_{3.25 / 1.08}$

ClF_3

(a) Y is ClF_3 and X is ClF

(b) 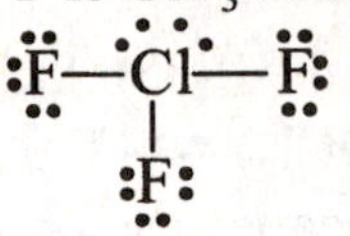 $:\!\overset{\displaystyle ..}{\underset{\displaystyle ..}{F}}\!\!-\!\overset{\displaystyle ..}{\underset{\displaystyle |}{Cl}}\!-\!\overset{\displaystyle ..}{\underset{\displaystyle ..}{F}}\!:$ There are five electron clouds around the Cl (3 bonding and 2 lone
$\qquad\quad :\!\overset{\displaystyle ..}{\underset{\displaystyle ..}{F}}\!:$ pairs). The geometry is T-shaped.

(c)   ΔH

$$Cl_2O(g) + 3 \ OF_2(g) \ \rightarrow \ 2 \ O_2(g) + 2 \ ClF_3(g) \qquad -532.8 \text{ kJ}$$
$$2 \ ClF(g) + O_2(g) \ \rightarrow \ Cl_2O(g) + OF_2(g) \qquad\qquad +205.4 \text{ kJ}$$
$$\underline{O_2(g) + 2 \ F_2(g) \ \rightarrow \ 2 \ OF_2(g) \qquad\qquad\qquad\quad 2(+24.5 \text{ kJ})}$$
$$2 \ ClF(g) + 2 \ F_2(g) \ \rightarrow \ 2 \ ClF_3(g) \qquad\qquad\quad -278.4 \text{ kJ}$$

Divide reaction coefficients and ΔH by 2.

$$ClF(g) \ + \ F_2(g) \ \rightarrow \ ClF_3(g) \qquad\qquad \Delta H = -278.4 \text{ kJ}/2 = -139.2 \text{ kJ/mol ClF}_3$$

(d) ClF, 54.45

$$q = 25.0 \text{ g ClF} \times \frac{1 \text{ mol ClF}}{54.45 \text{ g ClF}} \times \frac{-139.2 \text{ kJ}}{1 \text{ mol ClF}} \times 0.875 = -55.9 \text{ kJ}$$

55.9 kJ is released in this reaction.

9 Gases: Their Properties and Behavior

9.1   1.00 atm = 14.7 psi

$$1.00 \text{ mm Hg} \times \frac{1 \text{ atm}}{760 \text{ mm Hg}} \times \frac{14.7 \text{ psi}}{1 \text{ atm}} = 1.93 \times 10^{-2} \text{ psi}$$

9.2   1.00 atmosphere pressure can support a column of Hg 0.760 m high. Because the density of H_2O is 1.00 g/mL and that of Hg is 13.6 g/mL, 1.00 atmosphere pressure can support a column of H_2O 13.6 times higher than that of Hg. The column of H_2O supported by 1.00 atmosphere will be (0.760 m)(13.6) = 10.3 m.

9.3   The pressure in the flask is less than 0.975 atm because the liquid level is higher on the side connected to the flask. The 24.7 cm of Hg is the difference between the two pressures.

$$\text{Pressure difference} = 24.7 \text{ cm Hg} \times \frac{1.00 \text{ atm}}{76.0 \text{ cm Hg}} = 0.325 \text{ atm}$$

Pressure in flask = 0.975 atm − 0.325 atm = 0.650 atm

9.4   The pressure in the flask is greater than 750 mm Hg because the liquid level is lower on the side connected to the flask.

$$\text{Pressure difference} = 25 \text{ cm Hg} \times \frac{10 \text{ mm Hg}}{1 \text{ cm Hg}} = 250 \text{ mm Hg}$$

Pressure in flask = 750 mm Hg + 250 mm Hg = 1000 mm Hg

9.5   (a) Assume an initial volume of 1.00 L.
First consider the volume change resulting from a change in the number of moles with the pressure and temperature constant.

$$\frac{V_i}{n_i} = \frac{V_f}{n_f}; \quad V_f = \frac{V_i \, n_f}{n_i} = \frac{(1.00 \text{ L})(0.225 \text{ mol})}{0.3 \text{ mol}} = 0.75 \text{ L}$$

Now consider the volume change from 0.75 L as a result of a change in temperature with the number of moles and the pressure constant.

$$\frac{V_i}{T_i} = \frac{V_f}{T_f}; \quad V_f = \frac{V_i \, T_f}{T_i} = \frac{(0.75 \text{ L})(400 \text{ K})}{300 \text{ K}} = 1.0 \text{ L}$$

There is no net change in the volume as a result of the decrease in the number of moles of gas and a temperature increase.

(b) Assume an initial volume of 1.00 L.

First consider the volume change resulting from a change in the number of moles with the pressure and temperature constant.

$$\frac{V_i}{n_i} = \frac{V_f}{n_f}; \quad V_f = \frac{V_i\, n_f}{n_i} = \frac{(1.00\text{ L})(0.225\text{ mol})}{0.3\text{ mol}} = 0.75\text{ L}$$

Now consider the volume change from 0.75 L as a result of a change in temperature with the number of moles and the pressure constant.

$$\frac{V_i}{T_i} = \frac{V_f}{T_f}; \quad V_f = \frac{V_i\, T_f}{T_i} = \frac{(0.75\text{ L})(200\text{ K})}{300\text{ K}} = 0.5\text{ L}$$

The volume would be cut in half as a result of the decrease in the number of moles of gas and a temperature decrease.

9.6 $n = \dfrac{PV}{RT} = \dfrac{(1.000\text{ atm})(1.000 \times 10^5\text{ L})}{\left(0.082\ 06\ \dfrac{\text{L}\cdot\text{atm}}{\text{K}\cdot\text{mol}}\right)(273.15\text{ K})} = 4.461 \times 10^3\text{ mol CH}_4$

CH_4, 16.04;   mass $CH_4 = (4.461 \times 10^3\text{ mol})\left(\dfrac{16.04\text{ g}}{1\text{ mol}}\right) = 7.155 \times 10^4\text{ g CH}_4$

9.7 C_3H_8, 44.10;   V = 350 mL = 0.350 L;   T = 20 °C = 293 K

$n = 3.2\text{ g} \times \dfrac{1\text{ mol C}_3\text{H}_8}{44.10\text{ g C}_3\text{H}_8} = 0.073\text{ mol C}_3\text{H}_8$

$P = \dfrac{nRT}{V} = \dfrac{(0.073\text{ mol})\left(0.082\ 06\ \dfrac{\text{L}\cdot\text{atm}}{\text{K}\cdot\text{mol}}\right)(293\text{ K})}{0.350\text{ L}} = 5.0\text{ atm}$

9.8 $P = 1.51 \times 10^4\text{ kPa} \times \dfrac{1\text{ atm}}{101.325\text{ kPa}} = 149\text{ atm};$ T = 25.0 °C = 298 K

$n = \dfrac{PV}{RT} = \dfrac{(149\text{ atm})(43.8\text{ L})}{\left(0.082\ 06\ \dfrac{\text{L}\cdot\text{atm}}{\text{K}\cdot\text{mol}}\right)(298\text{ K})} = 267\text{ mol He}$

9.9   The volume and number of moles of gas remain constant.

$\dfrac{nR}{V} = \dfrac{P_i}{T_i} = \dfrac{P_f}{T_f}; \quad T_f = \dfrac{P_f T_i}{P_i} = \dfrac{(2.37\text{ atm})(273\text{ K})}{2.15\text{ atm}} = 301\text{ K} = 28\text{ °C}$

9.10 (a) The temperature has increased by about 10% (from 300 K to 325 K) while the amount and the pressure are unchanged. Thus, the volume should increase by about 10%.

(b) The temperature has increased by a factor of 1.5 (from 300 K to 450 K) and the pressure has increased by a factor of 3 (from 0.9 atm to 2.7 atm) while the amount is unchanged. Thus, the volume should decrease by half (1.5/3 = 0.5).

(c) Both the amount and the pressure have increased by a factor of 3 (from 0.075 mol to 0.22 mol and from 0.9 atm to 2.7 atm) while the temperature is unchanged. Thus, the volume is unchanged.

9.11 $CaCO_3(s) + 2 HCl(aq) \rightarrow CaCl_2(aq) + CO_2(g) + H_2O(l)$
$CaCO_3$, 100.1; CO_2, 44.01

$$\text{mole } CO_2 = 33.7 \text{ g } CaCO_3 \times \frac{1 \text{ mol } CaCO_3}{100.1 \text{ g } CaCO_3} \times \frac{1 \text{ mol } CO_2}{1 \text{ mol } CaCO_3} = 0.337 \text{ mol } CO_2$$

$$\text{mass } CO_2 = 0.337 \text{ mol } CO_2 \times \frac{44.01 \text{ g } CO_2}{1 \text{ mol } CO_2} = 14.8 \text{ g } CO_2$$

$$V = \frac{nRT}{P} = \frac{(0.337 \text{ mol})\left(0.082\ 06\ \dfrac{L \cdot atm}{K \cdot mol}\right)(273 \text{ K})}{1.00 \text{ atm}} = 7.55 \text{ L}$$

9.12 $C_3H_8(g) + 5 O_2(g) \rightarrow 3 CO_2(g) + 4 H_2O(l)$

$$n_{propane} = \frac{PV}{RT} = \frac{(4.5 \text{ atm})(15.0 \text{ L})}{\left(0.082\ 06\ \dfrac{L \cdot atm}{K \cdot mol}\right)(298 \text{ K})} = 2.76 \text{ mol } C_3H_8$$

$$2.76 \text{ mol } C_3H_8 \times \frac{3 \text{ mol } CO_2}{1 \text{ mol } C_3H_8} = 8.28 \text{ mol } CO_2$$

$$V = \frac{nRT}{P} = \frac{(8.28 \text{ mol})\left(0.082\ 06\ \frac{L \cdot atm}{K \cdot mol}\right)(273 \text{ K})}{1.00 \text{ atm}} = 186 \text{ L} = 190 \text{ L}$$

9.13 $n = \dfrac{PV}{RT} = \dfrac{(1.00 \text{ atm})(1.00 \text{ L})}{\left(0.082\ 06\ \frac{L \cdot atm}{K \cdot mol}\right)(273 \text{ K})} = 0.0446 \text{ mol}$

molar mass $= \dfrac{1.52 \text{ g}}{0.0446 \text{ mol}} = 34.1 \text{ g/mol};$ molecular mass $= 34.1$

$Na_2S(aq) + 2\ HCl(aq) \rightarrow H_2S(g) + 2\ NaCl(aq)$
The foul-smelling gas is H_2S, hydrogen sulfide.

9.14 $12.45 \text{ g } H_2 \ \times \ \dfrac{1 \text{ mol } H_2}{2.016 \text{ g } H_2} = 6.176 \text{ mol } H_2$

$60.67 \text{ g } N_2 \ \times \ \dfrac{1 \text{ mol } N_2}{28.01 \text{ g } N_2} = 2.166 \text{ mol } N_2$

$2.38 \text{ g } NH_3 \ \times \ \dfrac{1 \text{ mol } NH_3}{17.03 \text{ g } NH_3} = 0.140 \text{ mol } NH_3$

$n_{total} = n_{H_2} + n_{N_2} + n_{NH_3} = 6.176 \text{ mol} + 2.166 \text{ mol} + 0.140 \text{ mol} = 8.482 \text{ mol}$

$X_{H_2} = \dfrac{6.176 \text{ mol}}{8.482 \text{ mol}} = 0.7281;$ $X_{N_2} = \dfrac{2.166 \text{ mol}}{8.482 \text{ mol}} = 0.2554;$ $X_{NH_3} = \dfrac{0.140 \text{ mol}}{8.482 \text{ mol}} = 0.0165$

9.15 $n_{total} = 8.482 \text{ mol}$ (from Problem 9.14) $T = 90\ ^\circ C = 363 \text{ K}$

$$P_{total} = \frac{n_{total}RT}{V} = \frac{(8.482 \text{ mol})\left(0.082\ 06\ \frac{L \cdot atm}{K \cdot mol}\right)(363 \text{ K})}{10.00 \text{ L}} = 25.27 \text{ atm}$$

$P_{H_2} = X_{H_2} \cdot P_{total} = (0.7281)(25.27 \text{ atm}) = 18.4 \text{ atm}$

$P_{N_2} = X_{N_2} \cdot P_{total} = (0.2554)(25.27 \text{ atm}) = 6.45 \text{ atm}$

$P_{NH_3} = X_{NH_3} \cdot P_{total} = (0.0165)(25.27 \text{ atm}) = 0.417 \text{ atm}$

9.16 $P_{H_2O} = X_{H_2O} \cdot P_{Total} = (0.0287)(0.977 \text{ atm}) = 0.0280 \text{ atm}$

9.17 The number of moles of each gas is proportional to the number of each of the different gas molecules in the container.

$n_{total} = n_{red} + n_{yellow} + n_{green} = 6 + 2 + 4 = 12$

$X_{red} = \dfrac{n_{red}}{n_{total}} = \dfrac{6}{12} = 0.500;$ $X_{yellow} = \dfrac{n_{yellow}}{n_{total}} = \dfrac{2}{12} = 0.167;$ $X_{green} = \dfrac{n_{green}}{n_{total}} = \dfrac{4}{12} = 0.333$

$P_{red} = X_{red} \cdot P_{total} = (0.500)(600 \text{ mm Hg}) = 300 \text{ mm Hg}$
$P_{yellow} = X_{yellow} \cdot P_{total} = (0.167)(600 \text{ mm Hg}) = 100 \text{ mm Hg}$
$P_{green} = X_{green} \cdot P_{total} = (0.333)(600 \text{ mm Hg}) = 200 \text{ mm Hg}$

9.18 $u = \sqrt{\dfrac{3RT}{M}}$, M = molar mass, R = 8.314 J/(K · mol), 1 J = 1 kg · m²/s²

at 37 °C = 310 K, $u = \sqrt{\dfrac{3 \times 8.314 \text{ kg m}^2/(\text{s}^2 \text{ K mol}) \times 310 \text{ K}}{28.01 \times 10^{-3} \text{ kg/mol}}} = 525 \text{ m/s}$

at –25 °C = 248 K, $u = \sqrt{\dfrac{3 \times 8.314 \text{ kg m}^2/(\text{s}^2 \text{ K mol}) \times 248 \text{ K}}{28.01 \times 10^{-3} \text{ kg/mol}}} = 470 \text{ m/s}$

9.19 $u = \sqrt{\dfrac{3RT}{M}}$, M = molar mass, R = 8.314 J/(K · mol), 1 J = 1 kg · m²/s²

O_2, 32.00, 32.00 x 10⁻³ kg/mol

$u = 580 \text{ mi/h} \times \dfrac{1.6093 \text{ km}}{1 \text{ mi}} \times \dfrac{1000 \text{ m}}{1 \text{ km}} \times \dfrac{1 \text{ hr}}{60 \text{ min}} \times \dfrac{1 \text{ min}}{60 \text{ s}} = 259 \text{ m/s}$

$u = \sqrt{\dfrac{3RT}{M}}$; $u^2 = \dfrac{3RT}{M}$

$T = \dfrac{u^2 M}{3R} = \dfrac{(259 \text{ m/s})^2 (32.00 \times 10^{-3} \text{ kg/mol})}{(3)(8.314 \text{ kg} \cdot \text{m}^2/\text{s}^2 \cdot \text{K} \cdot \text{mol})} = 86.1 \text{ K}$

$T = 86.1 - 273.15 = -187.0 \text{ °C}$

9.20   (a) $\dfrac{\text{rate } O_2}{\text{rate Kr}} = \sqrt{\dfrac{M_{Kr}}{M_{O_2}}} = \sqrt{\dfrac{83.8}{32.0}}$; $\dfrac{\text{rate } O_2}{\text{rate Kr}} = 1.62$

O_2 diffuses 1.62 times faster than Kr.

(b) $\dfrac{\text{rate } C_2H_2}{\text{rate } N_2} = \sqrt{\dfrac{M_{N_2}}{M_{C_2H_2}}} = \sqrt{\dfrac{28.0}{26.0}}$; $\dfrac{\text{rate } C_2H_2}{\text{rate } N_2} = 1.04$

C_2H_2 diffuses 1.04 times faster than N_2.

9.21 $\dfrac{\text{rate } ^{20}\text{Ne}}{\text{rate}^{22}\text{Ne}} = \sqrt{\dfrac{M\ ^{22}\text{Ne}}{M\ ^{20}\text{Ne}}} = \sqrt{\dfrac{22}{20}} = 1.05$; $\dfrac{\text{rate } ^{21}\text{Ne}}{\text{rate } ^{22}\text{Ne}} = \sqrt{\dfrac{M\ ^{22}\text{Ne}}{M\ ^{21}\text{Ne}}} = \sqrt{\dfrac{22}{21}} = 1.02$

Thus, the relative rates of diffusion are $^{20}\text{Ne}(1.05) > {}^{21}\text{Ne}(1.02) > {}^{22}\text{Ne}(1.00)$.

9.22 $P = \dfrac{nRT}{V} = \dfrac{(0.500 \text{ mol})\left(0.082\ 06\ \dfrac{\text{L} \cdot \text{atm}}{\text{K} \cdot \text{mol}}\right)(300 \text{ K})}{(0.600 \text{ L})} = 20.5 \text{ atm}$

$$P = \frac{nRT}{V - nb} - \frac{an^2}{V^2}$$

$$P = \frac{(0.500 \text{ mol})\left(0.082\ 06\ \dfrac{L \cdot atm}{K \cdot mol}\right)(300 \text{ K})}{[(0.600 \text{ L}) - (0.500 \text{ mol})(0.0387 \text{ L/mol})]} - \frac{\left(1.35\ \dfrac{L^2 \cdot atm}{mol^2}\right)(0.500 \text{ mol})^2}{(0.600 \text{ L})^2} = 20.3 \text{ atm}$$

9.23   The amount of ozone is assumed to be constant.

Therefore $nR = \dfrac{P_i V_i}{T_i} = \dfrac{P_f V_f}{T_f}$

Because $V \propto h$, then $\dfrac{P_i h_i}{T_i} = \dfrac{P_f h_f}{T_f}$ where h is the thickness of the O_3 layer.

$$h_f = \frac{P_i}{P_f} \times \frac{T_f}{T_i} \times h_i = \left(\frac{1.6 \times 10^{-9} \text{ atm}}{1 \text{ atm}}\right)\left(\frac{273 \text{ K}}{230 \text{ K}}\right)(20 \times 10^3 \text{ m}) = 3.8 \times 10^{-5} \text{ m}$$

(Actually, $V = 4\pi r^2 h$, where r = the radius of Earth. When you go out ~30 km to get to the ozone layer, the change in r^2 is less than 1%. Therefore you can neglect the change in r^2 and assume that V is proportional to h.)

9.24   For ether, the MAC $= \dfrac{15 \text{ mm Hg}}{760 \text{ mm Hg}} \times 100\% = 2.0\%$

9.25   (a) Let X = partial pressure of chloroform.

MAC $= \dfrac{X}{760 \text{ mm Hg}} \times 100\% = 0.77\%$

Solve for X. $X = 760 \text{ mm Hg} \times \dfrac{0.77\%}{100\%} = 5.9 \text{ mm Hg}$

(b) $CHCl_3$, 119.4

$$PV = nRT; \quad n = \frac{PV}{RT} = \frac{\left(5.9 \text{ mm Hg} \times \dfrac{1.00 \text{ atm}}{760 \text{ mm Hg}}\right)(10.0 \text{ L})}{\left(0.082\ 06\ \dfrac{L \cdot atm}{K \cdot mol}\right)(273 \text{ K})} = 0.00347 \text{ mol } CHCl_3$$

mass $CHCl_3 = 0.00347 \text{ mol } CHCl_3 \times \dfrac{119.4 \text{ g } CHCl_3}{1 \text{ mol } CHCl_3} = 0.41 \text{ g } CHCl_3$

Conceptual Problems

9.26   The picture on the right will be the same as that on the left, apart from random scrambling of the He and Ar atoms.

9.28 (a)  The volume of a gas is proportional to the kelvin temperature at constant pressure. As the temperature increases from 300 K to 450 K, the volume will increase by a factor of 1.5.

(b) The volume of a gas is inversely proportional to pressure at constant temperature. As the pressure increases from 1 atm to 2 atm, the volume will decrease by a factor of 2.

(c) $PV = nRT$;   The amount of gas (n) is constant.

Therefore $nR = \dfrac{P_i V_i}{T_i} = \dfrac{P_f V_f}{T_f}$.

Assume $V_i = 1$ L and solve for V_f.

$\dfrac{P_i V_i T_f}{T_i P_f} = \dfrac{(3 \text{ atm})(1 \text{ L})(200 \text{ K})}{(300 \text{ K})(2 \text{ atm})} = V_f = 1 \text{ L}$

There is no change in volume.

9.30 The two gases should mix randomly and homogeneously and this is best represented by drawing (c).

9.32 The gas pressure in the bulb in mm Hg is equal to the difference in the height of the Hg in the two arms of the manometer.

9.34 (a) Because there are more yellow gas molecules than there are blue, the yellow gas molecules have the higher average speed.
(b) Each rate is proportional to the number of effused gas molecules of each type.
$M_{yellow} = 25$

$\dfrac{rate_{blue}}{rate_{yellow}} = \sqrt{\dfrac{M_{yellow}}{M_{blue}}}$; $\dfrac{5}{6} = \sqrt{\dfrac{25 \text{ amu}}{M_{blue}}}$; $\left(\dfrac{5}{6}\right)^2 = \dfrac{25 \text{ amu}}{M_{blue}}$; $M_{blue} = \dfrac{25 \text{ amu}}{\left(\dfrac{5}{6}\right)^2} = 36$

Section Problems
Gases and Gas Pressure (Section 9.1)

9.36 Temperature is a measure of the average kinetic energy of gas particles.

9.38 $P = 480 \text{ mm Hg} \times \dfrac{1.00 \text{ atm}}{760 \text{ mm Hg}} = 0.632 \text{ atm}$; $P = 480 \text{ mm Hg} \times \dfrac{101,325 \text{ Pa}}{760 \text{ mm Hg}} = 6.40 \times 10^4 \text{ Pa}$

9.40 $P_{flask} > 754.3 \text{ mm Hg}$; $P_{flask} = 754.3 \text{ mm Hg} + 176 \text{ mm Hg} = 930 \text{ mm Hg}$

9.42 $P_{flask} > 752.3$ mm Hg (see Figure 9.4); If the pressure in the flask can support a column of ethyl alcohol (d = 0.7893 g/mL) 55.1 cm high, then it can only support a column of Hg that is much shorter because of the higher density of Hg.

$$55.1 \text{ cm} \times \frac{0.7893 \text{ g/mL}}{13.546 \text{ g/mL}} = 3.21 \text{ cm Hg} = 32.1 \text{ mm Hg}$$

$$P_{flask} = 752.3 \text{ mm Hg} + 32.1 \text{ mm Hg} = 784.4 \text{ mm Hg}$$

$$P_{flask} = 784.4 \text{ mm Hg} \times \frac{101{,}325 \text{ Pa}}{760 \text{ mm Hg}} = 1.046 \times 10^5 \text{ Pa}$$

9.44   % Volume

	% Volume
N_2	78.08
O_2	20.95
Ar	0.93
CO_2	0.037

The % volume for a particular gas is proportional to the number of molecules of that gas in a mixture of gases.

Average molecular mass of air

= (0.7808)(mol. mass N_2) + (0.2095)(mol. mass O_2)
          + (0.0093)(at. mass Ar) + (0.000 37)(mol. mass CO_2)

= (0.7808)(28.01) + (0.2095)(32.00) + (0.0093)(39.95) + (0.000 37)(44.01) = 28.96

The Gas Laws (Sections 9.2 and 9.3)

9.46   (a) $\dfrac{nR}{V} = \dfrac{P_i}{T_i} = \dfrac{P_f}{T_f}$; $\dfrac{P_i T_f}{T_i} = P_f$

Let $P_i = 1$ atm, $T_i = 100$ K, $T_f = 300$ K

$$P_f = \frac{P_i T_f}{T_i} = \frac{(1 \text{ atm})(300 \text{ K})}{(100 \text{ K})} = 3 \text{ atm}$$

The pressure would triple.

(b) $\dfrac{RT}{V} = \dfrac{P_i}{n_i} = \dfrac{P_f}{n_f}$; $\dfrac{P_i n_f}{n_i} = P_f$

Let $P_i = 1$ atm, $n_i = 3$ mol, $n_f = 1$ mol

$$P_f = \frac{P_i n_f}{n_i} = \frac{(1 \text{ atm})(1 \text{ mol})}{(3 \text{ mol})} = \frac{1}{3} \text{ atm}$$

The pressure would be $\dfrac{1}{3}$ the initial pressure.

(c) $nRT = P_i V_i = P_f V_f$; $\dfrac{P_i V_i}{V_f} = P_f$

Let $P_i = 1$ atm, $V_i = 1$ L, $V_f = 1 - 0.45$ L $= 0.55$ L

$$P_f = \frac{P_i V_i}{V_f} = \frac{(1 \text{ atm})(1 \text{ L})}{(0.55 \text{ L})} = 1.8 \text{ atm}$$

The pressure would increase by 1.8 times.

(d) $$nR = \frac{P_i V_i}{T_i} = \frac{P_f V_f}{T_f}; \qquad \frac{P_i V_i T_f}{T_i V_f} = P_f$$

Let $P_i = 1$ atm, $V_i = 1$ L, $T_i = 200$ K, $V_f = 3$ L, $T_i = 100$ K

$$P_f = \frac{P_i V_i T_f}{T_i V_f} = \frac{(1 \text{ atm})(1 \text{ L})(100 \text{ K})}{(200 \text{ K})(3 \text{ L})} = 0.17 \text{ atm}$$

The pressure would be 0.17 times the initial pressure.

9.48   They all contain the same number of gas molecules.

9.50 n and T are constant; therefore $nRT = P_i V_i = P_f V_f$

$$V_f = \frac{P_i V_i}{P_f} = \frac{(150 \text{ atm})(49.0 \text{ L})}{(1.02 \text{ atm})} = 7210 \text{ L}$$

n and P are constant; therefore $\dfrac{nR}{P} = \dfrac{V_i}{T_i} = \dfrac{V_f}{T_f}$

$$V_f = \frac{V_i T_f}{T_i} = \frac{(49.0 \text{ L})(308 \text{ K})}{(293 \text{ K})} = 51.5 \text{ L}$$

9.52 $15.0 \text{ g } CO_2 \times \dfrac{1 \text{ mol } CO_2}{44.0 \text{ g } CO_2} = 0.341 \text{ mol } CO_2$

$$P = \frac{nRT}{V} = \frac{(0.341 \text{ mol})\left(0.082\ 06\ \dfrac{L \cdot atm}{K \cdot mol}\right)(300 \text{ K})}{(0.30 \text{ L})} = 27.98 \text{ atm}$$

$$27.98 \text{ atm} \times \frac{760 \text{ mm Hg}}{1 \text{ atm}} = 2.1 \times 10^4 \text{ mm Hg}$$

9.54 $\dfrac{1 \text{ H atom}}{cm^3} \times \dfrac{1 \text{ mol H}}{6.02 \times 10^{23} \text{ atoms}} \times \dfrac{1000 \text{ cm}^3}{1 \text{ L}} = 1.7 \times 10^{-21} \text{ mol H/L}$

$$P = \frac{nRT}{V} = \frac{(1.7 \times 10^{-21} \text{ mol})\left(0.082\ 06\ \dfrac{L \cdot atm}{K \cdot mol}\right)(100 \text{ K})}{(1 \text{ L})} = 1.4 \times 10^{-20} \text{ atm}$$

$$P = 1.4 \times 10^{-20} \text{ atm} \times \frac{760 \text{ mm Hg}}{1.0 \text{ atm}} = 1 \times 10^{-17} \text{ mm Hg}$$

9.56 $\quad n = \dfrac{PV}{RT} = \dfrac{\left(17{,}180 \text{ kPa} \times \dfrac{1000 \text{ Pa}}{1 \text{ kPa}} \times \dfrac{1 \text{ atm}}{101{,}325 \text{ Pa}}\right)(43.8 \text{ L})}{\left(0.082\ 06 \dfrac{\text{L} \cdot \text{atm}}{\text{K} \cdot \text{mol}}\right)(293\text{K})} = 308.9 \text{ mol}$

$\quad$ mass Ar $= 308.9 \text{ mol} \times \dfrac{39.948 \text{ g}}{1 \text{ mol}} = 12340 \text{ g} = 1.23 \times 10^4 \text{ g}$

Gas Stoichiometry (Section 9.4)

9.58 $\quad$ For steam, T = 123.0 °C = 396 K

$\quad n = \dfrac{PV}{RT} = \dfrac{(0.93 \text{ atm})(15.0 \text{ L})}{\left(0.082\ 06 \dfrac{\text{L} \cdot \text{atm}}{\text{K} \cdot \text{mol}}\right)(396 \text{ K})} = 0.43 \text{ mol steam}$

$\quad$ For ice, H_2O, 18.02; $\quad n = 10.5 \text{ g} \times \dfrac{1 \text{ mol}}{18.02 \text{ g}} = 0.583 \text{ mol ice}$

Because the number of moles of ice is larger than the number of moles of steam, the ice contains more H_2O molecules.

9.60 $\quad$ The containers are identical. Both containers contain the same number of gas molecules. Weigh the containers. Because the molecular mass for O_2 is greater than the molecular mass for H_2, the heavier container contains O_2.

9.62 $\quad$ room volume = 4.0 m x 5.0 m x 2.5 m x $\dfrac{1 \text{ L}}{10^{-3} \text{ m}^3} = 5.0 \times 10^4 \text{ L}$

$\quad n_{\text{total}} = \dfrac{PV}{RT} = \dfrac{(1.0 \text{ atm})(5.0 \times 10^4 \text{ L})}{\left(0.082\ 06 \dfrac{\text{L} \cdot \text{atm}}{\text{K} \cdot \text{mol}}\right)(273 \text{ K})} = 2.23 \times 10^3 \text{ mol}$

$\quad n_{O_2} = (0.2095)n_{\text{total}} = (0.2095)(2.23 \times 10^3 \text{ mol}) = 467 \text{ mol } O_2$

$\quad$ mass $O_2 = 467 \text{ mol} \times \dfrac{32.0 \text{ g}}{1 \text{ mol}} = 1.5 \times 10^4 \text{ g } O_2$

9.64 $\quad$ (a) CH_4, 16.04; $\qquad d = \dfrac{16.04 \text{ g}}{22.4 \text{ L}} = 0.716 \text{ g/L}$

$\qquad$ (b) CO_2, 44.01; $\qquad d = \dfrac{44.01 \text{ g}}{22.4 \text{ L}} = 1.96 \text{ g/L}$

$\qquad$ (c) O_2, 32.00; $\qquad d = \dfrac{32.00 \text{ g}}{22.4 \text{ L}} = 1.43 \text{ g/L}$

9.66 $\quad n = \dfrac{PV}{RT} = \dfrac{\left(356 \text{ mm Hg} \times \dfrac{1.00 \text{ atm}}{760 \text{ mm Hg}}\right)(1.500 \text{ L})}{\left(0.082\ 06 \dfrac{\text{L} \cdot \text{atm}}{\text{K} \cdot \text{mol}}\right)(295.5 \text{ K})} = 0.0290 \text{ mol}$

$\quad$ molar mass $= \dfrac{0.9847 \text{ g}}{0.0290 \text{ mol}} = 34.0 \text{ g/mol}; \quad$ molecular mass $= 34.0$

9.68 $\quad 2 \text{ HgO(s)} \rightarrow 2 \text{ Hg(l)} + \text{O}_2\text{(g)}; \qquad \text{HgO}, 216.59$

$\quad 10.57 \text{ g HgO} \times \dfrac{1 \text{ mol HgO}}{216.59 \text{ g HgO}} \times \dfrac{1 \text{ mol O}_2}{2 \text{ mol HgO}} = 0.024\ 40 \text{ mol O}_2$

$\quad V = \dfrac{nRT}{P} = \dfrac{(0.024\ 40 \text{ mol})\left(0.082\ 06 \dfrac{\text{L} \cdot \text{atm}}{\text{K} \cdot \text{mol}}\right)(273.15 \text{ K})}{1.000 \text{ atm}} = 0.5469 \text{ L}$

9.70 $\quad \text{Zn(s)} + 2 \text{ HCl(aq)} \rightarrow \text{ZnCl}_2\text{(aq)} + \text{H}_2\text{(g)}$

$\quad$ (a) $25.5 \text{ g Zn} \times \dfrac{1 \text{ mol Zn}}{65.39 \text{ g Zn}} \times \dfrac{1 \text{ mol H}_2}{1 \text{ mol Zn}} = 0.390 \text{ mol H}_2$

$\quad V = \dfrac{nRT}{P} = \dfrac{(0.390 \text{ mol})\left(0.082\ 06 \dfrac{\text{L} \cdot \text{atm}}{\text{K} \cdot \text{mol}}\right)(288 \text{ K})}{\left(742 \text{ mm Hg} \times \dfrac{1.00 \text{ atm}}{760 \text{ mm Hg}}\right)} = 9.44 \text{ L}$

$\quad$ (b) $n = \dfrac{PV}{RT} = \dfrac{\left(350 \text{ mm Hg} \times \dfrac{1.00 \text{ atm}}{760 \text{ mm Hg}}\right)(5.00 \text{ L})}{\left(0.082\ 06 \dfrac{\text{L} \cdot \text{atm}}{\text{K} \cdot \text{mol}}\right)(303.15 \text{ K})} = 0.092\ 56 \text{ mol H}_2$

$\quad 0.092\ 56 \text{ mol H}_2 \times \dfrac{1 \text{ mol Zn}}{1 \text{ mol H}_2} \times \dfrac{65.39 \text{ g Zn}}{1 \text{ mol Zn}} = 6.05 \text{ g Zn}$

9.72 $\quad$ (a) $V_{24h} = (4.50 \text{ L/min})(60 \text{ min/h})(24 \text{ h/day}) = 6480 \text{ L}$

$\qquad V_{CO_2} = (0.034)V_{24h} = (0.034)(6480 \text{ L}) = 220 \text{ L}$

$\qquad n = \dfrac{PV}{RT} = \dfrac{\left(735 \text{ mm Hg} \times \dfrac{1.00 \text{ atm}}{760 \text{ mm Hg}}\right)(220 \text{ L})}{\left(0.082\ 06 \dfrac{\text{L} \cdot \text{atm}}{\text{K} \cdot \text{mol}}\right)(298 \text{ K})} = 8.70 \text{ mol CO}_2$

$\qquad 8.70 \text{ mol CO}_2 \times \dfrac{44.01 \text{ g CO}_2}{1 \text{ mol CO}_2} = 383 \text{ g} = 380 \text{ g CO}_2$

(b) $2 Na_2O_2(s) + 2 CO_2(g) \rightarrow 2 Na_2CO_3(s) + O_2(g)$; Na_2O_2, 77.98

3.65 kg = 3650 g

$$3650 \text{ g } Na_2O_2 \times \frac{1 \text{ mol } Na_2O_2}{77.98 \text{ g } Na_2O_2} \times \frac{2 \text{ mol } CO_2}{2 \text{ mol } Na_2O_2} \times \frac{1 \text{ day}}{8.70 \text{ mol } CO_2} = 5.4 \text{ days}$$

Dalton's Law and Mole Fraction (Section 9.5)

9.74   Because of Avogadro's Law ($V \propto n$), the % volumes are also % moles.

	% mole
N_2	78.08
O_2	20.95
Ar	0.93
CO_2	0.038

In decimal form, % mole = mole fraction.

$P_{N_2} = X_{N_2} \cdot P_{total} = (0.7808)(1.000 \text{ atm}) = 0.7808 \text{ atm}$

$P_{O_2} = X_{O_2} \cdot P_{total} = (0.2095)(1.000 \text{ atm}) = 0.2095 \text{ atm}$

$P_{Ar} = X_{Ar} \cdot P_{total} = (0.0093)(1.000 \text{ atm}) = 0.0093 \text{ atm}$

$P_{CO_2} = X_{CO_2} \cdot P_{total} = (0.000\ 38)(1.000 \text{ atm}) = 0.000\ 38 \text{ atm}$

Pressures of the rest are negligible.

9.76   Assume a 100.0 g sample.   g CO_2 = 1.00 g and g O_2 = 99.0 g

$$\text{mol } CO_2 = 1.00 \text{ g } CO_2 \times \frac{1 \text{ mol } CO_2}{44.01 \text{ g } CO_2} = 0.0227 \text{ mol } CO_2$$

$$\text{mol } O_2 = 99.0 \text{ g } O_2 \times \frac{1 \text{ mol } O_2}{32.00 \text{ g } O_2} = 3.094 \text{ mol } O_2$$

n_{total} = 3.094 mol + 0.0227 mol = 3.117 mol

$$X_{O_2} = \frac{3.094 \text{ mol}}{3.117 \text{ mol}} = 0.993; \quad X_{CO_2} = \frac{0.0227 \text{ mol}}{3.117 \text{ mol}} = 0.007\ 28$$

$P_{O_2} = X_{O_2} \cdot P_{total} = (0.993)(0.977 \text{ atm}) = 0.970 \text{ atm}$

$P_{CO_2} = X_{CO_2} \cdot P_{total} = (0.007\ 28)(0.977 \text{ atm}) = 0.007\ 11 \text{ atm}$

9.78   Assume a 100.0 g sample.

$$\text{g HCl} = (0.0500)(100.0 \text{ g}) = 5.00 \text{ g}; \quad 5.00 \text{ g HCl} \times \frac{1 \text{ mol HCl}}{36.5 \text{ g HCl}} = 0.137 \text{ mol HCl}$$

$$\text{g } H_2 = (0.0100)(100.0 \text{ g}) = 1.00 \text{ g}; \quad 1.00 \text{ g } H_2 \times \frac{1 \text{ mol } H_2}{2.016 \text{ g } H_2} = 0.496 \text{ mol } H_2$$

$$\text{g Ne} = (0.94)(100.0 \text{ g}) = 94 \text{ g}; \quad 94 \text{ g Ne} \times \frac{1 \text{ mol Ne}}{20.18 \text{ g Ne}} = 4.66 \text{ mol Ne}$$

n_{total} = 0.137 + 0.496 + 4.66 = 5.3 mol

$$X_{HCl} = \frac{0.137 \text{ mol}}{5.3 \text{ mol}} = 0.026; \quad X_{H_2} = \frac{0.496 \text{ mol}}{5.3 \text{ mol}} = 0.094; \quad X_{Ne} = \frac{4.66 \text{ mol}}{5.3 \text{ mol}} = 0.88$$

9.80 (a) H_2, 2.016

$$P = \frac{nRT}{V} = \frac{\left(14.2 \text{ g} \times \frac{1 \text{ mol}}{2.016 \text{ g}}\right)\left(0.082 \ 06 \ \frac{L \cdot atm}{K \cdot mol}\right)(290 \text{ K})}{(100.0 \text{ L})} = 1.68 \text{ atm}$$

(b) $$P = \frac{nRT}{V} = \frac{\left(36.7 \text{ g} \times \frac{1 \text{ mol}}{39.95 \text{ g}}\right)\left(0.082 \ 06 \ \frac{L \cdot atm}{K \cdot mol}\right)(290 \text{ K})}{(100.0 \text{ L})} = 0.219 \text{ atm}$$

9.82 $P_{total} = P_{H_2} + P_{H_2O}; \quad P_{H_2} = P_{total} - P_{H_2O} = 747 \text{ mm Hg} - 23.8 \text{ mm Hg} = 723 \text{ mm Hg}$

$$n = \frac{PV}{RT} = \frac{\left(723 \text{ mm Hg} \times \frac{1.00 \text{ atm}}{760 \text{ mm Hg}}\right)(3.557 \text{ L})}{\left(0.082 \ 06 \ \frac{L \cdot atm}{K \cdot mol}\right)(298 \text{ K})} = 0.1384 \text{ mol } H_2$$

$$0.1384 \text{ mol } H_2 \times \frac{1 \text{ mol Mg}}{1 \text{ mol } H_2} \times \frac{24.3 \text{ g Mg}}{1 \text{ mol Mg}} = 3.36 \text{ g Mg}$$

Kinetic–Molecular Theory and Graham's Law (Sections 9.6 and 9.7)

9.84 The kinetic-molecular theory is based on the following assumptions:
1. A gas consists of tiny particles, either atoms or molecules, moving about at random.
2. The volume of the particles themselves is negligible compared with the total volume of the gas; most of the volume of a gas is empty space.
3. The gas particles act independently; there are no attractive or repulsive forces between particles.
4. Collisions of the gas particles, either with other particles or with the walls of the container, are elastic; that is, the total kinetic energy of the gas particles is constant at constant T.
5. The average kinetic energy of the gas particles is proportional to the Kelvin temperature of the sample.

9.86 Heat is the energy transferred from one object to another as the result of a temperature difference between them. Temperature is a measure of the kinetic energy of molecular motion.

9.88 $$u = \sqrt{\frac{3 \, RT}{M}} = \sqrt{\frac{3 \times 8.314 \text{ kg m}^2/(s^2 \, K \, mol) \times 220 \text{ K}}{28.0 \times 10^{-3} \text{ kg/mol}}} = 443 \text{ m/s}$$

9.90 For H_2, $u = \sqrt{\dfrac{3\,RT}{M}} = \sqrt{\dfrac{3 \times 8.314\ \text{kg m}^2/(\text{s}^2\,\text{K mol}) \times 150\ \text{K}}{2.02 \times 10^{-3}\ \text{kg/mol}}} = 1360\ \text{m/s}$

For He, $u = \sqrt{\dfrac{3 \times 8.314\ \text{kg m}^2/(\text{s}^2\,\text{K mol}) \times 648\ \text{K}}{4.00 \times 10^{-3}\ \text{kg/mol}}} = 2010\ \text{m/s}$

He at 375 °C has the higher average speed.

9.92 $\dfrac{\text{rate}_{H_2}}{\text{rate}_X} = \sqrt{\dfrac{M_X}{M_{H_2}}}$; $\dfrac{2.92}{1} = \dfrac{\sqrt{M_X}}{\sqrt{2.02}}$; $2.92\sqrt{2.02} = \sqrt{M_X}$

$M_X = (2.92\sqrt{2.02})^2 = 17.2\ \text{g/mol}$; molecular mass = 17.2

9.94 HCl, 36.5; F_2, 38.0; Ar, 39.9

$\dfrac{\text{rate HCl}}{\text{rate Ar}} = \sqrt{\dfrac{M_{Ar}}{M_{HCl}}} = \sqrt{\dfrac{39.9}{36.5}} = 1.05$; $\dfrac{\text{rate } F_2}{\text{rate Ar}} = \sqrt{\dfrac{M_{Ar}}{M_{F_2}}} = \sqrt{\dfrac{39.9}{38.0}} = 1.02$

The relative rates of diffusion are HCl(1.05) > F_2(1.02) > Ar(1.00).

9.96 $u = 45\ \text{m/s} = \sqrt{\dfrac{3 \times 8.314\ \text{kg m}^2/(\text{s}^2\,\text{K mol}) \times T}{4.00 \times 10^{-3}\ \text{kg/mol}}}$

Square both sides of the equation and solve for T.

$2025\ \text{m}^2/\text{s}^2 = \dfrac{3 \times 8.314\ \text{kg m}^2/(\text{s}^2\,\text{K mol}) \times T}{4.00 \times 10^{-3}\ \text{kg/mol}}$

$T = 0.325\ \text{K} = -272.83\ °\text{C}$ (near absolute zero)

Chapter Problems

9.98 (a) ozone depletion (b) acid rain (c) global warming (d) air pollution

9.100 $\dfrac{\text{rate }^{35}Cl_2}{\text{rate }^{37}Cl_2} = \sqrt{\dfrac{M\ ^{37}Cl_2}{M\ ^{35}Cl_2}} = \sqrt{\dfrac{74.0}{70.0}} = 1.03$

$\dfrac{\text{rate }^{35}Cl\,^{37}Cl}{\text{rate }^{37}Cl_2} = \sqrt{\dfrac{M\ ^{37}Cl_2}{M\ ^{35}Cl^{37}Cl}} = \sqrt{\dfrac{74.0}{72.0}} = 1.01$

The relative rates of diffusion are $^{35}Cl_2$(1.03) > $^{35}Cl^{37}Cl$(1.01) > $^{37}Cl_2$(1.00).

9.102 $V = \dfrac{nRT}{P} = \dfrac{(1.00 \text{ mol})\left(0.082\ 06\ \dfrac{L \cdot atm}{K \cdot mol}\right)(1050 \text{ K})}{(75 \text{ atm})} = 1.1 \text{ L}$

9.104 UF_6, 352.0; 70 °C = 70 + 273 = 343 K

(a) $P = \dfrac{nRT}{V} = \dfrac{\left(512.9 \text{ g} \times \dfrac{1 \text{ mol}}{352.0 \text{ g}}\right)\left(0.082\ 06\ \dfrac{L \cdot atm}{K \cdot mol}\right)(343 \text{ K})}{(22.9 \text{ L})} = 1.79 \text{ atm}$

(b) $P = \dfrac{nRT}{V - nb} - \dfrac{an^2}{V^2}$

$n = 512.9 \text{ g} \times \dfrac{1 \text{ mol } UF_6}{352.0 \text{ g } UF_6} = 1.457 \text{ mol } UF_6$

$P = \dfrac{(1.457 \text{ mol})\left(0.082\ 06\ \dfrac{L \cdot atm}{K \cdot mol}\right)(343 \text{ K})}{[(22.9 \text{ L}) - (1.457 \text{ mol})(0.1128 \text{ L/mol})]} - \dfrac{\left(15.80\ \dfrac{L^2 \cdot atm}{mol^2}\right)(1.457 \text{ mol})^2}{(22.9 \text{ L})^2} = 1.74 \text{ atm}$

9.106 Both tanks contain the same number of gas particles (atoms or molecules) at the same temperature with the same average kinetic energy. Because O_2 is lighter than Kr, the average O_2 velocity is greater than the average Kr velocity.
(a) $Kr < O_2$ (b) $O_2 < Kr$ (c) $Kr < O_2$ (d) Both are the same.

9.108 (a) $16{,}400 \text{ ft} \times \dfrac{12 \text{ in}}{1 \text{ ft}} \times \dfrac{2.54 \text{ cm}}{1 \text{ in}} \times \dfrac{1 \text{ m}}{100 \text{ cm}} = 5000 \text{ m}$

$P = e^{-h/7000} = e^{-(5000/7000)} = 0.490 \text{ atm}$

$0.490 \text{ atm} \times \dfrac{760 \text{ mm Hg}}{1 \text{ atm}} = 372 \text{ mm Hg}$

(b) $28{,}251 \text{ ft} \times \dfrac{12 \text{ in}}{1 \text{ ft}} \times \dfrac{2.54 \text{ cm}}{1 \text{ in}} \times \dfrac{1 \text{ m}}{100 \text{ cm}} = 8611 \text{ m}$

$P = e^{-h/7000} = e^{-(8611/7000)} = 0.292 \text{ atm}$

$0.292 \text{ atm} \times \dfrac{760 \text{ mm Hg}}{1 \text{ atm}} = 222 \text{ mm Hg}$

(c) $P_{O_2} = X_{O_2} \cdot P = (0.2095)(222 \text{ mm Hg}) = 46.5 \text{ mm Hg}$

9.110 $n = \dfrac{PV}{RT} = \dfrac{(2.15 \text{ atm})(7.35 \text{ L})}{\left(0.082\ 06\ \dfrac{L \cdot atm}{K \cdot mol}\right)(293 \text{ K})} = 0.657 \text{ mol Ar}$

$0.657 \text{ mol Ar} \times \dfrac{39.948 \text{ g Ar}}{1 \text{ mol Ar}} = 26.2 \text{ g Ar}$

$m_{total} = 478.1 \text{ g} + 26.2 \text{ g} = 504.3 \text{ g}$

9.112　(a)　Bulb A contains $CO_2(g)$ and $N_2(g)$;　Bulb B contains $CO_2(g)$, $N_2(g)$, and $H_2O(s)$.

(b)　Initial moles of gas $= n = \dfrac{PV}{RT} = \dfrac{\left(564 \text{ mm Hg} \times \dfrac{1.00 \text{ atm}}{760 \text{ mm Hg}}\right)(1.000 \text{ L})}{\left(0.082\ 06 \dfrac{\text{L} \cdot \text{atm}}{\text{K} \cdot \text{mol}}\right)(298 \text{ K})}$

Initial moles of gas = 0.030 35 mol

mol gas in Bulb A $= n = \dfrac{PV}{RT} = \dfrac{\left(219 \text{ mm Hg} \times \dfrac{1.00 \text{ atm}}{760 \text{ mm Hg}}\right)(1.000 \text{ L})}{\left(0.082\ 06 \dfrac{\text{L} \cdot \text{atm}}{\text{K} \cdot \text{mol}}\right)(298 \text{ K})} = 0.011\ 78 \text{ mol}$

mol gas in Bulb B $= n = \dfrac{PV}{RT} = \dfrac{\left(219 \text{ mm Hg} \times \dfrac{1.00 \text{ atm}}{760 \text{ mm Hg}}\right)(1.000 \text{ L})}{\left(0.082\ 06 \dfrac{\text{L} \cdot \text{atm}}{\text{K} \cdot \text{mol}}\right)(203 \text{ K})} = 0.017\ 30 \text{ mol}$

$n_{H_2O} = n_{initial} - n_A - n_B = 0.030\ 35 - 0.011\ 78 - 0.017\ 30 = 0.001\ 27 \text{ mol} = 0.0013 \text{ mol } H_2O$

(c)　Bulb A contains $N_2(g)$.
　　Bulb B contains $N_2(g)$ and $H_2O(s)$.
　　Bulb C contains $N_2(g)$ and $CO_2(s)$.

(d)　$n_A = \dfrac{PV}{RT} = \dfrac{\left(33.5 \text{ mm Hg} \times \dfrac{1.00 \text{ atm}}{760 \text{ mm Hg}}\right)(1.000 \text{ L})}{\left(0.082\ 06 \dfrac{\text{L} \cdot \text{atm}}{\text{K} \cdot \text{mol}}\right)(298 \text{ K})} = 0.001\ 803 \text{ mol}$

$n_B = \dfrac{PV}{RT} = \dfrac{\left(33.5 \text{ mm Hg} \times \dfrac{1.00 \text{ atm}}{760 \text{ mm Hg}}\right)(1.000 \text{ L})}{\left(0.082\ 06 \dfrac{\text{L} \cdot \text{atm}}{\text{K} \cdot \text{mol}}\right)(203 \text{ K})} = 0.002\ 646 \text{ mol}$

$n_C = \dfrac{PV}{RT} = \dfrac{\left(33.5 \text{ mm Hg} \times \dfrac{1.00 \text{ atm}}{760 \text{ mm Hg}}\right)(1.000 \text{ L})}{\left(0.082\ 06 \dfrac{\text{L} \cdot \text{atm}}{\text{K} \cdot \text{mol}}\right)(83 \text{ K})} = 0.006\ 472 \text{ mol}$

$n_{N_2} = n_A + n_B + n_C = 0.001\ 803 + 0.002\ 646 + 0.006\ 472 = 0.010\ 92 \text{ mol } N_2$

(e)　$n_{CO_2} = n_{initial} - n_{H_2O} - n_{N_2} = 0.030\ 35 - 0.0013 - 0.010\ 92 = 0.0181 \text{ mol } CO_2$

9.114 NH_3, 17.03;   mol NH_3 = 45.0 g x $\dfrac{1 \text{ mol}}{17.03 \text{ g}}$ = 2.64 mol

$$P = \frac{nRT}{V} \quad \text{or} \quad P = \frac{nRT}{(V-nb)} - \frac{an^2}{V^2}$$

(a) At T = 0 °C = 273 K

$$P = \frac{(2.64 \text{ mol})\left(0.082\ 06\ \dfrac{L\cdot atm}{K\cdot mol}\right)(273 \text{ K})}{(1.000 \text{ L})} = 59.1 \text{ atm}$$

$$P = \frac{(2.64 \text{ mol})\left(0.082\ 06\ \dfrac{L\cdot atm}{K\cdot mol}\right)(273 \text{ K})}{[(1.000 \text{ L}) - (2.64 \text{ mol})(0.0371 \text{ L/mol})]} - \frac{\left(4.17\ \dfrac{L^2\cdot atm}{mol^2}\right)(2.64 \text{ mol})^2}{(1.000 \text{ L})^2}$$

P = 65.6 atm – 29.1 atm = 36.5 atm

(b) At T = 50 °C = 323 K

$$P = \frac{(2.64 \text{ mol})\left(0.082\ 06\ \dfrac{L\cdot atm}{K\cdot mol}\right)(323 \text{ K})}{(1.000 \text{ L})} = 70.0 \text{ atm}$$

$$P = \frac{(2.64 \text{ mol})\left(0.082\ 06\ \dfrac{L\cdot atm}{K\cdot mol}\right)(323 \text{ K})}{[(1.000 \text{ L}) - (2.64 \text{ mol})(0.0371 \text{ L/mol})]} - \frac{\left(4.17\ \dfrac{L^2\cdot atm}{mol^2}\right)(2.64 \text{ mol})^2}{(1.000 \text{ L})^2}$$

P = 77.6 atm – 29.1 atm = 48.5 atm

(c) At T = 100 °C = 373 K

$$P = \frac{(2.64 \text{ mol})\left(0.082\ 06\ \dfrac{L\cdot atm}{K\cdot mol}\right)(373 \text{ K})}{(1.000 \text{ L})} = 80.8 \text{ atm}$$

$$P = \frac{(2.64 \text{ mol})\left(0.082\ 06\ \dfrac{L\cdot atm}{K\cdot mol}\right)(373 \text{ K})}{[(1.000 \text{ L}) - (2.64 \text{ mol})(0.0371 \text{ L/mol})]} - \frac{\left(4.17\ \dfrac{L^2\cdot atm}{mol^2}\right)(2.64 \text{ mol})^2}{(1.000 \text{ L})^2}$$

P = 89.6 atm – 29.1 atm = 60.5 atm

At the three temperatures, the van der Waals equation predicts a much lower pressure than does the ideal gas law. This is likely due to the fact that NH_3 can hydrogen bond leading to strong intermolecular forces.

9.116 CO_2, 44.01

$$\text{mol } CO_2 = 500.0 \text{ g } CO_2 \text{ x } \frac{1 \text{ mol } CO_2}{44.01 \text{ g } CO_2} = 11.36 \text{ mol } CO_2$$

$$PV = nRT; \quad P = \frac{nRT}{V} = \frac{(11.36 \text{ mol})\left(0.082\ 06\ \dfrac{L\cdot atm}{K\cdot mol}\right)(700 \text{ K})}{(0.800 \text{ L})} = 816 \text{ atm}$$

9.118   (a) average molecular mass for natural gas

$$= (0.915)(16.04) + (0.085)(30.07) = 17.2$$

$$\text{total moles of gas} = 15.50 \text{ g} \times \frac{1 \text{ mol gas}}{17.2 \text{ g gas}} = 0.901 \text{ mol gas}$$

    (b) $P = \dfrac{(0.901 \text{ mol})\left(0.082\ 06\ \dfrac{\text{L} \cdot \text{atm}}{\text{K} \cdot \text{mol}}\right)(293 \text{ K})}{(15.00 \text{ L})} = 1.44 \text{ atm}$

    (c) $P_{CH_4} = X_{CH_4} \cdot P_{total} = (1.44 \text{ atm})(0.915) = 1.32 \text{ atm}$

         $P_{C_2H_6} = X_{C_2H_6} \cdot P_{total} = (1.44 \text{ atm})(0.085) = 0.12 \text{ atm}$

    (d) $\Delta H_{combustion}(CH_4) = -890.3 \text{ kJ/mol}$ and $\Delta H_{combustion}(C_2H_6) = -1427.7 \text{ kJ/mol}$
        Heat liberated $= (0.915)(0.901 \text{ mol})(-890.3 \text{ kJ/mol})$
                           $+ (0.085)(0.901)(-1427.7 \text{ kJ/mol}) = -843 \text{ kJ}$

9.120   (a) $T = 0\ ^\circ C = 273 \text{ K}$; $PV = nRT$

      $n_Q = \dfrac{PV}{RT} = \dfrac{(0.229 \text{ atm})(0.0500 \text{ L})}{\left(0.082\ 06\ \dfrac{\text{L} \cdot \text{atm}}{\text{K} \cdot \text{mol}}\right)(273 \text{ K})} = 5.11 \times 10^{-4} \text{ mol Q}$

      Q molar mass $= \dfrac{0.100 \text{ g Q}}{5.11 \times 10^{-4} \text{ mol Q}} = 196 \text{ g/mol}$

      Xe molar mass = 131.3 g/mol

      O_n molar mass = 196 g/mol – 131.3 g/mol = 65 g/mol; $65/16 \approx 4$
      So, n = 4 and XeO_4 is the likely formula for Q.
    (b) The decomposition reaction is $XeO_4(g) \rightarrow Xe(g) + 2\ O_2(g)$.
      After decomposition $n_{Xe} = n_{XeO_4} = 5.11 \times 10^{-4}$ mol and $n_{O_2} = 2 \times n_{XeO_4} = 1.02 \times 10^{-3}$ mol

      $T = 100\ ^\circ C = 373 \text{ K}$

      $P_{Xe} = \dfrac{nRT}{V} = \dfrac{(5.11 \times 10^{-4} \text{ mol})\left(0.082\ 06\ \dfrac{\text{L} \cdot \text{atm}}{\text{K} \cdot \text{mol}}\right)(373 \text{ K})}{(0.0500 \text{ L})} = 0.313 \text{ atm}$

      $P_{O_2} = 2 \times P_{Xe} = 2 \times 0.313 \text{ atm} = 0.626 \text{ atm}$

      $P_{total} = P_{Xe} + P_{O_2} = 0.313 \text{ atm} + 0.626 \text{ atm} = 0.939 \text{ atm}$

9.122 PCl_3, 137.3; O_2, 32.00; $POCl_3$, 153.3
      $2\ PCl_3(g) + O_2(g) \rightarrow 2\ POCl_3(g)$

      $\text{mol } PCl_3 = 25.0 \text{ g} \times \dfrac{1 \text{ mol } PCl_3}{137.3 \text{ g } PCl_3} = 0.182 \text{ mol } PCl_3$

      $\text{mol } O_2 = 3.00 \text{ g} \times \dfrac{1 \text{ mol } O_2}{32.00 \text{ g } O_2} = 0.0937 \text{ mol } O_2$

Check for limiting reactant.

$$\text{mol } O_2 \text{ needed } = 0.182 \text{ mol } PCl_3 \text{ x } \frac{1 \text{ mol } O_2}{2 \text{ mol } PCl_3} = 0.0910 \text{ mol } O_2 \text{ needed}$$

There is a slight excess of O_2. PCl_3 is the limiting reactant.

$$\text{mol } POCl_3 = 0.182 \text{ mol } PCl_3 \text{ x } \frac{2 \text{ mol } POCl_3}{2 \text{ mol } PCl_3} = 0.182 \text{ mol } POCl_3$$

mol O_2 left over = 0.0937 mol − 0.0910 mol = 0.0027 mol O_2 left over
T = 200.0 °C = 200.0 + 273.15 = 473.1 K; PV = nRT

$$P = \frac{nRT}{V} = \frac{(0.182 \text{ mol} + 0.0027 \text{ mol})\left(0.082\ 06 \dfrac{L \cdot atm}{K \cdot mol}\right)(473.1 \text{ K})}{(5.00 \text{ L})} = 1.43 \text{ atm}$$

9.124 O_2, 32.00; O_3, 48.00

	$3\ O_2(g)$	$\rightarrow$	$2\ O_3(g)$
initial (atm)	32.00		0
change (atm)	−3x		+2x
after rxn (atm)	32.00 − 3x		2x

$$P_{Total} = P_{O_2} + P_{O_3} = 30.64 \text{ atm} = 32.00 \text{ atm} - 3x + 2x = 32.00 \text{ atm} - x$$

x = 32.00 atm − 30.64 atm = 1.36 atm

P_{O_2} = 32.00 − 3x = 32.00 − 3(1.36 atm) = 27.92 atm

P_{O_3} = 2x = 2(1.36 atm) = 2.72 atm

T = 25 °C = 25 + 273 = 298 K; PV = nRT

$$n_{O_2} = \frac{PV}{RT} = \frac{(27.92 \text{ atm})(10.00 \text{ L})}{\left(0.082\ 06 \dfrac{L \cdot atm}{K \cdot mol}\right)(298 \text{ K})} = 11.42 \text{ mol } O_2$$

$$n_{O_3} = \frac{PV}{RT} = \frac{(2.72 \text{ atm})(10.00 \text{ L})}{\left(0.082\ 06 \dfrac{L \cdot atm}{K \cdot mol}\right)(298 \text{ K})} = 1.11 \text{ mol } O_3$$

$$\text{mass } O_2 = 11.42 \text{ mol } O_2 \text{ x } \frac{32.00 \text{ g } O_2}{1 \text{ mol } O_2} = 365.4 \text{ g } O_2$$

$$\text{mass } O_3 = 1.11 \text{ mol } O_3 \text{ x } \frac{48.00 \text{ g } O_3}{1 \text{ mol } O_3} = 53.3 \text{ g } O_3$$

total mass = 365.4 g + 53.3 g = 418.7 g

$$\text{mass \% } O_3 = \frac{\text{mass } O_3}{\text{total mass}} = \frac{53.3 \text{ g}}{418.7 \text{ g}} \text{ x } 100\% = 12.7 \%$$

9.126 NO_2, 46.01

Calculate the NO_2 pressure before any NO_2 dimerization.

$$P = \frac{nRT}{V} = \frac{\left(9.66 \text{ g} \times \dfrac{1 \text{ mol}}{46.01 \text{ g}}\right)\left(0.082\ 06 \dfrac{L \cdot atm}{K \cdot mol}\right)(298 \text{ K})}{(6.51 \text{ L})} = 0.789 \text{ atm}$$

	$2 NO_2(g)$	$\rightarrow$	$N_2O_4(g)$
before reaction (atm)	0.789		0
change (atm)	$-2x$		$+x$
after reaction (atm)	$0.789 - 2x$		x

$0.487 = 0.789 - 2x + x$

$0.487 = 0.789 - x$

$x = 0.789 - 0.487 = 0.302 \text{ atm}$

$P_{NO_2} = 0.789 - 2x = 0.789 - 2(0.302) = 0.185 \text{ atm}$

$P_{N_2O_4} = x = 0.302 \text{ atm}$

$X_{NO_2} = \dfrac{0.185 \text{ atm}}{0.487 \text{ atm}} = 0.380$ and $X_{N_2O_4} = \dfrac{0.302 \text{ atm}}{0.487 \text{ atm}} = 0.620$

Multiconcept Problems

9.128 CO_2, 44.01

$CH_4(g) + 2 O_2(g) \rightarrow CO_2(g) + 2 H_2O(g)$ $\qquad \Delta H° = -802 \text{ kJ}$

(a) 1.00 atm of CH_4 only requires 2.00 atm O_2, therefore O_2 is in excess.

$T = 300 °C = 300 + 273 = 573 \text{ K}$; $PV = nRT$

$$n_{CH_4} = \frac{PV}{RT} = \frac{(1.00 \text{ atm})(4.00 \text{ L})}{\left(0.082\ 06 \dfrac{L \cdot atm}{K \cdot mol}\right)(573 \text{ K})} = 0.0851 \text{ mol } CH_4$$

$$n_{O_2} = \frac{PV}{RT} = \frac{(4.00 \text{ atm})(4.00 \text{ L})}{\left(0.082\ 06 \dfrac{L \cdot atm}{K \cdot mol}\right)(573 \text{ K})} = 0.340 \text{ mol } O_2$$

$$\text{mass } CO_2 = 0.0851 \text{ mol } CH_4 \times \frac{1 \text{ mol } CO_2}{1 \text{ mol } CH_4} \times \frac{44.01 \text{ g } CO_2}{1 \text{ mol } CO_2} = 3.75 \text{ g } CO_2$$

(b) $q_{rxn} = 0.0851 \text{ mol } CH_4 \times \dfrac{-802 \text{ kJ}}{1 \text{ mol } CH_4} = -68.3 \text{ kJ}$

	$CH_4(g)$	$+$	$2 O_2(g)$	$\rightarrow$	$CO_2(g)$	$+$	$2 H_2O(g)$
initial (mol)	0.0851		0.340		0		0
change (mol)	-0.0851		$-2(0.0851)$		$+0.0851$		$+2(0.0851)$
after rxn (mol)	0		$0.340 - 2(0.0851)$		0.0851		0.170

total moles of gas = 0.340 ml − 2(0.0851) mol + 0.0851 mol + 0.170 mol = 0.425 mol gas

$$q_{rxn} = -68.3 \text{ kJ} \times \frac{1000 \text{ J}}{1 \text{ kJ}} = -68,300 \text{ J}$$

$$q_{vessel} = -q_{rxn} = 68,300 \text{ J} = (0.425 \text{ mol})(21 \text{ J/(mol} \cdot {}^{\circ}\text{C}))(t_f - 300 \text{ }^{\circ}\text{C}) +$$

$$(14.500 \text{ kg})\left(\frac{1000 \text{ g}}{1 \text{ kg}}\right)(0.449 \text{ J/(g} \cdot {}^{\circ}\text{C}))(t_f - 300 \text{ }^{\circ}\text{C})$$

Solve for t_f.

$$68,300 \text{ J} = (8.925 \text{ J/}{}^{\circ}\text{C} + 6510 \text{ J/}{}^{\circ}\text{C})(t_f - 300 \text{ }^{\circ}\text{C}) = (6519 \text{ J/}{}^{\circ}\text{C})(t_f - 300 \text{ }^{\circ}\text{C})$$

$$\frac{68,300 \text{ J}}{6519 \text{ J/}{}^{\circ}\text{C}} = 10.5 \text{ }^{\circ}\text{C} = (t_f - 300 \text{ }^{\circ}\text{C})$$

$$300 \text{ }^{\circ}\text{C} + 10.5 \text{ }^{\circ}\text{C} = t_f$$

$$t_f = 310 \text{ }^{\circ}\text{C}$$

(c) $T = 310 \text{ }^{\circ}\text{C} = 310 + 273 = 583 \text{ K}$

$$P_{CO_2} = \frac{nRT}{V} = \frac{(0.0851 \text{ mol})\left(0.082 \ 06 \ \frac{\text{L} \cdot \text{atm}}{\text{K} \cdot \text{mol}}\right)(583 \text{ K})}{(4.00 \text{ L})} = 1.02 \text{ atm}$$

9.130 (a) $2 \text{ C}_8\text{H}_{18}(l) + 25 \text{ O}_2(g) \rightarrow 16 \text{ CO}_2(g) + 18 \text{ H}_2\text{O}(g)$

(b) $4.6 \times 10^{10} \text{ L C}_8\text{H}_{18} \times \dfrac{1000 \text{ mL}}{1 \text{ L}} \times \dfrac{0.792 \text{ g}}{1 \text{ mL}} = 3.64 \times 10^{13} \text{ g C}_8\text{H}_{18}$

$$3.64 \times 10^{13} \text{ g C}_8\text{H}_{18} \times \frac{1 \text{ mol C}_8\text{H}_{18}}{114.2 \text{ g C}_8\text{H}_{18}} \times \frac{16 \text{ mol CO}_2}{2 \text{ mol C}_8\text{H}_{18}} = 2.55 \times 10^{12} \text{ mol CO}_2$$

$$2.55 \times 10^{12} \text{ mol CO}_2 \times \frac{44.0 \text{ g CO}_2}{1 \text{ mol CO}_2} \times \frac{1 \text{ kg}}{1000 \text{ g}} = 1.1 \times 10^{11} \text{ kg CO}_2$$

(c) $V = \dfrac{nRT}{P} = \dfrac{(2.55 \times 10^{12} \text{ mol})\left(0.082 \ 06 \ \frac{\text{L} \cdot \text{atm}}{\text{K} \cdot \text{mol}}\right)(273 \text{ K})}{(1.00 \text{ atm})} = 5.7 \times 10^{13} \text{ L of CO}_2$

(d) 12.5 moles of O_2 are needed for each mole of isooctane (from part a).

$$12.5 \text{ mol O}_2 = (0.210)(n_{air}); \qquad n_{air} = \frac{12.5 \text{ mol}}{0.210} = 59.5 \text{ mol air}$$

$$V = \frac{nRT}{P} = \frac{(59.5 \text{ mol})\left(0.082 \ 06 \ \frac{\text{L} \cdot \text{atm}}{\text{K} \cdot \text{mol}}\right)(273 \text{ K})}{(1.00 \text{ atm})} = 1.33 \times 10^3 \text{ L}$$

9.132 $n = \dfrac{PV}{RT} = \dfrac{(1 \text{ atm})(1323 \text{ L})}{\left(0.082 \ 06 \ \frac{\text{L} \cdot \text{atm}}{\text{K} \cdot \text{mol}}\right)(2223 \text{ K})} = 7.25 \text{ mol of all gases}$

(a) $0.004 \ 00 \text{ mol "nitro"} \times \dfrac{7.25 \text{ mol gases}}{1 \text{ mol "nitro"}} = 0.0290 \text{ mol hot gases}$

(b) $n = \dfrac{PV}{RT} = \dfrac{\left(623 \text{ mm Hg} \times \dfrac{1.00 \text{ atm}}{760 \text{ mm Hg}}\right)(0.500 \text{ L})}{\left(0.082\ 06 \dfrac{\text{L} \cdot \text{atm}}{\text{K} \cdot \text{mol}}\right)(263 \text{ K})} = 0.0190 \text{ mol B + C + D}$

$n_A = n_{total} - n_{(B+C+D)} = 0.0290 - 0.0190 = 0.0100 \text{ mol A}; \quad A = H_2O$

(c) $n = \dfrac{PV}{RT} = \dfrac{\left(260 \text{ mm Hg} \times \dfrac{1.00 \text{ atm}}{760 \text{ mm Hg}}\right)(0.500 \text{ L})}{\left(0.082\ 06 \dfrac{\text{L} \cdot \text{atm}}{\text{K} \cdot \text{mol}}\right)(298 \text{ K})} = 0.007\ 00 \text{ mol C + D}$

$n_B = n_{(B+C+D)} - n_{(C+D)} = 0.0190 - 0.007\ 00 = 0.0120 \text{ mol B}; \quad B = CO_2$

(d) $n = \dfrac{PV}{RT} = \dfrac{\left(223 \text{ mm Hg} \times \dfrac{1.00 \text{ atm}}{760 \text{ mm Hg}}\right)(0.500 \text{ L})}{\left(0.082\ 06 \dfrac{\text{L} \cdot \text{atm}}{\text{K} \cdot \text{mol}}\right)(298 \text{ K})} = 0.006\ 00 \text{ mol D}$

$n_C = n_{(C+D)} - n_D = 0.007\ 00 - 0.006\ 00 = 0.001\ 00 \text{ mol C}; \quad C = O_2$

molar mass $D = \dfrac{0.168 \text{ g}}{0.006\ 00 \text{ mol}} = 28.0 \text{ g/mol}; \quad D = N_2$

(e) $0.004 \ C_3H_5N_3O_9(l) \rightarrow 0.0100 \ H_2O(g) + 0.012 \ CO_2(g) + 0.001 \ O_2(g) + 0.006 \ N_2(g)$
Multiply each coefficient by 1000 to obtain integers.
$4 \ C_3H_5N_3O_9(l) \rightarrow 10 \ H_2O(g) + 12 \ CO_2(g) + O_2(g) + 6 \ N_2(g)$

10

Liquids, Solids, and Phase Changes

10.1 $\mu = Q \times r = (1.60 \times 10^{-19}\ C)(92 \times 10^{-12}\ m)\left(\dfrac{1\ D}{3.336 \times 10^{-30}\ C \cdot m}\right) = 4.41\ D$

% ionic character for HF $= \dfrac{1.83\ D}{4.41\ D} \times 100\% = 41\%$

HF has more ionic character than HCl. HCl has only 18% ionic character.

10.2 (a) SF_6 has polar covalent bonds but the molecule is symmetrical (octahedral). The individual bond polarities cancel, and the molecule has no dipole moment.
(b) $H_2C{=}CH_2$ can be assumed to have nonpolar C–H bonds. In addition, the molecule is symmetrical. The molecule has no dipole moment.
(c)

The C–Cl bonds in $CHCl_3$ are polar covalent bonds, and the molecule is polar.
(d)

The C–Cl bonds in CH_2Cl_2 are polar covalent bonds, and the molecule is polar.

10.3

10.4 The N atom is electron rich (red) because of its high electronegativity. The C and H atoms are electron poor (blue) because they are less electronegative.

10.5 (a) Of the four substances, only HNO_3 has a net dipole moment.
(b) Only HNO_3 can hydrogen bond.
(c) Ar has fewer electrons than Cl_2 and CCl_4, and has the smallest dispersion forces.

10.6 H_2S dipole-dipole, dispersion
CH_3OH hydrogen bonding, dipole-dipole, dispersion
C_2H_6 dispersion
Ar   dispersion
$Ar < C_2H_6 < H_2S < CH_3OH$

10.7　(a)　$CO_2(s) \rightarrow CO_2(g)$, ΔS is positive

(b)　$H_2O(g) \rightarrow H_2O(l)$, ΔS is negative

(c)　ΔS is positive (more randomness)

10.8　$\Delta G = \Delta H - T\Delta S$; at the boiling point (phase change), $\Delta G = 0$.

$$\Delta H = T\Delta S; \quad T = \frac{\Delta H_{vap}}{\Delta S_{vap}} = \frac{29.2 \text{ kJ/mol}}{87.5 \times 10^{-3} \text{ kJ/(K} \cdot \text{mol)}} = 334 \text{ K}$$

10.9　The boiling point is the temperature where the vapor pressure of a liquid equals the external pressure.

$P_1 = 760$ mm Hg; $P_2 = 260$ mm Hg; $T_1 = 80.1\,^\circ$C

$\Delta H_{vap} = 30.7$ kJ/mol

$$\ln P_2 = \ln P_1 + \frac{\Delta H_{vap}}{R}\left(\frac{1}{T_1} - \frac{1}{T_2}\right)$$

$$(\ln P_2 - \ln P_1)\left(\frac{R}{\Delta H_{vap}}\right) = \frac{1}{T_1} - \frac{1}{T_2}$$

Solve for T_2 (the boiling point for benzene at 260 mm Hg).

$$\frac{1}{T_1} - (\ln P_2 - \ln P_1)\left(\frac{R}{\Delta H_{vap}}\right) = \frac{1}{T_2}$$

$$\frac{1}{353.2 \text{ K}} - [\ln(260) - \ln(760)]\left(\frac{8.3145 \dfrac{J}{K \cdot mol}}{30,700 \text{ J/mol}}\right) = \frac{1}{T_2}$$

$$\frac{1}{T_2} = 0.003\,122 \text{ K}^{-1}; \ T_2 = 320 \text{ K} = 47\,^\circ\text{C} \ \text{ (boiling point is lower at lower pressure)}$$

10.10　$$\Delta H_{vap} = \frac{(\ln P_2 - \ln P_1)(R)}{\left(\dfrac{1}{T_1} - \dfrac{1}{T_2}\right)}$$

$P_1 = 400$ mm Hg;　$T_1 = 41.0\,^\circ$C $= 314.2$ K

$P_2 = 760$ mm Hg;　$T_2 = 331.9$ K

$$\Delta H_{vap} = \frac{[\ln(760) - \ln(400)]\left(8.3145 \dfrac{J}{K \cdot mol}\right)}{\left(\dfrac{1}{314.2 \text{ K}} - \dfrac{1}{331.9 \text{ K}}\right)} = 31,442 \text{ J/mol} = 31.4 \text{ kJ/mol}$$

10.11　(a)　1/8 atom at 8 corners and 1 atom at body center = 2 atoms

(b)　1/8 atom at 8 corners and 1/2 atom at 6 faces = 4 atoms

10.12　For a simple cube, $d = 2r$;　$r = \dfrac{d}{2} = \dfrac{334 \text{ pm}}{2} = 167$ pm

10.13   For a simple cube, there is one atom per unit cell.

$$\text{mass of one Po atom} = 209 \text{ g/mol} \times \frac{1 \text{ mol}}{6.022 \times 10^{23} \text{ atoms}} = 3.4706 \times 10^{-22} \text{ g/atom}$$

unit cell edge = d = 334 pm = 334×10^{-12} m = 3.34×10^{-8} cm

unit cell volume = d^3 = $(3.34 \times 10^{-8} \text{ cm})^3$ = $3.7260 \times 10^{-23} \text{ cm}^3$

$$\text{density} = \frac{\text{mass}}{\text{volume}} = \frac{3.4706 \times 10^{-22} \text{ g}}{3.7260 \times 10^{-23} \text{ cm}^3} = 9.31 \text{ g/cm}^3$$

10.14   There are several possibilities. Here's one:

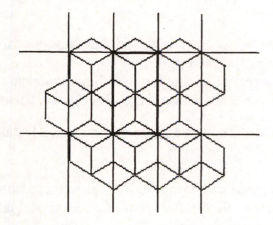

10.15   For CuCl:

1/8 Cl^- at 8 corners and 1/2 Cl^- at 6 faces = 4 Cl^- (4 minuses)

4 Cu^+ inside (4 pluses)

For $BaCl_2$:

1/8 Ba^{2+} at 8 corners and 1/2 Ba^{2+} at 6 faces = 4 Ba^{2+} (8 pluses)

8 Cl^- inside (8 minuses)

10.16   (a) In the unit cell there is a rhenium atom at each corner of the cube. The number of rhenium atoms in the unit cell = 1/8 Re at 8 corners = 1 Re atom.

In the unit cell there is an oxygen atom in the center of each edge of the cube. The number of oxygen atoms in the unit cell = 1/4 O on 12 edges = 3 O atoms.

(b) ReO_3

(c) Each oxide has a –2 charge and there are three of them for a total charge of –6. The charge (oxidation state) of rhenium must be +6 to balance the negative charge of the oxides.

(d) Each oxygen atom is surrounded by two rhenium atoms. The geometry is linear.

(e) Each rhenium atom is surrounded by six oxygen atoms. The geometry is octahedral.

10.17   The minimum pressure at which liquid CO_2 can exist is its triple point pressure of 5.11 atm.

10.18   (a) $CO_2(s) \rightarrow CO_2(g)$

(b) $CO_2(l) \rightarrow CO_2(g)$

(c) $CO_2(g) \rightarrow CO_2(l) \rightarrow$ supercritical CO_2

10.19 (a)

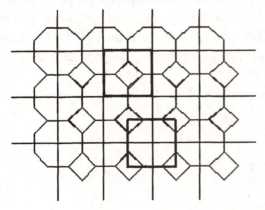

(b) Gallium has two triple points. The one below 1 atm is a solid, liquid, vapor triple point. The one at 10^4 atm is a solid(1), solid(2), liquid triple point.
(c) Increasing the pressure favors the liquid phase, giving the solid/liquid boundary a negative slope. At 1 atm pressure the liquid phase is more dense than the solid phase.

10.20 As the name implies, the constituent particles in an ionic liquid are cations and anions rather than molecules.

10.21 In ionic liquids the cation has an irregular shape and one or both of the ions are large and bulky to disperse charges over a large volume. Both factors minimize the crystal lattice energy, making the solid less stable and favoring the liquid.

Conceptual Problems

10.22 The electronegative O atoms are electron rich (red), while the rest of the molecule is electron poor (blue).

10.24 (a) cubic closest-packed
(b) $1/8$ S^{2-} at 8 corners and $1/2$ S^{2-} at 6 faces = 4 S^{2-};   4 Zn^{2+} inside

10.26 (a) normal boiling point ≈ 300 K; normal melting point ≈ 180 K
(b) (i) solid   (ii) gas   (iii) supercritical fluid

10.28 Here are two possibilities:

10.30

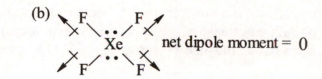

CH$_3$—C $\begin{matrix} \diagup O \cdots H — O \diagdown \\ \diagdown O — H \cdots O \diagup \end{matrix}$ C—CH$_3$

Section Problems
Dipole Moments and Intermolecular Forces (Sections 10.1–10.3)

10.32　If a molecule has polar covalent bonds, the molecular shape (and location of lone pairs of electrons) determines whether the bond dipoles cancel and thus whether the molecule has a dipole moment.

10.34　(a) $CHCl_3$ has a permanent dipole moment. Dipole-dipole intermolecular forces are important. London dispersion forces are also present.
(b) O_2 has no dipole moment. London dispersion intermolecular forces are important.
(c) Polyethylene, C_nH_{2n+2}. London dispersion intermolecular forces are important.
(d) CH_3OH has a permanent dipole moment. Dipole-dipole intermolecular forces and hydrogen bonding are important. London dispersion forces are also present.

10.36　For CH_3OH and CH_4, dispersion forces are small. CH_3OH can hydrogen bond; CH_4 cannot. This accounts for the large difference in boiling points.
For 1-decanol and decane, dispersion forces are comparable and relatively large along the C–H chain. 1-decanol can hydrogen bond; decane cannot. This accounts for the 57 °C higher boiling point for 1-decanol.

10.38　(a)

$\overset{\cdot\cdot}{O}$ with Cl and Cl — net (arrow up)

(b)

F, F, Xe, F, F — net dipole moment = 0

(c)

CH$_3$—C----H with Cl and H — net

(d)

F, B, F, F — net dipole moment = 0

10.40　$\mu = Q \times r = (1.60 \times 10^{-19} \text{ C})(213.9 \times 10^{-12} \text{ m})\left(\dfrac{1 \text{ D}}{3.336 \times 10^{-30} \text{ C} \cdot \text{m}} \right) = 10.26 \text{ D}$

% ionic character for BrCl = $\dfrac{0.518 \text{ D}}{10.26 \text{ D}} \times 100\% = 5.05\%$

10.42

$\overset{\cdot\cdot}{S}$ with O and O — net (arrow down)

O＝C＝O
net dipole moment = 0

SO_2 is bent and the individual bond dipole moments add to give the molecule a net dipole moment.
CO_2 is linear and the individual bond dipole moments point in opposite directions to cancel each other out. CO_2 has no net dipole moment.

10.44 (a)

$$\left[\begin{array}{c} Br \\ | \\ Cl—Pt—Cl \\ | \\ Br \end{array} \right]^{2-}$$

(b)

$$\left[\begin{array}{c} Br \\ | \\ Cl—Pt—Br \\ | \\ Cl \end{array} \right]^{2-}$$

10.46

:N—H---:N—H   (with H atoms above and below each N)
hydrogen bond

10.48   It's water's surface tension that keeps the needle on top.

Vapor Pressure and Phase Changes (Sections 10.4 and 10.5)

10.50 ΔH_{vap} is usually larger than ΔH_{fusion} because ΔH_{vap} is the heat required to overcome all intermolecular forces.

10.52   (a) $Hg(l) \rightarrow Hg(g)$
(b) no change of state, Hg remains a liquid
(c) $Hg(g) \rightarrow Hg(l) \rightarrow Hg(s)$

10.54   As the pressure over the liquid H_2O is lowered, H_2O vapor is removed by the pump. As H_2O vapor is removed, more of the liquid H_2O is converted to H_2O vapor. This conversion is an endothermic process and the temperature decreases. The combination of both a decrease in pressure and temperature takes the system across the liquid/solid boundary in the phase diagram so the H_2O that remains turns to ice.

10.56 H_2O, 18.02; $5.00 \text{ g } H_2O \times \dfrac{1 \text{ mol } H_2O}{18.02 \text{ g } H_2O} = 0.2775 \text{ mol } H_2O$

$q_1 = (0.2775 \text{ mol})[36.6 \times 10^{-3} \text{ kJ/(K} \cdot \text{mol)}](273 \text{ K} - 263 \text{ K}) = 0.1016 \text{ kJ}$
$q_2 = (0.2775 \text{ mol})(6.01 \text{ kJ/mol}) = 1.668 \text{ kJ}$
$q_3 = (0.2775 \text{ mol})[[75.3 \times 10^{-3} \text{ kJ/(K} \cdot \text{mol)}](303 \text{ K} - 273 \text{ K}) = 0.6269 \text{ kJ}$
$q_{total} = q_1 + q_2 + q_3 = 2.40 \text{ kJ};$ 2.40 kJ of heat is required.

10.58 H_2O, 18.02; $7.55 \text{ g } H_2O \times \dfrac{1 \text{ mol } H_2O}{18.02 \text{ g } H_2O} = 0.4190 \text{ mol } H_2O$

$q_1 = (0.4190 \text{ mol})[75.3 \times 10^{-3} \text{ kJ/(K} \cdot \text{mol)}](273.15 \text{ K} - 306.65 \text{ K}) = -1.057 \text{ kJ}$
$q_2 = -(0.4190 \text{ mol})(6.01 \text{ kJ/mol}) = -2.518 \text{ kJ}$
$q_3 = (0.4190 \text{ mol})[36.6 \times 10^{-3} \text{ kJ/(K} \cdot \text{mol)}](263.15 \text{ K} - 273.15 \text{ K}) = -0.1534 \text{ kJ}$
$q_{total} = q_1 + q_2 + q_3 = -3.73 \text{ kJ};$ 3.73 kJ of heat is released.

10.60

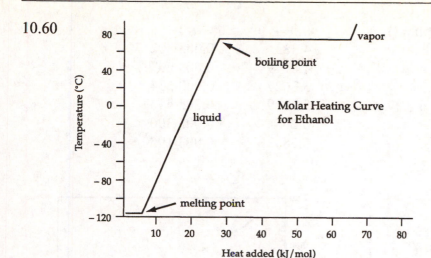

10.62 boiling point = 218 °C = 491 K

$\Delta G = \Delta H_{vap} - T\Delta S_{vap}$; At the boiling point (phase change), $\Delta G = 0$

$$\Delta H_{vap} = T\Delta S_{vap}; \quad \Delta S_{vap} = \frac{\Delta H_{vap}}{T} = \frac{43.3 \text{ kJ/mol}}{491 \text{ K}} = 0.0882 \text{ kJ/(K} \cdot \text{mol)} = 88.2 \text{ J/(K} \cdot \text{mol)}$$

10.64 $\Delta H_{vap} = \dfrac{(\ln P_2 - \ln P_1)(R)}{\left(\dfrac{1}{T_1} - \dfrac{1}{T_2}\right)}$

$T_1 = -5.1 \text{ °C} = 268.0 \text{ K}; \qquad P_1 = 100 \text{ mm Hg}$
$T_2 = 46.5 \text{ °C} = 319.6 \text{ K}; \qquad P_2 = 760 \text{ mm Hg}$

$$\Delta H_{vap} = \frac{[\ln(760) - \ln(100)][8.3145 \times 10^{-3} \text{ kJ/(K} \cdot \text{mol)}]}{\left(\dfrac{1}{268.0 \text{ K}} - \dfrac{1}{319.6 \text{ K}}\right)} = 28.0 \text{ kJ/mol}$$

10.66 $\ln P_2 = \ln P_1 + \dfrac{\Delta H_{vap}}{R}\left(\dfrac{1}{T_1} - \dfrac{1}{T_2}\right)$

$\Delta H_{vap} = 28.0 \text{ kJ/mol}$
$P_1 = 100 \text{ mm Hg}; \qquad T_1 = -5.1 \text{ °C} = 268.0 \text{ K}; \qquad T_2 = 20.0 \text{ °C} = 293.2 \text{ K}$
Solve for P_2.

$$\ln P_2 = \ln(100) + \frac{28.0 \text{ kJ/mol}}{[8.3145 \times 10^{-3} \text{ kJ/(K} \cdot \text{mol)}]}\left(\frac{1}{268.0 \text{ K}} - \frac{1}{293.2 \text{ K}}\right)$$

$\ln P_2 = 5.6852; \quad P_2 = e^{5.6852} = 294.5 \text{ mm Hg} = 294 \text{ mm Hg}$

10.68

T(K)	P_{vap}(mm Hg)	ln P_{vap}	1/T
263	80.1	4.383	0.003 802
273	133.6	4.8949	0.003 663
283	213.3	5.3627	0.003 534
293	329.6	5.7979	0.003 413
303	495.4	6.2054	0.003 300
313	724.4	6.5853	0.003 195

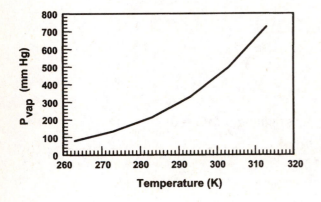

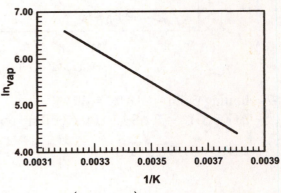

$$\ln P_{vap} = \left(-\frac{\Delta H_{vap}}{R} \right)\frac{1}{T} + C; \ \ C = 18.2$$

$$\text{slope} = -3628 \ K = -\frac{\Delta H_{vap}}{R}$$

$$\Delta H_{vap} = (3628 \ K)(R) = (3628 \ K)[8.3145 \times 10^{-3} \ kJ/(K \cdot mol)] = 30.1 \ kJ/mol$$

10.70 $\Delta H_{vap} = 30.1 \ kJ/mol$

10.72 $\Delta H_{vap} = \dfrac{(\ln P_2 - \ln P_1)(R)}{\left(\dfrac{1}{T_1} - \dfrac{1}{T_2} \right)}$

$P_1 = 80.1$ mm Hg; $\qquad T_1 = 263$ K
$P_2 = 724.4$ mm Hg; $\qquad T_2 = 313$ K

$$\Delta H_{vap} = \frac{[\ln(724.4) - \ln(80.1)][8.3145 \times 10^{-3} \ kJ/(K \cdot mol)]}{\left(\dfrac{1}{263 \ K} - \dfrac{1}{313 \ K} \right)} = 30.1 \ kJ/mol$$

The calculated ΔH_{vap} and that obtained from the plot in Problem 10.68 are the same.

Kinds and Structures of Solids (Sections 10.6–10.10)

10.74 molecular solid, CO_2, I_2; metallic solid, any metallic element;
covalent network solid, diamond; ionic solid, NaCl

10.76 (a) rubber   (b) Na_3PO_4 (c) CBr_4 (d) quartz   (e) Au

10.78 Silicon carbide is a covalent network solid.

10.80 $d = \dfrac{n\,\lambda}{2\sin\theta} = \dfrac{(1)(154.2\text{ pm})}{2\sin(22.5^\circ)} = 201$ pm

10.82 The unit cell is the smallest repeating unit in a crystal.

10.84 Cu is face-centered cubic. d = 362 pm; $r = \sqrt{\dfrac{d^2}{8}} = \sqrt{\dfrac{(362\text{ pm})^2}{8}} = 128$ pm

362 pm = 362 x 10^{-12} m = 3.62 x 10^{-8} cm
unit cell volume = (3.62 x 10^{-8} cm)^3 = 4.74 x 10^{-23} cm^3

mass of one Cu atom = 63.55 g/mol x $\dfrac{1\text{ mol}}{6.022 \times 10^{23}\text{ atom}}$ = 1.055 x 10^{-22} g/atom

Cu is face-centered cubic; there are, therefore four Cu atoms in the unit cell.
unit cell mass = (4 atoms)(1.055 x 10^{-22} g/atom) = 4.22 x 10^{-22} g

density = $\dfrac{\text{mass}}{\text{volume}}$ = $\dfrac{4.22 \times 10^{-22}\text{g}}{4.74 \times 10^{-23}\text{ cm}^3}$ = 8.90 g/cm^3

10.86 mass of one Al atom = 26.98 g/mol x $\dfrac{1\text{ mol}}{6.022 \times 10^{23}\text{ atom}}$ = 4.480 x 10^{-23} g/atom

Al is face-centered cubic; there are, therefore, four Al atoms in the unit cell.
unit cell mass = (4 atoms)(4.480 x 10^{-23} g/atom) = 1.792 x 10^{-22} g

density = $\dfrac{\text{mass}}{\text{volume}}$

unit cell volume = $\dfrac{\text{unit cell mass}}{\text{density}}$ = $\dfrac{1.792 \times 10^{-22}\text{ g}}{2.699\text{ g/cm}^3}$ = 6.640 x 10^{-23} cm^3

unit cell edge = d = $\sqrt[3]{6.640 \times 10^{-23}\text{ cm}^3}$ = 4.049 x 10^{-8} cm

d = 4.049 x 10^{-8} cm x $\dfrac{1\text{m}}{100\text{ cm}}$ = 4.049 x 10^{-10} m = 404.9 x 10^{-12} m = 404.9 pm

10.88 unit cell body diagonal = 4r = 549 pm
For W, r = $\dfrac{549\text{ pm}}{4}$ = 137 pm

10.90 mass of one Ti atom = 47.88 g/mol x $\dfrac{1\text{ mol}}{6.022 \times 10^{23}\text{ atoms}}$ = 7.951 x 10^{-23} g/atom

r = 144.8 pm = 144.8 x 10^{-12} m

r = 144.8 x 10^{-12} m x $\dfrac{100\text{ cm}}{1\text{ m}}$ = 1.448 x 10^{-8} cm

Calculate the volume and then the density for Ti assuming it is primitive cubic, body-centered cubic, and face-centered cubic. Compare the calculated density with the actual density to identify the unit cell.

For primitive cubic:

$d = 2r$; volume $= d^3 = [2(1.448 \times 10^{-8} \text{ cm})]^3 = 2.429 \times 10^{-23} \text{ cm}^3$

$$\text{density} = \frac{\text{unit cell mass}}{\text{volume}} = \frac{7.951 \times 10^{-23} \text{ g}}{2.429 \times 10^{-23} \text{ cm}^3} = 3.273 \text{ g/cm}^3$$

For face-centered cubic:

$d = 2\sqrt{2}\,r$; volume $= d^3 = [2\sqrt{2}(1.448 \times 10^{-8} \text{ cm})]^3 = 6.870 \times 10^{-23} \text{ cm}^3$

$$\text{density} = \frac{4(7.951 \times 10^{-23} \text{ g})}{6.870 \times 10^{-23} \text{ cm}^3} = 4.630 \text{ g/cm}^3$$

For body-centered cubic:

From Problem 10.87

$$d = \frac{4r}{\sqrt{3}}; \text{ volume} = d^3 = \left[\frac{4(1.448 \times 10^{-8} \text{ cm})}{\sqrt{3}}\right]^3 = 3.739 \times 10^{-23} \text{ cm}^3$$

$$\text{density} = \frac{2(7.951 \times 10^{-23} \text{ g})}{3.739 \times 10^{-23} \text{ cm}^3} = 4.253 \text{ g/cm}^3$$

The calculated density for a face-centered cube (4.630 g/cm^3) is closest to the actual density of 4.54 g/cm^3. Ti crystallizes in the face-centered cubic unit cell.

10.92   Six Na$^+$ ions touch each H$^-$ ion and six H$^-$ ions touch each Na$^+$ ion.

10.94   Na$^+$ H$^-$ Na$^+$
         $\leftarrow$ 488 pm $\rightarrow$ unit cell edge $= d = 488$ pm; Na–H bond $= d/2 = 244$ pm

Phase Diagrams (Section 10.11)

10.96   (a) gas   (b) liquid   (c) solid

10.98

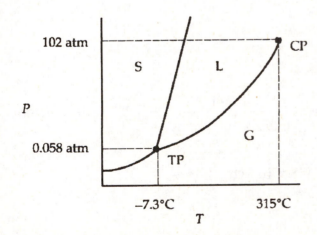

10.100 (a) $Br_2(s)$ (b) $Br_2(l)$

10.102 Solid O_2 does not melt when pressure is applied because the solid is denser than the liquid, and the solid/liquid boundary in the phase diagram slopes to the right.

10.104

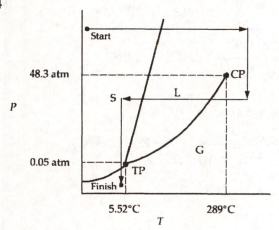

The starting phase is benzene as a solid, and the final phase is benzene as a gas.

10.106 solid → liquid → supercritical fluid → liquid → solid → gas

Chapter Problems

10.108 Because chlorine is larger than fluorine, the charge separation is larger in CH_3Cl compared to CH_3F, resulting in CH_3Cl having a slightly larger dipole moment.

10.110 $7.50 \text{ g} \times \dfrac{1 \text{ mol}}{200.6 \text{ g}} = 0.037\,39 \text{ mol Hg}$

$q_1 = (0.037\,39 \text{ mol})[28.2 \times 10^{-3} \text{ kJ/(K} \cdot \text{mol)}](234.3 \text{ K} - 223.1 \text{ K}) = 0.011\,81 \text{ kJ}$
$q_2 = (0.037\,39 \text{ mol})(2.33 \text{ kJ/mol}) = 0.087\,12 \text{ kJ}$
$q_3 = (0.037\,39 \text{ mol})[27.9 \times 10^{-3} \text{ kJ/(K} \cdot \text{mol)}](323.1 \text{ K} - 234.3 \text{ K}) = 0.092\,63 \text{ kJ}$
$q_{total} = q_1 + q_2 + q_3 = 0.192 \text{ kJ}$;   0.192 kJ of heat is required.

10.112 $\ln P_2 = \ln P_1 + \dfrac{\Delta H_{vap}}{R}\left(\dfrac{1}{T_1} - \dfrac{1}{T_2}\right)$

$\Delta H_{vap} = 40.67 \text{ kJ/mol}$
At 1 atm, H_2O boils at 100 °C; therefore set
$T_1 = 100 \text{ °C} = 373 \text{ K}$, and $P_1 = 1.00 \text{ atm}$.
Let $T_2 = 95 \text{ °C} = 368 \text{ K}$, and solve for P_2.   (P_2 is the atmospheric pressure in Denver.)

$\ln P_2 = \ln(1) + \dfrac{40.67 \text{ kJ/mol}}{[8.3145 \times 10^{-3} \text{ kJ/(K} \cdot \text{mol)}]}\left(\dfrac{1}{373 \text{ K}} - \dfrac{1}{368 \text{ K}}\right)$

$\ln P_2 = -0.1782$; $P_2 = e^{-0.1782} = 0.837 \text{ atm}$

10.114 Unit cell volume = $(7900 \times 10^{-12} \text{ m})^2(3800 \times 10^{-12} \text{ m})\left(\dfrac{100 \text{ cm}}{1 \text{ m}}\right)^3 = 2.37 \times 10^{-19} \text{ cm}^3$

total lysozyme mass = $8 \times 1.44 \times 10^4 \text{ u} \times \dfrac{1.660\,54 \times 10^{-27} \text{ kg}}{1 \text{ u}} \times \dfrac{1000 \text{ g}}{1 \text{ kg}} = 1.91 \times 10^{-19} \text{ g}$

total lysozyme volume = $1.91 \times 10^{-19} \text{ g} \times \dfrac{1 \text{ cm}^3}{1.35 \text{ g}} = 1.42 \times 10^{-19} \text{ cm3}$

% occupied unit cell volume = $\dfrac{1.42 \times 10^{-19} \text{ cm}^3}{2.37 \times 10^{-19} \text{ cm}^3} \times 100\% = 60\%$

10.116 $\Delta G = \Delta H - T\Delta S$; at the melting point (phase change), $\Delta G = 0$.

$\Delta H = T\Delta S$; $T = \dfrac{\Delta H_{fus}}{\Delta S_{fus}} = \dfrac{9.037 \text{ kJ/mol}}{9.79 \times 10^{-3} \text{ kJ/(K} \cdot \text{mol)}} = 923 \text{ K} = 650 \,^{\circ}\text{C}$

10.118 $\Delta H_{vap} = \dfrac{(\ln P_2 - \ln P_1)(R)}{\left(\dfrac{1}{T_1} - \dfrac{1}{T_2}\right)}$

$P_1 = 40.0 \text{ mm Hg}$; $T_1 = -81.6 \,^{\circ}\text{C} = 191.6 \text{ K}$
$P_2 = 400 \text{ mm Hg}$; $T_2 = -43.9 \,^{\circ}\text{C} = 229.2 \text{ K}$

$\Delta H_{vap} = \dfrac{[\ln(400) - \ln(40.0)]\left(8.3145 \times 10^{-3}\dfrac{\text{kJ}}{\text{K} \cdot \text{mol}}\right)}{\left(\dfrac{1}{191.6 \text{ K}} - \dfrac{1}{229.2 \text{ K}}\right)} = 22.36 \text{ kJ/mol}$

Using $\Delta H_{vap} = 22.36 \text{ kJ/mol}$

$\ln P_2 = \ln P_1 + \dfrac{\Delta H_{vap}}{R}\left(\dfrac{1}{T_1} - \dfrac{1}{T_2}\right)$

$(\ln P_2 - \ln P_1)\left(\dfrac{R}{\Delta H_{vap}}\right) = \dfrac{1}{T_1} - \dfrac{1}{T_2}$

$\dfrac{1}{T_1} - (\ln P_2 - \ln P_1)\left(\dfrac{R}{\Delta H_{vap}}\right) = \dfrac{1}{T_2}$

$P_1 = 40.0 \text{ mm Hg}$; $T_1 = 191.6 \text{ K}$
$P_2 = 760 \text{ mm Hg}$
Solve for T_2, the normal boiling point.

$\dfrac{1}{191.6 \text{ K}} - [\ln(760) - \ln(40.0)]\left(\dfrac{8.3145 \times 10^{-3}\dfrac{\text{kJ}}{\text{K} \cdot \text{mol}}}{22.36 \text{ kJ/mol}}\right) = \dfrac{1}{T_2}$

$\dfrac{1}{T_2} = 0.004\,124\,33$; $T_2 = 242.46 \text{ K} = -30.7 \,^{\circ}\text{C}$

10.120 $\Delta H_{vap} = \dfrac{(\ln P_2 - \ln P_1)(R)}{\left(\dfrac{1}{T_1} - \dfrac{1}{T_2}\right)}$

$P_1 = 100$ mm Hg; $\qquad$ $T_1 = -110.3\ °C = 162.85$ K
$P_2 = 760$ mm Hg; $\qquad$ $T_2 = -88.5\ °C = 184.65$ K

$\Delta H_{vap} = \dfrac{[\ln(760) - \ln(100)]\left(8.3145 \times 10^{-3}\dfrac{kJ}{K\cdot mol}\right)}{\left(\dfrac{1}{162.85\ K} - \dfrac{1}{184.65\ K}\right)} = 23.3$ kJ/mol

10.122

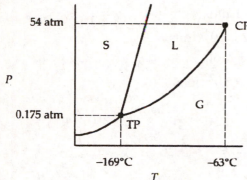

Kr cannot be liquified at room temperature because room temperature is above T_c (–63 °C).

10.124 For a body-centered cube:

$4r = \sqrt{3}$ edge; $\qquad$ edge $= \dfrac{4r}{\sqrt{3}}$

volume of sphere $= \dfrac{4}{3}\pi r^3$

volume of unit cell $= \left(\dfrac{4r}{\sqrt{3}}\right)^3 = \dfrac{64\,r^3}{3\sqrt{3}}$

volume of 2 spheres $= 2\left(\dfrac{4}{3}\pi r^3\right) = \dfrac{8}{3}\pi r^3$

% volume occupied $= \dfrac{\left(\dfrac{8}{3}\pi r^3\right)}{\left(\dfrac{64\,r^3}{3\sqrt{3}}\right)} \times 100\% = 68\%$

10.126 unit cell edge $= d = 287$ pm $= 287 \times 10^{-12}$ m $= 2.87 \times 10^{-8}$ cm
unit cell volume $= d^3 = (2.87 \times 10^{-8}$ cm$)^3 = 2.364 \times 10^{-23}$ cm^3
unit cell mass $= (2.364 \times 10^{-23}$ cm$^3)(7.86$ g/cm$^3) = 1.858 \times 10^{-22}$ g

Fe is body-centered cubic; therefore there are two Fe atoms per unit cell.

$$\text{mass of one Fe atom} = \frac{1.858 \times 10^{-22} \text{ g}}{2 \text{ Fe atoms}} = 9.290 \times 10^{-23} \text{ g/atom}$$

$$\text{Avogadro's number} = 55.85 \text{ g/mol} \times \frac{1 \text{ atom}}{9.290 \times 10^{-23} \text{ g}} = 6.01 \times 10^{23} \text{ atoms/mol}$$

10.128 (a) unit cell edge $= 2r_{Cl^-} + 2r_{Na^+} = 2(181 \text{ pm}) + 2(97 \text{ pm}) = 556 \text{ pm}$

(b) unit cell edge $= d = 556 \text{ pm} = 556 \times 10^{-12} \text{ m} = 5.56 \times 10^{-8} \text{ cm}$
unit cell volume $= (5.56 \times 10^{-8} \text{ cm})^3 = 1.719 \times 10^{-22} \text{ cm}^3$
The unit cell contains 4 Na^+ ions and 4 Cl^- ions.

$$\text{mass of one Na}^+ \text{ ion} = 22.99 \text{ g/mol} \times \frac{1 \text{ mol}}{6.022 \times 10^{23} \text{ ions}} = 3.818 \times 10^{-23} \text{ g/Na}^+$$

$$\text{mass of one Cl}^- \text{ ion} = 35.45 \text{ g/mol} \times \frac{1 \text{ mol}}{6.022 \times 10^{23} \text{ ions}} = 5.887 \times 10^{-23} \text{ g/Cl}^-$$

unit cell mass $= 4(3.818 \times 10^{-23} \text{ g}) + 4(5.887 \times 10^{-23} \text{ g}) = 3.882 \times 10^{-22} \text{ g}$

$$\text{density} = \frac{\text{unit cell mass}}{\text{unit cell volume}} = \frac{3.882 \times 10^{-22} \text{ g}}{1.719 \times 10^{-22} \text{ cm}^3} = 2.26 \text{ g/cm}^3$$

10.130 Al_2O_3, ionic (greater lattice energy than NaCl because of higher ion charges);
F_2, dispersion; H_2O, H–bonding, dipole-dipole; Br_2, dispersion (larger and more polarizable than F_2), ICl, dipole-dipole, NaCl, ionic
rank according to normal boiling points: $F_2 < Br_2 < ICl < H_2O < NaCl < Al_2O_3$

10.132 (a)

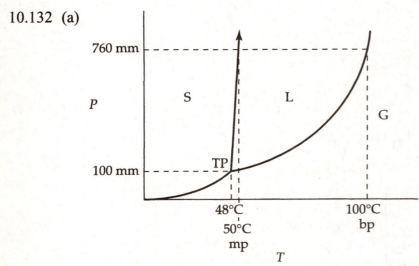

(b) (i) solid (ii) gas (iii) liquid (iv) liquid (v) solid

Multiconcept Problems

10.134 (a) Let the formula of magnetite be Fe_xO_y, then $Fe_xO_y + y\,CO \rightarrow x\,Fe + y\,CO_2$

$$n_{CO_2} = y = \frac{PV}{RT} = \frac{\left(751 \text{ mm Hg} \times \dfrac{1.00 \text{ atm}}{760 \text{ mm Hg}}\right)(1.136 \text{ L})}{\left(0.082\ 06\ \dfrac{L \cdot atm}{K \cdot mol}\right)(298 \text{ K})} = 0.04590 \text{ mol } CO_2$$

0.04590 mol CO_2 = mol of O in Fe_xO_y

mass of O in Fe_xO_y = 0.04590 mol O $\times \dfrac{16.0 \text{ g O}}{1 \text{ mol O}} = 0.7345$ g O

mass of Fe in Fe_xO_y = 2.660 g – 0.7345 g = 1.926 g Fe

(b) mol Fe in magnetite = 1.926 g Fe $\times \dfrac{1 \text{ mol Fe}}{55.85 \text{ g Fe}} = 0.0345$ mol Fe

formula of magnetite: $Fe_{0.0345}O_{0.0459}$ (divide each subscript by the smaller)

$Fe_{0.0345/0.0345}O_{0.0459/0.0345}$

$FeO_{1.33}$ (multiply both subscripts by 3)

$Fe_{(1 \times 3)}O_{(1.33 \times 3)};$ Fe_3O_4

(c) unit cell edge = d = 839 pm = 839×10^{-12} m

d = 839×10^{-12} m $\times \dfrac{100 \text{ cm}}{1 \text{ m}} = 8.39 \times 10^{-8}$ cm

unit cell volume = $d^3 = (8.39 \times 10^{-8} \text{ cm})^3 = 5.91 \times 10^{-22} \text{ cm}^3$

unit cell mass = $(5.91 \times 10^{-22} \text{ cm}^3)(5.20 \text{ g/cm}^3) = 3.07 \times 10^{-21}$ g

mass of Fe in unit cell = $\left(\dfrac{1.926 \text{ g Fe}}{2.660 \text{ g}}\right)(3.07 \times 10^{-21} \text{ g}) = 2.22 \times 10^{-21}$ g Fe

mass of O in unit cell = $\left(\dfrac{0.7345 \text{ g O}}{2.660 \text{ g}}\right)(3.07 \times 10^{-21} \text{ g}) = 8.47 \times 10^{-22}$ g O

Fe atoms in unit cell = 2.22×10^{-21} g $\times \dfrac{6.022 \times 10^{23} \text{ atoms/mol}}{55.847 \text{ g/mol}} = 24$ Fe atoms

O atoms in unit cell = 8.47×10^{-22} g $\times \dfrac{6.022 \times 10^{23} \text{ atoms/mol}}{16.00 \text{ g/mol}} = 32$ O atoms

10.136 (a) M = alkali metal; 500.0 mL = 0.5000 L; 802 °C = 1075 K

$$n_M = \frac{PV}{RT} = \frac{\left(12.5 \text{ mm Hg} \times \dfrac{1.00 \text{ atm}}{760 \text{ mm Hg}}\right)(0.5000 \text{ L})}{\left(0.082\ 06\ \dfrac{L \cdot atm}{K \cdot mol}\right)(1075 \text{ K})} = 9.32 \times 10^{-5} \text{ mol M}$$

1.62 mm = 1.62×10^{-3} m; crystal volume = $(1.62 \times 10^{-3} \text{ m})^3 = 4.25 \times 10^{-9} \text{ m}^3$

M atoms in crystal = $(9.32 \times 10^{-5} \text{ mol})(6.022 \times 10^{23} \text{ atoms/mol}) = 5.61 \times 10^{19}$ M atoms

Because M is body-centered cubic, only 68% (Table 10.10) of the total volume is occupied by M atoms.

$$\text{volume of M atom} = \frac{(0.68)(4.25 \times 10^{-9} \text{ m})}{5.61 \times 10^{19} \text{ M atoms}} = 5.15 \times 10^{-29} \text{ m}^3/\text{M atom}$$

$$\text{volume of a sphere} = \frac{4}{3}\pi r^3$$

$$r_M = \sqrt[3]{\frac{3(\text{volume})}{4\pi}} = \sqrt[3]{\frac{3(5.15 \times 10^{-29} \text{ m}^3)}{4\pi}} = 2.31 \times 10^{-10} \text{ m} = 231 \times 10^{-12} \text{ m} = 231 \text{ pm}$$

(b) The radius of 231 pm is closest to that of K.

(c) 1.62 mm = 0.162 cm

$$\text{density of solid} = \frac{(9.32 \times 10^{-5} \text{ mol})(39.1 \text{ g/mol})}{(0.162 \text{ cm})^3} = 0.857 \text{ g/cm}^3$$

$$\text{density of vapor} = \frac{(9.32 \times 10^{-5} \text{ mol})(39.1 \text{ g/mol})}{500.0 \text{ cm}^3} = 7.29 \times 10^{-6} \text{ g/cm}^3$$

11   Solutions and Their Properties

11.1   Toluene is nonpolar and is insoluble in water.
Br_2 is nonpolar but because of its size, is polarizable and is soluble in water.
KBr is an ionic compound and is very soluble in water.
toluene $<$ Br_2 $<$ KBr   (solubility in H_2O)

11.2   (a) Na^+ has the larger (more negative) hydration energy because the Na^+ ion is smaller than the Cs^+ ion and water molecules can approach more closely and bind more tightly to the Na^+ ion.
(b) Ba^{2+} has the larger (more negative) hydration energy because of its higher charge.

11.3   NaCl, 58.44; 1.00 mol NaCl = 58.44 g
1.00 L H_2O = 1000 mL = 1000 g (assuming a density of 1.00 g/mL)

$$\text{mass \% NaCl} = \frac{58.44 \text{ g}}{1000 \text{ g} + 58.44 \text{ g}} \times 100\% = 5.52 \text{ mass \%}$$

11.4 $$\text{ppm} = \frac{\text{mass of } CO_2}{\text{total mass of solution}} \times 10^6 \text{ ppm}$$

total mass of solution = density x volume = (1.3 g/L)(1.0 L) = 1.3 g

$$35 \text{ ppm} = \frac{\text{mass of } CO_2}{1.3 \text{ g}} \times 10^6 \text{ ppm}$$

$$\text{mass of } CO_2 = \frac{(35 \text{ ppm})(1.3 \text{ g})}{10^6 \text{ ppm}} = 4.6 \times 10^{-5} \text{ g } CO_2$$

11.5   Assume 1.00 L of sea water.
mass of 1.00 L = (1000 mL)(1.025 g/mL) = 1025 g

$$\frac{\text{mass NaCl}}{1025 \text{ g}} \times 100\% = 3.50 \text{ mass \%}; \quad \text{mass NaCl} = \frac{1025 \text{ g} \times 3.50}{100} = 35.88 \text{ g}$$

There are 35.88 g NaCl per 1.00 L of solution.

$$M = \frac{\left(35.88 \text{ g NaCl} \times \dfrac{1 \text{ mol NaCl}}{58.44 \text{ g NaCl}} \right)}{1.00 \text{ L}} = 0.614 \text{ M}$$

11.6 $C_{27}H_{46}O$, 386.7; $CHCl_3$, 119.4; $40.0 \text{ g} \times \dfrac{1 \text{ kg}}{1000 \text{ g}} = 0.0400 \text{ kg}$

$$\text{molality} = \frac{\text{mol } C_{27}H_{46}O}{\text{kg } CHCl_3} = \frac{\left(0.385 \text{ g} \times \dfrac{1 \text{ mol}}{386.7 \text{ g}} \right)}{0.0400 \text{ kg}} = 0.0249 \text{ mol/kg} = 0.0249 \ m$$

$$X_{C_{27}H_{46}O} = \frac{mol\ C_{27}H_{46}O}{mol\ C_{27}H_{46}O\ +\ mol\ CHCl_3}$$

$$X_{C_{27}H_{46}O} = \frac{\left(0.385\ g\ \times\ \dfrac{1\ mol}{386.7\ g}\right)}{\left[\left(0.385\ g\ \times\ \dfrac{1\ mol}{386.7\ g}\right)\ +\ \left(40.0\ g\ \times\ \dfrac{1\ mol}{119.4\ g}\right)\right]} = 2.96\ \times\ 10^{-3}$$

11.7 CH_3CO_2Na, 82.03

$$kg\ H_2O = (0.150\ mol\ CH_3CO_2Na)\left(\frac{1\ kg\ H_2O}{0.500\ mol\ CH_3CO_2Na}\right) = 0.300\ kg\ H_2O$$

$$mass\ CH_3CO_2Na = 0.150\ mol\ CH_3CO_2Na\ \times\ \frac{82.03\ g\ CH_3CO_2Na}{1\ mol\ CH_3CO_2Na} = 12.3\ g\ CH_3CO_2Na$$

mass of solution needed = 300 g + 12.3 g = 312 g

11.8   Assume you have a solution with 1.000 kg (1000 g) of H_2O. If this solution is 0.258 m, then it must also contain 0.258 mol glucose.

$$mass\ of\ glucose = 0.258\ mol\ \times\ \frac{180.2\ g}{1\ mol} = 46.5\ g\ glucose$$

mass of solution = 1000 g + 46.5 g = 1046.5 g
density = 1.0173 g/mL

$$volume\ of\ solution = 1046.5\ g\ \times\ \frac{1\ mL}{1.0173\ g} = 1028.7\ mL$$

$$volume = 1028.7\ mL\ \times\ \frac{1\ L}{1000\ mL} = 1.029\ L; \qquad molarity = \frac{0.258\ mol}{1.029\ L} = 0.251\ M$$

11.9   Assume 1.00 L of solution.
mass of 1.00 L = (1.0042 g/mL)(1000 mL) = 1004.2 g of solution

$$0.500\ mol\ CH_3CO_2H\ \times\ \frac{60.05\ g\ CH_3CO_2H}{1\ mol\ CH_3CO_2H} = 30.02\ g\ CH_3CO_2H$$

$$1004.2\ g - 30.02\ g = 974.2\ g = 0.9742\ kg\ of\ H_2O; \qquad molality = \frac{0.500\ mol}{0.9742\ kg} = 0.513\ m$$

11.10   Assume you have 100.0 g of seawater.
mass NaCl = (0.0350)(100.0 g) = 3.50 g NaCl
mass H_2O = 100.0 g – 3.50 g = 96.5 g H_2O

$$NaCl,\ 58.44; \quad mol\ NaCl = 3.50\ g\ \times\ \frac{1\ mol}{58.44\ g} = 0.0599\ mol\ NaCl$$

$$mass\ H_2O = 96.5\ g\ \times\ \frac{1\ kg}{1000\ g} = 0.0965\ kg\ H_2O; \qquad molality = \frac{0.0599\ mol}{0.0965\ kg} = 0.621\ m$$

11.11 $M = k \cdot P$; $k = \dfrac{M}{P} = \dfrac{3.2 \times 10^{-2} \, M}{1.0 \, atm} = 3.2 \times 10^{-2} \, mol/(L \cdot atm)$

11.12 (a) $M = k \cdot P = [3.2 \times 10^{-2} \, mol/(L \cdot atm)](2.5 \, atm) = 0.080 \, M$

(b) $M = k \cdot P = [3.2 \times 10^{-2} \, mol/(L \cdot atm)](4.0 \times 10^{-4} \, atm) = 1.3 \times 10^{-5} \, M$

11.13 $C_7H_6O_2$, 122.1; C_2H_6O, 46.07

$$X_{solv} = \frac{mol \, C_2H_6O}{mol \, C_2H_6O + mol \, C_7H_6O_2} = \frac{\left(100 \, g \times \dfrac{1 \, mol}{46.07 \, g}\right)}{\left(100 \, g \times \dfrac{1 \, mol}{46.07 \, g}\right) + \left(5.00 \, g \times \dfrac{1 \, mol}{122.1 \, g}\right)} = 0.981$$

$P_{soln} = P_{solv} \cdot X_{solv} = (100.5 \, mm \, Hg)(0.981) = 98.6 \, mm \, Hg$

11.14 $P_{soln} = P_{solv} \cdot X_{solv}$; $\quad X_{solv} = \dfrac{P_{soln}}{P_{solv}} = \dfrac{(55.3 - 1.30) \, mm \, Hg}{55.3 \, mm \, Hg} = 0.976$

NaBr dissociates into two ions in aqueous solution.

$$X_{solv} = \frac{mol \, H_2O}{mol \, H_2O + mol \, Na^+ + mol \, Br^-}$$

$$X_{solv} = 0.976 = \frac{\left(250 \, g \times \dfrac{1 \, mol}{18.02 \, g}\right)}{\left(250 \, g \times \dfrac{1 \, mol}{18.02 \, g}\right) + x \, mol \, Na^+ + x \, mol \, Br^-}$$

$0.976 = \dfrac{13.9 \, mol}{13.9 \, mol + 2x \, mol}$; $\qquad$ solve for x.

$0.976(13.9 \, mol + 2x \, mol) = 13.9 \, mol$

$13.566 \, mol + 1.952 \, x \, mol = 13.9 \, mol$

$1.952 \, x \, mol = 13.9 \, mol - 13.566 \, mol$

$x \, mol = \dfrac{13.9 \, mol - 13.566 \, mol}{1.952} = 0.171 \, mol$

$x = 0.171 \, mol \, Na^+ = 0.171 \, mol \, Br^- = 0.171 \, mol \, NaBr$

NaBr, 102.9; $\quad$ mass NaBr $= 0.171 \, mol \times \dfrac{102.9 \, g}{1 \, mol} = 17.6 \, g \, NaBr$

11.15 At any given temperature, the vapor pressure of a solution is lower than the vapor pressure of the pure solvent. The upper curve represents the vapor pressure of the pure solvent. The lower curve represents the vapor pressure of the solution.

11.16 C_2H_5OH, 46.07; H_2O, 18.02

(a) $25.0 \, g \, C_2H_5OH \times \dfrac{1 \, mol \, C_2H_5OH}{46.07 \, g \, C_2H_5OH} = 0.5426 \, mol \, C_2H_5OH$

$$100.0 \text{ g H}_2\text{O} \times \frac{1 \text{ mol H}_2\text{O}}{18.02 \text{ g H}_2\text{O}} = 5.549 \text{ mol H}_2\text{O}$$

$$X_{C_2H_5OH} = \frac{0.5426 \text{ mol}}{0.5426 \text{ mol} + 5.549 \text{ mol}} = 0.08907 \qquad X_{H_2O} = \frac{5.549 \text{ mol}}{0.5426 \text{ mol} + 5.549 \text{ mol}} = 0.9109$$

$$P_{soln} = X_{C_2H_5OH} P^o_{C_2H_5OH} + X_{H_2O} P^o_{H_2O}$$

$$P_{soln} = (0.08907)(61.2 \text{ mm Hg}) + (0.9109)(23.8 \text{ mm Hg}) = 27.1 \text{ mm Hg}$$

(b) $100 \text{ g C}_2\text{H}_5\text{OH} \times \dfrac{1 \text{ mol C}_2\text{H}_5\text{OH}}{46.07 \text{ g C}_2\text{H}_5\text{OH}} = 2.171 \text{ mol C}_2\text{H}_6\text{O}$

$$25.0 \text{ g H}_2\text{O} \times \frac{1 \text{ mol H}_2\text{O}}{18.02 \text{ g H}_2\text{O}} = 1.387 \text{ mol H}_2\text{O}$$

$$X_{C_2H_5OH} = \frac{2.171 \text{ mol}}{2.171 \text{ mol} + 1.387 \text{ mol}} = 0.6102 \qquad X_{H_2O} = \frac{1.387 \text{ mol}}{2.171 \text{ mol} + 1.387 \text{ mol}} = 0.3898$$

$$P_{soln} = X_{C_2H_5OH} P^o_{C_2H_5OH} + X_{H_2O} P^o_{H_2O}$$

$$P_{soln} = (0.6102)(61.2 \text{ mm Hg}) + (0.3898)(23.8 \text{ mm Hg}) = 46.6 \text{ mm Hg}$$

11.17 (a) Because the vapor pressure of the solution (red curve) is higher than that of the first liquid (green curve), the vapor pressure of the second liquid must be higher than that of the solution (red curve). Because the second liquid has a higher vapor pressure than the first liquid, the second liquid has a lower boiling point.

(b)

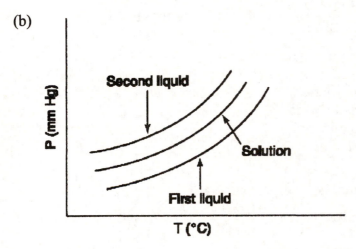

11.18 $C_9H_8O_4$, 180.2; $CHCl_3$ is the solvent. For $CHCl_3$, $K_b = 3.63 \dfrac{^\circ C \cdot kg}{mol}$

$$75.00 \text{ g} \times \frac{1 \text{ kg}}{1000 \text{ g}} = 0.075 \, 00 \text{ kg}$$

$$\Delta T_b = K_b \cdot m = \left(3.63 \frac{^\circ C \cdot kg}{mol} \right) \left(\frac{\left(1.50 \text{ g} \times \dfrac{1 \text{ mol}}{180.2 \text{ g}} \right)}{0.075 \, 00 \text{ kg}} \right) = 0.40 \, ^\circ C$$

Solution boiling point = $61.7 \, ^\circ C + \Delta T_b = 61.7 \, ^\circ C + 0.40 \, ^\circ C = 62.1 \, ^\circ C$

11.19 $MgCl_2$, 95.21

$$110 \text{ g} \times \frac{1 \text{ kg}}{1000 \text{ g}} = 0.110 \text{ kg}$$

$$\Delta T_f = K_f \cdot m \cdot i = \left(1.86 \frac{^\circ C \cdot kg}{mol} \right) \left(\frac{\left(7.40 \text{ g} \times \frac{1 \text{ mol}}{95.21 \text{ g}} \right)}{0.110 \text{ kg}} \right) (2.7) = 3.55 \ ^\circ C$$

Solution freezing point $= 0.00 \ ^\circ C - \Delta T_f = 0.00 \ ^\circ C - 3.55 \ ^\circ C = -3.55 \ ^\circ C$

11.20 $\Delta T_f = K_f \cdot m \cdot i$; For KBr, $i = 2$.

Solution freezing point $= -2.95 \ ^\circ C = 0.00 \ ^\circ C - \Delta T_f$; $\Delta T_f = 2.95 \ ^\circ C$

$$m = \frac{\Delta T_f}{K_f \cdot i} = \frac{2.95 \ ^\circ C}{\left(1.86 \frac{^\circ C \cdot kg}{mol} \right) (2)} = 0.793 \text{ mol/kg} = 0.793 \ m$$

11.21 HCl, 36.46; $\Delta T_f = K_f \cdot m \cdot i$

$$190 \text{ g} \times \frac{1 \text{ kg}}{1000 \text{ g}} = 0.190 \text{ kg}$$

Solution freezing point $= -4.65 \ ^\circ C = 0.00 \ ^\circ C - \Delta T_f$; $\Delta T_f = 4.65 \ ^\circ C$

$$i = \frac{\Delta T_f}{K_f \cdot m} = \frac{4.65 \ ^\circ C}{\left(1.86 \frac{^\circ C \cdot kg}{mol} \right) \left(\frac{9.12 \text{ g} \times \frac{1 \text{ mol}}{36.46 \text{ g}}}{0.190 \text{ kg}} \right)} = 1.9$$

11.22 The red curve represents the vapor pressure of pure chloroform.

(a) The normal boiling point for a liquid is the temperature where the vapor pressure of the liquid equals 1 atm (760 mm Hg). The approximate boiling point of pure chloroform is 62 °C.

(b) The approximate boiling point of the solution is 69 °C.

$\Delta T_b = 69 \ ^\circ C - 62 \ ^\circ C = 7 \ ^\circ C$

$$\Delta T_b = K_b \cdot m; \qquad m = \frac{\Delta T_b}{K_b} = \frac{7 \ ^\circ C}{3.63 \frac{^\circ C \cdot kg}{mol}} = 2 \text{ mol/kg} = 2 \ m$$

11.23 For $CaCl_2$ there are 3 ions (solute particles)/$CaCl_2$

$\Pi = MRT$; For $CaCl_2$, $\Pi = 3MRT$

$$\Pi = (3)(0.125 \text{ mol/L}) \left(0.082 \ 06 \frac{L \cdot atm}{K \cdot mol} \right) (310 \text{ K}) = 9.54 \text{ atm}$$

11.24 $\Pi = MRT$; $M = \dfrac{\Pi}{RT} = \dfrac{(3.85 \text{ atm})}{\left(0.082 \ 06 \frac{L \cdot atm}{K \cdot mol} \right) (300 \text{ K})} = 0.156 \text{ M}$

11.25 $\Delta T_f = K_f \cdot m$; $m = \dfrac{\Delta T_f}{K_f} = \dfrac{2.10\ ^\circ C}{37.7\ \dfrac{^\circ C \cdot kg}{mol}} = 0.0557\ mol/kg = 0.0557\ m$

$35.00\ g \times \dfrac{1\ kg}{1000\ g} = 0.03500\ kg$

$mol = 0.0557\ \dfrac{mol}{kg} \times 0.03500\ kg = 0.001\ 95\ mol\ naphthalene$

$molar\ mass\ of\ naphthalene = \dfrac{0.250\ g\ naphthalene}{0.001\ 95\ mol\ naphthalene} = 128\ g/mol$

11.26 $\Pi = MRT$; $M = \dfrac{\Pi}{RT} = \dfrac{\left(149\ mm\ Hg \times \dfrac{1\ atm}{760\ mm\ Hg}\right)}{\left(0.08206\ \dfrac{L \cdot atm}{K \cdot mol}\right)(298\ K)} = 8.02 \times 10^{-3}\ M$

$300.0\ mL = 0.3000\ L$

$(8.02 \times 10^{-3}\ mol/L)(0.3000\ L) = 0.002\ 406\ mol\ sucrose$

$molar\ mass\ of\ sucrose = \dfrac{0.822\ g\ sucrose}{0.002\ 406\ mol\ sucrose} = 342\ g/mol$

11.27   (a) and (c)

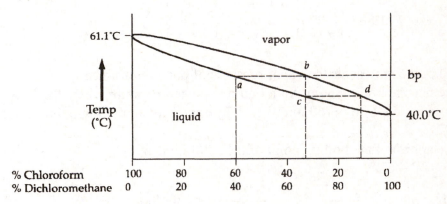

(b) The mixture will begin to boil at ~50 °C.
(d) After two cycles of boiling and condensing, the approximate composition of the liquid would be 90% dichloromethane and 10% chloroform.

11.28   Both solvent molecules and small solute particles can pass through a semipermeable dialysis membrane. Only large colloidal particles such as proteins can't pass through. Only solvent molecules can pass through a semipermeable membrane used for osmosis.

Conceptual Problems

11.30   (a) < (b) < (c)

11.32 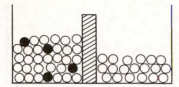 Assume that only the blue (open) spheres (solvent) can pass through the semipermeable membrane. There will be a net transfer of solvent from the right compartment (pure solvent) to the left compartment (solution) to achieve equilibrium.

11.34 The vapor pressure of the NaCl solution is lower than that of pure H_2O. More H_2O molecules will go into the vapor from the pure H_2O than from the NaCl solution. More H_2O vapor molecules will go into the NaCl solution than into pure H_2O. The result is represented by (b).

11.36 (a) Solution B (i = 3) is represented by the red line because its freezing point depressed the most. Solution A (i = 1) is represented by the blue line.
(b) There is a difference of 3.0 °C between the red and blue line. The 3.0 °C equates to 2 i units ($i_{red} - i_{blue} = 3 - 1 = 2$) so 1 i unit is worth 1.5 °C. The blue line (i = 1) is at 14.0 °C, so the freezing point of the pure liquid is 14.0 °C + 1.5 °C = 15.5 °C.
(c) Both the solutions are the same concentration.

$$\Delta T = K_f \cdot m; \quad m = \frac{\Delta T}{K_f} = \frac{1.5\ ^\circ C}{3.0\ ^\circ C/m} = 0.50\ m$$

Section Problems
Solutions and Energy Changes (Sections 11.1 and 11.2)

11.38 The surface area of a solid plays an important role in determining how rapidly a solid dissolves. The larger the surface area, the more solid-solvent interactions, and the more rapidly the solid will dissolve. Powdered NaCl has a much larger surface area than a large block of NaCl, and it will dissolve more rapidly.

11.40 Energy is required to overcome intermolecular forces holding solute particles together in the crystal. For an ionic solid, this is the lattice energy. Substances with higher lattice energies tend to be less soluble than substances with lower lattice energies.

11.42 Ethyl alcohol and water are both polar with small dispersion forces. They both can hydrogen bond, and are miscible.
Pentyl alcohol is slightly polar and can hydrogen bond. It has, however, a relatively large dispersion force because of its size, which limits its water solubility.

11.44 $CaCl_2$, 110.98
For a 1.00 m solution:
heat released = 81,300 J
mass of solution = 1000 g H_2O + 110.98 g $CaCl_2$ = 1110.98 g

$$\Delta T = \frac{q}{(\text{specific heat})(\text{mass of solution})} = \frac{81,300\ J}{[4.18\ J/(K \cdot g)](1110.98\ g)} = 17.5\ K = 17.5\ ^\circ C$$

Final temperature = 25.0 °C + 17.5 °C = 42.5 °C

Units of Concentration (Section 11.3)

11.46 $C_{16}H_{21}NO_2$, 259.3; 50 ng = 50 x 10^{-9} g
(a) mass of solution = (1.025 g/mL)(1000 mL) = 1025 g

$$ppb = \frac{50 \times 10^{-9} \text{ g}}{1025 \text{ g}} \times 10^9 = 0.049 \text{ ppb}$$

(b) 50 x 10^{-9} g x $\dfrac{1 \text{ mol } C_{16}H_{21}NO_2}{259.3 \text{ g } C_{16}H_{21}NO_2} = 1.9 \times 10^{-10}$ mol

$$M = \frac{1.9 \times 10^{-10} \text{ mol}}{1.0 \text{ L}} = 1.9 \times 10^{-10} \text{ mol/L}$$

11.48 (a) Dissolve 0.150 mol of glucose in water; dilute to 1.00 L.
(b) Dissolve 1.135 mol of KBr in 1.00 kg of H_2O.
(c) Mix together 0.15 mol of CH_3OH with 0.85 mol of H_2O.

11.50 $C_7H_6O_2$, 122.12; 165 mL = 0.165 L
mol $C_7H_6O_2$ = (0.0268 mol/L)(0.165 L) = 0.004 42 mol

mass $C_7H_6O_2$ = 0.004 42 mol x $\dfrac{122.12 \text{ g}}{1 \text{ mol}}$ = 0.540 g

Dissolve 4.42 x 10^{-3} mol (0.540 g) of $C_7H_6O_2$ in enough $CHCl_3$ to make 165 mL of solution.

11.52 (a) KCl, 74.6
A 0.500 M KCl solution contains 37.3 g of KCl per 1.00 L of solution.
A 0.500 mass % KCl solution contains 5.00 g of KCl per 995 g of water.
The 0.500 M KCl solution is more concentrated (that is, it contains more solute per amount of solvent).
(b) Both solutions contain the same amount of solute. The 1.75 M solution contains less solvent than the 1.75 m solution. The 1.75 M solution is more concentrated.

11.54 (a) $C_6H_8O_7$, 192.12

0.655 mol $C_6H_8O_7$ x $\dfrac{192.12 \text{ g } C_6H_8O_7}{1 \text{ mol } C_6H_8O_7}$ = 126 g $C_6H_8O_7$

mass % $C_6H_8O_7$ = $\dfrac{126 \text{ g}}{126 \text{ g} + 1000 \text{ g}}$ x 100% = 11.2 mass %

(b) 0.135 mg = 0.135 x 10^{-3} g
(5.00 mL H_2O)(1.00 g/mL) = 5.00 g H_2O

mass % KBr = $\dfrac{0.135 \times 10^{-3} \text{ g}}{(0.135 \times 10^{-3} \text{ g}) + 5.00 \text{ g}}$ x 100% = 0.002 70 mass % KBr

(c) mass % aspirin = $\dfrac{5.50 \text{ g}}{5.50 \text{ g} + 145 \text{ g}}$ x 100% = 3.65 mass % aspirin

11.56 $P_{O_3} = P_{total} \cdot X_{O_3}$

$X_{O_3} = \dfrac{P_{O_3}}{P_{total}} = \dfrac{1.6 \times 10^{-9} \text{ atm}}{1.3 \times 10^{-2} \text{ atm}} = 1.2 \times 10^{-7}$

Assume one mole of air (29 g/mol)

mol $O_3 = n_{air} \cdot X_{O_3} = (1 \text{ mol})(1.2 \times 10^{-7}) = 1.2 \times 10^{-7}$ mol O_3

O_3, 48.00;   mass $O_3 = 1.2 \times 10^{-7}$ mol $\times \dfrac{48.0 \text{ g}}{1 \text{ mol}} = 5.8 \times 10^{-6}$ g O_3

ppm $O_3 = \dfrac{5.8 \times 10^{-6} \text{ g}}{29 \text{ g}} \times 10^6 = 0.20$ ppm

11.58 (a) H_2SO_4, 98.08;   molality $= \dfrac{\left(25.0 \text{ g} \times \dfrac{1 \text{ mol}}{98.08 \text{ g}} \right)}{1.30 \text{ kg}} = 0.196$ mol/kg $= 0.196 \, m$

(b) $C_{10}H_{14}N_2$, 162.23; CH_2Cl_2, 84.93

2.25 g $C_{10}H_{14}N_2 \times \dfrac{1 \text{ mol } C_{10}H_{14}N_2}{162.23 \text{ g } C_{10}H_{14}N_2} = 0.0139$ mol $C_{10}H_{14}N_2$

80.0 g $CH_2Cl_2 \times \dfrac{1 \text{ mol } CH_2Cl_2}{84.93 \text{ g } CH_2Cl_2} = 0.942$ mol CH_2Cl_2

$X_{C_{10}H_{14}N_2} = \dfrac{0.0139 \text{ mol}}{0.942 \text{ mol} + 0.0139 \text{ mol}} = 0.0145$; $X_{CH_2Cl_2} = \dfrac{0.942 \text{ mol}}{0.942 \text{ mol} + 0.0139 \text{ mol}} = 0.985$

11.60 16.0 mass % $= \dfrac{16.0 \text{ g } H_2SO_4}{16.0 \text{ g } H_2SO_4 + 84.0 \text{ g } H_2O}$

H_2SO_4, 98.08;   density $= 1.1094$ g/mL

volume of solution $= 100.0$ g $\times \dfrac{1 \text{ mL}}{1.1094 \text{ g}} = 90.14$ mL $= 0.090 \, 14$ L

molarity $= \dfrac{\left(16.0 \text{ g} \times \dfrac{1 \text{ mol}}{98.08 \text{ g}} \right)}{0.090 \, 14 \text{ L}} = 1.81$ M

11.62 molality $= \dfrac{\left(40.0 \text{ g} \times \dfrac{1 \text{ mol}}{62.07 \text{ g}} \right)}{0.0600 \text{ kg}} = 10.7$ mol/kg $= 10.7 \, m$

11.64 $C_{19}H_{21}NO_3$, 311.38; 1.5 mg $= 1.5 \times 10^{-3}$ g

1.3×10^{-3} mol/kg $= \dfrac{\left(1.5 \times 10^{-3} \text{ g} \times \dfrac{1 \text{ mol}}{311.38 \text{ g}} \right)}{\text{kg of solvent}}$;   solve for kg of solvent.

$$\text{kg of solvent} = \frac{\left(1.5 \times 10^{-3} \text{ g} \times \dfrac{1 \text{ mol}}{311.38 \text{ g}}\right)}{1.3 \times 10^{-3} \text{ mol/kg}} = 0.0037 \text{ kg}$$

Because the solution is very dilute, kg of solvent ≈ kg of solution.

$$\text{g of solution} = (0.0037 \text{ kg})\left(\frac{1000 \text{ g}}{1 \text{ kg}}\right) = 3.7 \text{ g}$$

11.66 $C_6H_{12}O_6$, 180.16; H_2O, 18.02;   Assume 1.00 L of solution.

mass of solution = (1000 mL)(1.0624 g/mL) = 1062.4 g

$$\text{mass of solute} = 0.944 \text{ mol} \times \frac{180.16 \text{ g}}{1 \text{ mol}} = 170.1 \text{ g } C_6H_{12}O_6$$

mass of H_2O = 1062.4 g – 170.1 g = 892.3 g H_2O

$$\text{mol } C_6H_{12}O_6 = 0.944 \text{ mol}; \quad \text{mol } H_2O = 892.3 \text{ g} \times \frac{1 \text{ mol}}{18.02 \text{ g}} = 49.5 \text{ mol}$$

(a) $X_{C_6H_{12}O_6} = \dfrac{\text{mol } C_6H_{12}O_6}{\text{mol } C_6H_{12}O_6 + \text{mol } H_2O} = \dfrac{0.944 \text{ mol}}{0.944 \text{ mol} + 49.5 \text{ mol}} = 0.0187$

(b)   mass % = $\dfrac{\text{mass } C_6H_{12}O_6}{\text{total mass of solution}} \times 100\% = \dfrac{170.1 \text{ g}}{1062.4 \text{ g}} \times 100\% = 16.0\%$

(c)   molality = $\dfrac{\text{mol } C_6H_{12}O_6}{\text{kg } H_2O} = \dfrac{0.944 \text{ mol}}{0.8923 \text{ kg}} = 1.06 \text{ mol/kg} = 1.06 \ m$

Solubility and Henry's Law (Section 11.4)

11.68   From Figure 11.6, first determine the solubility, in g/100 mL, for each compound.

(a) $CuSO_4$, 159.6, ~42 g/100 mL; NH_4Cl, 53.5, ~56 g/100 mL

$$M = \frac{\left(42 \text{ g} \times \dfrac{1 \text{ mol } CuSO_4}{159.6 \text{ g}}\right)}{0.100 \text{ L}} = 2.6 \text{ mol/L}$$

$$M = \frac{\left(56 \text{ g} \times \dfrac{1 \text{ mol } NH_4Cl}{53.5 \text{ g}}\right)}{0.100 \text{ L}} = 10.5 \text{ mol/L}$$

NH_4Cl has the higher molar solubility.

(b) CH_3CO_2Na, 82.0, ~48 g/100 mL;   glucose ($C_6H_{12}O_6$), 180.2, ~90 g/100 mL

$$M = \frac{\left(48 \text{ g} \times \dfrac{1 \text{ mol } CH_3CO_2Na}{82.0 \text{ g}}\right)}{0.100 \text{ L}} = 5.9 \text{ mol/L}$$

$$M = \frac{\left(90 \text{ g} \times \dfrac{1 \text{ mol glucose}}{180.2 \text{ g}}\right)}{0.100 \text{ L}} = 5.0 \text{ mol/L}$$

CH_3CO_2Na has the higher molar solubility.

11.70 $M = k \cdot P = (0.091 \ \dfrac{mol}{L \cdot atm})(0.75 \ atm) = 0.068 \ M$

11.72 $M = k \cdot P$

$k = \dfrac{M}{P} = \dfrac{2.21 \ x \ 10^{-3} \ mol/L}{1.00 \ atm} = 2.21 \ x \ 10^{-3} \ \dfrac{mol}{L \cdot atm}$

Convert 4 mg/L to mol/L:

$4 \ mg = 4 \ x \ 10^{-3} \ g$

$O_2 \ molarity = \dfrac{\left(4 \ x \ 10^{-3} \ g \ x \ \dfrac{1 \ mol}{32.00 \ g}\right)}{1.00 \ L} = 1.25 \ x \ 10^{-4} \ M$

$P_{O_2} = \dfrac{M}{k} = \dfrac{1.25 \ x \ 10^{-4} \ \dfrac{mol}{L}}{2.21 \ x \ 10^{-3} \ \dfrac{mol}{L \cdot atm}} = 0.06 \ atm$

11.74 $k = 2.4 \ x \ 10^{-4} \ mol/(L \cdot atm)$

$M = k \cdot P = [2.4 \ x \ 10^{-4} \ mol/(L \cdot atm)](2.00 \ atm) = 4.8 \ x \ 10^{-4} \ mol/L$

Colligative Properties (Sections 11.5–11.8)

11.76 NaCl is a nonvolatile solute. Methyl alcohol is a volatile solute. When NaCl is added to water, the vapor pressure of the solution is decreased, which means that the boiling point of the solution will increase. When methyl alcohol is added to water, the vapor pressure of the solution is increased which means that the boiling point of the solution will decrease.

11.78

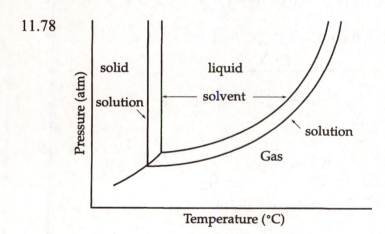

11.80 (a) CH_4N_2O, 60.06; H_2O, 18.02

$10.0 \ g \ CH_4N_2O \ x \ \dfrac{1 \ mol \ CH_4N_2O}{60.06 \ g \ CH_4N_2O} = 0.167 \ mol \ CH_4N_2O$

$$150.0 \text{ g H}_2\text{O} \times \frac{1 \text{ mol H}_2\text{O}}{18.02 \text{ g H}_2\text{O}} = 8.32 \text{ mol H}_2\text{O}$$

$$X_{\text{H}_2\text{O}} = \frac{8.32 \text{ mol}}{8.32 \text{ mol} + 0.167 \text{ mol}} = 0.980$$

$$P_{\text{soln}} = P^{\text{o}}_{\text{H}_2\text{O}} \cdot X_{\text{H}_2\text{O}} = (71.93 \text{ mm Hg})(0.980) = 70.5 \text{ mm Hg}$$

(b) LiCl, 42.39; $10.0 \text{ g LiCl} \times \dfrac{1 \text{ mol LiCl}}{42.39 \text{ g LiCl}} = 0.236 \text{ mol LiCl}$

LiCl dissociates into $\text{Li}^+(aq)$ and $\text{Cl}^-(aq)$ in H_2O.
mol Li^+ = mol Cl^- = mol LiCl = 0.236 mol

$$150.0 \text{ g H}_2\text{O} \times \frac{1 \text{ mol H}_2\text{O}}{18.02 \text{ g H}_2\text{O}} = 8.32 \text{ mol H}_2\text{O}$$

$$X_{\text{H}_2\text{O}} = \frac{8.32 \text{ mol}}{8.32 \text{ mol} + 0.236 \text{ mol} + 0.236 \text{ mol}} = 0.946$$

$$P_{\text{soln}} = P^{\text{o}}_{\text{H}_2\text{O}} \cdot X_{\text{H}_2\text{O}} = (71.93 \text{ mm Hg})(0.946) = 68.0 \text{ mm Hg}$$

11.82   For H_2O, $K_b = 0.51 \dfrac{^\circ\text{C} \cdot \text{kg}}{\text{mol}}$; 150.0 g = 0.1500 kg

(a) $\Delta T_b = K_b \cdot m = \left(0.51 \dfrac{^\circ\text{C} \cdot \text{kg}}{\text{mol}} \right)\left(\dfrac{0.167 \text{ mol}}{0.1500 \text{ kg}} \right) = 0.57\ ^\circ\text{C}$

Solution boiling point = 100.00 °C + ΔT_b = 100.00 °C + 0.57 °C = 100.57 °C

(b) $\Delta T_b = K_b \cdot m = \left(0.51 \dfrac{^\circ\text{C} \cdot \text{kg}}{\text{mol}} \right)\left(\dfrac{2(0.236 \text{ mol})}{0.1500 \text{ kg}} \right) = 1.6\ ^\circ\text{C}$

Solution boiling point = 100.00 °C + ΔT_b = 100.00 °C + 1.6 °C = 101.6 °C

11.84   Solution freezing point = -4.3 °C = 0.00 °C $- \Delta T_f$; $\Delta T_f = 4.3$ °C

$$\Delta T_f = K_f \cdot m \cdot i; \qquad i = \frac{\Delta T_f}{K_f \cdot m} = \frac{4.3\ ^\circ\text{C}}{\left(1.86 \dfrac{^\circ\text{C} \cdot \text{kg}}{\text{mol}} \right)(1.0 \text{ mol/kg})} = 2.3$$

11.86   Let $X_{\text{heptane}} = x$ and $X_{\text{octane}} = 1 - x$
(428 mm Hg)x + (175 mm Hg)(1 − x) = 305 mm Hg
(428 mm Hg)x + 175 mm Hg − (175 mm Hg)x = 305 mm Hg
(428 mm Hg − 175 mm Hg)x = 305 mm Hg − 175 mm Hg
(253 mm Hg)x = 130 mm Hg

$$x = X_{\text{heptane}} = \frac{130 \text{ mm Hg}}{253 \text{ mm Hg}} = 0.514$$

11.88   Acetone, $\text{C}_3\text{H}_6\text{O}$, 58.08, $P^{\text{o}}_{\text{C}_3\text{H}_6\text{O}} = 285 \text{ mm Hg}$

Ethyl acetate, $\text{C}_4\text{H}_8\text{O}_2$, 88.11, $P^{\text{o}}_{\text{C}_4\text{H}_8\text{O}_2} = 118 \text{ mm Hg}$

$$25.0 \text{ g } C_3H_6O \text{ x } \frac{1 \text{ mol } C_3H_6O}{58.08 \text{ g } C_3H_6O} = 0.430 \text{ mol } C_3H_6O$$

$$25.0 \text{ g } C_4H_8O_2 \text{ x } \frac{1 \text{ mol } C_4H_8O_2}{88.11 \text{ g } C_4H_8O_2} = 0.284 \text{ mol } C_4H_8O_2$$

$$X_{C_3H_6O} = \frac{0.430 \text{ mol}}{0.430 \text{ mol} + 0.284 \text{ mol}} = 0.602; \quad X_{C_4H_8O_2} = \frac{0.284 \text{ mol}}{0.430 \text{ mol} + 0.284 \text{ mol}} = 0.398$$

$$P_{soln} = P^o_{C_3H_6O} \cdot X_{C_3H_6O} + P^o_{C_4H_8O_2} \cdot X_{C_4H_8O_2}$$

$$P_{soln} = (285 \text{ mm Hg})(0.602) + (118 \text{ mm Hg})(0.398) = 219 \text{ mm Hg}$$

11.90 In the liquid, $X_{acetone} = 0.602$ and $X_{ethyl \, acetate} = 0.398$
In the vapor, $P_{Total} = 219$ mm Hg

$$P_{acetone} = P^o_{acetone} \cdot X_{acetone} = (285 \text{ mm Hg})(0.602) = 172 \text{ mm Hg}$$

$$P_{ethyl \, acetate} = P^o_{ethyl \, acetate} \cdot X_{ethyl \, acetate} = (118 \text{ mm Hg})(0.398) = 47 \text{ mm Hg}$$

$$X_{acetone} = \frac{P_{acetone}}{P_{total}} = \frac{172 \text{ mm Hg}}{219 \text{ mm Hg}} = 0.785; \quad X_{ethyl \, acetate} = \frac{P_{ethyl \, acetate}}{P_{total}} = \frac{47 \text{ mm Hg}}{219 \text{ mm Hg}} = 0.215$$

11.92 $C_9H_8O_4$, 180.16; 215 g = 0.215 kg

$$\Delta T_b = K_b \cdot m = 0.47 \, ^\circ C; \qquad K_b = \frac{\Delta T_b}{m} = \frac{0.47 \, ^\circ C}{\left(\dfrac{5.00 \text{ g x } \dfrac{1 \text{ mol}}{180.16 \text{ g}}}{0.215 \text{ kg}} \right)} = 3.6 \, \frac{^\circ C \cdot kg}{mol}$$

11.94 $\Delta T_b = K_b \cdot m = 1.76 \, ^\circ C; \qquad m = \dfrac{\Delta T_b}{K_b} = \dfrac{1.76 \, ^\circ C}{3.07 \, \dfrac{^\circ C \cdot kg}{mol}} = 0.573 \, m$

11.96 $\Pi = MRT$
(a) NaCl 58.44; 350.0 mL = 0.3500 L
There are 2 moles of ions/mole of NaCl

$$\Pi = (2) \left(\frac{5.00 \text{ g x } \dfrac{1 \text{ mol}}{58.44 \text{ g}}}{0.3500 \text{ L}} \right) \left(0.082 \, 06 \, \frac{L \cdot atm}{K \cdot mol} \right) (323 \text{ K}) = 13.0 \text{ atm}$$

(b) CH_3CO_2Na, 82.03; 55.0 mL = 0.0550 L
There are 2 moles of ions/mole of CH_3CO_2Na

$$\Pi = (2) \left(\frac{6.33 \text{ g x } \dfrac{1 \text{ mol}}{82.03 \text{ g}}}{0.0550 \text{ L}} \right) \left(0.082 \, 06 \, \frac{L \cdot atm}{K \cdot mol} \right) (283 \text{K}) = 65.2 \text{ atm}$$

11.98 $\Pi = MRT;$ $M = \dfrac{\Pi}{RT} = \dfrac{4.85 \text{ atm}}{\left(0.082\ 06\ \dfrac{L \cdot atm}{K \cdot mol}\right)(300 \text{ K})} = 0.197 \text{ M}$

Uses of Colligative Properties (Section 11.9)

11.100 K_f for snow (H_2O) is $1.86\ \dfrac{^\circ C \cdot kg}{mol}$.

Reasonable amounts of salt are capable of lowering the freezing point (ΔT_f) of the snow below an air temperature of −2 °C. Reasonable amounts of salt, however, are not capable of causing a ΔT_f of more than 30 °C which would be required if it is to melt snow when the air temperature is −30 °C.

11.102 $\Pi = 407.2 \text{ mm Hg} \times \dfrac{1 \text{ atm}}{760 \text{ mm Hg}} = 0.5358 \text{ atm}$

$\Pi = MRT;\ M = \dfrac{\Pi}{RT} = \dfrac{0.5358 \text{ atm}}{\left(0.082\ 06\ \dfrac{L \cdot atm}{K \cdot mol}\right)(298.15 \text{ K})} = 0.021\ 90 \text{ M}$

$200.0 \text{ mL} \times \dfrac{1 \text{ L}}{1000 \text{ mL}} = 0.2000 \text{ L}$

mol cellobiose = (0.2000 L)(0.021 90 mol/L) = 4.380×10^{-3} mol

molar mass of cellobiose = $\dfrac{1.500 \text{ g cellobiose}}{4.380 \times 10^{-3} \text{ mol cellobiose}} = 342.5 \text{ g/mol}$

molecular mass = 342.5

11.104   HCl is a strong electrolyte in H_2O and completely dissociates into two solute particles per each HCl.
HF is a weak electrolyte in H_2O. Only a few percent of the HF molecules dissociates into ions.

11.106   First, determine the empirical formula:
Assume 100.0 g of β-carotene.

10.51% H $10.51 \text{ g H} \times \dfrac{1 \text{ mol H}}{1.008 \text{ g H}} = 10.43 \text{ mol H}$

89.49% C $89.49 \text{ g C} \times \dfrac{1 \text{ mol C}}{12.01 \text{ g C}} = 7.45 \text{ mol C}$

$C_{7.45}H_{10.43}$; divide each subscript by the smaller, 7.45.
$C_{7.45 / 7.45}H_{10.43 / 7.45}$
$CH_{1.4}$
Multiply each subscript by 5 to obtain integers.
Empirical formula is C_5H_7, 67.1.

Second, calculate the molecular mass:

$$\Delta T_f = K_f \cdot m; \quad m = \frac{\Delta T_f}{K_f} = \frac{1.17\ ^\circ C}{37.7\ \dfrac{^\circ C \cdot kg}{mol}} = 0.0310\ mol/kg = 0.0310\ m$$

$$1.50\ g \times \frac{1\ kg}{1000\ g} = 1.50 \times 10^{-3}\ kg$$

mol β-carotene = $(1.50 \times 10^{-3}\ kg)(0.0310\ mol/kg) = 4.65 \times 10^{-5}\ mol$

molar mass of β-carotene = $\dfrac{0.0250\ g\ \beta\text{-carotene}}{4.65 \times 10^{-5}\ mol\ \beta\text{-carotene}} = 538\ g/mol$

molecular mass = 538

Finally, determine the molecular formula:
Divide the molecular mass by the empirical formula mass.

$$\frac{538\ amu}{67.1\ amu} = 8$$

molecular formula is $C_{(8 \times 5)}H_{(8 \times 7)}$, or $C_{40}H_{56}$

Chapter Problems

11.108 $C_2H_6O_2$, 62.07; $\Delta T_f = 22.0\ ^\circ C$

$$\Delta T_f = K_f \cdot m; \quad m = \frac{\Delta T_f}{K_f} = \frac{22.0\ ^\circ C}{1.86\ \dfrac{^\circ C \cdot kg}{mol}} = 11.8\ mol/kg = 11.8\ m$$

mol $C_2H_6O_2 = (3.55\ kg)(11.8\ mol/kg) = 41.9\ mol\ C_2H_6O_2$

mass $C_2H_6O_2 = 41.9\ mol\ C_2H_6O_2 \times \dfrac{62.07\ g\ C_2H_6O_2}{1\ mol\ C_2H_6O_2} = 2.60 \times 10^3\ g\ C_2H_6O_2$

11.110   When solid $CaCl_2$ is added to liquid water, the temperature rises because ΔH_{soln} for $CaCl_2$ is exothermic.
When solid $CaCl_2$ is added to ice at 0 °C, some of the ice will melt (an endothermic process) and the temperature will fall because the $CaCl_2$ lowers the freezing point of an ice/water mixture.

11.112   Sucrose ($C_{12}H_{22}O_{11}$), 342.3; fructose ($C_6H_{12}O_6$), 180.2

$$\Pi = MRT; \quad M = \frac{\Pi}{RT} = \frac{0.1843\ atm}{\left(0.082\ 06\ \dfrac{L \cdot atm}{K \cdot mol}\right)(298.0\ K)} = 0.007\ 537\ M$$

$n_{total} = (0.007\ 537\ mol/L)(1.50\ L) = 0.0113\ mol$
Let x = g of sucrose and 2.850 g − x = g of fructose
0.0113 mol = mol sucrose + mol fructose

$$0.0113\ mol = (x)\left(\frac{1\ mol\ sucrose}{342.3\ g}\right) + (2.850\ g - x)\left(\frac{1\ mol\ fructose}{180.2\ g}\right)$$

Solve for x.

$$0.0113 \text{ mol} = (x)\left(\frac{1 \text{ mol}}{342.3 \text{ g}}\right) + (2.850 \text{ g} - x)\left(\frac{1 \text{ mol}}{180.2 \text{ g}}\right)$$

$$0.0113 \text{ mol} = (x)\left(\frac{1 \text{ mol}}{342.3 \text{ g}}\right) + 0.0158 \text{ mol} - (x)\left(\frac{1 \text{ mol}}{180.2 \text{ g}}\right)$$

$$0.0113 \text{ mol} - 0.0158 \text{ mol} = (x)\left(\frac{1 \text{ mol}}{342.3 \text{ g}}\right) - (x)\left(\frac{1 \text{ mol}}{180.2 \text{ g}}\right)$$

$$-0.0045 \text{ mol} = (x)\left[\left(\frac{1 \text{ mol}}{342.3 \text{ g}}\right) - \left(\frac{1 \text{ mol}}{180.2 \text{ g}}\right)\right]$$

$$x = \frac{-0.0045 \text{ mol}}{\left[\left(\dfrac{1 \text{ mol}}{342.3 \text{ g}}\right) - \left(\dfrac{1 \text{ mol}}{180.2 \text{ g}}\right)\right]} = 1.71 \text{ g sucrose}$$

$$1.71 \text{ g sucrose} \times \frac{1 \text{ mol sucrose}}{342.3 \text{ g sucrose}} = 0.005\ 00 \text{ mol sucrose}$$

$$X_{sucrose} = \frac{n_{sucrose}}{n_{total}} = \frac{0.005\ 00 \text{ mol}}{0.0113 \text{ mol}} = 0.442$$

11.114 $C_{10}H_8$, 128.17; $\Delta T_f = 0.35 \text{ °C}$

$$\Delta T_f = K_f \cdot m; \qquad m = \frac{\Delta T_f}{K_f} = \frac{0.35 \text{ °C}}{5.12 \dfrac{\text{°C} \cdot \text{kg}}{\text{mol}}} = 0.0684 \text{ mol/kg} = 0.0684 \ m$$

$$150.0 \text{ g} \times \frac{1 \text{kg}}{1000 \text{ g}} = 0.1500 \text{ kg}$$

$$\text{mol } C_{10}H_8 = (0.1500 \text{ kg})(0.0684 \text{ mol/kg}) = 0.0103 \text{ mol } C_{10}H_8$$

$$\text{mass } C_{10}H_8 = 0.0103 \text{ mol } C_{10}H_8 \times \frac{128.17 \text{ g } C_{10}H_8}{1 \text{ mol } C_{10}H_8} = 1.3 \text{ g } C_{10}H_8$$

11.116 NaCl, 58.44; there are 2 ions/NaCl

A 3.5 mass % aqueous solution of NaCl contains 3.5 g NaCl and 96.5 g H_2O.

$$\text{molality} = \frac{\left(3.5 \text{ g} \times \dfrac{1 \text{ mol}}{58.44 \text{ g}}\right)}{0.0965 \text{ kg}} = 0.62 \text{ mol/kg} = 0.62 \ m$$

$$\Delta T_f = K_f \cdot 2 \cdot m = \left(1.86 \frac{\text{°C} \cdot \text{kg}}{\text{mol}}\right)(2)(0.62 \text{ mol/kg}) = 2.3 \text{ °C}$$

Solution freezing point = 0.0 °C $- \Delta T_f$ = 0.0 °C $-$ 2.3 °C = $-$2.3 °C

$$\Delta T_b = K_b \cdot 2 \cdot m = \left(0.51 \frac{\text{°C} \cdot \text{kg}}{\text{mol}}\right)(2)(0.62 \text{ mol/kg}) = 0.63 \text{ °C}$$

Solution boiling point = 100.00 °C $+ \Delta T_b$ = 100.00 °C + 0.63 °C = 100.63 °C

11.118 (a) 90 mass % isopropyl alcohol = $\dfrac{10.5 \text{ g}}{10.5 \text{ g} + \text{ mass of H}_2\text{O}}$ x 100%

Solve for the mass of H_2O.

mass of H_2O = $\left(10.5 \text{ g x } \dfrac{100}{90} \right)$ − 10.5 g = 1.2 g

mass of solution = 10.5 g + 1.2 g = 11.7 g

11.7 g of rubbing alcohol contains 10.5 g of isopropyl alcohol.

(b) C_3H_8O, 60.10

mass C_3H_8O = (0.90)(50.0 g) = 45 g

45 g C_3H_8O x $\dfrac{1 \text{ mol } C_3H_8O}{60.10 \text{ g } C_3H_8O}$ = 0.75 mol C_3H_8O

11.120 $\Delta T_f = K_f \cdot i \cdot m$; $CaCl_2$, 111.0

$m = \dfrac{\Delta T_f}{K_f \cdot i} = \dfrac{1.14 \text{ °C}}{\left(1.86 \dfrac{\text{°C} \cdot \text{kg}}{\text{mol}} \right)(2.71)}$ = 0.226 mol/kg

0.226 mol $CaCl_2$ x $\dfrac{111.0 \text{ g } CaCl_2}{1 \text{ mol } CaCl_2}$ = 25.1 g $CaCl_2$

mass % $CaCl_2$ = $\dfrac{\text{mass } CaCl_2}{\text{mass } CaCl_2 + \text{ mass } H_2O}$ x 100%

$= \dfrac{25.1 \text{ g}}{25.1 \text{ g} + 1000.0 \text{ g}}$ x 100% = 2.45%

11.122 H_2O, 18.02

A 0.62 m LiCl solution contains 0.62 mol of LiCl and 1.00 kg (= 1000 g) of H_2O.

(0.62 mol LiCl)(1.96) = 1.21 mol of solute particles

1000 g H_2O x $\dfrac{1 \text{ mol } H_2O}{18.02 \text{ g } H_2O}$ = 55.5 mol H_2O

$P_{soln} = P^o_{H_2O} \cdot X_{H_2O}$ = 23.76 mm Hg x $\dfrac{55.5 \text{ mol } H_2O}{55.5 \text{ mol } H_2O + 1.21 \text{ mol solute}}$ = 23.25 mm Hg

Vapor pressure depression = 23.76 mm Hg − 23.25 mm Hg = 0.51 mm Hg

11.124 First, determine the empirical formula.

3.47 mg = 3.47 x 10^{-3} g sample

10.10 mg = 10.10 x 10^{-3} g CO_2

2.76 mg = 2.76 x 10^{-3} g H_2O

mass C = 10.10 x 10^{-3} g CO_2 x $\dfrac{12.01 \text{ g C}}{44.01 \text{ g } CO_2}$ = 2.76 x 10^{-3} g C

mass H = 2.76 x 10^{-3} g H_2O x $\dfrac{2 \text{ x } 1.008 \text{ g H}}{18.02 \text{ g } H_2O}$ = 3.09 x 10^{-4} g H mass O = 3.47×10^{-3} g $-$ 2.76×10^{-3} g C $-$ 3.09×10^{-4} g H = 4.01×10^{-4} g O

$$2.76 \times 10^{-3} \text{ g C} \times \frac{1 \text{ mol C}}{12.01 \text{ g C}} = 2.30 \times 10^{-4} \text{ mol C}$$

$$3.09 \times 10^{-4} \text{ g H} \times \frac{1 \text{ mol H}}{1.008 \text{ g H}} = 3.07 \times 10^{-4} \text{ mol H}$$

$$4.01 \times 10^{-4} \text{ g O} \times \frac{1 \text{ mol O}}{16.00 \text{ g O}} = 2.51 \times 10^{-5} \text{ mol O} = 0.251 \times 10^{-4} \text{ mol O}$$

To simplify the empirical formula, divide each mol quantity by 10^{-4}.
$C_{2.30}H_{3.07}O_{0.251}$; divide all subscripts by the smallest, 0.251.
$C_{2.30/0.251}H_{3.07/0.251}O_{0.251/0.251}$
$C_{9.16}H_{12.23}O$
Empirical formula is $C_9H_{12}O$, 136.

Second, determine the molecular mass.

7.55 mg = 7.55×10^{-3} g estradiol; 0.500 g $\times \dfrac{1 \text{ kg}}{1000 \text{ g}} = 5.00 \times 10^{-4}$ kg camphor

$$\Delta T_f = K_f \cdot m; \quad m = \frac{\Delta T_f}{K_f} = \frac{2.10 \,^{\circ}\text{C}}{37.7 \,\dfrac{^{\circ}\text{C} \cdot \text{kg}}{\text{mol}}} = 0.0557 \text{ mol/kg} = 0.0557 \, m$$

$$m = \frac{\text{mol estradiol}}{\text{kg solvent}}$$

mol estradiol = m x (kg solvent) = (0.0557 mol/kg)(5.00×10^{-4} kg) = 2.79×10^{-5} mol

$$\text{molar mass} = \frac{7.55 \times 10^{-3} \text{ g estradiol}}{2.79 \times 10^{-5} \text{ mol estradiol}} = 271 \text{ g/mol}; \quad \text{molecular mass} = 271$$

Finally, determine the molecular formula:
Divide the molecular mass by the empirical formula mass.
$$\frac{271}{136} = 2$$
molecular formula is $C_{(2 \times 9)}H_{(2 \times 12)}O_{(2 \times 1)}$, or $C_{18}H_{24}O_2$

11.126 (a) H_2SO_4, 98.08; 2.238 mol H_2SO_4 x $\dfrac{98.08 \text{ g } H_2SO_4}{1 \text{ mol } H_2SO_4} = 219.50$ g H_2SO_4

mass of 2.238 m solution = 219.50 g H_2SO_4 + 1000 g H_2O = 1219.50 g

volume of 2.238 m solution = 1219.50 g x $\dfrac{1.0000 \text{ mL}}{1.1243 \text{ g}} = 1084.68$ mL = 1.0847 L

molarity of 2.238 m solution = $\dfrac{2.238 \text{ mol}}{1.0847 \text{ L}} = 2.063$ M

The molarity of the H_2SO_4 solution is less than the molarity of the $BaCl_2$ solution. Because equal volumes of the two solutions are mixed, H_2SO_4 is the limiting reactant and the number of moles of H_2SO_4 determines the number of moles of $BaSO_4$ produced as the white precipitate.

$$(0.05000 \text{ L}) \times (2.063 \text{ mol } H_2SO_4/L) \times \frac{1 \text{ mol } BaSO_4}{1 \text{ mol } H_2SO_4} \times \frac{233.39 \text{ g } BaSO_4}{1 \text{ mol } BaSO_4} = 24.07 \text{ g } BaSO_4$$

(b) More precipitate will form because of the excess $BaCl_2$ in the solution.

11.128 Let $x = X_{H_2O}$ and $y = X_{CH_3OH}$ and assume $n_{total} = 1.00$ mol

$(14.5 \text{ mm Hg})x + (82.5 \text{ mm Hg})y = 39.4 \text{ mm Hg}$

$(26.8 \text{ mm Hg})x + (140.3 \text{ mm Hg})y = 68.2 \text{ mm Hg}$

$$x = \frac{68.2 - 140.3y}{26.8}$$

$$\frac{14.5(68.2 - 140.3y)}{26.8} + 82.5y = 39.4$$

$$\frac{(988.9 - 2034.35y)}{26.8} + 82.5y = 39.4$$

$36.90 - 75.91y + 82.5y = 39.4; \quad 6.59y = 2.5; \quad y = \dfrac{2.5}{6.59} = 0.3794$

$$x = \frac{[68.2 - 140.3(0.3794)]}{26.8} = 0.5586$$

$X_{LiCl} = 1 - X_{H_2O} - X_{CH_3OH} = 1 - 0.5586 - 0.3794 = 0.0620$

The mole fraction equals the number of moles of each component because $n_{total} = 1.00$ mol.

$$\text{mass LiCl} = 0.0620 \text{ mol LiCl} \times \frac{42.39 \text{ g LiCl}}{1 \text{ mol LiCl}} = 2.6 \text{ g LiCl}$$

$$\text{mass } H_2O = 0.5588 \text{ mol } H_2O \times \frac{18.02 \text{ g } H_2O}{1 \text{ mol } H_2O} = 10.1 \text{ g } H_2O$$

$$\text{mass } CH_3OH = 0.3794 \text{ mol } CH_3OH \times \frac{32.04 \text{ g } CH_3OH}{1 \text{ mol } CH_3OH} = 12.2 \text{ g } CH_3OH$$

total mass = 2.6 g + 10.1 g + 12.2 g = 24.9 g

$$\text{mass \% LiCl} = \frac{2.6 \text{ g}}{24.9 \text{ g}} \times 100\% = 10\%$$

$$\text{mass \% } H_2O = \frac{10.1 \text{ g}}{24.9 \text{ g}} \times 100\% = 41\%$$

$$\text{mass \% } CH_3OH = \frac{12.2 \text{ g}}{24.9 \text{ g}} \times 100\% = 49\%$$

11.130 Solution freezing point = $-1.03\ °C = 0.00\ °C - \Delta T_f;\ \ \Delta T_f = 1.03\ °C$

$$\Delta T_f = K_f \cdot m;\ \ m = \frac{\Delta T_f}{K_f} = \frac{1.03\ °C}{1.86\ \dfrac{°C \cdot kg}{mol}} = 0.554\ mol/kg = 0.554\ m$$

$$\Pi = MRT;\ \ M = \frac{\Pi}{RT} = \frac{(12.16\ atm)}{\left(0.082\ 06\ \dfrac{L \cdot atm}{K \cdot mol}\right)(298\ K)} = 0.497\ \frac{mol}{L}$$

Assume 1.000 L = 1000 mL of solution.

mass of solution = (1000 mL)(1.063 g/mL) = 1063 g

$$\text{mass of } H_2O \text{ in 1000 mL of solution} = \frac{1000\ g\ H_2O}{0.554\ mol\ of\ solute} \times 0.497\ mol = 897\ g\ H_2O$$

mass of solute = total mass – mass of H_2O = 1063 g – 897 g = 166 g solute

$$\text{molar mass} = \frac{166\ g}{0.497\ mol} = 334\ g/mol$$

11.132 (a) NaCl, 58.44; $CaCl_2$, 110.98; H_2O, 18.02

$$\text{mol NaCl} = 100.0\ g\ NaCl \times \frac{1\ mol\ NaCl}{58.44\ g\ NaCl} = 1.711\ mol\ NaCl$$

$$\text{mol } CaCl_2 = 100.0\ g\ CaCl_2 \times \frac{1\ mol\ CaCl_2}{110.98\ g\ CaCl_2} = 0.9011\ mol\ CaCl_2$$

mass of solution = (1000 mL)(1.15 g/mL) = 1150 g

mass of H_2O in solution = mass of solution – mass NaCl – mass $CaCl_2$

$$= 1150\ g - 100.0\ g - 100.0\ g = 950\ g$$

$$= 950\ g \times \frac{1\ kg}{1000\ g} = 0.950\ kg$$

$$\Delta T_b = K_b \cdot (m_{NaCl} \cdot i + m_{CaCl_2} \cdot i)$$

$$\Delta T_b = \left(0.51\ \frac{°C \cdot kg}{mol}\right)\left(\frac{(1.711\ mol\ NaCl \cdot 2) + (0.9011\ mol\ CaCl_2 \cdot 3)}{0.950\ kg}\right) = 3.3\ °C$$

solution boiling point = 100.0 °C + ΔT_b = 100.0 °C + 3.3 °C = 103.3 °C

(b) $\text{mol } H_2O = 950\ g\ H_2O \times \dfrac{1\ mol\ H_2O}{18.02\ g\ H_2O} = 52.7\ mol\ H_2O$

$$P_{Solution} = P° \cdot X_{H_2O}$$

$$P_{Solution} = P° \cdot \left(\frac{52.7\ mol\ H_2O}{(52.7\ mol\ H_2O) + (1.711\ mol\ NaCl \cdot 2) + (0.9011\ mol\ CaCl_2 \cdot 3)}\right)$$

$$P_{Solution} = (23.8\ mm\ Hg)(0.896) = 21.3\ mm\ Hg$$

11.134 (a) KI, 166.00

Assume you have 1.000 L of 1.24 M solution.

mass of solution = (1000 mL)(1.15 g/mL) = 1150 g

mass of KI in solution = 1.24 mol KI x $\dfrac{166.00 \text{ g KI}}{1 \text{ mol KI}}$ = 206 g KI

mass of H_2O in solution = mass of solution – mass KI = 1150 g – 206 g = 944 g

$$= 944 \text{ g} \times \dfrac{1 \text{ kg}}{1000 \text{ g}} = 0.944 \text{ kg}$$

molality = $\dfrac{1.24 \text{ mol KI}}{0.944 \text{ kg } H_2O}$ = 1.31 m

(b) For KI, i = 2 assuming complete dissociation.

$$\Delta T_f = K_f \cdot m \cdot i = \left(1.86 \ \dfrac{^{\circ}C \cdot kg}{mol} \right)(1.31 \ m)(2) = 4.87 \ ^{\circ}C$$

Solution freezing point = 0.00 °C – ΔT_f = 0.00 °C – 4.87 °C = – 4.87 °C

(c) i = $\dfrac{\Delta T_f}{K_f \cdot m}$ = $\dfrac{4.46 \ ^{\circ}C}{\left(1.86 \ \dfrac{^{\circ}C \cdot kg}{mol} \right)(1.31 \text{ mol/kg})}$ = 1.83

Because the calculated i is only 1.83 and not 2, the percent dissociation for KI is 83%.

11.136 NaCl, 58.44; $C_{12}H_{22}O_{11}$, 342.3

Let X = mass NaCl and Y = mass $C_{12}H_{22}O_{11}$, then X + Y = 100.0 g.

$$500.0 \text{ g} \ \times \ \dfrac{1 \text{ kg}}{1000 \text{ g}} = 0.5000 \text{ kg}$$

Solution freezing point = –2.25 °C = 0.00 °C – ΔT_f; ΔT_f = 0.00 °C + 2.25 °C = 2.25 °C

$$\Delta T_f = K_f \cdot (m_{NaCl} \cdot i + m_{C_{12}H_{22}O_{11}})$$

$$\Delta T_b = \left(1.86 \ \dfrac{^{\circ}C \cdot kg}{mol} \right) \left(\dfrac{(\text{mol NaCl} \cdot 2) + (\text{mol } C_{12}H_{22}O_{11})}{0.5000 \text{ kg}} \right) = 2.25 \ ^{\circ}C$$

mol NaCl = X g NaCl x $\dfrac{1 \text{ mol NaCl}}{58.44 \text{ g NaCl}}$ = X/58.44 mol

mol $C_{12}H_{22}O_{11}$ = Y g $C_{12}H_{22}O_{11}$ x $\dfrac{1 \text{ mol } C_{12}H_{22}O_{11}}{342.3 \text{ g } C_{12}H_{22}O_{11}}$ = Y/342.3 mol

$$\Delta T_b = \left(1.86 \ \dfrac{^{\circ}C \cdot kg}{mol} \right) \left(\dfrac{((X/58.44) \cdot 2 \text{ mol}) + ((Y/342.3) \text{ mol})}{0.5000 \text{ kg}} \right) = 2.25 \ ^{\circ}C$$

X = 100 – Y

$$\left(1.86 \ \dfrac{^{\circ}C \cdot kg}{mol} \right) \left(\dfrac{\{[(100 - Y)/58.44] \cdot 2 \text{ mol}\} + [(Y/342.3) \text{ mol}]}{0.5000 \text{ kg}} \right) = 2.25 \ ^{\circ}C$$

$$\left(\frac{[(200/58.44) - (2Y/58.44) + (Y/342.3)]\ \text{mol}}{0.5000\ \text{kg}} \right) = \frac{2.25\ ^\circ\text{C}}{\left(1.86\ \dfrac{^\circ\text{C}\cdot\text{kg}}{\text{mol}} \right)} = 1.21\ \text{mol/kg}$$

$$\left(\frac{[(3.42) - (0.0313Y)]\ \text{mol}}{0.5000\ \text{kg}} \right) = 1.21\ \text{mol/kg}$$

$[(3.42) - (0.0313Y)] = (0.5000\ \text{kg})(1.21) = 0.605$

$-0.0313\ Y = 0.605 - 3.42 = -2.81$

$Y = (-2.81)/(-0.0313) = 89.8\ \text{g of }C_{12}H_{22}O_{11}$

$X = 100.0\ \text{g} - Y = 100.0\ \text{g} - 89.8\ \text{g} = 10.2\ \text{g of NaCl}$

Multiconcept Problems

11.138 (a) 20.00 mL = 0.02000 L

mol NaOH = (0.02000 L)(2.00 mol/L) = 0.0400 mol NaOH

$$\text{mol } CO_2 = 0.0400\ \text{mol NaOH} \times \frac{1\ \text{mol } CO_2}{2\ \text{mol NaOH}} = 0.0200\ \text{mol } CO_2$$

$$\text{mol } C = 0.0200\ \text{mol } CO_2 \times \frac{1\ \text{mol } C}{1\ \text{mol } CO_2} = 0.0200\ \text{mol } C$$

$$\text{mass } C = 0.0200\ \text{mol } C \times \frac{12.011\ \text{g } C}{1\ \text{mol } C} = 0.240\ \text{g } C$$

mass H = mass of compound – mass of C = 0.270 g – 0.240 g = 0.030 g H

$$\text{mol } H = 0.030\ \text{g } H \times \frac{1\ \text{mol } H}{1.008\ \text{g } H} = 0.030\ \text{mol } H$$

The mole ratio of C and H in the molecule is $C_{0.0200}H_{0.030}$.

$C_{0.0200}H_{0.030}$; divide both subscripts by the smaller of the two, 0.0200.

$C_{0.0200 / 0.0200}H_{0.030 / 0.0200}$

$C_1H_{1.5}$, multiply both subscripts by 2.

$C_{(2 \times 1)}H_{(2 \times 1.5)}$

C_2H_3 (27.05) is the empirical formula.

(b) $\Delta T_f = K_f \cdot m$; $m = \dfrac{\Delta T_f}{K_f} = \dfrac{(179.8\ ^\circ\text{C} - 177.9\ ^\circ\text{C})}{37.7\dfrac{^\circ\text{C}\cdot\text{kg}}{\text{mol}}} = 0.050\ \text{mol/kg} = 0.050\ m$

$$50.0\ \text{g} \times \frac{1\ \text{kg}}{1000\ \text{g}} = 0.0500\ \text{kg}$$

mol solute = (0.050 mol/kg)(0.0500 kg) = 0.0025 mol

$$\text{molar mass} = \frac{0.270\ \text{g}}{0.0025\ \text{mol}} = 108\ \text{g/mol}; \quad \text{molecular mass} = 108$$

(c) To find the molecular formula, first divide the molecular mass by the mass of the empirical formula unit.

$$\frac{108}{27} = 4$$

Multiply the subscripts in the empirical formula by the result of this division, 4.

$C_{(4 \times 2)} H_{(4 \times 3)}$

C_8H_{12} is the molecular formula of the compound.

11.140   AgCl, 143.32

Solution freezing point $= -4.42 \, °C = 0.00 \, °C - \Delta T_f;$ $\Delta T_f = 0.00 \, °C + 4.42 \, °C = 4.42 \, °C$

$\Delta T_f = K_f \cdot m$

total ion $m = \dfrac{\Delta T_f}{K_f} = \dfrac{4.42 \, °C}{1.86 \, \dfrac{°C \cdot kg}{mol}} = 2.376 \, mol/kg = 2.376 \, m$

$150.0 \, g \times \dfrac{1 \, kg}{1000 \, g} = 0.1500 \, kg$

total mol of ions $= (2.376 \, mol/kg)(0.1500 \, kg) = 0.3564$ mol of ions

An excess of $AgNO_3$ reacts with all Cl^- to produce 27.575 g AgCl.

total mol $Cl^- = 27.575 \, g \, AgCl \times \dfrac{1 \, mol \, AgCl}{143.32 \, g \, AgCl} \times \dfrac{1 \, mol \, Cl^-}{1 \, mol \, AgCl} = 0.1924$ mol Cl^-

Let P = mol XCl and Q = mol YCl_2.

0.3564 mol ions $= 2 \times$ mol XCl $+ 3 \times$ mol $YCl_2 = (2 \times P) + (3 \times Q)$

0.1924 mol $Cl^- =$ mol XCl $+ 2 \times$ mol $YCl_2 = P + (2 \times Q)$

P = 0.1924 − (2 × Q)

0.3564 = 2 × [0.1924 − (2 × Q)] + (3 × Q) = 0.3848 − (4 × Q) + (3 × Q)

Q = 0.3848 − 0.3564 = 0.0284 mol YCl_2

P = 0.1924 − (2 × Q) = 0.1924 − (2 × 0.0284) = 0.1356 mol XCl

mass Cl in XCl $= 0.1356 \, mol \, XCl \times \dfrac{1 \, mol \, Cl}{1 \, mol \, XCl} \times \dfrac{35.453 \, g \, Cl}{1 \, mol \, Cl} = 4.81 \, g \, Cl$

mass Cl in $YCl_2 = 0.0284 \, mol \, YCl_2 \times \dfrac{2 \, mol \, Cl}{1 \, mol \, YCl_2} \times \dfrac{35.453 \, g \, Cl}{1 \, mol \, Cl} = 2.01 \, g \, Cl$

total mass of XCl and YCl_2 = 8.900 g

mass of X + Y = total mass − mass Cl = 8.900 g − 4.81 g − 2.01 g = 2.08 g

X is an alkali metal and there are 0.1356 mol of X in XCl.

If X = Li, then mass of X = (0.1356 mol)(6.941 g/mol) = 0.941 g

If X = Na, then mass of X = (0.1356 mol)(22.99 g/mol) = 3.12 g but this is not possible because 3.12 g is greater than the total mass of X + Y. Therefore, X is Li.

mass of Y = 2.08 − mass of X = 2.08 g − 0.941 g = 1.14 g

Y is an alkaline earth metal and there are 0.0284 mol of Y in YCl_2.

molar mass of Y = 1.14 g/0.0284 mol = 40.1 g/mol. Therefore, Y is Ca.

mass LiCl $= 0.1356 \, mol \, LiCl \times \dfrac{42.39 \, g \, LiCl}{1 \, mol \, LiCl} = 5.75 \, g \, LiCl$

mass $CaCl_2 = 0.0284 \, mol \, CaCl_2 \times \dfrac{110.98 \, g \, CaCl_2}{1 \, mol \, CaCl_2} = 3.15 \, g \, CaCl_2$

12

The Rates and Mechanisms of Chemical Reactions

12.1 $3\,I^-(aq) + H_3AsO_4(aq) + 2\,H^+(aq) \rightarrow I_3^-(aq) + H_3AsO_3(aq) + H_2O(l)$

(a) $-\dfrac{\Delta[I^-]}{\Delta t} = 4.8 \times 10^{-4}\ M/s$

$\dfrac{\Delta[I_3^-]}{\Delta t} = \dfrac{1}{3}\left(-\dfrac{\Delta[I^-]}{\Delta t}\right) = \left(\dfrac{1}{3}\right)(4.8 \times 10^{-4}\ M/s) = 1.6 \times 10^{-4}\ M/s$

(b) $-\dfrac{\Delta[H^+]}{\Delta t} = 2\left(\dfrac{\Delta[I_3^-]}{\Delta t}\right) = (2)(1.6 \times 10^{-4}\ M/s) = 3.2 \times 10^{-4}\ M/s$

12.2 $2\,N_2O_5(g) \rightarrow 4\,NO_2(g) + O_2(g)$

time	$[N_2O_5]$	$[O_2]$
200 s	0.0142 M	0.0029 M
300 s	0.0120 M	0.0040 M

Rate of decomposition of $N_2O_5 = -\dfrac{\Delta[N_2O_5]}{\Delta t} = -\dfrac{0.0120\ M - 0.0142\ M}{300\ s - 200\ s} = 2.2 \times 10^{-5}\ M/s$

Rate of formation of $O_2 = \dfrac{\Delta[O_2]}{\Delta t} = \dfrac{0.0040\ M - 0.0029\ M}{300\ s - 200\ s} = 1.1 \times 10^{-5}\ M/s$

12.3 Rate $= k[BrO_3^-][Br^-][H^+]^2$,
1st order in BrO_3^-, 1st order in Br^-, 2nd order in H^+, 4th order overall
Rate $= k[H_2][I_2]$, 1st order in H_2, 1st order in I_2, 2nd order overall

12.4 $H_2O_2(aq) + 3\,I^-(aq) + 2\,H^+(aq) \rightarrow I_3^-(aq) + 2\,H_2O(l)$

Rate $= \dfrac{\Delta[I_3^-]}{\Delta t} = k[H_2O_2]^m[I^-]^n$

(a) $\dfrac{Rate_3}{Rate_1} = \dfrac{2.30 \times 10^{-4}\ M/s}{1.15 \times 10^{-4}\ M/s} = 2$ $\dfrac{[H_2O_2]_3}{[H_2O_2]_1} = \dfrac{0.200\ M}{0.100\ M} = 2$

Because both ratios are the same, m = 1.

$\dfrac{Rate_2}{Rate_1} = \dfrac{2.30 \times 10^{-4}\ M/s}{1.15 \times 10^{-4}\ M/s} = 2$ $\dfrac{[I^-]_2}{[I^-]_1} = \dfrac{0.200\ M}{0.100\ M} = 2$

Because both ratios are the same, n = 1.
The rate law is: Rate $= k[H_2O_2][I^-]$

(b) $k = \dfrac{\text{Rate}}{[H_2O_2][I^-]}$

Using data from Experiment 1: $k = \dfrac{1.15 \times 10^{-4}\ M/s}{(0.100\ M)(0.100\ M)} = 1.15 \times 10^{-2}\ /(M \cdot s)$

(c) Rate $= k[H_2O_2][I^-] = [1.15 \times 10^{-2}/(M \cdot s)](0.300\ M)(0.400\ M) = 1.38 \times 10^{-3}\ M/s$

12.5

Rate Law	Units of k
Rate $= k[(CH_3)_3CBr]$	$1/s$
Rate $= k[Br_2]$	$1/s$
Rate $= k[BrO_3^-][Br^-][H^+]^2$	$1/(M^3 \cdot s)$
Rate $= k[H_2][I_2]$	$1/(M \cdot s)$

12.6 (a) The reactions in vessels (a) and (b) have the same rate, the same number of B molecules, but different numbers of A molecules. Therefore, the rate does not depend on A and its reaction order is zero. The same conclusion can be drawn from the reactions in vessels (c) and (d).

The rate for the reaction in vessel (c) is four times the rate for the reaction in vessel (a). Vessel (c) has twice as many B molecules than does vessel (a). Because the rate quadruples when the concentration of B doubles, the reaction order for B is two.

(b) rate $= k[B]^2$

12.7 (a)

$$\ln \frac{[Co(NH_3)_5Br^{2+}]_t}{[Co(NH_3)_5Br^{2+}]_o} = -kt$$

$k = 6.3 \times 10^{-6}/s;$ $\qquad t = 10.0\ h \times \dfrac{3600\ s}{1\ h} = 36{,}000\ s$

$\ln[Co(NH_3)_5Br^{2+}]_t = -kt + \ln[Co(NH_3)_5Br^{2+}]_o$

$\ln[Co(NH_3)_5Br^{2+}]_t = -(6.3 \times 10^{-6}/s)(36{,}000\ s) + \ln(0.100)$

$\ln[Co(NH_3)_5Br^{2+}]_t = -2.5294;$ $\qquad$ After 10.0 h, $[Co(NH_3)_5Br^{2+}] = e^{-2.5294} = 0.080\ M$

(b) $[Co(NH_3)_5Br^{2+}]_o = 0.100\ M$

If 75% of the $Co(NH_3)_5Br^{2+}$ reacts then 25% remains.

$[Co(NH_3)_5Br^{2+}]_t = (0.25)(0.100\ M) = 0.025\ M$

$$\ln \frac{[Co(NH_3)_5Br^{2+}]_t}{[Co(NH_3)_5Br^{2+}]_o} = -kt; \quad t = \frac{\ln \dfrac{[Co(NH_3)_5Br^{2+}]_t}{[Co(NH_3)_5Br^{2+}]_o}}{-k}$$

$t = \dfrac{\ln\left(\dfrac{0.025}{0.100}\right)}{-(6.3 \times 10^{-6}/s)} = 2.2 \times 10^5\ s;$ $\quad t = 2.2 \times 10^5\ s \times \dfrac{1\ h}{3600\ s} = 61\ h$

12.8

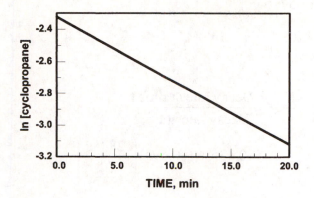

Slope = –0.03989/min = –6.6 x 10^{-4}/s and k = –slope

A plot of ln[cyclopropane] versus time is linear, indicating that the data fit the equation for a first-order reaction. k = 6.6 x 10^{-4}/s (0.040/min)

12.9 (a) k = 1.8 x 10^{-5}/s

$$t_{1/2} = \frac{0.693}{k} = \frac{0.693}{1.8 \times 10^{-5}/s} = 38{,}500 \text{ s}; \qquad t_{1/2} = 38{,}500 \text{ s} \times \frac{1 \text{ h}}{3600 \text{ s}} = 11 \text{ h}$$

(b) $0.30 \text{ M} \xrightarrow{t_{1/2}} 0.15 \text{ M} \xrightarrow{t_{1/2}} 0.075 \text{ M} \xrightarrow{t_{1/2}} 0.0375 \text{ M} \xrightarrow{t_{1/2}} 0.019 \text{ M}$

(c) Because 25% of the initial concentration corresponds to 1/4 or $(1/2)^2$ of the initial concentration, the time required is two half-lives: $t = 2t_{1/2} = 2(11 \text{ h}) = 22 \text{ h}$

12.10 After one half-life, there would be four A molecules remaining. After two half-lives, there would be two A molecules remaining. This is represented by the drawing at t = 10 min. 10 min is equal to two half-lives, therefore, $t_{1/2}$ = 5 min for this reaction. After 15 min (three half-lives) only one A molecule would remain.

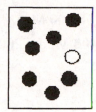

○ red   ● blue

12.11 $t_{1/2} = \frac{0.693}{k} = \frac{0.693}{1.08 \times 10^{-2} \text{ h}^{-1}} = 64.2 \text{ h}$

12.12 $k = \frac{0.693}{t_{1/2}} = \frac{0.693}{30.2 \text{ y}} = 0.0229 \text{ y}^{-1}$

12.13 $\ln\left(\dfrac{N}{N_0}\right) = -0.693\left(\dfrac{t}{t_{1/2}}\right) = -0.693\left(\dfrac{16{,}230\text{ y}}{5715\text{ y}}\right) = -1.968$

$\dfrac{N}{N_0} = e^{-1.968} = 0.140;\qquad \dfrac{N}{100\%} = 0.140;\qquad N = 14.0\%$

12.14 $\ln\left(\dfrac{N}{N_0}\right) = (-0.693)\left(\dfrac{t}{t_{1/2}}\right);\qquad \dfrac{N}{N_0} = \dfrac{\text{Decay rate at time t}}{\text{Decay rate at t} = 0}$

$\ln\left(\dfrac{10{,}860}{16{,}800}\right) = (-0.693)\left(\dfrac{28.0\text{ d}}{t_{1/2}}\right);\qquad -0.436 = (-0.693)\left(\dfrac{28.0\text{ d}}{t_{1/2}}\right)$

$t_{1/2} = \dfrac{(-0.693)(28.0\text{ d})}{(-0.436)} = 44.5\text{ d}$

12.15

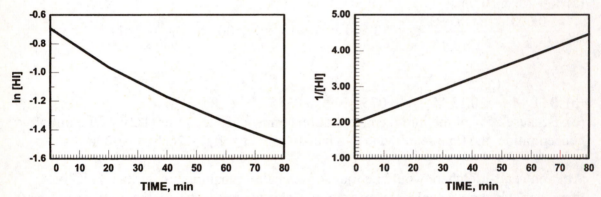

(a) A plot of 1/[HI] versus time is linear. The reaction is second-order.

(b) $k = \text{slope} = 0.0308/(\text{M}\cdot\text{min})$

(c) $t = \dfrac{1}{k}\left[\dfrac{1}{[\text{HI}]_t} - \dfrac{1}{[\text{HI}]_0}\right] = \dfrac{1}{0.0308/(\text{M}\cdot\text{min})}\left[\dfrac{1}{0.100\text{ M}} - \dfrac{1}{0.500\text{ M}}\right] = 260\text{ min}$

(d) It requires one half-life ($t_{1/2}$) for the [HI] to drop from 0.400 M to 0.200 M.

$t_{1/2} = \dfrac{1}{k[\text{HI}]_0} = \dfrac{1}{[0.0308/(\text{M}\cdot\text{min})](0.400\text{ M})} = 81.2\text{ min}$

12.16 (a) $E_a = 100\text{ kJ/mol} - 20\text{ kJ/mol} = 80\text{ kJ/mol}$

(b) The reaction is endothermic because the energy of the products is higher than the energy of the reactants.

(c) A--·C
     B--·D

(d) The reaction rates of all chemical reactions increase as the temperature increases. This is explained by collision theory.

12.17 (a) $\ln\left(\dfrac{k_2}{k_1}\right) = \left(\dfrac{-E_a}{R}\right)\left(\dfrac{1}{T_2} - \dfrac{1}{T_1}\right)$

$k_1 = 3.7 \times 10^{-5}/s,\ T_1 = 25\ ^\circ C = 298\ K$
$k_2 = 1.7 \times 10^{-3}/s,\ T_2 = 55\ ^\circ C = 328\ K$

$E_a = -\dfrac{[\ln k_2 - \ln k_1]R}{\left(\dfrac{1}{T_2} - \dfrac{1}{T_1}\right)}$

$E_a = -\dfrac{[\ln(1.7 \times 10^{-3}) - \ln(3.7 \times 10^{-5})][8.314 \times 10^{-3}\ kJ/(K \cdot mol)]}{\left(\dfrac{1}{328\ K} - \dfrac{1}{298\ K}\right)} = 104\ kJ/mol$

(b) $k_1 = 3.7 \times 10^{-5}/s,\ T_1 = 25\ ^\circ C = 298\ K$
   solve for k_2, $T_2 = 35\ ^\circ C = 308\ K$

$\ln k_2 = \left(\dfrac{-E_a}{R}\right)\left(\dfrac{1}{T_2} - \dfrac{1}{T_1}\right) + \ln k_1$

$\ln k_2 = \left(\dfrac{-104\ kJ/mol}{8.314 \times 10^{-3}\ kJ/(K \cdot mol)}\right)\left(\dfrac{1}{308\ K} - \dfrac{1}{298\ K}\right) + \ln(3.7 \times 10^{-5})$

$\ln k_2 = -8.84;\ k_2 = e^{-8.84} = 1.4 \times 10^{-4}/s$

12.18 (a)
$$NO_2(g) + F_2(g) \rightarrow NO_2F(g) + F(g)$$
$$\underline{F(g) + NO_2(g) \rightarrow NO_2F(g)}$$

Overall reaction $\quad 2\ NO_2(g) + F_2(g) \rightarrow 2\ NO_2F(g)$

Because F(g) is produced in the first reaction and consumed in the second, it is a reaction intermediate.

(b) In each reaction there are two reactants, so each elementary reaction is bimolecular.

12.19 (a) Rate = $k[O_3][O]$ (b) Rate = $k[Br]^2[Ar]$ (c) Rate = $k[Co(CN)_5(H_2O)^{2-}]$

12.20
$$Co(CN)_5(H_2O)^{2-}(aq) \rightarrow Co(CN)_5^{2-}(aq) + H_2O(l) \qquad \text{(slow)}$$
$$\underline{Co(CN)_5^{2-}(aq) + I^-(aq) \rightarrow Co(CN)_5I^{3-}(aq)} \qquad \text{(fast)}$$

Overall reaction $\quad Co(CN)_5(H_2O)^{2-}(aq) + I^-(aq) \rightarrow Co(CN)_5I^{3-}(aq) + H_2O(l)$

The predicted rate law for the overall reaction is the rate law for the first (slow) elementary reaction: Rate = $k[Co(CN)_5(H_2O)^{2-}]$
The predicted rate law is in accord with the observed rate law.

12.21 (a) $2\ NO(g) + O_2(g) \rightarrow 2\ NO_2(g)$
(b) $\text{Rate}_{forward} = k_1[NO][O_2]$ and $\text{Rate}_{reverse} = k_{-1}[NO_3]$
Because of the equilibrium, $\text{Rate}_{forward} = \text{Rate}_{reverse}$, and $k_1[NO][O_2] = k_{-1}[NO_3]$.

$$[NO_3] = \frac{k_1}{k_{-1}}[NO][O_2]$$

The rate law for the rate determining step is Rate $= k_2[NO_3][NO]$. Because 2 NO disappear in the overall reaction for every NO that reacts in the second step, the rate law for the overall reaction is Rate $= -\Delta[NO]/\Delta t = 2k_2[NO_3][NO]$. In this rate law substitute for $[NO_3]$.

Rate $= 2k_2 \dfrac{k_1}{k_{-1}}[NO]^2[O_2]$, which is consistent with the experimental rate law.

(c) $k = \dfrac{2k_2 k_1}{k_{-1}}$

12.22   Assume that concentration is proportional to the number of each molecule in a box.
(a) Comparing boxes (1) and (2), the concentration of A doubles, B and C_2 remain the same and the rate does not change. This means the reaction is zeroth-order in A. Comparing boxes (1) and (3), the concentration of C_2 doubles, A and B remain the same and the rate doubles. This means the reaction is first-order in C_2. Comparing boxes (1) and (4), the concentration of B triples, A and C_2 remain the same and the rate triples. This means the reaction is first-order in B.
(b) Rate $= k[B][C_2]$
(c)

$$
\begin{array}{lll}
B + C_2 & \rightarrow & BC_2 \qquad\qquad (slow) \\
A + BC_2 & \rightarrow & AC + BC \\
\underline{A + BC} & \rightarrow & \underline{AC + B} \\
2A + C_2 & \rightarrow & 2AC \qquad\qquad (overall)
\end{array}
$$

(d) B doesn't appear in the overall reaction because it is consumed in the first step and regenerated in the third step. B is therefore a catalyst. BC_2 and BC are intermediates because they are formed in one step and then consumed in a subsequent step in the reaction.

12.23

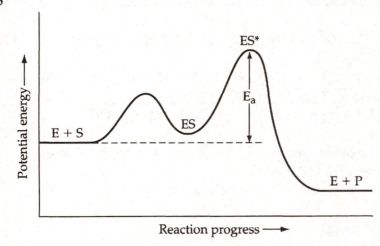

Conceptual Problems

12.24 (a) Because Rate = k[A][B], the rate is proportional to the product of the number of A molecules and the number of B molecules. The relative rates of the reaction in vessels (a) – (d) are 2 : 1 : 4 : 2.
(b) Because the same reaction takes place in each vessel, the k's are all the same.

12.26 (a) For the first-order reaction, half of the A molecules are converted to B molecules each minute.

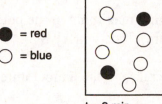
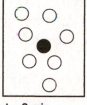

● = red
○ = blue

t = 2 min   t = 3 min

(b) Because half of the A molecules are converted to B molecules in 1 min, the half-life is 1 minute.

12.28 (a) Because the half-life is inversely proportional to the concentration of A molecules, the reaction is second-order in A.
(b) Rate = k[A]^2
(c) The second box represents the passing of one half-life, and the third box represents the passing of a second half-life for a second-order reaction. A relative value of k can be calculated.

$$k = \frac{1}{t_{1/2}[A]} = \frac{1}{(1)(16)} = 0.0625$$

$t_{1/2}$ in going from box 3 to box 4 is: $t_{1/2} = \dfrac{1}{k[A]} = \dfrac{1}{(0.0625)(4)} = 4$ min

(For fourth box, t = 7 min)

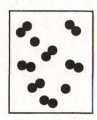

t = 3 min + 4 min = 7 min

12.30 Assume that concentration is proportional to the number of each molecule in a box.
(a) Comparing boxes (1) and (2), the concentration of B doubles, A remains the same and the rate does not change. This means the reaction is zeroth-order in B.
Comparing boxes (3) and (2), the concentration of A doubles, B remains the same and the rate quadruples. This means the reaction is second-order in A.
(b) Rate = k [A]^2.

(c) $2 A \rightarrow A_2$ (slow)

$\dfrac{A_2 + B \rightarrow AB + A}{A + B \rightarrow AB}$ (overall)

(d) A_2 is an intermediate because it is formed in one step and then consumed in a subsequent step in the reaction.

12.32   (a) $BC + D \rightarrow B + CD$

(b) 1. B–C + D (reactants), A (catalyst); 2. B---C---A (transition state), D (reactant); 3. A–C (intermediate), B (product), D (reactant); 4. A---C---D (transition state), B (product); 5. A (catalyst), C–D + B (products)

(c) The first step is rate determining because the first maximum in the potential energy curve is greater than the second (relative) maximum; Rate = k[A][BC]

(d) Endothermic

(e) The reaction rates of all chemical reactions increase as the temperature increases. This is explained by collision theory.

Section Problems
Reaction Rates (Section 12.1)

12.34   (a) Rate $= \dfrac{\Delta[NO_2]}{\Delta t} = \dfrac{(0.0278\ M) - (0\ M)}{700\ s - 0\ s} = 4.0 \times 10^{-5}\ M/s$

(b) Rate $= \dfrac{\Delta[NO_2]}{\Delta t} = \dfrac{(0.0256\ M) - (0.0063\ M)}{600\ s - 100\ s} = 3.9 \times 10^{-5}\ M/s$

(c) Rate $= \dfrac{\Delta[NO_2]}{\Delta t} = \dfrac{(0.0229\ M) - (0.0115\ M)}{500\ s - 200\ s} = 3.8 \times 10^{-5}\ M/s$

(d) Rate $= \dfrac{\Delta[NO_2]}{\Delta t} = \dfrac{(0.0197\ M) - (0.0160\ M)}{400\ s - 300\ s} = 3.7 \times 10^{-5}\ M/s$

Rate (d) is the best estimate of the instantaneous rate because it is determined from measurements taken over the smallest time interval.

12.36

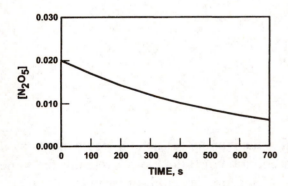

(a) The instantaneous rate of decomposition of N_2O_5 at t = 200 s is determined from the slope of the curve at t = 200 s.

$$\text{Rate} = -\frac{\Delta[N_2O_5]}{\Delta t} = -\text{slope} = -\frac{(1.20 \times 10^{-2} \text{ M}) - (1.69 \times 10^{-2} \text{ M})}{300 \text{ s} - 100 \text{ s}} = 2.5 \times 10^{-5} \text{ M/s}$$

(b) The initial rate of decomposition of N_2O_5 is determined from the slope of the curve at $t = 0$ s. This is equivalent to the slope of the curve from 0 s to 100 s because in this time interval the curve is almost linear.

$$\text{Initial rate} = -\frac{\Delta[N_2O_5]}{\Delta t} = -\text{slope} = -\frac{(1.69 \times 10^{-2} \text{ M}) - (2.00 \times 10^{-2} \text{ M})}{100 \text{ s} - 0 \text{ s}} = 3.1 \times 10^{-5} \text{ M/s}$$

12.38 (a) $-\dfrac{\Delta[H_2]}{\Delta t} = -3\,\dfrac{\Delta[N_2]}{\Delta t}$; The rate of consumption of H_2 is 3 times faster.

(b) $\dfrac{\Delta[NH_3]}{\Delta t} = -2\,\dfrac{\Delta[N_2]}{\Delta t}$; The rate of formation of NH_3 is 2 times faster.

12.40 (a) $-\dfrac{1}{2}\dfrac{\Delta[Br_2]}{\Delta t} = \dfrac{\Delta[ClO_2^-]}{\Delta t} = -2.4 \times 10^{-6}$ M/s

(b) Rate $= -\dfrac{\Delta[Br^-]}{\Delta t} = -4\,\dfrac{\Delta[ClO_2^-]}{\Delta t} = -4(-2.4 \times 10^{-6} \text{ M/s}) = 9.6 \times 10^{-6}$ M/s

Rate Laws (Sections 12.2 and 12.3)

12.42 Rate $= k[H_2][ICl]$; units for k are $\dfrac{L}{mol \cdot s}$ or $1/(M \cdot s)$

12.44 (a) Rate $= k[CH_3Br][OH^-]$
(b) Because the reaction is first-order in OH^-, if the $[OH^-]$ is decreased by a factor of 5, the rate will also decrease by a factor of 5.
(c) Because the reaction is first-order in each reactant, if both reactant concentrations are doubled, the rate will increase by a factor of $2 \times 2 = 4$.

12.46 (a) Rate $= k[Cu(C_{10}H_8N_2)_2^+]^2[O_2]$
(b) The overall reaction order is $2 + 1 = 3$.
(c) Because the reaction is second-order in $Cu(C_{10}H_8N_2)_2^+$, if the $[Cu(C_{10}H_8N_2)_2^+]$ is decreased by a factor of four, the rate will decrease by a factor of 16 ($1/4 \times 1/4 = 1/16$).

12.48 (a) Rate $= k[N_2O_5]^m$

$$m = \frac{\ln\left(\dfrac{\text{Rate}_2}{\text{Rate}_1}\right)}{\ln\left(\dfrac{[N_2O_5]_2}{[N_2O_5]_1}\right)} = \frac{\ln\left(\dfrac{6.8 \times 10^{-5}}{2.4 \times 10^{-5}}\right)}{\ln\left(\dfrac{0.040}{0.014}\right)} = 1$$

Rate $= k[N_2O_5]$

(b) From Experiment 1: $k = \dfrac{\text{Rate}}{[N_2O_5]} = \dfrac{2.4 \times 10^{-5} \text{ M/s}}{0.014 \text{ M}} = 1.7 \times 10^{-3} \text{ /s}$

(c) Rate $= k[N_2O_5] = (1.7 \times 10^{-3} \text{ /s})(0.030 \text{ M}) = 5.1 \times 10^{-5} \text{ M/s}$

12.50 (a) Rate $= k[A]^m[B]^n[C]^p$

$$m = \dfrac{\ln\left(\dfrac{\text{Rate}_4}{\text{Rate}_3}\right)}{\ln\left(\dfrac{[A]_4}{[A]_3}\right)} = \dfrac{\ln\left(\dfrac{7.2 \times 10^{-2}}{1.8 \times 10^{-2}}\right)}{\ln\left(\dfrac{0.20}{0.10}\right)} = 2; \quad n = \dfrac{\ln\left(\dfrac{\text{Rate}_3}{\text{Rate}_1}\right)}{\ln\left(\dfrac{[B]_3}{[B]_1}\right)} = \dfrac{\ln\left(\dfrac{1.8 \times 10^{-2}}{6.0 \times 10^{-3}}\right)}{\ln\left(\dfrac{0.36}{0.12}\right)} = 1$$

$$p = \dfrac{\ln\left(\left(\dfrac{\text{Rate}_2}{\text{Rate}_1}\right)\bigg/\left(\dfrac{[B]_2}{[B]_1}\right)\right)}{\ln\left(\dfrac{[C]_2}{[C]_1}\right)} = \dfrac{\ln\left(\left(\dfrac{1.2 \times 10^{-2}}{6.0 \times 10^{-3}}\right)\bigg/\left(\dfrac{0.24}{0.12}\right)\right)}{\ln\left(\dfrac{0.075}{0.050}\right)} = 0$$

Rate $= k[A]^2[B]$

(b) From Experiment 1: $k = \dfrac{\text{Rate}}{[A]^2[B]} = \dfrac{6.0 \times 10^{-3} \text{ M/s}}{(0.10 \text{ M})^2(0.12 \text{ M})} = 5.0/(M^2 \cdot s)$

(c) Rate $= k[A]^2[B] = [5.0/(M^2 \cdot s)](0.020 \text{ M})^2(0.020 \text{ M}) = 4.0 \times 10^{-5} \text{ M/s}$

Integrated Rate Law; Half-Life (Sections 12.4, 12.5, 12.7 and 12.8)

12.52 $\ln \dfrac{[C_3H_6]_t}{[C_3H_6]_0} = -kt$, $k = 6.7 \times 10^{-4}$/s

(a) $t = 30 \text{ min} \times \dfrac{60 \text{ s}}{1 \text{ min}} = 1800 \text{ s}$

$\ln[C_3H_6]_t = -kt + \ln[C_3H_6]_0 = -(6.7 \times 10^{-4}\text{/s})(1800 \text{ s}) + \ln(0.0500) = -4.202$

$[C_3H_6]_t = e^{-4.202} = 0.015 \text{ M}$

(b) $t = \dfrac{\ln \dfrac{[C_3H_6]_t}{[C_3H_6]_0}}{-k} = \dfrac{\ln\left(\dfrac{0.0100}{0.0500}\right)}{-(6.7 \times 10^{-4}\text{/s})} = 2402 \text{ s}$; $t = 2402 \text{ s} \times \dfrac{1 \text{ min}}{60 \text{ s}} = 40 \text{ min}$

(c) $[C_3H_6]_0 = 0.0500 \text{ M}$; If 25% of the C_3H_6 reacts then 75% remains.

$[C_3H_6]_t = (0.75)(0.0500 \text{ M}) = 0.0375 \text{ M}$

$t = \dfrac{\ln \dfrac{[C_3H_6]_t}{[C_3H_6]_0}}{-k} = \dfrac{\ln\left(\dfrac{0.0375}{0.0500}\right)}{-(6.7 \times 10^{-4}\text{/s})} = 429 \text{ s}$; $t = 429 \text{ s} \times \dfrac{1 \text{ min}}{60 \text{ s}} = 7.2 \text{ min}$

12.54 $t_{1/2} = \dfrac{0.693}{k} = \dfrac{0.693}{6.7 \times 10^{-4}/s} = 1034\ s = 17\ min$

$t = \dfrac{\ln\dfrac{[C_3H_6]_t}{[C_3H_6]_o}}{-k} = \dfrac{\ln\dfrac{(0.0625)(0.0500)}{(0.0500)}}{-6.7 \times 10^{-4}/s} = 4140\ s$

$t = 4140\ s \times \dfrac{1\ min}{60\ s} = 69\ min$

This is also 4 half-lives. $100 \overset{t_{1/2}}{\to} 50 \overset{t_{1/2}}{\to} 25 \overset{t_{1/2}}{\to} 12.5 \overset{t_{1/2}}{\to} 6.25$

12.56 $kt = \dfrac{1}{[C_4H_6]_t} - \dfrac{1}{[C_4H_6]_o}, \qquad k = 4.0 \times 10^{-2}/(M \cdot s)$

(a) $t = 1.00\ h \times \dfrac{60\ min}{1\ hr} \times \dfrac{60\ s}{1\ min} = 3600\ s$

$\dfrac{1}{[C_4H_6]_t} = kt + \dfrac{1}{[C_4H_6]_o} = (4.0 \times 10^{-2}/(M \cdot s))(3600\ s) + \dfrac{1}{0.0200\ M}$

$\dfrac{1}{[C_4H_6]_t} = 194/M \ \text{ and } \ [C_4H_6] = 5.2 \times 10^{-3}\ M$

(b) $t = \dfrac{1}{k}\left[\dfrac{1}{[C_4H_6]_t} - \dfrac{1}{[C_4H_6]_o} \right]$

$t = \dfrac{1}{4.0 \times 10^{-2}/(M \cdot s)}\left[\dfrac{1}{(0.0020\ M)} - \dfrac{1}{(0.0200\ M)} \right] = 11,250\ s$

$t = 11,250\ s \times \dfrac{1\ min}{60\ s} \times \dfrac{1\ hr}{60\ min} = 3.1\ h$

12.58 $t_{1/2} = \dfrac{1}{k[C_4H_6]_o} = \dfrac{1}{[4.0 \times 10^{-2}/(M \cdot s)](0.0200\ M)} = 1250\ s = 21\ min$

$t = t_{1/2} = \dfrac{1}{k[C_4H_6]_o} = \dfrac{1}{[4.0 \times 10^{-2}/(M \cdot s)](0.0100\ M)} = 2500\ s = 42\ min$

12.60

time (min)	[N$_2$O]	ln[N$_2$O]	1/[N$_2$O]
0	0.250	−1.386	4.00
60	0.228	−1.478	4.39
90	0.216	−1.532	4.63
300	0.128	−2.056	7.81
600	0.0630	−2.765	15.9

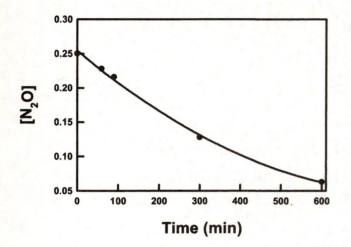

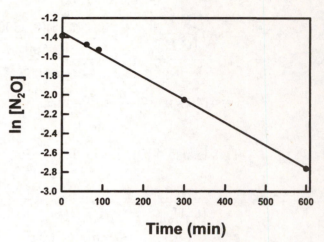

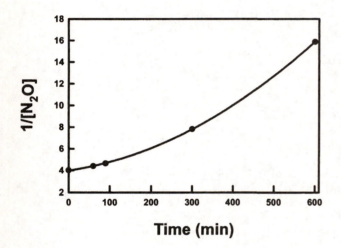

A plot of $\ln[N_2O]$ versus time is linear. The reaction is first-order in N_2O.

$k = -\text{slope} = -(-2.35 \times 10^{-3}/\text{min}) = 2.35 \times 10^{-3}/\text{min}$

$k = 2.35 \times 10^{-3}/\text{min} \times \dfrac{1 \text{ min}}{60 \text{ s}} = 3.92 \times 10^{-5}/\text{s}$

12.62 $k = \dfrac{0.693}{t_{1/2}} = \dfrac{0.693}{248 \text{ s}} = 2.79 \times 10^{-3}/\text{s}$

12.64 (a) The units for the rate constant, k, indicate the reaction is zeroth-order.

(b) For a zeroth-order reaction, $[A]_t - [A]_o = -kt$

$t = 30 \text{ min} \times \dfrac{60 \text{ s}}{1 \text{ min}} = 1800 \text{ s}$

$[A]_t = -kt + [A]_o = -(3.6 \times 10^{-5} \text{ M/s})(1800 \text{ s}) + 0.096 \text{ M} = 0.031 \text{ M}$

(c) Let $[A]_t = [A]_o/2$, $t_{1/2} = \dfrac{[A]_o/2 - [A]_o}{-k} = \dfrac{0.096/2 \text{ M} - 0.096 \text{ M}}{-3.6 \times 10^{-5} \text{ M/s}} = 1333 \text{ s}$

$t_{1/2} = 1333 \text{ s} \times \dfrac{1 \text{ min}}{60 \text{ s}} = 22 \text{ min}$

Radioactive Decay Rates (Section 12.6)

12.66 $\quad k = \dfrac{0.693}{t_{1/2}} = \dfrac{0.693}{2.805 \text{ d}} = 0.247 \text{ d}^{-1}$

12.68 $\quad t_{1/2} = \dfrac{0.693}{k} = \dfrac{0.693}{7.95 \times 10^{-3} \text{ d}^{-1}} = 87.17 \text{d}$

$\ln\left(\dfrac{N}{N_0}\right) = (-0.693)\left(\dfrac{t}{t_{1/2}}\right) = (-0.693)\left(\dfrac{185 \text{ d}}{87.17 \text{ d}}\right) = -1.4707$

$\dfrac{N}{N_0} = e^{-1.4707} = 0.2298; \qquad \dfrac{N}{100\%} = 0.2298; \quad N = 23.0\%$

12.70 $\quad t_{1/2} = (102 \text{ y})(365 \text{ d/y})(24 \text{ h/d})(3600 \text{ s/h}) = 3.2167 \times 10^9 \text{ s}$

$k = \dfrac{0.693}{t_{1/2}} = \dfrac{0.693}{3.2167 \times 10^9 \text{ s}} = 2.1544 \times 10^{-10} \text{ s}^{-1}$

$N = (1.0 \times 10^{-9} \text{ g})\left(\dfrac{1 \text{ mol Po}}{209 \text{ g Po}}\right)(6.022 \times 10^{23} \text{ atoms/mol}) = 2.881 \times 10^{12} \text{ atoms}$

Decay rate = $kN = (2.1544 \times 10^{-10} \text{ s}^{-1})(2.881 \times 10^{12} \text{ atoms}) = 6.21 \times 10^2 \text{ s}^{-1}$
621 α particles are emitted in 1.0 s.

12.72 $\quad$ Decay rate = kN

$N = (1.0 \times 10^{-3} \text{ g})\left(\dfrac{1 \text{ mol } ^{79}\text{Se}}{79 \text{ g}}\right)(6.022 \times 10^{23} \text{ atoms/mol}) = 7.6 \times 10^{18} \text{ atoms}$

$k = \dfrac{\text{Decay rate}}{N} = \dfrac{1.5 \times 10^5 /\text{s}}{7.6 \times 10^{18}} = 2.0 \times 10^{-14} \text{ s}^{-1}$

$t_{1/2} = \dfrac{0.693}{k} = \dfrac{0.693}{2.0 \times 10^{-14} \text{ s}^{-1}} = 3.5 \times 10^{13} \text{ s}$

$t_{1/2} = (3.5 \times 10^{13} \text{ s})\left(\dfrac{1 \text{ h}}{3600 \text{ s}}\right)\left(\dfrac{1 \text{ d}}{24 \text{ h}}\right)\left(\dfrac{1 \text{ y}}{365 \text{ d}}\right) = 1.1 \times 10^6 \text{ y}$

12.74 $\quad \ln\left(\dfrac{N}{N_0}\right) = (-0.693)\left(\dfrac{t}{t_{1/2}}\right); \qquad \dfrac{N}{N_0} = \dfrac{\text{Decay rate at time } t}{\text{Decay rate at time } t = 0}$

$\ln\left(\dfrac{6990}{8540}\right) = (-0.693)\left(\dfrac{10.0 \text{ d}}{t_{1/2}}\right); \qquad t_{1/2} = 34.6 \text{ d}$

The Arrhenius Equation (Sections 12.9 and 12.10)

12.76 $\quad$ Very few collisions involve a collision energy greater than or equal to the activation energy, and only a fraction of those have the proper orientation for reaction.

12.78   Plot ln k versus 1/T to determine the activation energy, E_a.

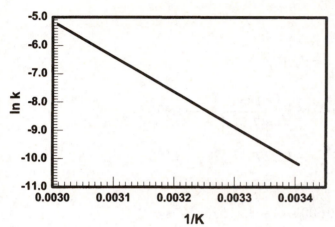

Slope = -1.25×10^4 K

$E_a = -R(\text{slope}) = -[8.314 \times 10^{-3}\ \text{kJ/(K} \cdot \text{mol})](-1.25 \times 10^4\ \text{K}) = 104\ \text{kJ/mol}$

12.80   (a) $\ln\left(\dfrac{k_2}{k_1}\right) = \left(\dfrac{-E_a}{R}\right)\left(\dfrac{1}{T_2} - \dfrac{1}{T_1}\right)$

$k_1 = 1.3/(\text{M} \cdot \text{s}),\ T_1 = 700\ \text{K}$
$k_2 = 23.0/(\text{M} \cdot \text{s}),\ T_2 = 800\ \text{K}$

$E_a = -\dfrac{[\ln k_2 - \ln k_1](R)}{\left(\dfrac{1}{T_2} - \dfrac{1}{T_1}\right)}$

$E_a = -\dfrac{[\ln(23.0) - \ln(1.3)][8.314 \times 10^{-3}\ \text{kJ/(K} \cdot \text{mol})]}{\left(\dfrac{1}{800\ \text{K}} - \dfrac{1}{700\ \text{K}}\right)} = 134\ \text{kJ/mol}$

(b) $k_1 = 1.3/(\text{M} \cdot \text{s}),\ T_1 = 700\ \text{K}$
solve for $k_2,\ T_2 = 750\ \text{K}$

$\ln k_2 = \left(\dfrac{-E_a}{R}\right)\left(\dfrac{1}{T_2} - \dfrac{1}{T_1}\right) + \ln k_1$

$\ln k_2 = \left(\dfrac{-133.8\ \text{kJ/mol}}{8.314 \times 10^{-3}\ \text{kJ/(K} \cdot \text{mol})}\right)\left(\dfrac{1}{750\ \text{K}} - \dfrac{1}{700\ \text{K}}\right) + \ln(1.3) = 1.795$

$k_2 = e^{1.795} = 6.0/(\text{M} \cdot \text{s})$

12.82 $\ln\left(\dfrac{k_2}{k_1}\right) = \left(\dfrac{-E_a}{R}\right)\left(\dfrac{1}{T_2} - \dfrac{1}{T_1}\right)$

assume $k_1 = 1.0/(\text{M} \cdot \text{s})$ at $T_1 = 25\ ^\circ\text{C} = 298\ \text{K}$
assume $k_2 = 15/(\text{M} \cdot \text{s})$ at $T_2 = 50\ ^\circ\text{C} = 323\ \text{K}$

$$E_a = -\frac{[\ln k_2 - \ln k_1](R)}{\left(\dfrac{1}{T_2} - \dfrac{1}{T_1}\right)}$$

$$E_a = -\frac{[\ln(15) - \ln(1.0)][8.314 \times 10^{-3}\ kJ/(K \cdot mol)]}{\left(\dfrac{1}{323\ K} - \dfrac{1}{298\ K}\right)} = 87\ kJ/mol$$

12.84

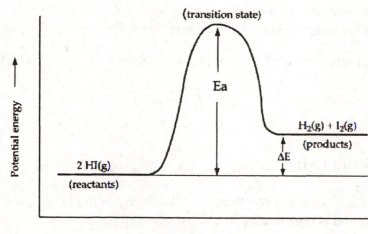

Reaction Mechanisms (Sections 12.11–12.13)

12.86 There is no relationship between the coefficients in a balanced chemical equation for an overall reaction and the exponents in the rate law unless the overall reaction occurs in a single elementary step, in which case the coefficients in the balanced equation are the exponents in the rate law.

12.88 (a)

$$H_2(g) + ICl(g) \rightarrow HI(g) + HCl(g)$$
$$\underline{HI(g) + ICl(g) \rightarrow I_2(g) + HCl(g)}$$

Overall reaction $H_2(g) + 2\ ICl(g) \rightarrow I_2(g) + 2\ HCl(g)$

(b) Because HI(g) is produced in the first step and consumed in the second step, it is a reaction intermediate.

(c) In each reaction there are two reactant molecules, so each elementary reaction is bimolecular.

12.90 (a) bimolecular, Rate = $k[O_3][Cl]$ (b) unimolecular, Rate = $k[NO_2]$
    (c) bimolecular, Rate = $k[ClO][O]$ (d) termolecular, Rate = $k[Cl]^2[N_2]$

12.92 (a)

$$NO_2Cl(g) \rightarrow NO_2(g) + Cl(g)$$
$$\underline{Cl(g) + NO_2Cl(g) \rightarrow NO_2(g) + Cl_2(g)}$$

Overall reaction $2\ NO_2Cl(g) \rightarrow 2\ NO_2(g) + Cl_2(g)$

(b) 1. unimolecular; 2. bimolecular

(c) Rate = $k[NO_2Cl]$

12.94 $NO_2(g) + F_2(g) \rightarrow NO_2F(g) + F(g)$ (slow)
$F(g) + NO_2(g) \rightarrow NO_2F(g)$ (fast)

12.96 (a) $2 NO(g) + O_2(g) \rightarrow 2 NO_2(g)$
(b) $Rate_{forward} = k_1[NO]^2$ and $Rate_{reverse} = k_{-1}[N_2O_2]$
Because of the equilibrium, $Rate_{forward} = Rate_{reverse}$, and $k_1[NO]^2 = k_{-1}[N_2O_2]$.

$$[N_2O_2] = \frac{k_1}{k_{-1}}[NO]^2$$

The rate law for the rate determining step is Rate = $-\Delta[NO]/\Delta t = 2k_2[N_2O_2][O_2]$ because two NO molecules are consumed in the overall reaction for every N_2O_2 that reacts in the second step. In this rate law substitute for $[N_2O_2]$. Rate = $2k_2 \dfrac{k_1}{k_{-1}}[NO]^2[O_2]$

(c) $k = \dfrac{2k_2 k_1}{k_{-1}}$

Catalysis (Sections 12.14 and 12.15)

12.98 A catalyst does participate in the reaction, but it is not consumed because it reacts in one step of the reaction and is regenerated in a subsequent step.

12.100 (a) $O_3(g) + O(g) \rightarrow 2 O_2(g)$
(b) Cl acts as a catalyst.
(c) ClO is a reaction intermediate.
(d) A catalyst reacts in one step and is regenerated in a subsequent step. A reaction intermediate is produced in one step and consumed in another.

12.102 (a)
$NH_2NO_2(aq) + OH^-(aq) \rightarrow NHNO_2^-(aq) + H_2O(l)$
$\underline{NHNO_2^-(aq) \rightarrow N_2O(g) + OH^-(aq)}$
Overall reaction $NH_2NO_2(aq) \rightarrow N_2O(g) + H_2O(l)$
(b) OH^- acts as a catalyst because it is used in the first step and regenerated in the second. $NHNO_2^-$ is a reaction intermediate because it is produced in the first step and consumed in the second.
(c) The rate will decrease because added acid decreases the concentration of OH^-, which appears in the rate law since it is a catalyst.

Chapter Problems

12.104 $2 AB_2 \rightarrow A_2 + 2 B_2$
(a) Measure the change in the concentration of AB_2 as a function of time.
(b) and (c) If a plot of $[AB_2]$ versus time is linear, the reaction is zeroth-order and $k = -$slope. If a plot of $\ln [AB_2]$ versus time is linear, the reaction is first-order and $k = -$slope. If a plot of $1/[AB_2]$ versus time is linear, the reaction is second-order and $k = $slope.

12.106 (a) Rate = $k[B_2][C]$

(b) $B_2 + C \rightarrow CB + B$ (slow)

$CB + A \rightarrow AB + C$ (fast)

(c) C is a catalyst. C does not appear in the chemical equation because it is consumed in the first step and regenerated in the second step.

12.108 The first maximum represents the potential energy of the transition state for the first step. The second maximum represents the potential energy of the transition state for the second step. The saddle point between the two maxima represents the potential energy of the intermediate products.

12.110 (a) The reaction rate will increase with an increase in temperature at constant volume.

(b) The reaction rate will decrease with an increase in volume at constant temperature because reactant concentrations will decrease.

(c) The reaction rate will increase with the addition of a catalyst.

(d) Addition of an inert gas at constant volume will not affect the reaction rate.

12.112 (a) Rate = $k[C_2H_4Br_2]^m[I^-]^n$

$$m = \frac{\ln\left(\dfrac{Rate_2}{Rate_1}\right)}{\ln\left(\dfrac{[C_2H_4Br_2]_2}{[C_2H_4Br_2]_1}\right)} = \frac{\ln\left(\dfrac{1.74 \times 10^{-4}}{6.45 \times 10^{-5}}\right)}{\ln\left(\dfrac{0.343}{0.127}\right)} = 1$$

$$n = \frac{\ln\left(\dfrac{Rate_3 \cdot [C_2H_4Br_2]_2}{Rate_2 \cdot [C_2H_4Br_2]_3}\right)}{\ln\left(\dfrac{[I^-]_3}{[I^-]_2}\right)} = \frac{\ln\left(\dfrac{(1.26 \times 10^{-4})(0.343)}{(1.74 \times 10^{-4})(0.203)}\right)}{\ln\left(\dfrac{0.125}{0.102}\right)} = 1$$

Rate = $k[C_2H_4Br_2][I^-]$

(b) From Experiment 1:

$$k = \frac{Rate}{[C_2H_4Br_2][I^-]} = \frac{6.45 \times 10^{-5} \text{ M/s}}{(0.127 \text{ M})(0.102 \text{ M})} = 4.98 \times 10^{-3}/(\text{M} \cdot \text{s})$$

(c) Rate = $k[C_2H_4Br_2][I^-]$ = [$4.98 \times 10^{-3}/(\text{M} \cdot \text{s})$](0.150 M)(0.150 M) = 1.12×10^{-4} M/s

12.114 $\ln \dfrac{[11-cis-retinal]_t}{[11-cis-retinal]_0} = -kt$, $k = 1.02 \times 10^{-5}/s$

(a) $t = 6 \text{ h} \times \dfrac{60 \text{ min}}{1 \text{ h}} \times \dfrac{60 \text{ s}}{1 \text{ min}} = 21,600 \text{ s}$

$\ln[\text{11-cis-retinal}]_t = -kt + \ln[\text{11-cis-retinal}]_0$

$\ln[\text{11-cis-retinal}]_t = -(1.02 \times 10^{-5}/s)(21{,}600\ s) + \ln(3.50 \times 10^{-3}) = -5.875$

$[\text{11-cis-retinal}]_t = e^{-5.875} = 2.81 \times 10^{-3}\ M$

(b) $[\text{11-cis-retinal}]_0 = 3.50 \times 10^{-3}\ M$

If 25% of the 11-cis-retinal reacts then 75% remains.

$[\text{11-cis-retinal}]_t = (0.75)(3.50 \times 10^{-3}\ M) = 2.625 \times 10^{-3}\ M$

$$t = \frac{\ln \dfrac{[\text{11-cis-retinal}]_t}{[\text{11-cis-retinal}]_0}}{-k} = \frac{\ln\left(\dfrac{2.625 \times 10^{-3}\ M}{3.50 \times 10^{-3}\ M}\right)}{-(1.02 \times 10^{-5}/s)} = 28{,}200\ s$$

$$t = 28{,}200\ s \times \frac{1\ \min}{60\ s} = 470\ \min$$

(c) $[\text{11-cis-retinal}]_t = [\text{11-cis-retinal}]_0 - [\text{11-trans-retinal}]_t$

$[\text{11-cis-retinal}]_t = (3.50 \times 10^{-3}) - (3.15 \times 10^{-3}) = 3.50 \times 10^{-4}\ M$

$$t = \frac{\ln \dfrac{[\text{11-cis-retinal}]_t}{[\text{11-cis-retinal}]_0}}{-k} = \frac{\ln\left(\dfrac{3.50 \times 10^{-4}\ M}{3.50 \times 10^{-3}\ M}\right)}{-(1.02 \times 10^{-5}/s)} = 225{,}744\ s$$

$$t = 225{,}744\ s \times \frac{1\ \min}{60\ s} \times \frac{1\ h}{60\ \min} = 63\ h$$

12.116   For $E_a = 50$ kJ/mol

$$f = e^{-E_a/RT} = \exp\left\{\frac{-50\ \text{kJ/mol}}{[8.314 \times 10^{-3}\ \text{kJ/(K}\cdot\text{mol)}](300\ K)}\right\} = 2.0 \times 10^{-9}$$

For $E_a = 100$ kJ/mol

$$f = e^{-E_a/RT} = \exp\left\{\frac{-100\ \text{kJ/mol}}{[8.314 \times 10^{-3}\ \text{kJ/(K}\cdot\text{mol)}](300\ K)}\right\} = 3.9 \times 10^{-18}$$

12.118 $[A] = -kt + [A]_0$

$[A]_0/2 = -kt_{1/2} + [A]_0$

$[A]_0/2 - [A]_0 = -kt_{1/2}$

$-[A]_0/2 = -kt_{1/2}$

$[A]_0/2 = kt_{1/2}$

For a zeroth-order reaction, $t_{1/2} = \dfrac{[A]_0}{2k}$

For a zeroth-order reaction, each half-life is half of the previous one.
For a first-order reaction, each half-life is the same as the previous one.
For a second-order reaction, each half-life is twice the previous one.

12.120 (a) $\text{Rate}_f = k_f[A]$ and $\text{Rate}_r = k_r[B]$

(b)

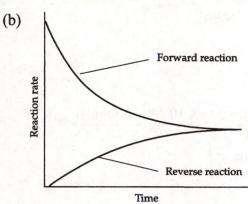

Reaction rate (y-axis) vs Time (x-axis), with curves labeled "Forward reaction" and "Reverse reaction"

(c) When $\text{Rate}_f = \text{Rate}_r$, $k_f[A] = k_r[B]$, and $\dfrac{[B]}{[A]} = \dfrac{k_f}{k_r} = \dfrac{(3.0 \times 10^{-3})}{(1.0 \times 10^{-3})} = 3$

12.122 $k = \dfrac{0.693}{t_{1/2}} = \dfrac{0.693}{5715 \text{ y}} = 1.21 \times 10^{-4}/\text{y}$

$$t = \dfrac{\ln \dfrac{(^{14}C)_t}{(^{14}C)_o}}{-k} = \dfrac{\ln \dfrac{(2.3)}{(15.3)}}{-1.21 \times 10^{-4}/\text{y}} = 1.6 \times 10^4 \text{ y}$$

12.124 $X \rightarrow$ products   is a first-order reaction

$t = 60 \text{ min} \times \dfrac{60 \text{ s}}{1 \text{ min}} = 3600 \text{ s}$

$\ln \dfrac{[X]_t}{[X]_o} = -kt; \qquad k = \dfrac{\ln \dfrac{[X]_t}{[X]_o}}{-t}$

At 25 °C, calculate k_1: $k_1 = \dfrac{\ln\left(\dfrac{0.600 \text{ M}}{1.000 \text{ M}}\right)}{-3600 \text{ s}} = 1.42 \times 10^{-4} \text{ s}^{-1}$

At 35 °C, calculate k_2: $k_2 = \dfrac{\ln\left(\dfrac{0.200 \text{ M}}{0.600 \text{ M}}\right)}{-3600 \text{ s}} = 3.05 \times 10^{-4} \text{ s}^{-1}$

At an unknown temperature calculate k_3: $k_3 = \dfrac{\ln\left(\dfrac{0.010 \text{ M}}{0.200 \text{ M}}\right)}{-3600 \text{ s}} = 8.32 \times 10^{-4} \text{ s}^{-1}$

$T_1 = 25 \text{ °C} = 25 + 273 = 298 \text{ K}$
$T_2 = 35 \text{ °C} = 35 + 273 = 308 \text{ K}$

Calculate E_a using k_1 and k_2.

$$\ln\left(\frac{k_2}{k_1}\right) = \left(\frac{-E_a}{R}\right)\left(\frac{1}{T_2} - \frac{1}{T_1}\right)$$

$$E_a = -\frac{[\ln k_2 - \ln k_1](R)}{\left(\frac{1}{T_2} - \frac{1}{T_1}\right)}$$

$$E_a = -\frac{[\ln(3.05 \times 10^{-4}) - \ln(1.42 \times 10^{-4})][8.314 \times 10^{-3} \text{ kJ/(K·mol)}]}{\left(\frac{1}{308 \text{ K}} - \frac{1}{298 \text{ K}}\right)} = 58.3 \text{ kJ/mol}$$

Use E_a, k_1, and k_3 to calculate T_3.

$$\frac{1}{T_3} = \frac{\ln\left(\frac{k_3}{k_1}\right)}{\left(\frac{-E_a}{R}\right)} + \frac{1}{T_1} = \frac{\ln\left(\frac{8.32 \times 10^{-4}}{1.42 \times 10^{-4}}\right)}{\left(\frac{-58.3 \text{ kJ/mol}}{8.314 \times 10^{-3} \text{ kJ/(K·mol)}}\right)} + \frac{1}{298 \text{ K}} = 0.003104/\text{K}$$

$$T_3 = \frac{1}{0.003104/\text{K}} = 322 \text{ K} = 322 - 273 = 49 \text{ °C}$$

At 3:00 p.m. raise the temperature to 49 °C to finish the reaction by 4:00 p.m.

12.126 (a) When equal volumes of two solutions are mixed, both concentrations are cut in half.
$[H_3O^+]_o = [OH^-]_o = 1.0$ M
When 99.999% of the acid is neutralized, $[H_3O^+] = [OH^-] = 1.0$ M $- (1.0$ M x 0.99999$)$
$= 1.0 \times 10^{-5}$ M

Using the 2nd order integrated rate law:

$$kt = \frac{1}{[H_3O^+]_t} - \frac{1}{[H_3O^+]_o}; \qquad t = \frac{1}{k}\left[\frac{1}{[H_3O^+]_t} - \frac{1}{[H_3O^+]_o}\right]$$

$$t = \frac{1}{(1.3 \times 10^{11} \text{ M}^{-1}\text{s}^{-1})}\left[\frac{1}{(1.0 \times 10^{-5} \text{ M})} - \frac{1}{(1.0 \text{ M})}\right] = 7.7 \times 10^{-7} \text{ s}$$

(b) The rate of an acid-base neutralization reaction would be limited by the speed of mixing, which is much slower than the intrinsic rate of the reaction itself.

12.128 Looking at the two experiments at 600 K, when the NO_2 concentration is doubled, the rate increased by a factor of 4. Therefore, the reaction is 2nd order.
Rate = k $[NO_2]^2$
Calculate k_1 at 600 K: $k_1 = $ Rate$/[NO_2]^2 = 5.4 \times 10^{-7}$ M s$^{-1}/(0.0010$ M$)^2 = 0.54$ M^{-1} s^{-1}
Calculate k_2 at 700 K: $k_2 = $ Rate$/[NO_2]^2 = 5.2 \times 10^{-5}$ M s$^{-1}/(0.0020$ M$)^2 = 13$ M^{-1} s^{-1}
Calculate E_a using k_1 and k_2.

$$\ln\left(\frac{k_2}{k_1}\right) = \left(\frac{-E_a}{R}\right)\left(\frac{1}{T_2} - \frac{1}{T_1}\right)$$

$$E_a = -\frac{[\ln k_2 - \ln k_1](R)}{\left(\dfrac{1}{T_2} - \dfrac{1}{T_1}\right)}$$

$$E_a = -\frac{[\ln(13) - \ln(0.54)][8.314 \times 10^{-3} \text{ kJ/(K} \cdot \text{mol)}]}{\left(\dfrac{1}{700 \text{ K}} - \dfrac{1}{600 \text{ K}}\right)} = 111 \text{ kJ/mol}$$

Calculate k_3 at 650 K using E_a and k_1.

Solve for k_3.

$$\ln k_3 = \frac{-E_a}{R}\left(\frac{1}{T_3} - \frac{1}{T_1}\right) + \ln k_1$$

$$\ln k_3 = \frac{-111 \text{ kJ/mol}}{[8.314 \times 10^{-3} \text{ kJ/(K} \cdot \text{mol)}]}\left(\frac{1}{650 \text{ K}} - \frac{1}{600 \text{ K}}\right) + \ln(0.54) = 1.0955$$

$$k_3 = e^{1.0955} = 3.0 \text{ M}^{-1} \text{ s}^{-1}$$

$$k_3 t = \frac{1}{[NO_2]_t} - \frac{1}{[NO_2]_o}; \qquad t = \frac{1}{k_3}\left[\frac{1}{[NO_2]_t} - \frac{1}{[NO_2]_o}\right]$$

$$t = \frac{1}{(3.0 \text{ M}^{-1}\text{s}^{-1})}\left[\frac{1}{(0.0010 \text{ M})} - \frac{1}{(0.0050 \text{ M})}\right] = 2.7 \times 10^2 \text{ s}$$

12.130 $A \rightarrow C$ is a first-order reaction.

The reaction is complete at 200 s when the absorbance of C reaches 1.200.

Because there is a one to one stoichiometry between A and C, the concentration of A must be proportional to 1.200 – absorbance of C. Any two data points can be used to find k. Let $[A]_o \propto 1.200$ and at 100 s, $[A]_t \propto 1.200 - 1.188 = 0.012$

$$\ln\frac{[A]_t}{[A]_o} = -kt; \qquad k = \frac{\ln\dfrac{[A]_t}{[A]_o}}{-t}; \qquad k = \frac{\ln\left(\dfrac{0.012 \text{ M}}{1.200 \text{ M}}\right)}{-100 \text{ s}} = 0.0461 \text{ s}^{-1}$$

$$t_{1/2} = \frac{0.693}{k} = \frac{0.693}{0.0461 \text{ s}^{-1}} = 15 \text{ s}$$

12.132   For radioactive decay, $\ln\dfrac{N}{N_o} = -kt$

For ^{235}U, $k_1 = \dfrac{0.693}{t_{1/2}} = \dfrac{0.693}{7.04 \times 10^8 \text{ y}} = 9.84 \times 10^{-10} \text{ y}^{-1}$

For ^{238}U, $k_2 = \dfrac{0.693}{t_{1/2}} = \dfrac{0.693}{4.47 \times 10^9 \text{ y}} = 1.55 \times 10^{-10} \text{ y}^{-1}$

For ^{235}U, $\ln\dfrac{N_1}{N_{o1}} = -k_1 t$ and $\ln\dfrac{N_1}{N_{o1}} + k_1 t = 0$

For ^{238}U,　$\ln \dfrac{N_2}{N_{o2}} = -k_2 t$ and $\ln \dfrac{N_2}{N_{o2}} + k_2 t = 0$

Set the two equations that are equal to zero equal to each other and solve for t.

$$\ln \dfrac{N_1}{N_{o1}} + k_1 t = \ln \dfrac{N_2}{N_{o2}} + k_2 t$$

$$\ln \dfrac{N_1}{N_{o1}} - \ln \dfrac{N_2}{N_{o2}} = k_2 t - k_1 t = (k_2 - k_1)t$$

$$\ln \dfrac{\left(\dfrac{N_1}{N_{o1}}\right)}{\left(\dfrac{N_2}{N_{o2}}\right)} = (k_2 - k_1)t, \text{ now } N_{o1} = N_{o2}, \text{ so } \ln\dfrac{N_1}{N_2} = (k_2 - k_1)t$$

$\dfrac{N_1}{N_2} = 7.25 \times 10^{-3}$, so $\ln(7.25 \times 10^{-3}) = (1.55 \times 10^{-10}\ y^{-1} - 9.84 \times 10^{-10}\ y^{-1})t$

$$t = \dfrac{-4.93}{-8.29 \times 10^{-10}\ y^{-1}} = 5.9 \times 10^9\ y$$

The age of the elements is 5.9×10^9 y (6 billion years).

12.134　$k = \dfrac{0.693}{t_{1/2}}$;　$500\ y \times \dfrac{365\ d}{1\ y} \times \dfrac{24\ h}{1\ d} \times \dfrac{60\ min}{1\ h} \times \dfrac{60\ s}{1\ min} = 1.58 \times 10^{10}\ s$

$k_1 = \dfrac{0.693}{1.58 \times 10^{10}\ s} = 4.39 \times 10^{-11}\ /s$ and $k_2 = \dfrac{0.693}{0.010\ s} = 69.3\ /s$

$\dfrac{k_2}{k_1} = \dfrac{69.3\ /s}{4.39 \times 10^{-11}\ /s} = 1.6 \times 10^{12}$

The enzyme increases the rate of the peptide bond breaking reaction by a factor of 1.6×10^{12}.

12.136　(a)

$$\ln\left(\dfrac{k_2}{k_1}\right) = \left(\dfrac{-E_a}{R}\right)\left(\dfrac{1}{T_2} - \dfrac{1}{T_1}\right)$$

$k_1 = 2.60 \times 10^{-4}\ /s$,　$T_1 = 530\ ^{\circ}C = 803\ K$
$k_2 = 9.45 \times 10^{-3}\ /s$,　$T_2 = 620\ ^{\circ}C = 893\ K$

$$E_a = -\dfrac{[\ln k_2 - \ln k_1](R)}{\left(\dfrac{1}{T_2} - \dfrac{1}{T_1}\right)}$$

$$E_a = -\frac{[\ln(9.45 \times 10^{-3}) - \ln(2.60 \times 10^{-4})][8.314 \times 10^{-3} \text{ kJ}/(\text{K} \cdot \text{mol})]}{\left(\dfrac{1}{893 \text{ K}} - \dfrac{1}{803 \text{ K}}\right)} = 238 \text{ kJ/mol}$$

(b) Calculate k_3 at 580 °C = 853 K using E_a and k_1.
Solve for k_3.

$$\ln k_3 = \frac{-E_a}{R}\left(\frac{1}{T_3} - \frac{1}{T_1}\right) + \ln k_1$$

$$\ln k_3 = \frac{-238 \text{ kJ/mol}}{[8.314 \times 10^{-3} \text{ kJ}/(\text{K} \cdot \text{mol})]}\left(\frac{1}{853 \text{ K}} - \frac{1}{803 \text{ K}}\right) + \ln(2.60 \times 10^{-4}) = -6.165$$

$$k_3 = e^{-6.165} = 2.10 \times 10^{-3} \text{ /s}$$

$$t_{1/2} = \frac{0.693}{k_3} = \frac{0.693}{2.10 \times 10^{-3} \text{ /s}} = 330 \text{ s}$$

Multiconcept Problems

12.138 $2 \text{ HI}(g) \rightarrow \text{H}_2(g) + \text{I}_2(g)$

(a) mass HI = $1.50 \text{ L} \times \dfrac{1000 \text{ mL}}{1 \text{ L}} \times \dfrac{0.0101 \text{ g}}{1 \text{ mL}} = 15.15 \text{ g HI}$

$15.15 \text{ g HI} \times \dfrac{1 \text{ mol HI}}{127.91 \text{ g HI}} = 0.118 \text{ mol HI}$

$[\text{HI}] = \dfrac{0.118 \text{ mol}}{1.50 \text{ L}} = 0.0787 \text{ mol/L}$

$-\dfrac{\Delta[\text{HI}]}{\Delta t} = k[\text{HI}]^2 = (0.031/(\text{M} \cdot \text{min}))(0.0787 \text{ M})^2 = 1.92 \times 10^{-4} \text{ M/min}$

$2 \text{ HI}(g) \rightarrow \text{H}_2(g) + \text{I}_2(g)$

$\dfrac{\Delta[\text{I}_2]}{\Delta t} = \dfrac{1}{2}\left(-\dfrac{\Delta[\text{HI}]}{\Delta t}\right) = \dfrac{1.92 \times 10^{-4} \text{ M/min}}{2} = 9.60 \times 10^{-5} \text{ M/min}$

$(9.60 \times 10^{-5} \text{ M/min})(1.50 \text{ L})(6.022 \times 10^{23} \text{ molecules/mol}) = 8.7 \times 10^{19} \text{ molecules/min}$

(b) Rate = $k[\text{HI}]^2$

$\dfrac{1}{[\text{HI}]_t} = kt + \dfrac{1}{[\text{HI}]_o} = (0.031/(\text{M} \cdot \text{min}))\left(8.00 \text{ h} \times \dfrac{60.0 \text{ min}}{1 \text{ h}}\right) + \dfrac{1}{0.0787 \text{ M}} = 27.59/\text{M}$

$[\text{HI}]_t = \dfrac{1}{27.59/\text{M}} = 0.0362 \text{ M}$

From stoichiometry, $[\text{H}_2]_t = 1/2 ([\text{HI}]_o - [\text{HI}]_t) = 1/2 (0.0787 \text{ M} - 0.0362 \text{ M}) = 0.0213 \text{ M}$
410 °C = 683 K
PV = nRT

$$P_{\text{H}_2} = \left(\frac{n}{V}\right)RT = (0.0213 \text{ mol/L})\left(0.082\ 06 \frac{\text{L} \cdot \text{atm}}{\text{K} \cdot \text{mol}}\right)(683 \text{ K}) = 1.2 \text{ atm}$$

12.140 (a) N_2O_5, 108.01

$$[N_2O_5]_o = \frac{\left(2.70 \text{ g } N_2O_5 \times \dfrac{1 \text{ mol } N_2O_5}{108.01 \text{ g } N_2O_5}\right)}{2.00 \text{ L}} = 0.0125 \text{ mol/L}$$

$$\ln [N_2O_5]_t = -kt + \ln [N_2O_5]_o = -(1.7 \times 10^{-3} \text{ s}^{-1})\left(13.0 \text{ min} \times \frac{60.0 \text{ s}}{1 \text{ min}}\right) + \ln (0.0125) = -5.71$$

$[N_2O_5]_t = e^{-5.71} = 3.31 \times 10^{-3}$ mol/L
After 13.0 min, mol $N_2O_5 = (3.31 \times 10^{-3}$ mol/L$)(2.00$ L$) = 6.62 \times 10^{-3}$ mol N_2O_5

$$N_2O_5(g) \rightarrow 2 \text{ NO}_2(g) + 1/2 \text{ O}_2(g)$$

before reaction (mol)	0.0250	0	0
change (mol)	$-x$	$+2x$	$+1/2x$
after reaction (mol)	$0.0250 - x$	$2x$	$1/2x$

After 13.0 min, mol $N_2O_5 = 6.62 \times 10^{-3} = 0.0250 - x$
$x = 0.0184$ mol

After 13.0 min, $n_{total} = n_{N_2O_5} + n_{NO_2} + n_{O_2} = (6.62 \times 10^{-3}) + 2(0.0184) + 1/2(0.0184)$

$n_{total} = 0.0526$ mol
55 °C = 328 K

$$PV = nRT; \quad P_{total} = \frac{nRT}{V} = \frac{(0.0526 \text{ mol})\left(0.082\ 06 \dfrac{\text{L} \cdot \text{atm}}{\text{K} \cdot \text{mol}}\right)(328 \text{ K})}{2.00 \text{ L}} = 0.71 \text{ atm}$$

(b) $N_2O_5(g) \rightarrow 2 \text{ NO}_2(g) + 1/2 \text{ O}_2(g)$
$\Delta H°_{rxn} = 2 \Delta H°_f(NO_2) - \Delta H°_f(N_2O_5)$
$\Delta H°_{rxn} = (2 \text{ mol})(33.2 \text{ kJ/mol}) - (1 \text{ mol})(11 \text{ kJ/mol}) = 55.4 \text{ kJ} = 5.54 \times 10^4$ J
initial rate $= k[N_2O_5]_o = (1.7 \times 10^{-3} \text{ s}^{-1})(0.0125 \text{ mol/L}) = 2.125 \times 10^{-5}$ mol/(L · s)
initial rate absorbing heat $= [2.125 \times 10^{-5}$ mol/(L · s)$](2.00 \text{ L})(5.54 \times 10^4 \text{ J/mol}) = 2.4$ J/s
(c)

$$\ln [N_2O_5]_t = -kt + \ln [N_2O_5]_o = -(1.7 \times 10^{-3} \text{ s}^{-1})\left(10.0 \text{ min} \times \frac{60.0 \text{ s}}{1 \text{ min}}\right) + \ln (0.0125) = -5.40$$

$[N_2O_5]_t = e^{-5.40} = 4.52 \times 10^{-3}$ mol/L
After 10.0 min, mol $N_2O_5 = (4.52 \times 10^{-3}$ mol/L$)(2.00$ L$) = 9.04 \times 10^{-3}$ mol N_2O_5

$$N_2O_5(g) \rightarrow 2 \text{ NO}_2(g) + 1/2 \text{ O}_2(g)$$

before reaction (mol)	0.0250	0	0
change (mol)	$-x$	$+2x$	$+1/2x$
after reaction (mol)	$0.0250 - x$	$2x$	$1/2x$

After 10.0 min, mol $N_2O_5 = 9.04 \times 10^{-3} = 0.0250 - x$
$x = 0.0160$ mol
heat absorbed $= (0.0160 \text{ mol})(55.4 \text{ kJ/mol}) = 0.89$ kJ

12.142 H_2O_2, 34.01

mass H_2O_2 = (0.500 L)(1000 mL/1 L)(1.00 g/ 1 mL)(0.0300) = 15.0 g H_2O_2

mol H_2O_2 = 15.0 g H_2O_2 x $\dfrac{1 \text{ mol } H_2O_2}{34.01 \text{ g } H_2O_2}$ = 0.441 H_2O_2

$[H_2O_2]_o$ = $\dfrac{0.441 \text{ mol}}{0.500 L}$ = 0.882 mol /L

k = $\dfrac{0.693}{t_{1/2}}$ = $\dfrac{0.693}{10.7 \text{ h}}$ = 6.48 x 10^{-2}/h

$\ln [H_2O_2]_t$ = $-kt$ + $\ln [H_2O_2]_o$
$\ln [H_2O_2]_t$ = $-(6.48 \times 10^{-2}$/h)(4.02 h) + $\ln (0.882)$
$\ln [H_2O_2]_t$ = -0.386; $[H_2O_2]_t$ = $e^{-0.386}$ = 0.680 mol/L
mol H_2O_2 = (0.680 mol/L)(0.500 L) = 0.340 mol

	2 H_2O_2(aq)	→	2 H_2O(l)	+	O_2(g)
before reaction (mol)	0.441		0		0
change (mol)	$- 2x$		$+2x$		$+x$
after reaction (mol)	$0.441 - 2x$		$2x$		x

After 4.02 h, mol H_2O_2 = 0.340 mol = 0.441 $- 2x$; solve for x.
$2x$ = 0.101
x = 0.0505 mol = mol O_2

P = 738 mm Hg x $\dfrac{1.00 \text{ atm}}{760 \text{ mm Hg}}$ = 0.971 atm

PV = nRT

$V = \dfrac{nRT}{P} = \dfrac{(0.0505 \text{ mol})\left(0.082\ 06\ \dfrac{\text{L}\cdot\text{atm}}{\text{K}\cdot\text{mol}}\right)(293 \text{ K})}{0.971 \text{ atm}}$ = 1.25 L

$P\Delta V$ = (0.971 atm)(1.25 L) = 1.21 L · atm

$w = -P\Delta V = -1.21 \text{ L·atm} = (-1.21 \text{ L} \cdot \text{atm})\left(101\ \dfrac{\text{J}}{\text{L·atm}}\right)$ = -122 J

13 Chemical Equilibrium: The Extent of Chemical Reactions

13.1 (a) $K_c = \dfrac{[SO_3]^2}{[SO_2]^2[O_2]}$ (b) $K_c = \dfrac{[SO_2]^2[O_2]}{[SO_3]^2}$

13.2 (a) $K_c = \dfrac{[SO_3]^2}{[SO_2]^2[O_2]} = \dfrac{(5.0 \times 10^{-2})^2}{(3.0 \times 10^{-3})^2(3.5 \times 10^{-3})} = 7.9 \times 10^4$

(b) $K_c = \dfrac{[SO_2]^2[O_2]}{[SO_3]^2} = \dfrac{(3.0 \times 10^{-3})^2(3.5 \times 10^{-3})}{(5.0 \times 10^{-2})^2} = 1.3 \times 10^{-5}$

13.3 (a) $K_c = \dfrac{[H^+][C_3H_5O_3^-]}{[C_3H_6O_3]}$

(b) $K_c = \dfrac{[(0.100)(0.0365)]^2}{[0.100 - (0.100)(0.0365)]} = 1.38 \times 10^{-4}$

13.4 From (1), $K_c = \dfrac{[AB][B]}{[A][B_2]} = \dfrac{(1)(2)}{(1)(2)} = 1$

For a mixture to be at equilibrium, $\dfrac{[AB][B]}{[A][B_2]}$ must be equal to 1.

For (2), $\dfrac{[AB][B]}{[A][B_2]} = \dfrac{(2)(1)}{(2)(1)} = 1$. This mixture is at equilibrium.

For (3), $\dfrac{[AB][B]}{[A][B_2]} = \dfrac{(1)(1)}{(4)(2)} = 0.125$. This mixture is not at equilibrium.

For (4), $\dfrac{[AB][B]}{[A][B_2]} = \dfrac{(2)(2)}{(4)(1)} = 1$. This mixture is at equilibrium.

13.5 (a) Because K_c is so large, k_f is larger than k_r.

(b) $K_c = \dfrac{k_f}{k_r}$; $k_r = \dfrac{k_f}{K_c} = \dfrac{8.5 \times 10^6 \text{ M}^{-1}\text{s}^{-1}}{3.4 \times 10^{34}} = 2.5 \times 10^{-28} \text{ M}^{-1}\text{ s}^{-1}$

(c) Because the reaction is exothermic, E_a (forward) is less than E_a (reverse). Consequently, as the temperature decreases, k_r decreases more than k_f decreases, and

therefore $K_c = \dfrac{k_f}{k_r}$ increases.

13.6 $\quad K_p = \dfrac{(P_{CO_2})(P_{H_2})}{(P_{CO})(P_{H_2O})} = \dfrac{(6.12)(20.3)}{(1.31)(10.0)} = 9.48$

13.7 $\quad 2\,NO(g) + O_2 \rightleftharpoons 2\,NO_2(g); \Delta n = 2 - 3 = -1$

$K_p = K_c(RT)^{\Delta n}, \qquad K_c = K_p(1/RT)^{\Delta n}$

at 500 K: $\quad K_p = (6.9 \times 10^5)[(0.082\,06)(500)]^{-1} = 1.7 \times 10^4$

at 1000 K: $\quad K_c = (1.3 \times 10^{-2})\left(\dfrac{1}{(0.082\,06)(1000)} \right)^{-1} = 1.1$

13.8 $\quad$ (a) $\;K_c = \dfrac{[H_2]^3}{[H_2O]^3}, \quad K_p = \dfrac{(P_{H_2})^3}{(P_{H_2O})^3}, \quad \Delta n = (3) - (3) = 0$ and $K_p = K_c$

$\quad$ (b) $\;K_c = [H_2]^2[O_2], \quad K_p = (P_{H_2})^2(P_{O_2}), \quad \Delta n = (3) - (0) = 3$ and $K_p = K_c(RT)^3$

$\quad$ (c) $\;K_c = \dfrac{[HCl]^4}{[SiCl_4][H_2]^2}, \quad K_p = \dfrac{(P_{HCl})^4}{(P_{SiCl_4})(P_{H_2})^2}, \quad \Delta n = (4) - (3) = 1$ and $K_p = K_c(RT)$

$\quad$ (d) $\;K_c = \dfrac{1}{[Hg_2^{2+}][Cl^-]^2}$

13.9 $\quad K_c = 1.2 \times 10^{-42}$. Because K_c is very small, the equilibrium mixture contains mostly H_2 molecules. H is in periodic group 1A. A very small value of K_c is consistent with strong bonding between 2 H atoms, each with one valence electron.

13.10 $\quad$ The container volume of 5.0 L must be included to calculate molar concentrations.

$\quad$ (a) $\quad Q_c = \dfrac{[NO_2]_t^2}{[NO]_t^2[O_2]_t} = \dfrac{(0.80\ mol/5.0\ L)^2}{(0.060\ mol/5.0\ L)^2(1.0\ mol/5.0\ L)} = 890$

Because $Q_c < K_c$, the reaction is not at equilibrium. The reaction will proceed to the right to reach equilibrium.

$\quad$ (b) $\quad Q_c = \dfrac{[NO_2]_t^2}{[NO]_t^2[O_2]_t} = \dfrac{(4.0\ mol/5.0\ L)^2}{(5.0 \times 10^{-3}\ mol/5.0\ L)^2(0.20\ mol/5.0\ L)} = 1.6 \times 10^7$

Because $Q_c > K_c$, the reaction is not at equilibrium. The reaction will proceed to the left to reach equilibrium.

13.11 $\quad K_c = \dfrac{[AB]^2}{[A_2][B_2]} = 4$; For a mixture to be at equilibrium, $\dfrac{[AB]^2}{[A_2][B_2]}$ must be equal to 4.

For (1), $Q_c = \dfrac{[AB]^2}{[A_2][B_2]} = \dfrac{(6)^2}{(1)(1)} = 36, Q_c > K_c$

For (2), $Q_c = \dfrac{[AB]^2}{[A_2][B_2]} = \dfrac{(4)^2}{(2)(2)} = 4, Q_c = K_c$

For (3), $Q_c = \dfrac{[AB]^2}{[A_2][B_2]} = \dfrac{(2)^2}{(3)(3)} = 0.44$, $Q_c < K_c$

(a) (2)   (b) (1), reverse; (3), forward

13.12 $K_c = \dfrac{[H]^2}{[H_2]} = 1.2 \times 10^{-42}$

(a) $[H] = \sqrt{K_c[H_2]} = \sqrt{(1.2 \times 10^{-42})(0.10)} = 3.5 \times 10^{-22}$ M

(b) H atoms $= (3.5 \times 10^{-22}$ mol/L$)(1.0$ L$)(6.022 \times 10^{23}$ atoms/mol$) = 210$ H atoms

H_2 molecules $= (0.10$ mol/L$)(1.0$ L$)(6.022 \times 10^{23}$ molecules/mol$) = 6.0 \times 10^{22}$ H_2 molecules

13.13

	$CO(g)$ +	$H_2O(g)$ ⇌	$CO_2(g)$ +	$H_2(g)$
initial (M)	0.150	0.150	0	0
change (M)	–x	–x	+x	+x
equil (M)	0.150 – x	0.150 – x	x	x

$K_c = 4.24 = \dfrac{[CO_2][H_2]}{[CO][H_2O]} = \dfrac{x^2}{(0.150 - x)^2}$

Take the square root of both sides and solve for x.

$\sqrt{4.24} = \sqrt{\dfrac{x^2}{(0.150 - x)^2}}$; $2.06 = \dfrac{x}{0.150 - x}$; $x = 0.101$

At equilibrium, $[CO_2] = [H_2] = x = 0.101$ M

$[CO] = [H_2O] = 0.150 - x = 0.150 - 0.101 = 0.049$ M

13.14

	$N_2O_4(g)$ ⇌	$2\,NO_2(g)$
initial (M)	0.0500	0
change (M)	–x	+2x
equil (M)	0.0500 – x	2x

$K_c = 4.64 \times 10^{-3} = \dfrac{[NO_2]^2}{[N_2O_4]} = \dfrac{(2x)^2}{(0.0500 - x)}$

$4x^2 + (4.64 \times 10^{-3})x - (2.32 \times 10^{-4}) = 0$

Use the quadratic formula to solve for x.

$x = \dfrac{-(4.64 \times 10^{-3}) \pm \sqrt{(4.64 \times 10^{-3})^2 - 4(4)(-2.32 \times 10^{-4})}}{2(4)} = \dfrac{-0.00464 \pm 0.06110}{8}$

$x = -0.008\ 22$ and $0.007\ 06$

Discard the negative solution (–0.008 22) because it leads to a negative concentration of NO_2 and that is impossible.

$[N_2O_4] = 0.0500 - x = 0.0500 - 0.007\ 06 = 0.0429$ M

$[NO_2] = 2x = 2(0.007\ 06) = 0.0141$ M

13.15 $N_2O_4(g) \rightleftharpoons 2 NO_2(g)$

$$Q_c = \frac{[NO_2]_t^2}{[N_2O_4]_t} = \frac{(0.0300 \text{ mol/L})^2}{(0.0200 \text{ mol/L})} = 0.0450; \quad Q_c > K_c$$

The reaction will approach equilibrium by going from right to left.

	$N_2O_4(g)$	$\rightleftharpoons$	$2 NO_2(g)$
initial (M)	0.0200		0.0300
change (M)	+x		−2x
equil (M)	0.0200 + x		0.0300 − 2x

$$K_c = 4.64 \times 10^{-3} = \frac{[NO_2]^2}{[N_2O_4]} = \frac{(0.0300 - 2x)^2}{(0.0200 + x)}$$

$4x^2 - 0.1246x + (8.072 \times 10^{-4}) = 0$

Use the quadratic formula to solve for x.

$$x = \frac{-(-0.1246) \pm \sqrt{(-0.1246)^2 - 4(4)(8.072 \times 10^{-4})}}{2(4)} = \frac{0.1246 \pm 0.05109}{8}$$

x = 0.0220 and 0.009 19

Discard the larger solution (0.0220) because it leads to a negative concentration of NO_2, and that is impossible.

$[N_2O_4] = 0.0200 + x = 0.0200 + 0.009\ 19 = 0.0292$ M

$[NO_2] = 0.0300 - 2x = 0.0300 - 2(0.009\ 19) = 0.0116$ M

13.16 $K_p = \dfrac{(P_{CO})(P_{H_2})}{(P_{H_2O})} = 2.44, \quad Q_p = \dfrac{(1.00)(1.40)}{(1.20)} = 1.17, \quad Q_p < K_p$ and the reaction goes to

the right to reach equilibrium.

	$C(s)$	+	$H_2O(g)$	$\rightleftharpoons$	$CO(g)$	+	$H_2(g)$
initial (atm)			1.20		1.00		1.40
change (atm)			−x		+x		+x
equil (atm)			1.20 − x		1.00 + x		1.40 + x

$$K_p = \frac{(P_{CO})(P_{H_2})}{(P_{H_2O})} = 2.44 = \frac{(1.00 + x)(1.40 + x)}{(1.20 - x)}$$

$x^2 + 4.84x - 1.53 = 0$

Use the quadratic formula to solve for x.

$$x = \frac{-(4.84) \pm \sqrt{(4.84)^2 - 4(1)(-1.53)}}{2(1)} = \frac{-4.84 \pm 5.44}{2}$$

x = −5.14 and 0.300

Discard the negative solution (−5.14) because it leads to negative partial pressures and that is impossible.

$P_{H_2O} = 1.20 - x = 1.20 - 0.300 = 0.90$ atm

$P_{CO} = 1.00 + x = 1.00 + 0.300 = 1.30$ atm

$P_{H_2} = 1.40 + x = 1.40 + 0.300 = 1.70$ atm

13.17 (a) CO (reactant) added, H_2 concentration increases.
(b) CO_2 (product) added, H_2 concentration decreases.
(c) H_2O (reactant) removed, H_2 concentration decreases.
(d) CO_2 (product) removed, H_2 concentration increases.

At equilibrium, $Q_c = K_c = \dfrac{[CO_2][H_2]}{[CO][H_2O]}$. If some CO_2 is removed from the

equilibrium mixture, the numerator in Q_c is decreased, which means that
$Q_c < K_c$ and the reaction will shift to the right, increasing the H_2 concentration.

13.18 (a) Because there are 2 mol of gas on both sides of the balanced equation, the
composition of the equilibrium mixture is unaffected by a change in pressure. The
number of moles of reaction products remains the same.
(b) Because there are 2 mol of gas on the left side and 1 mol of gas on the right side of
the balanced equation, the stress of an increase in pressure is relieved by a shift in the
reaction to the side with fewer moles of gas (in this case, to products). The number of
moles of reaction products increases.
(c) Because there is 1 mol of gas on the left side and 2 mol of gas on the right side of the
balanced equation, the stress of an increase in pressure is relieved by a shift in the
reaction to the side with fewer moles of gas (in this case, to reactants). The number of
moles of reaction product decreases.

13.19

13.20 Le Châtelier's principle predicts that a stress of added heat will be relieved by net
reaction in the direction that absorbs the heat. Since the reaction is endothermic, the
equilibrium will shift from left to right (K_c will increase) with an increase in temperature.
Therefore, the equilibrium mixture will contain more of the offending NO, the higher the
temperature.

13.21 The reaction is exothermic. As the temperature is increased, the reaction shifts from
right to left. The amount of ethyl acetate decreases.

$K_c = \dfrac{[CH_3CO_2C_2H_5][H_2O]}{[CH_3CO_2H][C_2H_5OH]}$

As the temperature is decreased, the reaction shifts from left to right. The product
concentrations increase, and the reactant concentrations decrease. This corresponds to
an increase in K_c.

13.22 There are more AB(g) molecules at the higher temperature. The equilibrium shifted to
the right at the higher temperature, which means the reaction is endothermic.

13.23   (a) A catalyst does not affect the equilibrium composition. The amount of CO remains the same.
(b) The reaction is exothermic. An increase in temperature shifts the reaction toward reactants. The amount of CO increases.
(c) Because there are 3 mol of gas on the left side and 2 mol of gas on the right side of the balanced equation, the stress of an increase in pressure is relieved by a shift in the reaction to the side with fewer moles of gas (in this case, to products). The amount of CO decreases.
(d) An increase in pressure as a result of the addition of an inert gas (with no volume change) does not affect the equilibrium composition. The amount of CO remains the same.
(e) Adding O_2 increases the O_2 concentration and shifts the reaction toward products. The amount of CO decreases.

13.24 $Hb + O_2 \rightleftharpoons Hb(O_2)$
If CO binds to Hb, Hb is removed from the reaction and the reaction will shift to the left, resulting in O_2 being released from $Hb(O_2)$. This will decrease the effectiveness of Hb for carrying O_2.

13.25   The partial pressure of O_2 in the atmosphere is 0.2095 atm.
$$PV = nRT$$
$$n = \frac{PV}{RT} = \frac{(0.2095 \text{ atm})(0.500 \text{ L})}{\left(0.082\ 06\ \frac{\text{L} \cdot \text{atm}}{\text{K} \cdot \text{mol}}\right)(298 \text{ K})} = 4.28 \times 10^{-3} \text{ mol } O_2$$

$$4.28 \times 10^{-3} \text{ mol } O_2 \times \frac{6.022 \times 10^{23} \ O_2 \text{ molecules}}{1 \text{ mol } O_2} = 2.58 \times 10^{21} \ O_2 \text{ molecules}$$

Conceptual Problems

13.26   (a) (1) and (3) because the number of A's and B's are the same in the third and fourth box.
(b) $K_c = \dfrac{[B]}{[A]} = \dfrac{6}{4} = 1.5$

(c) Because the same number of molecules appear on both sides of the equation, the volume terms in K_c all cancel. Therefore, we can calculate K_c without including the volume.

13.28   (a) Only reaction (3), $K_c = \dfrac{[A][AB]}{[A_2][B]} = \dfrac{(2)(4)}{(2)(2)} = 2$, is at equilibrium.

(b) $Q_c = \dfrac{[A][AB]}{[A_2][B]} = \dfrac{(3)(5)}{(1)(1)} = 15$ for reaction (1). Because $Q_c > K_c$, the reaction will go

in the reverse direction to reach equilibrium.

$Q_c = \dfrac{[A][AB]}{[A_2][B]} = \dfrac{(1)(3)}{(3)(3)} = 1/3$ for reaction (2). Because $Q_c < K_c$, the reaction will go in

the forward direction to reach equilibrium.

13.30   When the stopcock is opened, the reaction will go in the reverse direction because there will be initially an excess of AB molecules.

13.32   (a) AB → A + B
(b) The reaction is endothermic because a stress of added heat (higher temperature) shifts the AB ⇌ A + B equilibrium to the right.
(c) If the volume is increased, the pressure is decreased. The stress of decreased pressure will be relieved by a shift in the equilibrium from left to right, thus increasing the number of A atoms.

13.34   (a)   (b)   (c)

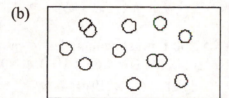
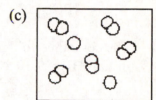

13.36   (a) A → 2 B
(b) (1) The reaction is exothermic. As the temperature is increased, the reaction shifts to the left. A increases, B decreases, and K_c decreases.
(2) When the volume decreases, the reaction shifts to the side with fewer gas molecules, which is towards the reactant. The amount of A increases.
(3) If there is no volume change, there is no change in the equilibrium composition, and the amount of A remains the same.
(4) A catalyst does not change the equilibrium composition, and the amount of A remains the same.

Section Problems
Equilibrium-Constant Expressions and Equilibrium Constants (Sections 13.1, 13.2, 13.4, and 13.5)

13.38   (a) $K_c = \dfrac{[CO][H_2]^3}{[CH_4][H_2O]}$ (b) $K_c = \dfrac{[ClF_3]^2}{[F_2]^3[Cl_2]}$ (c) $K_c = \dfrac{[HF]^2}{[H_2][F_2]}$

13.40   (a) $K_p = \dfrac{(P_{CO})(P_{H_2})^3}{(P_{CH_4})(P_{H_2O})}$, $\Delta n = 2$ and $K_p = K_c(RT)^2$

(b) $K_p = \dfrac{(P_{ClF_3})^2}{(P_{F_2})^3(P_{Cl_2})}$, $\Delta n = -2$ and $K_p = K_c(RT)^{-2}$

(c) $K_p = \dfrac{(P_{HF})^2}{(P_{H_2})(P_{F_2})}$, $\Delta n = 0$ and $K_p = K_c$

13.42 $K_c = \dfrac{[C_2H_5OC_2H_5][H_2O]}{[C_2H_5OH]^2}$

13.44 The two reactions are the reverse of each other.

$K_c(\text{reverse}) = \dfrac{1}{K_c(\text{forward})} = \dfrac{1}{7.5 \times 10^{-9}} = 1.3 \times 10^8$

13.46 $K_c = \dfrac{[PCl_3][Cl_2]}{[PCl_5]} = \dfrac{(1.5 \times 10^{-2})(3.2 \times 10^{-2})}{(8.3 \times 10^{-3})} = 0.058$

13.48 The container volume of 2.00 L must be included to calculate molar concentrations.
Initial [HI] = 9.30×10^{-3} mol/2.00 L = 4.65×10^{-3} M = 0.004 65 M

	$H_2(g)$	+	$I_2(g)$	$\rightleftarrows$	$2\ HI(g)$
initial (M)	0		0		0.004 65
change (M)	+x		+x		$-2x$
equil (M)	x		x		0.004 65 $-$ 2x

x = [H_2] = [I_2] = 6.29×10^{-4} M = 0.000 629 M
[HI] = 0.004 65 $-$ 2x = 0.004 65 $-$ 2(0.000 629) = 0.003 39 M

$K_c = \dfrac{[HI]^2}{[H_2][I_2]} = \dfrac{(0.003\ 39)^2}{(0.000\ 629)^2} = 29.0$

13.50 (a) $K_c = \dfrac{[CH_3CO_2C_2H_5][H_2O]}{[CH_3CO_2H][C_2H_5OH]}$

(b)

	$CH_3CO_2H(soln)$	+	$C_2H_5OH(soln)$	$\rightleftarrows$	$CH_3CO_2C_2H_5(soln)$	+	$H_2O(soln)$
initial (mol)	1.00		1.00		0		0
change (mol)	$-x$		$-x$		$+x$		$+x$
equil (mol)	1.00 $-$ x		1.00 $-$ x		x		x

x = 0.65 mol; 1.00 $-$ x = 0.35 mol; $K_c = \dfrac{(0.65)^2}{(0.35)^2} = 3.4$

Because there are the same number of molecules on both sides of the equation, the volume terms in K_c cancel. Therefore, we can calculate K_c without including the volume.

13.52 $\Delta n = (1) - (1 + 1) = -1$
$K_p = K_c(RT)^{\Delta n} = (2.2 \times 10^5)[(0.08206)(298)]^{-1} = 9.0 \times 10^3$

13.54 $K_p = P_{H_2O} = 0.0313$ atm; $\Delta n = 1$

$K_c = K_p\left(\dfrac{1}{RT}\right)^{\Delta n} = (0.0313)\left(\dfrac{1}{(0.082\ 06)(298)}\right) = 1.28 \times 10^{-3}$

13.56 (a) $K_c = \dfrac{[CO_2]^3}{[CO]^3}, \quad K_p = \dfrac{(P_{CO_2})^3}{(P_{CO})^3}$ (b) $K_c = \dfrac{1}{[O_2]^3}, \quad K_p = \dfrac{1}{(P_{O_2})^3}$

(c) $K_c = [SO_3], \quad K_p = P_{SO_3}$ (d) $K_c = [Ba^{2+}][SO_4^{2-}]$

Chemical Equilibrium and Chemical Kinetics (Section 13.3)

13.58 $A + B \rightleftharpoons C$

rate$_f$ = $k_f[A][B]$ and rate$_r$ = $k_r[C]$; at equilibrium, rate$_f$ = rate$_r$

$k_f[A][B] = k_r[C]; \quad \dfrac{k_f}{k_r} = \dfrac{[C]}{[A][B]} = K_c$

13.60 $K_c = \dfrac{k_f}{k_r} = \dfrac{0.13}{6.2 \times 10^{-4}} = 210$

13.62 k_r increases more than k_f, this means that E_a (reverse) is greater than E_a (forward). The reaction is exothermic when E_a (reverse) > E_a (forward).

13.64 Because the rate of the forward reaction is greater than the rate of the reverse reaction, the reaction will proceed in the forward direction to attain equilibrium.

Using the Equilibrium Constant (Section 13.6)

13.66 (a) Because K_c is very large, the equilibrium mixture contains mostly product.
(b) Because K_c is very small, the equilibrium mixture contains mostly reactants.

13.68 $K_c = 1.2 \times 10^{82}$ is very large. When equilibrium is reached, very little if any ethanol will remain because the reaction goes to completion.

13.70 The container volume of 10.0 L must be included to calculate molar concentrations.

$Q_c = \dfrac{[CS_2]_t[H_2]_t^4}{[CH_4]_t[H_2S]_t^2} = \dfrac{(3.0 \text{ mol}/10.0 \text{ L})(3.0 \text{ mol}/10.0 \text{ L})^4}{(2.0 \text{ mol}/10.0 \text{ L})(4.0 \text{ mol}/10.0 \text{ L})^2} = 7.6 \times 10^{-2}; \quad K_c = 2.5 \times 10^{-3}$

The reaction is not at equilibrium because $Q_c > K_c$. The reaction will proceed in the reverse direction to attain equilibrium.

13.72 $Q_p = \dfrac{(P_{P_2})(P_{H_2})^3}{(P_{PH_3})^2} = \dfrac{(0.871)(0.517)^3}{(0.0260)^2} = 178; \quad K_p = 398$

Because $Q_p < K_p$, the reaction will proceed in the forward direction to attain equilibrium.

13.74   (a) $Q_p = \dfrac{(P_{InH_2})}{(P_{In})(P_{H_2})} = \dfrac{(0.0760)}{(0.0600)(0.0350)} = 36.2; \quad K_p = 1.48$

Because $Q_p > K_p$, the reaction will proceed in the reverse direction to attain equilibrium.

(b)	$In(g)$	+	$H_2(g)$	$\rightleftharpoons$	$InH_2(g)$
initial (atm)	0.0600		0.0350		0.0760
change (atm)	+x		+x		−x
equil (atm)	0.0600 + x		0.0350 + x		0.0760 − x

$K_p = \dfrac{(P_{InH_2})}{(P_{In})(P_{H_2})} = 1.48 = \dfrac{(0.0760 - x)}{(0.0600 + x)(0.0350 + x)}$

$1.48 = \dfrac{(0.0760 - x)}{[(0.0600)(0.0350) + (0.0600x) + (0.0350x) + x^2]}$

$1.48x^2 + 1.1406x - 0.07289 = 0$

Use the quadratic formula to solve for x.

$x = \dfrac{-(1.1406) \pm \sqrt{(1.1406)^2 - 4(1.48)(-0.07289)}}{2(1.48)} = \dfrac{-1.1406 \pm 1.3162}{2.96}$

$x = -0.830$ and 0.0593

Discard the negative solution (−0.830) because it gives negative In and H_2 partial pressures and that is impossible.

$P_{In} = 0.0600 + 0.0593 = 0.1193$ atm

$P_{H_2} = 0.0350 + 0.0593 = 0.0943$ atm

$P_{InH_2} = 0.0760 - 0.0593 = 0.0167$ atm

13.76 $K_c = \dfrac{[NH_3]^2}{[N_2][H_2]^3} = 0.29; \quad$ At equilibrium, $[N_2] = 0.036$ M and $[H_2] = 0.15$ M

$[NH_3] = \sqrt{[N_2] \times [H_2]^3 \times K_c} = \sqrt{(0.036)(0.15)^3(0.29)} = 5.9 \times 10^{-3}$ M

13.78

	$N_2(g)$	+	$O_2(g)$	$\rightleftharpoons$	$2 NO(g)$
initial (M)	1.40		1.40		0
change (M)	−x		−x		+2x
equil (M)	1.40 − x		1.40 − x		2x

$K_c = \dfrac{[NO]^2}{[N_2][O_2]} = 1.7 \times 10^{-3} = \dfrac{(2x)^2}{(1.40 - x)^2}$

Take the square root of both sides and solve for x.

$\sqrt{1.7 \times 10^{-3}} = \sqrt{\dfrac{(2x)^2}{(1.40 - x)^2}}; \quad 4.1 \times 10^{-2} = \dfrac{2x}{1.40 - x}; \quad x = 2.8 \times 10^{-2}$

At equilibrium, $[NO] = 2x = 2(2.8 \times 10^{-2}) = 0.056$ M

$[N_2] = [O_2] = 1.40 - x = 1.40 - (2.8 \times 10^{-2}) = 1.37$ M

13.80

	L-α-lysine	⇌	L-β-lysine
initial (M)	0.003 00		0
change (M)	−x		+x
equil (M)	0.003 00 − x		x

$$K_c = \frac{[\text{L-}\beta\text{-lysine}]}{[\text{L-}\alpha\text{-lysine}]} = 7.20 = \frac{x}{(0.003\ 00 - x)}$$

$$0.0216 - 7.20x = x; \quad 0.0216 = 8.20x; \quad x = \frac{0.0216}{8.20} = 0.002\ 63\ \text{M}$$

[L-α-lysine] = 0.003 00 − x = 0.003 00 − 0.002 63 = 0.000 37 M

[L-β-lysine] = x = 0.002 63 M

13.82 (a) $$K_c = \frac{[CH_3CO_2C_2H_5][H_2O]}{[CH_3CO_2H][C_2H_5OH]} = 3.4 = \frac{(x)(12.0)}{(4.0)(6.0)}; \quad x = 6.8 \text{ moles } CH_3CO_2C_2H_5$$

Note that the volume cancels because the same number of molecules appear on both sides of the chemical equation.

(b)

	CH$_3$CO$_2$H(soln) +	C$_2$H$_5$OH(soln)	⇌	CH$_3$CO$_2$C$_2$H$_5$(soln) +	H$_2$O(soln)
initial (mol)	1.00	10.00		0	0
change (mol)	−x	−x		+x	+x
equil (mol)	1.00 − x	10.00 − x		x	x

$$K_c = 3.4 = \frac{x^2}{(1.00 - x)(10.00 - x)}$$

$$2.4x^2 - 37.4x + 34 = 0$$

Use the quadratic formula to solve for x.

$$x = \frac{-(-37.4) \pm \sqrt{(-37.4)^2 - 4(2.4)(34)}}{2(2.4)} = \frac{37.4 \pm 32.75}{4.8}$$

x = 0.969 and 14.6

Discard the larger solution (14.6) because it leads to negative concentrations and that is impossible.

mol CH$_3$CO$_2$H = 1.00 − x = 1.00 − 0.969 = 0.03 mol

mol C$_2$H$_5$OH = 10.00 − x = 10.00 − 0.969 = 9.03 mol

mol CH$_3$CO$_2$C$_2$H$_5$ = mol H$_2$O = x = 0.97 mol

13.84

	ClF$_3$(g)	⇌	ClF(g) +	F$_2$(g)
initial (atm)	1.47		0	0
change (atm)	−x		+x	+x
equil (atm)	1.47 − x		x	x

$$K_p = \frac{(P_{ClF})(P_{F_2})}{(P_{ClF_3})} = 0.140 = \frac{(x)(x)}{1.47 - x}; \quad \text{solve for x.}$$

$$x^2 + 0.140x - 0.2058 = 0$$

Use the quadratic formula to solve for x.

$$x = \frac{-(0.140) \pm \sqrt{(0.140)^2 - (4)(1)(-0.2058)}}{2(1)}$$

$$x = \frac{-0.140 \pm 0.918}{2}$$

$x = 0.389$ and -0.529

Discard the negative solution (-0.529) because it gives negative partial pressures and that is impossible.

$P_{ClF} = P_{F_2} = x = 0.389$ atm

$P_{ClF_3} = 1.47 - x = 1.47 - 0.389 = 1.08$ atm

Le Châtelier's Principle (Sections 13.7–13.11)

13.86   (a) Cl^- (reactant) added, $AgCl(s)$ increases
        (b) Ag^+ (reactant) added, $AgCl(s)$ increases
        (c) Ag^+ (reactant) removed, $AgCl(s)$ decreases
        (d) Cl^- (reactant) removed, $AgCl(s)$ decreases

Disturbing the equilibrium by decreasing $[Cl^-]$ increases Q_c $\left(Q_c = \dfrac{1}{[Ag^+]_t[Cl^-]_t} \right)$ to a

value greater than K_c. To reach a new state of equilibrium, Q_c must decrease, which means that the denominator must increase; that is, the reaction must go from right to left, thus decreasing the amount of solid $AgCl$.

13.88   (a) Because there are 2 mol of gas on the left side and 3 mol of gas on the right side of the balanced equation, the stress of an increase in pressure is relieved by a shift in the reaction to the side with fewer moles of gas (in this case, to reactants). The number of moles of reaction products decreases.
        (b) Because there are 2 mol of gas on both sides of the balanced equation, the composition of the equilibrium mixture is unaffected by a change in pressure. The number of moles of reaction product remains the same.
        (c) Because there are 2 mol of gas on the left side and 1 mol of gas on the right side of the balanced equation, the stress of an increase in pressure is relieved by a shift in the reaction to the side with fewer moles of gas (in this case, to products). The number of moles of reaction products increases.

13.90 $CO(g) + H_2O(g) \rightleftharpoons CO_2(g) + H_2(g)$ $\qquad\qquad$ $\Delta H° = -41.2$ kJ
        The reaction is exothermic. $[H_2]$ decreases when the temperature is increased.
        As the temperature is decreased, the reaction shifts to the right. $[CO_2]$ and $[H_2]$ increase, $[CO]$ and $[H_2O]$ decrease, and K_c increases.

13.92   (a) HCl is a source of Cl^- (product), the reaction shifts left, the equilibrium $[CoCl_4^{2-}]$ increases.
        (b) $Co(NO_3)_2$ is a source of $Co(H_2O)_6^{2+}$ (product), the reaction shifts left, the equilibrium $[CoCl_4^{2-}]$ increases.

(c) All concentrations will initially decrease and the reaction will shift to the right; the equilibrium $[CoCl_4^{2-}]$ decreases.

(d) For an exothermic reaction, the reaction shifts to the left when the temperature is increased; the equilibrium $[CoCl_4^{2-}]$ increases.

13.94 (a) The reaction is exothermic. The amount of CH_3OH (product) decreases as the temperature increases.

(b) When the volume decreases, the reaction shifts to the side with fewer gas molecules. The amount of CH_3OH increases.

(c) Addition of an inert gas (He) does not affect the equilibrium composition. There is no change.

(d) Addition of CO (reactant) shifts the reaction toward product. The amount of CH_3OH increases.

(e) Addition or removal of a catalyst does not affect the equilibrium composition. There is no change.

13.96 (a) add Au (a solid); no shift

(b) OH^- (product) increased; shift toward reactants.

(c) O_2 (reactant) partial pressure increased; shift toward products.

(d) Fe^{3+} decreases CN^- (reactant) by forming $Fe(CN)_6^{3-}$; shift toward reactants.

Chapter Problems

13.98 $2\ HI(g) \rightleftharpoons H_2(g) + I_2(g)$

Calculate K_c. $K_c = \dfrac{[H_2][I_2]}{[HI]^2} = \dfrac{(0.13)(0.70)}{(2.1)^2} = 0.0206$

$[HI] = \dfrac{0.20\ mol}{0.5000\ L} = 0.40\ M$

	$2\ HI(g)$	$\rightleftharpoons$	$H_2(g)$	+	$I_2(g)$
initial (M)	0.40		0		0
change (M)	$-2x$		$+x$		$+x$
equil (M)	$0.40 - 2x$		x		x

$K_c = 0.0206 = \dfrac{[H_2][I_2]}{[HI]^2} = \dfrac{x^2}{(0.40 - 2x)^2}$

Take the square root of both sides, and solve for x.

$\sqrt{0.0206} = \sqrt{\dfrac{x^2}{(0.40 - 2x)^2}}$

$0.144 = \dfrac{x}{0.40 - 2x}$; $x = 0.045$

At equilibrium, $[H_2] = [I_2] = x = 0.045\ M$; $[HI] = 0.40 - 2x = 0.40 - 2(0.045) = 0.31\ M$

13.100 $[H_2O] = \dfrac{6.00 \text{ mol}}{5.00 \text{ L}} = 1.20 \text{ M}$

	$C(s)$	+	$H_2O(g)$	$\rightleftarrows$	$CO(g)$	+	$H_2(g)$
initial (M)			1.20		0		0
change (M)			$-x$		$+x$		$+x$
equil (M)			$1.20 - x$		x		x

$K_c = \dfrac{[CO][H_2]}{[H_2O]} = 3.0 \times 10^{-2} = \dfrac{x^2}{1.20 - x}$

$x^2 + (3.0 \times 10^{-2})x - 0.036 = 0$

Use the quadratic formula to solve for x.

$x = \dfrac{-(0.030) \pm \sqrt{(0.030)^2 - 4(-0.036)}}{2(1)} = \dfrac{-0.030 \pm 0.381}{2}$

$x = 0.176$ and -0.206

Discard the negative solution (-0.206) because it leads to negative concentrations and that is impossible.

$[CO] = [H_2] = x = 0.18 \text{ M};$ $[H_2O] = 1.20 - x = 1.20 - 0.18 = 1.02 \text{ M}$

13.102 A decrease in volume (a) and the addition of reactants (c) will affect the composition of the equilibrium mixture, but leave the value of K_c unchanged.
A change in temperature (b) affects the value of K_c.
Addition of a catalyst (d) or an inert gas (e) affects neither the composition of the equilibrium mixture nor the value of K_c.

13.104 (a) $[PCl_5] = 1.000 \text{ mol}/5.000 \text{ L} = 0.2000 \text{ M}$

	$PCl_5(g)$	$\rightleftarrows$	$PCl_3(g)$	+	$Cl_2(g)$
initial (M)	0.2000		0		0
change (M)	$-(0.2000)(0.7850)$		$+(0.2000)(0.7850)$		$+(0.2000)(0.7850)$
equil (M)	0.0430		0.1570		0.1570

$K_c = \dfrac{[PCl_3][Cl_2]}{[PCl_5]} = \dfrac{(0.1570)(0.1570)}{(0.0430)} = 0.573$

$\Delta n = 1$ and $K_p = K_c(RT) = (0.573)(0.082\,06)(500) = 23.5$

(b) $Q_c = \dfrac{[PCl_3][Cl_2]}{[PCl_5]} = \dfrac{(0.150)(0.600)}{(0.500)} = 0.18$

Because $Q_c < K_c$, the reaction proceeds to the right to reach equilibrium.

	$PCl_5(g)$	$\rightleftarrows$	$PCl_3(g)$	+	$Cl_2(g)$
initial (M)	0.500		0.150		0.600
change (M)	$-x$		$+x$		$+x$
equil (M)	$0.500 - x$		$0.150 + x$		$0.600 + x$

$K_c = \dfrac{[PCl_3][Cl_2]}{[PCl_5]} = 0.573 = \dfrac{(0.150 + x)(0.600 + x)}{(0.500 - x)}$; solve for x.

$x^2 + 1.323x - 0.1965 = 0$

$$x = \frac{-(1.323) \pm \sqrt{(1.323)^2 - (4)(1)(-0.1965)}}{2(1)} = \frac{-1.323 \pm 1.593}{2}$$

$x = -1.458$ and 0.135

Discard the negative solution (-1.458) because it will lead to negative concentrations and that is impossible.

$[PCl_5] = 0.500 - x = 0.500 - 0.135 = 0.365$ M

$[PCl_3] = 0.150 + x = 0.150 + 0.135 = 0.285$ M

$[Cl_2] = 0.600 + x = 0.600 + 0.135 = 0.735$ M

13.106 (a) $K_c = \dfrac{[C_2H_6][C_2H_4]}{[C_4H_{10}]}$ $\qquad$ $K_p = \dfrac{(P_{C_2H_6})(P_{C_2H_4})}{P_{C_4H_{10}}}$

(b) $K_p = 12$; $\Delta n = 1$; $K_c = K_p\left(\dfrac{1}{RT}\right) = (12)\left(\dfrac{1}{(0.082\ 06)(773)}\right) = 0.19$

(c) $\qquad\qquad\quad$ $C_4H_{10}(g)$ $\rightleftarrows$ $C_2H_6(g)$ + $C_2H_4(g)$

initial (atm)	50	0	0
change (atm)	$-x$	$+x$	$+x$
equil (atm)	$50 - x$	x	x

$K_p = 12 = \dfrac{x^2}{50 - x}$; $\qquad x^2 + 12x - 600 = 0$

Use the quadratic formula to solve for x.

$$x = \frac{(-12) \pm \sqrt{(12)^2 - 4(1)(-600)}}{2(1)} = \frac{-12 \pm 50.44}{2}$$

$x = -31.22$ and 19.22

Discard the negative solution (-31.22) because it leads to negative concentrations and that is impossible.

% C_4H_{10} converted $= \dfrac{19.22}{50} \times 100\% = 38\%$

$P_{total} = P_{C_4H_{10}} + P_{C_2H_6} + P_{C_2H_4} = (50 - x) + x + x = (50 - 19) + 19 + 19 = 69$ atm

(d) A decrease in volume would decrease the % conversion of C_4H_{10}.

13.108 (a) $K_p = 3.45$; $\Delta n = 1$; $K_c = K_p\left(\dfrac{1}{RT}\right) = (3.45)\left(\dfrac{1}{(0.082\ 06)(500)}\right) = 0.0840$

(b) $[(CH_3)_3CCl] = 1.00$ mol/5.00 L $= 0.200$ M

$\qquad\qquad\qquad$ $(CH_3)_3CCl(g)$ $\rightleftarrows$ $(CH_3)_2C{=}CH_2(g)$ + $HCl(g)$

initial (M)	0.200	0	0
change (M)	$-x$	$+x$	$+x$
equil (M)	$0.200 - x$	x	x

$K_c = 0.0840 = \dfrac{x^2}{0.200 - x}$; $\qquad x^2 + 0.0840x - 0.0168 = 0$

Use the quadratic formula to solve for x.

$$x = \frac{(-0.0840) \pm \sqrt{(0.0840)^2 - 4(1)(-0.0168)}}{2(1)} = \frac{-0.0840 \pm 0.272}{2}$$

$x = -0.178$ and 0.094

Discard the negative solution (-0.178) because it leads to negative concentrations and that is impossible.

$[(CH_3)_2C=CCH_2] = [HCl] = x = 0.094$ M

$[(CH_3)_3CCl] = 0.200 - x = 0.200 - 0.094 = 0.106$ M

(c) $K_p = 3.45$

	$(CH_3)_3CCl(g)$	$\rightleftarrows$	$(CH_3)_2C=CH_2(g)$	+	$HCl(g)$
initial (atm)	0		0.400		0.600
change (atm)	$+x$		$-x$		$-x$
equil (atm)	x		$0.400 - x$		$0.600 - x$

$$K_p = 3.45 = \frac{(0.400 - x)(0.600 - x)}{x}$$

$x^2 - 4.45x + 0.240 = 0$

Use the quadratic formula to solve for x.

$$x = \frac{-(-4.45) \pm \sqrt{(-4.45)^2 - 4(1)(0.240)}}{2(1)} = \frac{4.45 \pm 4.34}{2}$$

$x = 0.055$ and 4.40

Discard the larger solution (4.40) because it leads to negative partial pressures and that is impossible.

$P_{t-butyl\ chloride} = x = 0.055$ atm; $P_{isobutylene} = 0.400 - x = 0.400 - 0.055 = 0.345$ atm

$P_{HCl} = 0.600 - x = 0.600 - 0.055 = 0.545$ atm

13.110 The activation energy (E_a) is positive, and for an exothermic reaction, $E_{a,r} > E_{a,f}$.

$$k_f = A_f\, e^{-E_{a,f}/RT}, \quad k_r = A_r\, e^{-E_{a,r}/RT}$$

$$K_c = \frac{k_f}{k_r} = \frac{A_f e^{-E_{a,f}/RT}}{A_r e^{-E_{a,r}/RT}} = \frac{A_f}{A_r} e^{(E_{a,r} - E_{a,f})/RT}$$

$(E_{a,r} - E_{a,f})$ is positive, so the exponent is always positive. As the temperature increases, the exponent, $(E_{a,r} - E_{a,f})/RT$, decreases and the value for K_c decreases as well.

13.112 (a) $PV = nRT;$ $n_{total} = \dfrac{PV}{RT} = \dfrac{(0.588\ atm)(1.00\ L)}{\left(0.082\ 06\ \dfrac{L \cdot atm}{K \cdot mol}\right)(300\ K)} = 0.0239$ mol

	$2\ NOBr(g)$	$\rightleftarrows$	$2\ NO(g)$	+	$Br_2(g)$
initial (mol)	0.0200		0		0
change (mol)	$-2x$		$+2x$		$+x$
equil (mol)	$0.0200 - 2x$		$2x$		x

$n_{total} = 0.0239 \text{ mol} = (0.0200 - 2x) + 2x + x = 0.0200 + x$

$x = 0.0239 - 0.0200 = 0.0039 \text{ mol}$

Because the volume is 1.00 L, the molarity equals the number of moles.

$[NOBr] = 0.0200 - 2x = 0.0200 - 2(0.0039) = 0.0122 \text{ M}$

$[NO] = 2x = 2(0.0039) = 0.0078 \text{ M}$

$[Br_2] = x = 0.0039 \text{ M}$

$$K_c = \frac{[NO]^2[Br_2]}{[NOBr]^2} = \frac{(0.0078)^2(0.0039)}{(0.0122)^2} = 1.6 \times 10^{-3}$$

(b) $\Delta n = (3) - (2) = 1$, $K_p = K_c(RT) = (1.6 \times 10^{-3})(0.082\ 06)(300) = 0.039$

13.114 (a) $W(s) + 4 Br(g) \rightleftarrows WBr_4(g)$

$$K_p = \frac{P_{WBr_4}}{(P_{Br})^4} = 100, \quad P_{WBr_4} = (P_{Br})^4(100) = (0.010 \text{ atm})^4(100) = 1.0 \times 10^{-6} \text{ atm}$$

(b) Because K_p is smaller at the higher temperature, the reaction has shifted toward reactants at the higher temperature, which means the reaction is exothermic.

(c) At 2800 K, $Q_p = \dfrac{(1.0 \times 10^{-6})}{(0.010)^4} = 100$, $Q_p > K_p$ so the reaction will go from products

to reactants, depositing tungsten back onto the filament.

13.116 $2 NO_2(g) \rightleftarrows N_2O_4(g)$

$\Delta n = (1) - (2) = -1$ and $K_p = K_c(RT)^{-1} = (216)[(0.082\ 06)(298)]^{-1} = 8.83$

$$K_p = \frac{P_{N_2O_4}}{(P_{NO_2})^2} = 8.83; \qquad \text{Let } X = P_{N_2O_4} \text{ and } Y = P_{NO_2}$$

$P_{total} = 1.50 \text{ atm} = X + Y$ and $\dfrac{X}{Y^2} = 8.83$. Use these two equations to solve for X and Y.

$X = 1.50 - Y$

$\dfrac{1.50 - Y}{Y^2} = 8.83$

$8.83Y^2 + Y - 1.50 = 0$

Use the quadratic formula to solve for Y.

$$Y = \frac{-(1) \pm \sqrt{(1)^2 - 4(8.83)(-1.50)}}{2(8.83)} = \frac{-1 \pm 7.35}{17.7}$$

$Y = -0.472$ and 0.359

Discard the negative solution (−0.472) because it leads to a negative partial pressure of NO_2 and that is impossible.

$Y = P_{NO_2} = 0.359 \text{ atm}$

$X = P_{N_2O_4} = 1.50 \text{ atm} - Y = 1.50 \text{ atm} - 0.359 \text{ atm} = 1.14 \text{ atm}$

13.118 (a) $P_{NO} = X_{NO} \cdot P_{total} = \dfrac{2.0 \text{ mol NO}}{(2.0 \text{ mol NO} + 3.0 \text{ mol NO}_2)} \times (1.65 \text{ atm}) = 0.66 \text{ atm}$

$P_{NO_2} = X_{NO_2} \cdot P_{total} = \dfrac{3.0 \text{ mol NO}_2}{(2.0 \text{ mol NO} + 3.0 \text{ mol NO}_2)} \times (1.65 \text{ atm}) = 0.99 \text{ atm}$

$$NO(g) \ + \ NO_2(g) \ \rightleftharpoons \ N_2O_3(g)$$

initial (atm)   0.66   0.99   0

change (atm)   –x   –x   +x

equil (atm)   0.66 – x   0.99 – x   x

$K_c = 13$; $\Delta n = (1) - (1 + 1) = -1$; $K_p = K_c(RT)^{-1} = (13)[(0.082\ 06)(298)]^{-1} = 0.53$

$K_p = \dfrac{P_{N_2O_3}}{(P_{NO})(P_{NO_2})} = 0.53 = \dfrac{x}{(0.66 - x)(0.99 - x)}$

$0.53x^2 - 1.87x + 0.35 = 0$

Use the quadratic formula to solve for x.

$x = \dfrac{-(-1.87) \pm \sqrt{(-1.87)^2 - 4(0.53)(0.35)}}{2(0.53)} = \dfrac{1.87 \pm 1.66}{1.06}$

$x = 3.33$ and 0.20

Discard the larger solution (3.33) because it leads to a negative concentration of NO and NO_2 and that is impossible.

$P_{NO} = 0.66 - x = 0.66 - 0.20 = 0.46 \text{ atm}$

$P_{NO_2} = 0.99 - x = 0.99 - 0.20 = 0.79 \text{ atm}$

$P_{N_2O_3} = x = 0.20 \text{ atm}$

(b) $PV = nRT$; $V = \dfrac{nRT}{P} = \dfrac{(2.0 \text{ mol} + 3.0 \text{ mol})\left(0.082\ 06 \dfrac{L \cdot atm}{K \cdot mol}\right)(298 \text{ K})}{1.65 \text{ atm}} = 74 \text{ L}$

13.120 $$N_2(g) \ + \ 3\,H_2(g) \ \rightleftharpoons \ 2\,NH_3(g)$$

initial (mol)   0   0   X

change (mol)   +y   +3y   –2y

equil (mol)   y   3y   X – 2y

$y = 0.200 \text{ mol}$

Because the volume is 1.00 L, the molarity equals the number of moles.

$[N_2] = y = 0.200 \text{ M}$; $[H_2] = 3y = 3(0.200) = 0.600 \text{ M}$

$K_c = \dfrac{[NH_3]^2}{[N_2][H_2]^3} = \dfrac{[NH_3]^2}{(0.200)(0.600)^3} = 4.20$, solve for $[NH_3]_{eq}$

$[NH_3]_{eq}^2 = [N_2][H_2]^3(4.20) = (0.200)(0.600)^3(4.20)$

$[NH_3]_{eq} = \sqrt{[N_2][H_2]^3(4.20)} = \sqrt{(0.200)(0.600)^3(4.20)} = 0.426 \text{ M}$

$[NH_3]_{eq} = 0.426 \text{ M} = X - 2(0.200) = [NH_3]_o - 2(0.200)$

$[NH_3]_o = 0.426 + 2(0.200) = 0.826 \text{ M}$

0.826 mol of NH_3 was placed in the 1.00 L reaction vessel.

13.122 ClF_3, 92.45

(a) mol ClF_3 = 9.25 g x $\dfrac{1 \text{ mol } ClF_3}{92.45 \text{ g}}$ = 0.100 mol ClF_3

$[ClF_3]$ = $\dfrac{0.100 \text{ mol } ClF_3}{2.00 \text{ L}}$ = 0.0500 M

$$
\begin{array}{lccc}
 & ClF_3(g) & \rightleftharpoons \quad ClF(g) & + \quad F_2(g) \\
\text{initial (M)} & 0.0500 & 0 & 0 \\
\text{change (M)} & -x & +x & +x \\
\text{equil (M)} & 0.0500 - x & x & x \\
 & 0.0401 & 0.009\ 90 & 0.009\ 90
\end{array}
$$

where x = 0.0500 x 0.198 = 0.009 90

K_c = $\dfrac{[ClF][F_2]}{[ClF_3]}$ = $\dfrac{(0.009\ 90)^2}{0.0401}$ = 0.002 44

(b) $K_p = K_c(RT)^{\Delta n}$; $\Delta n = 2 - 1 = 1$; $K_p = K_c(RT) = (0.002\ 44)(0.082\ 06)(700) = 0.140$

(c) mol ClF_3 = 39.4 g x $\dfrac{1 \text{ mol } ClF_3}{92.45 \text{ g}}$ = 0.426 mol ClF_3

$[ClF_3]$ = $\dfrac{0.426 \text{ mol } ClF_3}{2.00 \text{ L}}$ = 0.213 M

$$
\begin{array}{lccc}
 & ClF_3(g) & \rightleftharpoons \quad ClF(g) & + \quad F_2(g) \\
\text{initial (M)} & 0.213 & 0 & 0 \\
\text{change (M)} & -x & +x & +x \\
\text{equil (M)} & 0.213 - x & x & x
\end{array}
$$

K_c = $\dfrac{[ClF][F_2]}{[ClF_3]}$ = 0.00 244 = $\dfrac{x^2}{0.213 - x}$

$(0.002\ 44)(0.213 - x) = x^2$

$5.20 \times 10^{-4} - 0.002\ 44x = x^2$

$x^2 + 0.002\ 44x - (5.20 \times 10^{-4}) = 0$

Use the quadratic formula to solve for x.

$x = \dfrac{-(0.002\ 44) \pm \sqrt{(0.002\ 44)^2 - 4(1)(-5.20 \times 10^{-4})}}{2(1)} = \dfrac{-0.002\ 44 \pm 0.0457}{2}$

x = −0.0241 and 0.0216

Discard the negative solution (−0.0241) because it leads to a negative partial pressure and that is impossible.

$[ClF_3]$ = 0.213 − x = 0.213 − 0.0216 = 0.191 M

$[ClF]$ = $[F_2]$ = x = 0.0216 M

13.124

	Fumarate	$\rightleftharpoons$	L-Malate
initial (M)	0.001 56		0.002 27
change (M)	$-x$		$+x$
equil (M)	0.001 56 – x		0.002 27 + x

$$K_c = \frac{[\text{L-Malate}]}{[\text{Fumarate}]} = 3.3 = \frac{0.002\ 27 + x}{0.001\ 56 - x}$$

$(3.3)(0.001\ 56 - x) = 0.002\ 27 + x$

$0.005\ 15 - 3.3x = 0.002\ 27 + x$

$0.002\ 88 = 4.3x$

$$x = \frac{0.002\ 88}{4.3} = 6.7 \times 10^{-4} = 0.000\ 67$$

$[\text{Fumarate}] = 0.001\ 56 - x = 0.001\ 56 - 0.000\ 67 = 8.9 \times 10^{-4}\ \text{M}$

$[\text{L-Malate}] = 0.002\ 27 + x = 0.002\ 27 + 0.000\ 67 = 2.94 \times 10^{-3}\ \text{M}$

Multiconcept Problems

13.126 (a) $[\text{N}_2\text{O}_4] = \dfrac{0.500\ \text{mol}}{4.00\ \text{L}} = 0.125\ \text{M}$

	$\text{N}_2\text{O}_4(g)$	$\rightleftharpoons$	$2\ \text{NO}_2(g)$
initial (M)	0.125		0
change (M)	$-(0.793)(0.125)$		$+(2)(0.793)(0.125)$
equil (M)	$0.125 - (0.793)(0.125)$		$(2)(0.793)(0.125)$

At equilibrium, $[\text{N}_2\text{O}_4] = 0.125 - (0.793)(0.125) = 0.0259\ \text{M}$

$[\text{NO}_2] = (2)(0.793)(0.125) = 0.198\ \text{M}$

$$K_c = \frac{[\text{NO}_2]^2}{[\text{N}_2\text{O}_4]} = \frac{(0.198)^2}{(0.0259)} = 1.51$$

$\Delta n = 2 - 1 = 1$ and $K_p = K_c(RT)^{\Delta n}$; $K_p = K_c(RT) = (1.51)(0.082\ 06)(400) = 49.6$

(b)

13.128 2 monomer $\rightleftharpoons$ dimer

(a) In benzene, $K_c = 1.51 \times 10^2$

	2 monomer	$\rightleftharpoons$	dimer
initial (M)	0.100		0
change (M)	$-2x$		$+x$
equil (M)	0.100 – 2x		x

$$K_c = \frac{[\text{dimer}]}{[\text{monomer}]^2} = 1.51 \times 10^2 = \frac{x}{(0.100 - 2x)^2}$$

$604x^2 - 61.4x + 1.51 = 0$

Use the quadratic formula to solve for x.

$$x = \frac{-(-61.4) \pm \sqrt{(-61.4)^2 - (4)(604)(1.51)}}{2(604)} = \frac{61.4 \pm 11.04}{1208}$$

$x = 0.0600$ and 0.0417

Discard the larger solution (0.0600) because it gives a negative concentration of the monomer and that is impossible.

[monomer] = $0.100 - 2x = 0.100 - 2(0.0417) = 0.017$ M;   [dimer] = $x = 0.0417$ M

$$\frac{[dimer]}{[monomer]} = \frac{0.0417 \text{ M}}{0.017 \text{ M}} = 2.5$$

(b) In H_2O, $K_c = 3.7 \times 10^{-2}$

	2 monomer	$\rightleftharpoons$	dimer
initial (M)	0.100		0
change (M)	$-2x$		$+x$
equil (M)	$0.100 - 2x$		x

$$K_c = \frac{[dimer]}{[monomer]^2} = 3.7 \times 10^{-2} = \frac{x}{(0.100 - 2x)^2}$$

$0.148x^2 - 1.0148x + 0.000\,37 = 0$

Use the quadratic formula to solve for x.

$$x = \frac{-(-1.0148) \pm \sqrt{(-1.0148)^2 - (4)(0.148)(0.00037)}}{2(0.148)} = \frac{1.0148 \pm 1.0147}{0.296}$$

$x = 6.86$ and 3.4×10^{-4}

Discard the larger solution (6.86) because it gives a negative concentration of the monomer and that is impossible.

[monomer] = $0.100 - 2x = 0.100 - 2(3.4 \times 10^{-4}) = 0.099$ M;   [dimer] = $x = 3.4 \times 10^{-4}$ M

$$\frac{[dimer]}{[monomer]} = \frac{3.4 \times 10^{-4} \text{ M}}{0.099 \text{ M}} = 0.0034$$

(c) K_c for the water solution is so much smaller than K_c for the benzene solution because H_2O can hydrogen bond with acetic acid, thus preventing acetic acid dimer formation. Benzene cannot hydrogen bond with acetic acid.

13.130 (a) CO_2, 44.01; CO, 28.01

$$79.2 \text{ g } CO_2 \times \frac{1 \text{ mol } CO_2}{44.01 \text{ g } CO_2} = 1.80 \text{ mol } CO_2$$

	$CO_2(g)$	+	C(s)	$\rightleftharpoons$	2 CO(g)
initial (mol)	1.80				0
change (mol)	$-x$				$+2x$
equil (mol)	$1.80 - x$				2x

total mass of gas in flask = (16.3 g/L)(5.00 L) = 81.5 g

$81.5 = (1.80 - x)(44.01) + (2x)(28.01)$

$81.5 = 79.22 - 44.01x + 56.02x$; $2.28 = 12.01x$; $x = 2.28/12.01 = 0.19$

$n_{CO_2} = 1.80 - x = 1.80 - 0.19 = 1.61$ mol CO_2; $n_{CO} = 2x = 2(0.19) = 0.38$ mol CO

$$P_{CO_2} = \frac{nRT}{V} = \frac{(1.61\ mol)\left(0.082\ 06\ \dfrac{L \cdot atm}{K \cdot mol}\right)(1000\ K)}{5.0\ L} = 26.4\ atm$$

$$P_{CO} = \frac{nRT}{V} = \frac{(0.38\ mol)\left(0.082\ 06\ \dfrac{L \cdot atm}{K \cdot mol}\right)(1000\ K)}{5.0\ L} = 6.24\ atm$$

$$K_p = \frac{(P_{CO})^2}{(P_{CO_2})} = \frac{(6.24)^2}{(26.4)} = 1.47$$

(b) At 1100K, the total mass of gas in flask = (16.9 g/L)(5.00 L) = 84.5 g
84.5 = (1.80 − x)(44.01) + (2x)(28.01)
84.5 = 79.22 − 44.01x + 56.02x;   5.28 = 12.01x;   x = 5.28/12.01 = 0.44
n_{CO_2} = 1.80 − x = 1.80 − 0.44 = 1.36 mol CO_2; n_{CO} = 2x = 2(0.44) = 0.88 mol CO

$$P_{CO_2} = \frac{nRT}{V} = \frac{(1.36\ mol)\left(0.082\ 06\ \dfrac{L \cdot atm}{K \cdot mol}\right)(1100\ K)}{5.0\ L} = 24.6\ atm$$

$$P_{CO} = \frac{nRT}{V} = \frac{(0.88\ mol)\left(0.082\ 06\ \dfrac{L \cdot atm}{K \cdot mol}\right)(1100\ K)}{5.0\ L} = 15.9\ atm$$

$$K_p = \frac{(P_{CO})^2}{(P_{CO_2})} = \frac{(15.9)^2}{(24.6)} = 10.3$$

(c) In agreement with Le Châtelier's principle, the reaction is endothermic because K_p increases with increasing temperature.

13.132 (a) N_2O_4, 92.01

$$14.58\ g\ N_2O_4 \times \frac{1\ mol\ N_2O_4}{92.01\ g\ N_2O_4} = 0.1585\ mol\ N_2O_4$$

$$PV = nRT \qquad P_{N_2O_4} = \frac{nRT}{V} = \frac{(0.1585\ mol)\left(0.082\ 06\ \dfrac{L \cdot atm}{K \cdot mol}\right)(400\ K)}{1.000\ L} = 5.20\ atm$$

$$N_2O_4(g) \rightleftharpoons 2\ NO_2(g)$$

initial (atm)   5.20   0
change (atm)   −x   +2x
equil (atm)   5.20 − x   2x
$P_{total} = P_{N_2O_4} + P_{NO_2} = (5.20 − x) + (2x) = 9.15\ atm$
5.20 + x = 9.15 atm
x = 3.95 atm

$P_{N_2O_4} = 5.20 - x = 5.20 - 3.95 = 1.25$ atm

$P_{NO_2} = 2x = 2(3.95) = 7.90$ atm

$K_p = \dfrac{(P_{NO_2})^2}{(P_{N_2O_4})} = \dfrac{(7.90)^2}{(1.25)} = 49.9$

$\Delta n = 1$ and $K_c = K_p\left(\dfrac{1}{RT}\right) = \dfrac{(49.9)}{(0.082\ 06)(400)} = 1.52$

(b) $\Delta H°_{rxn} = [2\ \Delta H°_f(NO_2)] - \Delta H°_f(N_2O_4)$

$\Delta H°_{rxn} = [(2\ mol)(33.2\ kJ/mol)] - [(1mol)(11.1\ kJ/mol)] = 55.3$ kJ

moles N_2O_4 reacted $= n = \dfrac{PV}{RT} = \dfrac{(3.95\ atm)(1.000\ L)}{\left(0.082\ 06\ \dfrac{L \cdot atm}{K \cdot mol}\right)(400\ K)} = 0.1203$ mol N_2O_4

$q = (55.3\ kJ/mol\ N_2O_4)(0.1203\ mol\ N_2O_4) = 6.65$ kJ

13.134 The atmosphere is 21% (0.21) O_2; $P_{O_2} = (0.21)\left(720\ mm\ Hg\ x\ \dfrac{1\ atm}{760\ mm\ Hg}\right) = 0.199$ atm

$2\ O_3(g) \rightleftharpoons 3\ O_2(g)$

$K_p = \dfrac{(P_{O_2})^3}{(P_{O_3})^2}$; $P_{O_3} = \sqrt{\dfrac{(P_{O_2})^3}{K_p}} = \sqrt{\dfrac{(0.199)^3}{1.3\ x\ 10^{57}}} = 2.46\ x\ 10^{-30}$ atm

vol $= 10\ x\ 10^6\ m^3\ x\ \left(\dfrac{100\ cm}{1\ m}\right)^3\ x\ \dfrac{1\ L}{1000\ cm^3} = 1.0\ x\ 10^{10}$ L

$n_{O_3} = \dfrac{PV}{RT} = \dfrac{(2.46\ x\ 10^{-30}\ atm)(1.0\ x\ 10^{10}\ L)}{\left(0.082\ 06\ \dfrac{L \cdot atm}{K \cdot mol}\right)(298\ K)} = 1.0\ x\ 10^{-21}$ mol O_3

O_3 molecules $= 1.0\ x\ 10^{-21}$ mol $O_3\ x\ \dfrac{6.022\ x\ 10^{23}\ O_3\ molecules}{1\ mol\ O_3} = 6.0\ x\ 10^2\ O_3$ molecules

13.136 $PCl_5(g) \rightleftharpoons PCl_3(g) + Cl_2(g)$

$\Delta n = (2) - (1) = 1$ and at 700 K, $K_p = K_c(RT) = (46.9)(0.082\ 06)(700) = 2694$

(a) Because K_p is larger at the higher temperature, the reaction has shifted toward products at the higher temperature, which means the reaction is endothermic. Because the reaction involves breaking two P–Cl bonds and forming just one Cl–Cl bond, it should be endothermic.

(b) PCl_5, 208.24

mol $PCl_5 = 1.25$ g $PCl_5\ x\ \dfrac{1\ mol\ PCl_5}{208.24\ g\ PCl_5} = 6.00\ x\ 10^{-3}$ mol

$PV = nRT$, $P_{PCl_5} = \dfrac{nRT}{V} = \dfrac{(6.00\ x\ 10^{-3}\ mol)\left(0.082\ 06\ \dfrac{L \cdot atm}{K \cdot mol}\right)(700\ K)}{0.500\ L} = 0.689$ atm

Because K_p is so large, first assume the reaction goes to completion and then allow for a small back reaction.

	$PCl_5(g)$	$\rightleftarrows$	$PCl_3(g)$	+	$Cl_2(g)$
before rxn (atm)	0.689		0		0
100% rxn (atm)	−0.689		+0.689		+0.689
after rxn (atm)	0		0.689		0.689
back rxn (atm)	+x		−x		−x
equil (atm)	x		0.689 − x		0.689 − x

$$K_p = \frac{(P_{PCl_3})(P_{Cl_2})}{P_{PCl_5}} = 2694 = \frac{(0.689 - x)^2}{x} \approx \frac{(0.689)^2}{x}$$

$$x = P_{PCl_5} = \frac{(0.689)^2}{2694} = 1.76 \times 10^{-4} \text{ atm}$$

$$P_{total} = P_{PCl_5} + P_{PCl_3} + P_{Cl_2}$$

$$P_{total} = x + (0.689 - x) + (0.689 - x) = 0.689 + 0.689 - 1.76 \times 10^{-4} = 1.38 \text{ atm}$$

$$\% \text{ dissociation} = \frac{(P_{PCl_5})_o - (P_{PCl_5})}{(P_{PCl_5})_o} \times 100\% = \frac{0.689 - (1.76 \times 10^{-4})}{0.689} \times 100\% = 99.97\%$$

(c)

The molecular geometry is trigonal bipyramidal. There is no dipole moment because of a symmetrical distribution of Cl's around the central P.

The molecular geometry is trigonal pyramidal. There is a dipole moment because of the lone pair of electrons on the P and an unsymmetrical distribution of Cl's around the central P.

Aqueous Equilibria: Acids and Bases

14.1 (a) $H_2SO_4(aq) + H_2O(l) \rightleftarrows H_3O^+(aq) + HSO_4^-(aq)$
conjugate base

(b) $HSO_4^-(aq) + H_2O(l) \rightleftarrows H_3O^+(aq) + SO_4^{2-}(aq)$
conjugate base

(c) $H_3O^+(aq) + H_2O(l) \rightleftarrows H_3O^+(aq) + H_2O(l)$
conjugate base

(d) $NH_4^+(aq) + H_2O(l) \rightleftarrows H_3O^+(aq) + NH_3(aq)$
conjugate base

14.2 (a) $HCO_3^-(aq) + H_2O(l) \rightleftarrows H_2CO_3(aq) + OH^-(aq)$
conjugate acid

(b) $CO_3^{2-}(aq) + H_2O(l) \rightleftarrows HCO_3^-(aq) + OH^-(aq)$
conjugate acid

(c) $OH^-(aq) + H_2O(l) \rightleftarrows H_2O(l) + OH^-(aq)$
conjugate acid

(d) $H_2PO_4^-(aq) + H_2O(l) \rightleftarrows H_3PO_4(aq) + OH^-(aq)$
conjugate acid

14.3 $HCl(aq) + NH_3(aq) \rightleftarrows NH_4^+(aq) + Cl^-(aq)$
 acid   base   acid   base

conjugate acid-base pairs

14.4 (a) $H_9O_4^+$, $H_{19}O_9^+$, $H_{43}O_{21}^+$ (b) $H_9O_4^+$ (4 H_2O), $H_{19}O_9^+$ (9 H_2O), $H_{43}O_{21}^+$ (21 H_2O)

14.5 (a) $HF(aq) + NO_3^-(aq) \rightleftarrows HNO_3(aq) + F^-(aq)$
HNO_3 is a stronger acid than HF, and F^- is a stronger base than NO_3^- (see Table 14.1). Because proton transfer occurs from the stronger acid to the stronger base, the reaction proceeds from right to left.
(b) $NH_4^+(aq) + CO_3^{2-}(aq) \rightleftarrows HCO_3^-(aq) + NH_3(aq)$
NH_4^+ is a stronger acid than HCO_3^-, and CO_3^{2-} is a stronger base than NH_3 (see Table 14.1). Because proton transfer occurs from the stronger acid to the stronger base, the reaction proceeds from left to right.

14.6 (a) Both HX and HY have the same initial concentration. HY is more dissociated than HX. Therefore, HY is the stronger acid.
(b) The conjugate base (X^-) of the weaker acid (HX) is the stronger base.
(c) $HX + Y^- \rightleftarrows HY + X^-$; Proton transfer occurs from the stronger acid to the stronger base. The reaction proceeds to the left.

14.7   (a) H_2Se is a stronger acid than H_2S because Se is below S in the 6A group and the H–Se bond is weaker than the H–S bond.
(b) HI is a stronger acid than H_2Te because I is to the right of Te in the same row of the periodic table, I is more electronegative than Te, and the H–I bond is more polar.
(c) HNO_3 is a stronger acid than HNO_2 because acid strength increases with increasing oxidation number of N. The oxidation number for N is +5 in HNO_3 and +3 in HNO_2.
(d) H_2SO_3 is a stronger acid than H_2SeO_3 because acid strength increases with increasing electronegativity of the central atom. S is more electronegative than Se.

14.8 $[OH^-] = \dfrac{K_w}{[H_3O^+]} = \dfrac{1.0 \times 10^{-14}}{1.4 \times 10^{-4}} = 7.1 \times 10^{-11}$ M

Because $[H_3O^+] > [OH^-]$, the solution is acidic.

14.9 $[H_3O^+] = \dfrac{K_w}{[OH^-]} = \dfrac{1.0 \times 10^{-14}}{2.0 \times 10^{-6}} = 5.0 \times 10^{-9}$ M

Because $[OH^-] > [H_3O^+]$, the solution is basic.

14.10 $K_w = [H_3O^+][OH^-]$;   In a neutral solution, $[H_3O^+] = [OH^-]$
At 50 °C, $[H_3O^+] = [OH^-] = \sqrt{K_w} = \sqrt{5.5 \times 10^{-14}} = 2.3 \times 10^{-7}$ M

14.11   (a) $[H_3O^+] = \dfrac{K_w}{[OH^-]} = \dfrac{1.0 \times 10^{-14}}{1.58 \times 10^{-6}} = 6.3 \times 10^{-9}$ M
pH $= -\log[H_3O^+] = -\log(6.3 \times 10^{-9}) = 8.20$
(b) pH $= -\log[H_3O^+] = -\log(6.0 \times 10^{-5}) = 4.22$

14.12   (a) $[H_3O^+] = 10^{-pH} = 10^{-7.40} = 4.0 \times 10^{-8}$ M

$[OH^-] = \dfrac{K_w}{[H_3O^+]} = \dfrac{1.0 \times 10^{-14}}{4.0 \times 10^{-8}} = 2.5 \times 10^{-7}$ M

(b) $[H_3O^+] = 10^{-pH} = 10^{-2.8} = 2 \times 10^{-3}$ M

$[OH^-] = \dfrac{K_w}{[H_3O^+]} = \dfrac{1.0 \times 10^{-14}}{2 \times 10^{-3}} = 5 \times 10^{-12}$ M

14.13   (a) Because $HClO_4$ is a strong acid, $[H_3O^+] = 0.050$ M.
pH $= -\log[H_3O^+] = -\log(0.050) = 1.30$
(b) Because HCl is a strong acid, $[H_3O^+] = 6.0$ M.
pH $= -\log[H_3O^+] = -\log(6.0) = -0.78$
(c) Because KOH is a strong base, $[OH^-] = 4.0$ M.

$[H_3O^+] = \dfrac{K_w}{[OH^-]} = \dfrac{1.0 \times 10^{-14}}{4.0} = 2.5 \times 10^{-15}$ M

pH $= -\log[H_3O^+] = -\log(2.5 \times 10^{-15}) = 14.60$

(d) Because $Ba(OH)_2$ is a strong base, $[OH^-] = 2(0.010 \text{ M}) = 0.020$ M.

$$[H_3O^+] = \frac{K_w}{[OH^-]} = \frac{1.0 \times 10^{-14}}{0.020} = 5.0 \times 10^{-13} \text{ M}$$

$$pH = -\log[H_3O^+] = -\log(5.0 \times 10^{-13}) = 12.30$$

14.14 $BaO(s) + H_2O(l) \rightarrow Ba(OH)_2(aq);$ BaO, 153.33

$$0.25 \text{ g BaO} \times \frac{1 \text{ mol BaO}}{153.33 \text{ g BaO}} \times \frac{1 \text{ mol Ba(OH)}_2}{1 \text{ mol BaO}} \times \frac{2 \text{ mol OH}^-}{1 \text{ mol Ba(OH)}_2} = 3.26 \times 10^{-3} \text{ mol OH}^-$$

$$[OH^-] = \frac{3.26 \times 10^{-3} \text{ mol OH}^-}{0.500 \text{ L}} = 6.52 \times 10^{-3} \text{ M}$$

$$[H_3O^+] = \frac{K_w}{[OH^-]} = \frac{1.0 \times 10^{-14}}{6.52 \times 10^{-3}} = 1.53 \times 10^{-12} \text{ M}$$

$$pH = -\log[H_3O^+] = -\log(1.53 \times 10^{-12}) = 11.82$$

14.15
	$HOCl(aq) + H_2O(l)$	$\rightleftarrows$	$H_3O^+(aq)$	$+ OCl^-(aq)$
initial (M)	0.10		~0	0
change (M)	–x		+x	+x
equil (M)	0.10 – x		x	x

$x = [H_3O^+] = 10^{-pH} = 10^{-4.23} = 5.9 \times 10^{-5}$ M

$[OCl^-] = x = 5.9 \times 10^{-5}$ M; $[HOCl] = 0.10 - x = (0.10 - 5.9 \times 10^{-5})$ M

$$K_a = \frac{[H_3O^+][OCl^-]}{[HOCl]} = \frac{(5.9 \times 10^{-5})(5.9 \times 10^{-5})}{(0.10 - 5.9 \times 10^{-5})} = 3.5 \times 10^{-8}$$

This value of K_a agrees with the value in Table 14.2.

14.16 (a) HZ is completely dissociated. HX and HY are at the same concentration and HX is more dissociated than HY. The strongest acid is HZ, the weakest is HY.
K_a (HY) $<$ K_a (HX) $<$ K_a (HZ)
(b) HZ
(c) HY has the highest pH; HZ has the lowest pH (highest $[H_3O^+]$).

14.17 (a)
	$CH_3CO_2H(aq) + H_2O(l)$	$\rightleftarrows$	$H_3O^+(aq)$	$+ CH_3CO_2^-(aq)$
initial (M)	1.00		~0	0
change (M)	–x		+x	+x
equil (M)	1.00 – x		x	x

$$K_a = \frac{[H_3O^+][CH_3CO_2^-]}{[CH_3CO_2H]} = 1.8 \times 10^{-5} = \frac{x^2}{1.00 - x} \approx \frac{x^2}{1.00}$$

Solve for x. $x = [H_3O^+] = 4.2 \times 10^{-3}$ M

$pH = -\log[H_3O^+] = -\log(4.2 \times 10^{-3}) = 2.38$

$[CH_3CO_2^-] = x = 4.2 \times 10^{-3}$ M; $[CH_3CO_2H] = 1.00 - x = 1.00$ M

$$[OH^-] = \frac{K_w}{[H_3O^+]} = \frac{1.0 \times 10^{-14}}{4.2 \times 10^{-3}} = 2.4 \times 10^{-12} \text{ M}$$

(b) $\qquad$ $CH_3CO_2H(aq) + H_2O(l) \rightleftarrows H_3O^+(aq) + CH_3CO_2^-(aq)$

initial (M)	0.0100	~0	0
change (M)	$-x$	$+x$	$+x$
equil (M)	$0.0100 - x$	x	x

$$K_a = \frac{[H_3O^+][CH_3CO_2^-]}{[CH_3CO_2H]} = 1.8 \times 10^{-5} = \frac{x^2}{0.0100 - x}$$

$x^2 + (1.8 \times 10^{-5})x - (1.8 \times 10^{-7}) = 0$

Use the quadratic formula to solve for x.

$$x = \frac{-(1.8 \times 10^{-5}) \pm \sqrt{(1.8 \times 10^{-5})^2 - (4)(-1.8 \times 10^{-7})}}{2(1)} = \frac{(-1.8 \times 10^{-5}) \pm (8.5 \times 10^{-4})}{2}$$

$x = 4.2 \times 10^{-4}$ and -4.3×10^{-4}

Of the two solutions for x, only the positive value of x has physical meaning because x is the $[H_3O^+]$.

$x = [H_3O^+] = 4.2 \times 10^{-4}$ M

$pH = -\log[H_3O^+] = -\log(4.2 \times 10^{-4}) = 3.38$

$[CH_3CO_2^-] = x = 4.2 \times 10^{-4}$ M

$[CH_3CO_2H] = 0.0100 - x = 0.0100 - (4.2 \times 10^{-4}) = 0.0096$ M

$$[OH^-] = \frac{K_w}{[H_3O^+]} = \frac{1.0 \times 10^{-14}}{4.2 \times 10^{-4}} = 2.4 \times 10^{-11} \text{ M}$$

14.18 $\quad C_6H_8O_6$, 176.13; 250 mg = 0.250 g; 250 mL = 0.250 L

$$[C_6H_8O_6] = \frac{\left(0.250 \text{ g} \times \dfrac{1 \text{ mol}}{176.13 \text{ g}}\right)}{0.250 \text{ L}} = 5.68 \times 10^{-3} \text{ M}$$

$\qquad$ $C_6H_8O_6(aq) + H_2O(l) \rightleftarrows H_3O^+(aq) + C_6H_7O_6^-(aq)$

initial (M)	5.68×10^{-3}	~0	0
change (M)	$-x$	$+x$	$+x$
equil (M)	$(5.68 \times 10^{-3}) - x$	x	x

$$K_a = \frac{[H_3O^+][C_6H_7O_6^-]}{[C_6H_8O_6]} = 8.0 \times 10^{-5} = \frac{x^2}{(5.68 \times 10^{-3}) - x}$$

$x^2 + (8.0 \times 10^{-5})x - (4.54 \times 10^{-7}) = 0$

Use the quadratic formula to solve for x.

$$x = \frac{-(8.0 \times 10^{-5}) \pm \sqrt{(8.0 \times 10^{-5})^2 - (4)(-4.54 \times 10^{-7})}}{2(1)} = \frac{(-8.0 \times 10^{-5}) \pm 0.001\ 35}{2}$$

$x = 6.35 \times 10^{-4}$ and -7.15×10^{-4}

Of the two solutions for x, only the positive value of x has physical meaning because x is the $[H_3O^+]$.

$x = [H_3O^+] = 6.35 \times 10^{-4}$ M

$pH = -\log[H_3O^+] = -\log(6.35 \times 10^{-4}) = 3.20$

14.19 (a) From Example 14.10 in the text:

$[H_3O^+] = [HF]_{diss} = 4.0 \times 10^{-3}$ M

% dissociation $= \dfrac{[HF]_{diss}}{[HF]_{initial}} \times 100\% = \dfrac{4.0 \times 10^{-3} \text{ M}}{0.050 \text{ M}} \times 100\% = 8.0\%$ dissociation

(b)

	$HF(aq)$	$+ \ H_2O(l)$	$\rightleftarrows$	$H_3O^+(aq)$	$+$	$F^-(aq)$
initial (M)	0.50			~0		0
change (M)	$-x$			$+x$		$+x$
equil (M)	$0.50 - x$			x		x

$K_a = \dfrac{[H_3O^+][F^-]}{[HF]} = 3.5 \times 10^{-4} = \dfrac{x^2}{0.50 - x}$

$x^2 + (3.5 \times 10^{-4})x - (1.75 \times 10^{-4}) = 0$

Use the quadratic formula to solve for x.

$x = \dfrac{-(3.5 \times 10^{-4}) \pm \sqrt{(3.5 \times 10^{-4})^2 - (4)(-1.75 \times 10^{-4})}}{2(1)} = \dfrac{(-3.5 \times 10^{-4}) \pm 0.0265}{2}$

$x = 0.0131$ and -0.0134

Of the two solutions for x, only the positive value of x has physical meaning, because x is the $[H_3O^+]$.

$[H_3O^+] = [HF]_{diss} = 0.013$ M

% dissociation $= \dfrac{[HF]_{diss}}{[HF]_{initial}} \times 100\% = \dfrac{0.013 \text{ M}}{0.50 \text{ M}} \times 100\% = 2.6\%$ dissociation

14.20

	$H_2SO_3(aq)$	$+ \ H_2O(l)$	$\rightleftarrows$	$H_3O^+(aq)$	$+$	$HSO_3^-(aq)$
initial (M)	0.10			~0		0
change (M)	$-x$			$+x$		$+x$
equil (M)	$0.10 - x$			x		x

$K_{a1} = \dfrac{[H_3O^+][HSO_3^-]}{[H_2SO_3]} = 1.5 \times 10^{-2} = \dfrac{x^2}{0.10 - x}$

$x^2 + 0.015x - 0.0015 = 0$

Use the quadratic formula to solve for x.

$x = \dfrac{-(0.015) \pm \sqrt{(0.015)^2 - (4)(-0.0015)}}{2(1)} = \dfrac{-0.015 \pm 0.079}{2}$

$x = 0.032$ and -0.047

Of the two solutions for x, only the positive value of x has physical meaning since x is the $[H_3O^+]$.

$x = [H_3O^+] = [HSO_3^-] = 0.032$ M; $\quad [H_2SO_3] = 0.10 - x = 0.10 - 0.032 = 0.07$ M

The second dissociation of H_2SO_3 produces a negligible amount of H_3O^+ compared with that from the first dissociation.

$HSO_3^-(aq) + H_2O(l) \rightleftarrows H_3O^+(aq) + SO_3^{2-}(aq)$

$$K_{a2} = \frac{[H_3O^+][SO_3^{2-}]}{[HSO_3^-]} = 6.3 \times 10^{-8} = \frac{(0.032)[SO_3^{2-}]}{(0.032)}$$

$[SO_3^{2-}] = K_{a2} = 6.3 \times 10^{-8}$ M

$$[OH^-] = \frac{K_w}{[H_3O^+]} = \frac{1.0 \times 10^{-14}}{0.032} = 3.1 \times 10^{-13} \text{ M}$$

$pH = -\log[H_3O^+] = -\log(0.032) = 1.49$

14.21   From the complete dissociation of the first proton, $[H_3O^+] = [HSeO_4^-] = 0.50$ M.
        For the dissociation of the second proton, the following equilibrium must be considered:

$$HSeO_4^-(aq) \; + \; H_2O(l) \; \rightleftarrows \; H_3O^+(aq) \; + \; SeO_4^{2-}(aq)$$

initial (M)	0.50	0.50	0
change (M)	−x	+x	+x
equil (M)	0.50 − x	0.50 + x	x

$$K_{a2} = \frac{[H_3O^+][SeO_4^{2-}]}{[HSeO_4^-]} = 1.2 \times 10^{-2} = \frac{(0.50 + x)(x)}{0.50 - x}$$

$x^2 + 0.512x - 0.0060 = 0$
Use the quadratic formula to solve for x.

$$x = \frac{-(0.512) \pm \sqrt{(0.512)^2 - (4)(-0.0060)}}{2(1)} = \frac{-0.512 \pm 0.535}{2}$$

$x = 0.011$ and -0.524
Of the two solutions for x, only the positive value of x has physical meaning, since x is
the $[SeO_4^{2-}]$.
$[H_2SeO_4] = 0$ M; $[HSeO_4^-] = 0.50 - x = 0.49$ M; $[SeO_4^{2-}] = x = 0.011$ M
$[H_3O^+] = 0.50 + x = 0.51$ M
$pH = -\log[H_3O^+] = -\log(0.51) = 0.29$

$$[OH^-] = \frac{K_w}{[H_3O^+]} = \frac{1.0 \times 10^{-14}}{0.51} = 2.0 \times 10^{-14} \text{ M}$$

14.22

$$NH_3(aq) \; + \; H_2O(l) \; \rightleftarrows \; NH_4^+(aq) \; + \; OH^-(aq)$$

initial (M)	0.40	0	~0
change (M)	−x	+x	+x
equil (M)	0.40 − x	x	x

$$K_b = \frac{[NH_4^+][OH^-]}{[NH_3]} = 1.8 \times 10^{-5} = \frac{x^2}{0.40 - x} \approx \frac{x^2}{0.40}$$

Solve for x. $x = [OH^-] = 2.7 \times 10^{-3}$ M
$[NH_4^+] = x = 2.7 \times 10^{-3}$ M; $[NH_3] = 0.40 - x = 0.40$ M

$$[H_3O^+] = \frac{K_w}{[OH^-]} = \frac{1.0 \times 10^{-14}}{2.7 \times 10^{-3}} = 3.7 \times 10^{-12} \text{ M}$$

$pH = -\log[H_3O^+] = -\log(3.7 \times 10^{-12}) = 11.43$

14.23 $C_{21}H_{22}N_2O_2$, 334.42; 16 mg = 0.016 g

$$\text{molarity} = \frac{\left(0.016\ \text{g} \times \dfrac{1\ \text{mol}}{334.42\ \text{g}}\right)}{0.100\ \text{L}} = 4.8 \times 10^{-4}\ \text{M}$$

$$C_{21}H_{22}N_2O_2(aq)\ +\ H_2O(l)\ \rightleftharpoons\ C_{21}H_{23}N_2O_2^+(aq)\ +\ OH^-(aq)$$

initial (M)	4.8×10^{-4}	0	~0
change (M)	$-x$	$+x$	$+x$
equil (M)	$(4.8 \times 10^{-4}) - x$	x	x

$$K_b = \frac{[C_{21}H_{23}N_2O_2^+][OH^-]}{[C_{21}H_{22}N_2O_2]} = 1.8 \times 10^{-6} = \frac{x^2}{(4.8 \times 10^{-4}) - x}$$

$x^2 + (1.8 \times 10^{-6})x - (8.6 \times 10^{-10}) = 0$

Use the quadratic formula to solve for x.

$$x = \frac{-(1.8 \times 10^{-6}) \pm \sqrt{(1.8 \times 10^{-6})^2 - (4)(-8.6 \times 10^{-10})}}{2(1)} = \frac{(-1.8 \times 10^{-6}) \pm (5.87 \times 10^{-5})}{2}$$

$x = 2.85 \times 10^{-5}$ and -3.03×10^{-5}

Of the two solutions for x, only the positive value of x has physical meaning, because x is the [OH$^-$].

$[OH^-] = 2.84 \times 10^{-5}$ M

$$[H_3O^+] = \frac{K_w}{[OH^-]} = \frac{1.0 \times 10^{-14}}{2.85 \times 10^{-5}} = 3.51 \times 10^{-10}\ \text{M}$$

$pH = -\log[H_3O^+] = -\log(3.51 \times 10^{-10}) = 9.45$

14.24 (a) $K_a = \dfrac{K_w}{K_b \text{ for } C_5H_{11}N} = \dfrac{1.0 \times 10^{-14}}{1.3 \times 10^{-3}} = 7.7 \times 10^{-12}$

(b) $K_b = \dfrac{K_w}{K_a \text{ for HOCl}} = \dfrac{1.0 \times 10^{-14}}{3.5 \times 10^{-8}} = 2.9 \times 10^{-7}$

(c) pK_b for $HCO_2^- = 14.00 - pK_a = 14.00 - 3.74 = 10.26$

14.25 (a) 0.25 M NH_4Br

NH_4^+ is an acidic cation. Br$^-$ is a neutral anion. The salt solution is acidic.

For NH_4^+, $K_a = \dfrac{K_w}{K_b \text{ for } NH_3} = \dfrac{1.0 \times 10^{-14}}{1.8 \times 10^{-5}} = 5.6 \times 10^{-10}$

$$NH_4^+(aq)\ +\ H_2O(l)\ \rightleftharpoons\ H_3O^+(aq)\ +\ NH_3(aq)$$

initial (M)	0.25	~0	0
change (M)	$-x$	$+x$	$+x$
equil (M)	$0.25 - x$	x	x

$$K_a = \frac{[H_3O^+][NH_3]}{[NH_4^+]} = 5.6 \times 10^{-10} = \frac{x^2}{0.25 - x} \approx \frac{x^2}{0.25}$$

Solve for x. $x = [H_3O^+] = 1.2 \times 10^{-5}$ M

$pH = -\log[H_3O^+] = -\log(1.2 \times 10^{-5}) = 4.92$

(b) 0.40 M $ZnCl_2$

Zn^{2+} is an acidic cation. Cl^- is a neutral anion. The salt solution is acidic.

$$Zn(H_2O)_6^{2+}(aq) + H_2O(l) \rightleftharpoons H_3O^+(aq) + Zn(H_2O)_5(OH)^+(aq)$$

initial (M)	0.40	~0	0
change (M)	−x	+x	+x
equil(M)	0.40 − x	x	x

$$K_a = \frac{[H_3O^+][Zn(H_2O)_5(OH)^+]}{[Zn(H_2O)_6^{2+}]} = 2.5 \times 10^{-10} = \frac{x^2}{0.40-x} \approx \frac{x^2}{0.40}$$

Solve for x. $x = [H_3O^+] = 1.0 \times 10^{-5}$ M

$pH = -\log[H_3O^+] = -\log(1.0 \times 10^{-5}) = 5.00$

14.26 For NO_2^-, $K_b = \dfrac{K_w}{K_a \text{ for } HNO_2} = \dfrac{1.0 \times 10^{-14}}{4.6 \times 10^{-4}} = 2.2 \times 10^{-11}$

$$NO_2^-(aq) + H_2O(l) \rightleftharpoons HNO_2(aq) + OH^-(aq)$$

initial (M)	0.20	0	~0
change (M)	−x	+x	+x
equil (M)	0.20 − x	x	x

$$K_b = \frac{[HNO_2][OH^-]}{[NO_2^-]} = 2.2 \times 10^{-11} = \frac{x^2}{0.20-x} \approx \frac{x^2}{0.20}$$

Solve for x. $x = [OH^-] = 2.1 \times 10^{-6}$ M

$$[H_3O^+] = \frac{K_w}{[OH^-]} = \frac{1.0 \times 10^{-14}}{2.1 \times 10^{-6}} = 4.8 \times 10^{-9} \text{ M}$$

$pH = -\log[H_3O^+] = -\log(4.8 \times 10^{-9}) = 8.32$

14.27 For NH_4^+, $K_a = \dfrac{K_w}{K_b \text{ for } NH_3} = \dfrac{1.0 \times 10^{-14}}{1.8 \times 10^{-5}} = 5.6 \times 10^{-10}$

For CN^-, $K_b = \dfrac{K_w}{K_a \text{ for } HCN} = \dfrac{1.0 \times 10^{-14}}{4.9 \times 10^{-10}} = 2.0 \times 10^{-5}$

Because $K_b > K_a$, the solution is basic.

14.28 (a) KBr: K^+, neutral cation; Br^-, neutral anion; solution is neutral
(b) $NaNO_2$: Na^+, neutral cation; NO_2^-, basic anion; solution is basic
(c) NH_4Br: NH_4^+, acidic cation; Br^-, neutral anion; solution is acidic
(d) $ZnCl_2$: Zn^{2+}, acidic cation; Cl^-, neutral anion; solution is acidic
(e) NH_4F

For NH_4^+, $K_a = \dfrac{K_w}{K_b \text{ for } NH_3} = \dfrac{1.0 \times 10^{-14}}{1.8 \times 10^{-5}} = 5.6 \times 10^{-10}$

For F^-, $K_b = \dfrac{K_w}{K_a \text{ for } HF} = \dfrac{1.0 \times 10^{-14}}{3.5 \times 10^{-4}} = 2.9 \times 10^{-11}$

Because $K_a > K_b$, the solution is acidic.

14.29   (a) Lewis acid, $AlCl_3$; Lewis base, Cl^- (b) Lewis acid, Ag^+; Lewis base, NH_3
(c) Lewis acid, SO_2; Lewis base, OH^- (d) Lewis acid, Cr^{3+}; Lewis base, H_2O

14.30

14.31   Lewis acids include not only H^+ but also other cations and neutral molecules having vacant valence orbitals that can accept a share in a pair of electrons donated by a Lewis base. The O^{2-} from CaO is the Lewis base and SO_2 is the Lewis acid.

14.32 NO_2, 46.01

$$\text{mol } NO_2 = 5.47 \text{ mg} \times \frac{1.00 \times 10^{-3} \text{ g}}{1 \text{ mg}} \times \frac{1 \text{ mol } NO_2}{46.01 \text{ g } NO_2} = 1.19 \times 10^{-4} \text{ mol } NO_2$$

$$3 \, NO_2(g) + H_2O(l) \rightarrow 2 \, HNO_3(aq) + NO(g)$$

$$\text{mol } HNO_3 = 1.19 \times 10^{-4} \text{ mol } NO_2 \times \frac{2 \text{ mol } HNO_3}{3 \text{ mol } NO_2} = 7.93 \times 10^{-5} \text{ mol } HNO_3$$

$$[HNO_3] = \frac{7.93 \times 10^{-5} \text{ mol } HNO_3}{1.00 \text{ L}} = 7.93 \times 10^{-5} \text{ M}$$

HNO_3 is a strong acid and is completely dissociated therefore $[H_3O^+] = 7.93 \times 10^{-5}$ M
$pH = -\log [H_3O^+] = -\log (7.93 \times 10^{-5} \text{ M}) = 4.101$

Conceptual Problems

14.34   (a) X^-, Y^-, Z^- (b) $HX < HZ < HY$ (c) HY (d) HX (e) (2/10) x 100% = 20%

14.36   For H_2SO_4, there is complete dissociation of only the first H^+. At equilibrium, there should be H_3O^+, HSO_4^-, and a small amount of SO_4^{2-}. This is best represented by (b).

14.38   (a) $Y^- < Z^- < X^-$
(b) The weakest base, Y^-, has the strongest conjugate acid.
(c) X^- is the strongest conjugate base and has the smallest pK_b.
(d) The numbers of HA molecules and OH^- ions are equal because the reaction of A^- with water has a 1:1 stoichiometry: $A^- + H_2O \rightleftharpoons HA + OH^-$

14.40

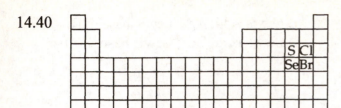

(a) H_2S, weakest; HBr, strongest. Acid strength for H_nX increases with increasing polarity of the H–X bond and with increasing size of X.

(b) H_2SeO_3, weakest; $HClO_3$, strongest. Acid strength for H_nYO_3 increases with increasing electronegativity of Y.

14.42 (a) $H_3BO_3(aq) + H_2O(l) \rightleftharpoons H_3O^+(aq) + H_2BO_3^-(aq)$

(b) $H_3BO_3(aq) + 2 H_2O(l) \rightleftharpoons H_3O^+(aq) + B(OH)_4^-(aq)$

Section Problems
Acid–Base Concepts (Sections 14.1 and 14.2)

14.44 NH_3, CN^-, and NO_2^-

14.46 (a) SO_4^{2-} (b) HSO_3^- (c) HPO_4^{2-} (d) NH_3 (e) OH^- (f) NH_2^-

14.48 (a) $CH_3CO_2H(aq) + NH_3(aq) \rightleftharpoons NH_4^+(aq) + CH_3CO_2^-(aq)$
           acid   base ———— acid   base

(b) $CO_3^{2-}(aq) + H_3O^+(aq) \rightleftharpoons H_2O(l) + HCO_3^-(aq)$
       base   acid ———— base   acid

(c) $HSO_3^-(aq) + H_2O(l) \rightleftharpoons H_3O^+(aq) + SO_3^{2-}(aq)$
       acid   base ———— acid   base

(d) $HSO_3^-(aq) + H_2O(l) \rightleftharpoons H_2SO_3(aq) + OH^-(aq)$
       base   acid   acid   base

14.50 From data in Table 14.1:   Strong acids: HNO_3 and H_2SO_4;   Strong bases: H^- and O^{2-}

14.52 In this reaction, H_3PO_4 is an acid and $H_2PO_4^-$ is a base. In the conjugate acid-base pair of HCl/Cl^-, according to Table 14.1, HCl is a stronger acid than H_3PO_4 and $H_2PO_4^-$ is a stronger base than Cl^-. Therefore, $H_2PO_4^-$ gets the proton and the reaction proceeds from left to right and $K_c > 1$. $HCl + H_2PO_4^- \rightarrow Cl^- + H_3PO_4$

Factors That Affect Acid Strength (Section 14.3)

14.54   (a) $PH_3 < H_2S < HCl$; electronegativity increases from P to Cl
      (b) $NH_3 < PH_3 < AsH_3$; X–H bond strength decreases from N to As (down a group)
      (c) $HBrO < HBrO_2 < HBrO_3$; acid strength increases with the number of O atoms

14.56   (a) HCl; The strength of a binary acid H_nA increases as A moves from left to right and from top to bottom in the periodic table.
      (b) $HClO_3$; The strength of an oxoacid increases with increasing electronegativity and increasing oxidation state of the central atom.
      (c) HBr; The strength of a binary acid H_nA increases as A moves from left to right and from top to bottom in the periodic table.

14.58   (a) H_2Te, weaker X–H bond
      (b) H_3PO_4, P has higher electronegativity
      (c) $H_2PO_4^-$, lower negative charge
      (d) NH_4^+, higher positive charge and N is more electronegative than C

Dissociation of Water; pH (Sections 14.4–14.6)

14.60   If $[H_3O^+] > 1.0 \times 10^{-7}$ M, solution is acidic.
      If $[H_3O^+] < 1.0 \times 10^{-7}$ M, solution is basic.
      If $[H_3O^+] = [OH^-] = 1.0 \times 10^{-7}$ M, solution is neutral.
      If $[OH^-] > 1.0 \times 10^{-7}$ M, solution is basic
      If $[OH^-] < 1.0 \times 10^{-7}$ M, solution is acidic.

(a) $[OH^-] = \dfrac{K_w}{[H_3O^+]} = \dfrac{1.0 \times 10^{-14}}{3.4 \times 10^{-9}} = 2.9 \times 10^{-6}$ M, basic

(b) $[H_3O^+] = \dfrac{K_w}{[OH^-]} = \dfrac{1.0 \times 10^{-14}}{0.010} = 1.0 \times 10^{-12}$ M, basic

(c) $[H_3O^+] = \dfrac{K_w}{[OH^-]} = \dfrac{1.0 \times 10^{-14}}{1.0 \times 10^{-10}} = 1.0 \times 10^{-4}$ M, acidic

(d) $[OH^-] = \dfrac{K_w}{[H_3O^+]} = \dfrac{1.0 \times 10^{-14}}{1.0 \times 10^{-7}} = 1.0 \times 10^{-7}$ M, neutral

(e) $[OH^-] = \dfrac{K_w}{[H_3O^+]} = \dfrac{1.0 \times 10^{-14}}{8.6 \times 10^{-5}} = 1.2 \times 10^{-10}$ M, acidic

14.62   At 200 °C and 750 atm, $K_w = [H_3O^+][OH^-] = 1.5 \times 10^{-11}$ and in pure water $[H_3O^+] = [OH^-]$.
      $[H_3O^+]^2 = K_w$

      $[H_3O^+] = [OH^-] = \sqrt{K_w} = \sqrt{1.5 \times 10^{-11}} = 3.9 \times 10^{-6}$ M

      Because $[H_3O^+] = [OH^-]$, the solution is neutral.

14.64   (a) $pH = -\log[H_3O^+] = -\log(2.0 \times 10^{-5}) = 4.70$

(b) $[H_3O^+] = \dfrac{K_w}{[OH^-]} = \dfrac{1.0 \times 10^{-14}}{4 \times 10^{-3}} = 2.5 \times 10^{-12}$ M

$pH = -\log[H_3O^+] = -\log(2.5 \times 10^{-12}) = 11.60$

(c) $pH = -\log[H_3O^+] = -\log(3.56 \times 10^{-9}) = 8.449$

(d) $pH = -\log[H_3O^+] = -\log(10^{-3}) = 3$

(e) $[H_3O^+] = \dfrac{K_w}{[OH^-]} = \dfrac{1.0 \times 10^{-14}}{12} = 8.3 \times 10^{-16}$ M

$pH = -\log[H_3O^+] = -\log(8.3 \times 10^{-16}) = 15.08$

14.66 $[H_3O^+] = 10^{-pH};$ (a) 8×10^{-5} M   (b) 1.5×10^{-11} M   (c) 1.0 M

(d) 5.6×10^{-15} M   (e) 10 M   (f) 5.78×10^{-6} M

14.68   (a) chlorphenol red   (b) thymol blue   (c) methyl orange

Strong Acids and Strong Bases (Section 14.7)

14.70   CaO, 56.08

$[H_3O^+] = 10^{-pH} = 10^{-(10.50)} = 3.16 \times 10^{-11}$ M

$[OH^-] = \dfrac{K_w}{[H_3O^+]} = \dfrac{1.0 \times 10^{-14}}{3.16 \times 10^{-11}} = 3.16 \times 10^{-4}$ M $= 3.16 \times 10^{-4}$ mol/L

$CaO(s) + H_2O(l) \rightarrow Ca^{2+}(aq) + 2\,OH^-(aq)$

3.16×10^{-4} mol OH^- x $\dfrac{1 \text{ mol CaO}}{2 \text{ mol OH}^-}$ x $\dfrac{56.08 \text{ g CaO}}{1 \text{ mol CaO}} = 0.0089$ g CaO

14.72   (a) LiOH, 23.95; 250 mL = 0.250 L

molarity of LiOH(aq) $= \dfrac{\left(4.8 \text{ g } \times \dfrac{1 \text{ mol}}{23.95 \text{ g}}\right)}{0.250 \text{ L}} = 0.80$ M

LiOH is a strong base; therefore $[OH^-] = 0.80$ M

$[H_3O^+] = \dfrac{K_w}{[OH^-]} = \dfrac{1.0 \times 10^{-14}}{0.80} = 1.25 \times 10^{-14}$ M

$pH = -\log[H_3O^+] = -\log(1.25 \times 10^{-14}) = 13.90$

(b) HCl, 36.46

molarity of HCl(aq) $= \dfrac{\left(0.93 \text{ g } \times \dfrac{1 \text{ mol}}{36.46 \text{ g}}\right)}{0.40 \text{ L}} = 0.064$ M

HCl is a strong acid; therefore $[H_3O^+] = 0.064$ M

$pH = -\log[H_3O^+] = -\log(0.064) = 1.19$

(c) $M_f \cdot V_f = M_i \cdot V_i$

$$M_f = \frac{M_i \cdot V_i}{V_f} = \frac{(0.10 \text{ M})(50 \text{ mL})}{(1000 \text{ mL})} = 5.0 \times 10^{-3} \text{ M}$$

$$pH = -\log[H_3O^+] = -\log(5.0 \times 10^{-3}) = 2.30$$

(d) For HCl, $M_f = \dfrac{M_i \cdot V_i}{V_f} = \dfrac{(2.0 \times 10^{-3} \text{ M})(100 \text{ mL})}{(500 \text{ mL})} = 4.0 \times 10^{-4} \text{ M}$

For $HClO_4$, $M_f = \dfrac{M_i \cdot V_i}{V_f} = \dfrac{(1.0 \times 10^{-3} \text{ M})(400 \text{ mL})}{(500 \text{ mL})} = 8.0 \times 10^{-4} \text{ M}$

$$[H_3O^+] = (4.0 \times 10^{-4} \text{ M}) + (8.0 \times 10^{-4} \text{ M}) = 1.2 \times 10^{-3} \text{ M}$$
$$pH = -\log[H_3O^+] = -\log(1.2 \times 10^{-3}) = 2.92$$

Weak Acids (Sections 14.8–14.10)

14.74 (a) The larger the K_a, the stronger the acid.
$C_6H_5OH < HOCl < CH_3CO_2H < HNO_3$
(b) The larger the K_a, the larger the percent dissociation for the same concentration.
$HNO_3 > CH_3CO_2H > HOCl > C_6H_5OH$
1 M HNO_3, $[H_3O^+] = 1$ M

1 M CH_3CO_2H, $[H_3O^+] = \sqrt{[HA] \times K_a} = \sqrt{(1 \text{ M})(1.8 \times 10^{-5})} = 4 \times 10^{-3}$ M

1 M $HOCl$, $[H_3O^+] = \sqrt{[HA] \times K_a} = \sqrt{(1 \text{ M})(3.5 \times 10^{-8})} = 2 \times 10^{-4}$ M

1 M C_6H_5OH, $[H_3O^+] = \sqrt{[HA] \times K_a} = \sqrt{(1 \text{ M})(1.3 \times 10^{-10})} = 1 \times 10^{-5}$ M

14.76

	$HOBr(aq)$	+	$H_2O(l)$	$\rightleftharpoons$	$H_3O^+(aq)$	+	$OBr^-(aq)$
initial (M)	0.040				~0		0
change (M)	−x				+x		+x
equil (M)	0.040 − x				x		x

$x = [H_3O^+] = 10^{-pH} = 10^{-5.05} = 8.9 \times 10^{-6}$ M

$$K_a = \frac{[H_3O^+][OBr^-]}{[HOBr]} = \frac{x^2}{0.040 - x} = \frac{(8.9 \times 10^{-6})^2}{0.040 - (8.9 \times 10^{-6})} = 2.0 \times 10^{-9}$$

14.78 $K_a = 10^{-pK_a} = 10^{-4.25} = 5.6 \times 10^{-5}$

(a)

	$HC_3H_3O_2(aq)$	+	$H_2O(l)$	$\rightleftharpoons$	$H_3O^+(aq)$	+	$C_3H_3O_2^-(aq)$
initial (M)	0.150				~0		0
change (M)	−x				+x		+x
equil (M)	0.150 − x				x		x

$$K_a = \frac{[H_3O^+][C_3H_3O_2^-]}{[HC_3H_3O_2]} = 5.6 \times 10^{-5} = \frac{x^2}{0.150 - x} \approx \frac{x^2}{0.150}$$

Solve for x. $x = 0.0029$ M $= [H_3O^+] = [C_3H_3O_2^-]$
$[HC_3H_3O_2] = 0.150 - x = 0.150 - 0.0029 = 0.147$ M

$$pH = -\log[H_3O^+] = -\log(0.0029) = 2.54$$

$$[OH^-] = \frac{K_w}{[H_3O^+]} = \frac{1.0 \times 10^{-14}}{0.0029} = 3.4 \times 10^{-12} \text{ M}$$

(b) $HC_3H_3O_2(aq) + H_2O(l) \rightleftharpoons H_3O^+(aq) + C_3H_3O_2^-(aq)$

initial (M)	0.0500	~0	0
change (M)	–x	+x	+x
equil (M)	0.0500 – x	x	x

$$K_a = \frac{[H_3O^+][C_3H_3O_2^-]}{[HC_3H_3O_2]} = 5.6 \times 10^{-5} = \frac{x^2}{0.0500 - x} \approx \frac{x^2}{0.0500}$$

Solve for x. $x = 0.001\ 67 \text{ M} = [H_3O^+] = [HC_3H_3O_2]_{diss}$

$$\% \text{ dissociation} = \frac{[HC_3H_3O_2]_{diss}}{[HC_3H_3O_2]_{initial}} \times 100\% = \frac{0.001\ 67 \text{ M}}{0.0500 \text{ M}} \times 100\% = 3.3\%$$

14.80 $HNO_2(aq) + H_2O(l) \rightleftharpoons H_3O^+(aq) + NO_2^-(aq)$

initial (M)	1.5	~0	0
change (M)	–x	+x	+x
equil (M)	1.5 – x	x	x

$$K_a = \frac{[H_3O^+][NO_2^-]}{[HNO_2]} = 4.5 \times 10^{-4} = \frac{x^2}{1.5 - x} \approx \frac{x^2}{1.5}$$

Solve for x. $x = 0.026 \text{ M} = [H_3O^+]$

$$pH = -\log[H_3O^+] = -\log(0.026) = 1.59$$

$$\% \text{ dissociation} = \frac{[HNO_2]_{diss}}{[HNO_2]_{initial}} \times 100\% = \frac{0.026 \text{ M}}{1.5 \text{ M}} \times 100\% = 1.7\%$$

Polyprotic Acids (Section 14.11)

14.82 $H_2SeO_4(aq) + H_2O(l) \rightleftharpoons H_3O^+(aq) + HSeO_4^-(aq)$; $K_{a1} = \dfrac{[H_3O^+][HSeO_4^-]}{[H_2SeO_4]}$

$HSeO_4^-(aq) + H_2O(l) \rightleftharpoons H_3O^+(aq) + SeO_4^{2-}(aq)$; $K_{a2} = \dfrac{[H_3O^+][SeO_4^{2-}]}{[HSeO_4^-]}$

14.84 $H_2CO_3(aq) + H_2O(l) \rightleftharpoons H_3O^+(aq) + HCO_3^-(aq)$

initial (M)	0.010	~0	0
change (M)	–x	+x	+x
equil (M)	0.010 – x	x	x

$$K_{a1} = \frac{[H_3O^+][HCO_3^-]}{[H_2CO_3]} = 4.3 \times 10^{-7} = \frac{x^2}{0.010 - x} \approx \frac{x^2}{0.010}$$

Solve for x. $x = 6.6 \times 10^{-5}$

$[H_3O^+] = [HCO_3^-] = x = 6.6 \times 10^{-5} \text{ M};$ $[H_2CO_3] = 0.010 - x = 0.010 \text{ M}$

The second dissociation of H_2CO_3 produces a negligible amount of H_3O^+ compared with that from the first dissociation.

$$HCO_3^-(aq) + H_2O(l) \rightleftarrows H_3O^+(aq) + CO_3^{2-}(aq)$$

$$K_{a2} = \frac{[H_3O^+][CO_3^{2-}]}{[HCO_3^-]} = 5.6 \times 10^{-11} = \frac{(6.6 \times 10^{-5})[CO_3^{2-}]}{(6.6 \times 10^{-5})}$$

$$[CO_3^{2-}] = K_{a2} = 5.6 \times 10^{-11} \text{ M}$$

$$[OH^-] = \frac{K_w}{[H_3O^+]} = \frac{1.0 \times 10^{-14}}{6.6 \times 10^{-5}} = 1.5 \times 10^{-10} \text{ M}$$

$$pH = -\log[H_3O^+] = -\log(6.6 \times 10^{-5}) = 4.18$$

14.86 For the dissociation of the first proton, the following equilibrium must be considered:

$$H_2C_2O_4(aq) + H_2O(l) \rightleftarrows H_3O^+(aq) + HC_2O_4^-(aq)$$

initial (M)	0.20	~0	0
change (M)	−x	+x	+x
equil (M)	0.20 − x	x	x

$$K_{a1} = \frac{[H_3O^+][HC_2O_4^-]}{[H_2C_2O_4]} = 5.9 \times 10^{-2} = \frac{x^2}{0.20 - x}$$

$$x^2 + 0.059x - 0.0118 = 0$$

Use the quadratic formula to solve for x.

$$x = \frac{-(0.059) \pm \sqrt{(0.059)^2 - 4(1)(-0.0118)}}{2(1)} = \frac{-0.059 \pm 0.225}{2}$$

$$x = 0.083 \text{ and } -0.142$$

Of the two solutions for x, only the positive value of x has physical meaning, because x is the $[H_3O^+]$.

$$[H_3O^+] = [HC_2O_4^-] = 0.083 \text{ M}$$

For the dissociation of the second proton, the following equilibrium must be considered:

$$HC_2O_4^-(aq) + H_2O(l) \rightleftarrows H_3O^+(aq) + C_2O_4^{2-}(aq)$$

initial (M)	0.083	0.083	0
change (M)	−x	+x	+x
equil (M)	0.083 − x	0.083 + x	x

$$K_{a2} = \frac{[H_3O^+][C_2O_4^{2-}]}{[HC_2O_4^-]} = 6.4 \times 10^{-5} = \frac{(0.083 + x)(x)}{0.083 - x} \approx \frac{(0.083)(x)}{0.083} = x$$

$$[H_3O^+] = 0.083 + x = 0.083 \text{ M}$$
$$pH = -\log[H_3O^+] = -\log(0.083) = 1.08$$
$$[C_2O_4^{2-}] = x = 6.4 \times 10^{-5} \text{ M}$$

Weak Bases; Relation Between K_a and K_b (Sections 14.12 and 14.13)

14.88 (a) $(CH_3)_2NH(aq) + H_2O(l) \rightleftharpoons (CH_3)_2NH_2^+(aq) + OH^-(aq)$; $K_b = \dfrac{[(CH_3)_2NH_2^+][OH^-]}{[(CH_3)_2NH]}$

(b) $C_6H_5NH_2(aq) + H_2O(l) \rightleftharpoons C_6H_5NH_3^+(aq) + OH^-(aq)$; $K_b = \dfrac{[C_6H_5NH_3^+][OH^-]}{[C_6H_5NH_2]}$

(c) $CN^-(aq) + H_2O(l) \rightleftharpoons HCN(aq) + OH^-(aq)$; $K_b = \dfrac{[HCN][OH^-]}{[CN^-]}$

14.90 $[H_3O^+] = 10^{-pH} = 10^{-9.5} = 3.16 \times 10^{-10}$ M

$[OH^-] = \dfrac{K_w}{[H_3O^+]} = \dfrac{1.0 \times 10^{-14}}{3.16 \times 10^{-10}} = 3.16 \times 10^{-5}$ M

$$\begin{array}{lccc}
 & C_{17}H_{19}NO_3(aq) + H_2O(l) & \rightleftharpoons & C_{17}H_{20}NO_3^+(aq) + OH^-(aq) \\
\text{initial (M)} & 7.0 \times 10^{-4} & 0 & \sim 0 \\
\text{change (M)} & -x & +x & +x \\
\text{equil (M)} & (7.0 \times 10^{-4}) - x & x & x
\end{array}$$

$x = [OH^-] = 3.16 \times 10^{-5}$ M

$K_b = \dfrac{[C_{17}H_{20}NO_3^+][OH^-]}{[C_{17}H_{19}NO_3]} = \dfrac{x^2}{(7.0 \times 10^{-4}) - x} = \dfrac{(3.16 \times 10^{-5})^2}{(7.0 \times 10^{-4}) - (3.16 \times 10^{-5})} = 1.49 \times 10^{-6}$

$K_b = 1 \times 10^{-6}$

$pK_b = -\log K_b = -\log(1.49 \times 10^{-6}) = 5.827 = 5.8$

14.92 $K_b = 10^{-pK_b} = 10^{-5.47} = 3.4 \times 10^{-6}$

$$\begin{array}{lccc}
 & C_{18}H_{21}NO_4(aq) + H_2O(l) & \rightleftharpoons & HC_{18}H_{21}NO_4^+(aq) + OH^-(aq) \\
\text{initial (M)} & 2.50 \times 10^{-3} & 0 & \sim 0 \\
\text{change (M)} & -x & +x & +x \\
\text{equil (M)} & (2.50 \times 10^{-3}) - x & x & x
\end{array}$$

$K_b = \dfrac{[HC_{18}H_{21}NO_4^+][OH^-]}{[C_{18}H_{21}NO_4]} = 3.4 \times 10^{-6} = \dfrac{x^2}{(2.50 \times 10^{-3}) - x}$

$x^2 + (3.4 \times 10^{-6})x - (8.5 \times 10^{-9}) = 0$

Use the quadratic formula to solve for x.

$x = \dfrac{-(3.4 \times 10^{-6}) \pm \sqrt{(3.4 \times 10^{-6})^2 - 4(1)(-8.5 \times 10^{-9})}}{2(1)}$

$x = \dfrac{-(3.4 \times 10^{-6}) \pm (1.84 \times 10^{-4})}{2}$

$x = 9.0 \times 10^{-5}$ and -9.4×10^{-5}

Of the two solutions for x, only the positive value of x has physical meaning, because x is the $[OH^-]$.

$x = 9.0 \times 10^{-5}$ M $= [OH^-] = [HC_{18}H_{21}NO_4^+]$

$[C_{18}H_{21}NO_4] = (2.50 \times 10^{-3}) - x = (2.50 \times 10^{-3}) - (9.0 \times 10^{-5}) = 0.0024$ M

$$[H_3O^+] = \frac{K_w}{[OH^-]} = \frac{1.0 \times 10^{-14}}{9.0 \times 10^{-5}} = 1.1 \times 10^{-10} \text{ M}$$

$$pH = -\log[H_3O^+] = -\log(1.1 \times 10^{-10}) = 9.96$$

14.94 (a) $K_a = \dfrac{K_w}{K_b \text{ for } C_3H_7NH_2} = \dfrac{1.0 \times 10^{-14}}{5.1 \times 10^{-4}} = 2.0 \times 10^{-11}$

(b) $K_a = \dfrac{K_w}{K_b \text{ for } NH_2OH} = \dfrac{1.0 \times 10^{-14}}{9.1 \times 10^{-9}} = 1.1 \times 10^{-6}$

(c) $K_a = \dfrac{K_w}{K_b \text{ for } C_6H_5NH_2} = \dfrac{1.0 \times 10^{-14}}{4.3 \times 10^{-10}} = 2.3 \times 10^{-5}$

(d) $K_a = \dfrac{K_w}{K_b \text{ for } C_5H_5N} = \dfrac{1.0 \times 10^{-14}}{1.8 \times 10^{-9}} = 5.6 \times 10^{-6}$

Acid–Base Properties of Salts (Section 14.14)

14.96 (a) $CH_3NH_3^+(aq) + H_2O(l) \rightleftarrows H_3O^+(aq) + CH_3NH_2(aq)$

       acid   base ——— acid   base

(b) $Cr(H_2O)_6^{3+}(aq) + H_2O(l) \rightleftarrows H_3O^+(aq) + Cr(H_2O)_5(OH)^{2+}(aq)$

       acid   base ——— acid   base

(c) $CH_3CO_2^-(aq) + H_2O(l) \rightleftarrows CH_3CO_2H(aq) + OH^-(aq)$

       base   acid   acid   base

(d) $PO_4^{3-}(aq) + H_2O(l) \rightleftarrows HPO_4^{2-}(aq) + OH^-(aq)$

       base   acid   acid   base

14.98 (a) F^- (conjugate base of a weak acid), basic solution

(b) Br^- (anion of a strong acid), neutral solution

(c) NH_4^+ (conjugate acid of a weak base), acidic solution

(d) $K(H_2O)_6^+$ (neutral cation), neutral solution

(e) SO_3^{2-} (conjugate base of a weak acid), basic solution

(f) $Cr(H_2O)_6^{3+}$ (acidic cation), acidic solution

14.100 (a) $(C_2H_5NH_3)NO_3$: $C_2H_5NH_3^+$, acidic cation; NO_3^-, neutral anion
$C_2H_5NH_2$, $K_b = 6.4 \times 10^{-4}$

$$C_2H_5NH_3^+, \quad K_a = \frac{K_w}{K_b \text{ for } C_2H_5NH_2} = \frac{1.0 \times 10^{-14}}{6.4 \times 10^{-4}} = 1.56 \times 10^{-11}$$

$$C_2H_5NH_3^+(aq) + H_2O(l) \rightleftarrows H_3O^+(aq) + C_2H_5NH_2(aq)$$

initial (M)	0.10	~0	0
change (M)	–x	+x	+x
equil (M)	0.10 – x	x	x

$$K_a = \frac{[H_3O^+][C_2H_5NH_2]}{[C_2H_5NH_3^+]} = 1.56 \times 10^{-11} = \frac{x^2}{0.10-x} \approx \frac{x^2}{0.10}$$

Solve for x. $x = 1.25 \times 10^{-6}$ M $= 1.2 \times 10^{-6}$ M $= [H_3O^+] = [C_2H_5NH_2]$
pH $= -\log[H_3O^+] = -\log(1.25 \times 10^{-6}) = 5.90$
$[C_2H_5NH_3^+] = 0.10 - x = 0.10$ M; $[NO_3^-] = 0.10$ M

$$[OH^-] = \frac{K_w}{[H_3O^+]} = \frac{1.0 \times 10^{-14}}{1.25 \times 10^{-6} \text{ M}} = 8.0 \times 10^{-9}$$

(b) $Na(CH_3CO_2)$: Na^+, neutral cation; $CH_3CO_2^-$, basic anion
CH_3CO_2H, $K_a = 1.8 \times 10^{-5}$

$$CH_3CO_2^-, \quad K_b = \frac{K_w}{K_a \text{ for } CH_3CO_2H} = \frac{1.0 \times 10^{-14}}{1.8 \times 10^{-5}} = 5.6 \times 10^{-10}$$

$$CH_3CO_2^-(aq) + H_2O(aq) \rightleftarrows CH_3CO_2H(aq) + OH^-(aq)$$

initial (M)	0.10	0	~0
change (M)	–x	+x	+x
equil (M)	0.10 – x	x	x

$$K_b = \frac{[CH_3CO_2H][OH^-]}{[CH_3CO_2^-]} = 5.6 \times 10^{-10} = \frac{x^2}{0.10-x} \approx \frac{x^2}{0.10}$$

Solve for x. $x = 7.5 \times 10^{-6}$ M $= [CH_3CO_2H] = [OH^-]$
$[CH_3CO_2^-] = 0.10 - x = 0.10$ M; $[Na^+] = 0.10$ M

$$[H_3O^+] = \frac{K_w}{[OH^-]} = \frac{1.0 \times 10^{-14}}{7.5 \times 10^{-6}} = 1.3 \times 10^{-9} \text{ M}$$

pH $= -\log[H_3O^+] = -\log(1.3 \times 10^{-9}) = 8.89$
(c) $NaNO_3$: Na^+, neutral cation; NO_3^-, neutral anion
$[Na^+] = [NO_3^-] = 0.10$ M
$[H_3O^+] = [OH^-] = 1.0 \times 10^{-7}$ M; pH $= 7.00$

Lewis Acids and Bases (Section 14.15)

14.102 (a) Lewis acid, SiF_4; Lewis base, F^- (b) Lewis acid, Zn^{2+}; Lewis base, NH_3
(c) Lewis acid, $HgCl_2$; Lewis base, Cl^- (d) Lewis acid, CO_2; Lewis base, H_2O

14.104 (a)

$$2\ :\!\ddot{\underset{..}{F}}\!:^{-} \ +\ SiF_4 \longrightarrow SiF_6{}^{2-}$$

(b)

$$4\ \ddot{N}H_3 \ +\ Zn^{2+} \longrightarrow Zn(NH_3)_4{}^{2+}$$

(c)

$$2\ :\!\ddot{\underset{..}{Cl}}\!:^{-} \ +\ HgCl_2 \longrightarrow HgCl_4{}^{2-}$$

(d)

$$H_2\ddot{\underset{..}{O}}\!: \ +\ CO_2 \longrightarrow H_2CO_3$$

14.106 (a) CN^-, Lewis base   (b) H^+, Lewis acid   (c) H_2O, Lewis base
(d) Fe^{3+}, Lewis acid   (e) OH^-, Lewis base   (f) CO_2, Lewis acid
(g) $P(CH_3)_3$, Lewis base   (h) $B(CH_3)_3$, Lewis acid

Chapter Problems

14.108 In aqueous solution:
H_2S acts as an acid only.
HS^- can act as both an acid and a base.
S^{2-} can act as a base only.
H_2O can act as both an acid and a base.
H_3O^+ acts as an acid only.
OH^- acts as a base only.

14.110 $HCO_3^-(aq) + Al(H_2O)_6{}^{3+}(aq) \rightarrow H_2O(l) + CO_2(g) + Al(H_2O)_5(OH)^{2+}(aq)$

14.112 H_2O, 18.02

$$\text{at 0 °C, } [H_2O] = \frac{\left(0.9998 \text{ g} \times \dfrac{1 \text{ mol}}{18.02 \text{ g}}\right)}{0.001 \text{ L}} = 55.48 \text{ M}$$

$K_w = [H_3O^+][OH^-]$, for a neutral solution $[H_3O^+] = [OH^-]$

$[H_3O^+] = \sqrt{K_w} = \sqrt{1.14 \times 10^{-15}} = 3.376 \times 10^{-8} \text{ M}$

$pH = -\log[H_3O^+] = -\log(3.376 \times 10^{-8}) = 7.472$

$$\text{fraction dissociated} = \frac{[H_2O]_{diss}}{[H_2O]_{initial}} = \frac{3.376 \times 10^{-8} \text{ M}}{55.48 \text{ M}} = 6.09 \times 10^{-10}$$

$$\text{\% dissociation} = \frac{[H_2O]_{diss}}{[H_2O]_{initial}} \times 100\% = \frac{3.376 \times 10^{-8} \text{ M}}{55.48 \text{ M}} \times 100\% = 6.09 \times 10^{-8}\ \%$$

14.114

	HA(aq)	+	H₂O(l)	⇌	H₃O⁺(aq)	+	A⁻(aq)

$$\begin{array}{lccc}
 & HA(aq) + H_2O(l) \rightleftharpoons & H_3O^+(aq) + & A^-(aq) \\
\text{initial (M)} & 0.050 & \sim 0 & 0 \\
\text{change (M)} & -x & +x & +x \\
\text{equil (M)} & 0.050 - x & x & x
\end{array}$$

$x = [H_3O^+] = 10^{-pH} = 10^{-2.86} = 1.38 \times 10^{-3} \text{ M}$

$$K_a = \frac{[H_3O^+][A^-]}{[HA]} = \frac{x^2}{0.050 - x} = \frac{(1.38 \times 10^{-3})^2}{0.050 - (1.38 \times 10^{-3})} = 3.92 \times 10^{-5} = 3.9 \times 10^{-5}$$

$pK_a = -\log K_a = -\log(3.92 \times 10^{-5}) = 4.41$

14.116 For $C_{10}H_{14}N_2H^+$, $K_{a1} = \dfrac{K_w}{K_{b1} \text{ for } C_{10}H_{14}N_2} = \dfrac{1.0 \times 10^{-14}}{1.0 \times 10^{-6}} = 1.0 \times 10^{-8}$

For $C_{10}H_{14}N_2H_2^{2+}$, $K_{a2} = \dfrac{K_w}{K_{b2} \text{ for } C_{10}H_{14}N_2H^+} = \dfrac{1.0 \times 10^{-14}}{1.3 \times 10^{-11}} = 7.7 \times 10^{-4}$

14.118 (a) $A^-(aq) + H_2O(l) \rightleftharpoons HA(aq) + OH^-(aq)$; basic

(b) $M(H_2O)_6^{3+}(aq) + H_2O(l) \rightleftharpoons H_3O^+(aq) + M(H_2O)_5(OH)^{2+}(aq)$; acidic

(c) $2 H_2O(l) \rightleftharpoons H_3O^+(aq) + OH^-(aq)$; neutral

(d) $M(H_2O)_6^{3+}(aq) + A^-(aq) \rightleftharpoons HA(aq) + M(H_2O)_5(OH)^{2+}(aq)$;
acidic because K_a for $M(H_2O)_6^{3+}$ (10^{-4}) is greater than K_b for A^- (10^{-9})

14.120

	$HIO_3(aq)$	$+ H_2O(l) \rightleftharpoons$	$H_3O^+(aq)$	$+ IO_3^-(aq)$
initial (M)	0.0500		~0	0
change (M)	–x		+x	+x
equil (M)	0.0500 – x		x	x

$K_a = \dfrac{[H_3O^+][IO_3^-]}{[HIO_3]} = 1.7 \times 10^{-1} = \dfrac{x^2}{0.0500 - x}$

$x^2 + 0.17x - 0.0085 = 0$

Use the quadratic formula to solve for x.

$x = \dfrac{-(0.17) \pm \sqrt{(0.17)^2 - (4)(1)(-0.0085)}}{2(1)} = \dfrac{(-0.17) \pm 0.251}{2}$

$x = -0.210$ and 0.0405

Of the two solutions for x, only the positive value of x has physical meaning because x is the $[H_3O^+]$.

$x = [H_3O^+] = 0.0405$ M $= 0.040$ M

$pH = -\log[H_3O^+] = -\log(0.040) = 1.39$

$[HIO_3] = 0.0500 - x = 0.0500 - 0.040 = 0.010$ M

$[IO_3^-] = x = 0.040$ M

$[OH^-] = \dfrac{K_w}{[H_3O^+]} = \dfrac{1.0 \times 10^{-14}}{0.040} = 2.5 \times 10^{-13}$ M

14.122 $K_{a1} = 10^{-pK_{a1}} = 10^{-2.89} = 1.3 \times 10^{-3}$; $K_{a2} = 10^{-pK_{a2}} = 10^{-5.51} = 3.1 \times 10^{-6}$

	$H_2C_8H_4O_4(aq)$	$+ H_2O(l) \rightleftharpoons$	$H_3O^+(aq)$	$+ HC_8H_4O_4^-(aq)$
initial (M)	0.0250		~0	0
change (M)	–x		+x	+x
equil (M)	0.0250 – x		x	x

$K_{a1} = \dfrac{[H_3O^+][HC_8H_4O_4^-]}{[H_2C_8H_4O_4]} = 1.3 \times 10^{-3} = \dfrac{x^2}{0.0250 - x}$

$x^2 + (1.3 \times 10^{-3})x - (3.25 \times 10^{-5}) = 0$

Use the quadratic formula to solve for x.

$$x = \frac{-(1.3 \times 10^{-3}) \pm \sqrt{(1.3 \times 10^{-3})^2 - (4)(1)(-3.25 \times 10^{-5})}}{2(1)} = \frac{-(1.3 \times 10^{-3}) \pm 0.0115}{2}$$

$x = 0.0051$ and -0.0064

Of the two solutions for x, only the positive value of x has physical meaning because x is the $[H_3O^+]$.

$x = 0.0051 \; M = [H_3O^+] = [HC_8H_4O_4^-]$

$[H_2C_8H_4O_4] = 0.0250 - x = 0.0250 - 0.0051 = 0.020 \; M$

The second dissociation of $H_2C_7H_3NO_4$ produces a negligible amount of H_3O^+ compared with that from the first dissociation.

$HC_8H_4O_4^-(aq) + H_2O(l) \rightleftarrows H_3O^+(aq) + C_8H_4O_4^{2-}(aq)$

$$K_{a2} = \frac{[H_3O^+][C_8H_4O_4^{2-}]}{[HC_8H_4O_4^-]} = 3.1 \times 10^{-6} = \frac{(0.0051)[C_8H_4O_4^{2-}]}{(0.0051)}$$

$[C_8H_4O_4^{2-}] = K_{a2} = 3.1 \times 10^{-6} \; M$

$$[OH^-] = \frac{K_w}{[H_3O^+]} = \frac{1.0 \times 10^{-14}}{0.0051} = 2.0 \times 10^{-12} \; M$$

$pH = -\log[H_3O^+] = -\log(0.0051) = 2.29$

14.124   (a) NH_4F;   For NH_4^+, $K_a = 5.6 \times 10^{-10}$ and for F^-, $K_b = 2.9 \times 10^{-11}$
Because $K_a > K_b$, the salt solution is acidic.
(b) $(NH_4)_2SO_3$;   For NH_4^+, $K_a = 5.6 \times 10^{-10}$ and for SO_3^{2-}, $K_b = 1.6 \times 10^{-7}$
Because $K_b > K_a$, the salt solution is basic.

14.126   Fraction dissociated $= \dfrac{[HA]_{diss}}{[HA]_{initial}}$

For a weak acid, $[HA]_{diss} = [H_3O^+] = [A^-]$

$$K_a = \frac{[H_3O^+][A^-]}{[HA]} = \frac{[H_3O^+]^2}{[HA]}$$

$[H_3O^+] = \sqrt{K_a [HA]}$

Fraction dissociated $= \dfrac{[HA]_{diss}}{[HA]} = \dfrac{[H_3O^+]}{[HA]} = \dfrac{\sqrt{K_a [HA]}}{[HA]} = \sqrt{\dfrac{K_a}{[HA]}}$

When the concentration of HA that dissociates is negligible compared with its initial concentration, the equilibrium concentration, [HA], equals the initial concentration, $[HA]_{initial}$.

% dissociation $= \sqrt{\dfrac{K_a}{[HA]_{initial}}} \times 100\%$

14.128 Both reactions occur together.

Let $x = [H_3O^+]$ from CH_3CO_2H and $y = [H_3O^+]$ from $C_6H_5CO_2H$

The following two equilibria must be considered:

$$CH_3CO_2H(aq) + H_2O(l) \rightleftharpoons H_3O^+(aq) + CH_3CO_2^-(aq)$$

initial (M)	0.10	y	0
change (M)	−x	+x	+x
equil (M)	0.10 − x	x + y	x

$$C_6H_5CO_2H(aq) + H_2O(l) \rightleftharpoons H_3O^+(aq) + C_6H_5CO_2^-(aq)$$

initial (M)	0.10	x	0
change (M)	−y	+y	+y
equil (M)	0.10 − y	x + y	y

$$K_a \text{(for } CH_3CO_2H) = \frac{[H_3O^+][CH_3CO_2^-]}{[CH_3CO_2H]} = 1.8 \times 10^{-5} = \frac{(x+y)(x)}{0.10-x} \approx \frac{(x+y)(x)}{0.10}$$

$$1.8 \times 10^{-6} = (x+y)(x)$$

$$K_a \text{(for } C_6H_5CO_2H) = \frac{[H_3O^+][C_6H_5CO_2^-]}{[C_6H_5CO_2H]} = 6.5 \times 10^{-5} = \frac{(x+y)(y)}{0.10-y} \approx \frac{(x+y)(y)}{0.10}$$

$$6.5 \times 10^{-6} = (x+y)(y)$$

$$1.8 \times 10^{-6} = (x+y)(x)$$

$$6.5 \times 10^{-6} = (x+y)(y)$$

These two equations must be solved simultaneously for x and y. Divide the first equation by the second.

$$\frac{x}{y} = \frac{1.8 \times 10^{-6}}{6.5 \times 10^{-6}}; \quad x = 0.277y$$

$6.5 \times 10^{-6} = (x+y)(y)$; substitute $x = 0.277y$ into this equation and solve for y.

$6.5 \times 10^{-6} = (0.277y + y)(y) = 1.277y^2$

$y = 0.002\ 256$

$x = 0.277y = (0.277)(0.002\ 256) = 0.000\ 624\ 9$

$[H_3O^+] = (x + y) = (0.000\ 624\ 9 + 0.002\ 256) = 0.002\ 881$ M

$pH = -\log[H_3O^+] = -\log(0.002\ 881) = 2.54$

14.130 $2\ NO_2(g) + H_2O(l) \rightarrow HNO_3(aq) + HNO_2(aq)$

$$\text{mol } HNO_3 = 0.0500 \text{ mol } NO_2 \times \frac{1 \text{ mol } HNO_3}{2 \text{ mol } NO_2} = 0.0250 \text{ mol } HNO_3$$

$$\text{mol } HNO_2 = 0.0500 \text{ mol } NO_2 \times \frac{1 \text{ mol } HNO_2}{2 \text{ mol } NO_2} = 0.0250 \text{ mol } HNO_2$$

Because the volume is 1.00 L, mol and molarity are the same.

HNO_3 is a strong acid and completely dissociated.

From HNO_3, $[NO_3^-] = [H_3O^+] = 0.0250$ M.

$$HNO_2(aq) + H_2O(l) \rightleftharpoons H_3O^+(aq) + NO_2^-(aq)$$

initial (M)	0.0250	0.0250	0
change (M)	−x	+x	+x
equil (M)	0.0250 − x	0.0250 + x	x

$$K_a = \frac{[H_3O^+][NO_2^-]}{[HNO_2]} = 4.5 \times 10^{-4} = \frac{(0.0250 + x)x}{0.0250 - x}$$

$$x^2 + 0.02545x - 1.125 \times 10^{-5} = 0$$

Solve for x using the quadratic formula.

$$x = \frac{-(0.025\ 45) \pm \sqrt{(0.025\ 45)^2 - (4)(-1.125 \times 10^{-5})}}{2(1)} = \frac{(-0.025\ 45) \pm (0.026\ 32)}{2}$$

$x = -0.0259$ and 4.35×10^{-4}

Of the two solutions for x, only the positive value of x has physical meaning, because x is the $[NO_2^-]$.

$[H_3O^+] = (0.0250) + x = (0.0250) + (4.35 \times 10^{-4}) = 0.0254$ M

$pH = -\log[H_3O^+] = -\log(0.02543) = 1.59$

$$[OH^-] = \frac{K_w}{[H_3O^+]} = \frac{1.0 \times 10^{-14}}{0.0254} = 3.9 \times 10^{-13} \text{ M}$$

$[NO_3^-] = 0.0250$ M

$[HNO_2] = 0.0250 - x = 0.0250 - 4.35 \times 10^{-4} = 0.0246$ M; $[NO_2^-] = x = 4.3 \times 10^{-4}$ M

Multiconcept Problems

14.132 H_3PO_4, 98.00

Assume 1.000 L of solution.

Mass of solution $= 1.000 \text{ L} \times \dfrac{1000 \text{ mL}}{1 \text{ L}} \times \dfrac{1.0353 \text{ g}}{1 \text{ mL}} = 1035.3$ g

Mass $H_3PO_4 = (0.070)(1035.3 \text{ g}) = 72.47 \text{ g } H_3PO_4$

mol $H_3PO_4 = 72.47 \text{ g } H_3PO_4 \times \dfrac{1 \text{ mol } H_3PO_4}{98.00 \text{ g } H_3PO_4} = 0.740 \text{ mol } H_3PO_4$

$$[H_3PO_4] = \frac{0.740 \text{ mol } H_3PO_4}{1.000 \text{ L}} = 0.740 \text{ M}$$

For the dissociation of the first proton, the following equilibrium must be considered:

$$H_3PO_4(aq) + H_2O(l) \rightleftharpoons H_3O^+(aq) + H_2PO_4^-(aq)$$

initial (M)	0.740	~0	0
change (M)	−x	+x	+x
equil (M)	0.740 − x	x	x

$$K_{a1} = \frac{[H_3O^+][H_2PO_4^-]}{[H_3PO_4]} = 7.5 \times 10^{-3} = \frac{x^2}{0.740 - x}$$

$$x^2 + (7.5 \times 10^{-3})x - (5.55 \times 10^{-3}) = 0$$

Solve for x using the quadratic formula.

$$x = \frac{-(7.5 \times 10^{-3}) \pm \sqrt{(7.5 \times 10^{-3})^2 - (4)(-5.55 \times 10^{-3})}}{2(1)} = \frac{(-7.5 \times 10^{-3}) \pm 0.149}{2}$$

$x = 0.0708$ and -0.0783

Of the two solutions for x, only the positive value of x has physical meaning, because x is the $[H_3O^+]$.

$x = 0.0708 \text{ M} = [H_2PO_4^-] = [H_3O^+]$

For the dissociation of the second proton, the following equilibrium must be considered:

$$H_2PO_4^-(aq) + H_2O(l) \rightleftharpoons H_3O^+(aq) + HPO_4^{2-}(aq)$$

initial (M)	0.0708	0.0708	0
change (M)	$-y$	$+y$	$+y$
equil (M)	$0.0708 - y$	$0.0708 + y$	y

$$K_{a2} = \frac{[H_3O^+][HPO_4^{2-}]}{[H_2PO_4^-]} = 6.2 \times 10^{-8} = \frac{(0.0708 + y)(y)}{0.0708 - y} \approx \frac{(0.0708)(y)}{0.0708} = y$$

$y = 6.2 \times 10^{-8} \text{ M} = [HPO_4^{2-}]$

For the dissociation of the third proton, the following equilibrium must be considered:

$$HPO_4^{2-}(aq) + H_2O(l) \rightleftharpoons H_3O^+(aq) + PO_4^{3-}(aq)$$

initial (M)	6.2×10^{-8}	0.0708	0
change (M)	$-z$	$+z$	$+z$
equil (M)	$(6.2 \times 10^{-8}) - z$	$0.0708 + z$	z

$$K_{a3} = \frac{[H_3O^+][PO_4^{3-}]}{[HPO_4^{2-}]} = 4.8 \times 10^{-13} = \frac{(0.0708 + z)(z)}{(6.2 \times 10^{-8}) - z} \approx \frac{(0.0708)(z)}{6.2 \times 10^{-8}}$$

$z = 4.2 \times 10^{-19} \text{ M} = [PO_4^{3-}]$

$[H_3PO_4] = 0.740 - x = 0.740 - 0.0708 = 0.67 \text{ M}$

$[H_2PO_4^-] = [H_3O^+] = 0.0708 \text{ M} = 0.071 \text{ M}$

$[HPO_4^{2-}] = 6.2 \times 10^{-8} \text{ M}$

$[PO_4^{3-}] = 4.2 \times 10^{-19} \text{ M}$

$$[OH^-] = \frac{K_w}{[H_3O^+]} = \frac{1.0 \times 10^{-14}}{0.0708} = 1.4 \times 10^{-13} \text{ M}$$

$pH = -\log[H_3O^+] = -\log(0.0708) = 1.15$

14.134 CH_3CO_2Na, 82.034; $[H_3O^+] = 10^{-pH} = 10^{-9.07} = 8.51 \times 10^{-10} \text{ M}$

$[H_3O^+][OH^-] = K_w$

$$[OH^-] = \frac{K_w}{[H_3O^+]} = \frac{1.0 \times 10^{-14}}{8.51 \times 10^{-10}} = 1.18 \times 10^{-5} \text{ M} = x \text{ below.}$$

$$K_a = 1.8 \times 10^{-5} \text{ for } CH_3CO_2H \text{ and } K_b = \frac{K_w}{K_a} = \frac{1.0 \times 10^{-14}}{1.8 \times 10^{-5}} = 5.56 \times 10^{-10}$$

Use the equilibrium associated with a weak base to solve for $[CH_3CO_2^-] = y$ below.

$$CH_3CO_2^-(aq) + H_2O(l) \rightleftharpoons CH_3CO_2H(aq) + OH^-(aq)$$

initial (M)	y	~0	0
change (M)	−x	+x	+x
equil (M)	y − x	x	x

$$K_b = \frac{[CH_3CO_2H][OH^-]}{[CH_3CO_2^-]} = 5.56 \times 10^{-10} = \frac{(1.18 \times 10^{-5})^2}{[y - (1.18 \times 10^{-5})]}$$

Solve for y.

$(5.56 \times 10^{-10})[y - (1.18 \times 10^{-5})] = (1.18 \times 10^{-5})^2$

$(5.56 \times 10^{-10})y - 6.56 \times 10^{-15} = 1.39 \times 10^{-10}$

$(5.56 \times 10^{-10})y = 1.39 \times 10^{-10}$

$y = (1.39 \times 10^{-10})/(5.56 \times 10^{-10}) = [CH_3CO_2^-] = 0.25 \text{ M}$

In 1.00 L of solution, the mass of CH_3CO_2Na solute $= (0.25 \text{ mol/L})\left(\dfrac{82.034 \text{ g } CH_3CO_2Na}{1 \text{ mol } CH_3CO_2Na}\right) = 20.5 \text{ g}$

mass of solution $= (1000 \text{ mL})\left(\dfrac{1.0085 \text{ g}}{1 \text{ mL}}\right) = 1008.5 \text{ g}$

mass of solvent $= 1008.5 \text{ g} - 20.5 \text{ g} = 988 \text{ g} = 0.988 \text{ kg}$

$$m = \frac{0.25 \text{ mol } CH_3CO_2Na}{0.988 \text{ kg}} = 0.25 \ m$$

Because CH_3CO_2Na is a strong electrolyte, the ionic compound is completely dissociated and $[CH_3CO_2^-] = [Na^+]$. The contribution of CH_3CO_2H and OH^- to the total molality of the solution is negligible.

$\Delta T_f = K_f \cdot (2 \cdot m) = (1.86 \ ^\circ C/m)(2)(0.25 \ m) = 0.93 \ ^\circ C$

Solution freezing point $= 0.00 \ ^\circ C - \Delta T_f = 0.00 \ ^\circ C - 0.93 \ ^\circ C = -0.93 \ ^\circ C$

14.136 Na_3PO_4, 163.94

$3.28 \text{ g } Na_3PO_4 \times \dfrac{1 \text{ mol } Na_3PO_4}{163.94 \text{ g } Na_3PO_4} = 0.0200 \text{ mol} = 20.0 \text{ mmol } Na_3PO_4$

$300.0 \text{ mL} \times 0.180 \text{ mmol/mL} = 54.0 \text{ mmol HCl}$

$$H_3O^+(aq) + PO_4^{3-}(aq) \rightleftharpoons HPO_4^{2-}(aq) + H_2O(l)$$

before (mmol)	54.0	20.0	0
change (mmol)	−20.0	−20.0	+20.0
after (mmol)	34.0	0	20.0

$$H_3O^+(aq) + HPO_4^{2-}(aq) \rightleftharpoons H_2PO_4^-(aq) + H_2O(l)$$

before (mmol)	34.0	20.0	0
change (mmol)	−20.0	−20.0	+20.0
after (mmol)	14.0	0	20.0

$$H_3O^+(aq) \ + \ H_2PO_4^-(aq) \ \rightleftharpoons \ H_3PO_4(aq) \ + \ H_2O(l)$$

before (mmol)	14.0	20.0	0
change (mmol)	−14.0	−14.0	+14.0
after (mmol)	0	6.0	14.0

$$[H_3PO_4] = \frac{14.0 \text{ mmol}}{300.0 \text{ mL}} = 0.047 \text{ M}; \quad [H_2PO_4^-] = \frac{6.0 \text{ mmol}}{300.0 \text{ mL}} = 0.020 \text{ M}$$

$$H_3PO_4(aq) \ + \ H_2O(l) \ \rightleftharpoons \ H_3O^+(aq) \ + \ H_2PO_4^-(aq)$$

initial (M)	0.047	~0	0.020
change (M)	−x	+x	+x
equil (M)	0.047 − x	x	0.020 + x

$$K_a = \frac{[H_3O^+][H_2PO_4^{2-}]}{[H_3PO_4]} = 7.5 \times 10^{-3} = \frac{x(0.020 + x)}{(0.047 - x)}$$

$x^2 + 0.0275x - (3.525 \times 10^{-4}) = 0$

Solve for x using the quadratic formula.

$$x = \frac{-(0.0275) \pm \sqrt{(0.0275)^2 - (4)(-3.525 \times 10^{-4})}}{2(1)} = \frac{-0.0275 \pm 0.0465}{2}$$

$x = 0.0095$ and -0.0370

Of the two solutions for x, only the positive value of x has physical meaning, because x is the $[H_3O^+]$.

$pH = -\log[H_3O^+] = -\log(0.0095) = 2.02$

14.138 (a) $PV = nRT$; $n = \dfrac{PV}{RT} = \dfrac{(0.601 \text{ atm})(1.000 \text{ L})}{\left(0.082 \ 06 \ \dfrac{L \cdot atm}{K \cdot mol}\right)(293.1 \text{ K})} = 0.0250 \text{ mol HF}$

$$50.0 \text{ mL} \times \frac{1.00 \text{ L}}{1000 \text{ mL}} = 0.0500 \text{ L}$$

$$[HF] = \frac{0.0250 \text{ mol HF}}{0.0500 \text{ L}} = 0.500 \text{ M}$$

$$HF(aq) \ + \ H_2O(l) \ \rightleftharpoons \ H_3O^+(aq) \ + \ F^-(aq)$$

initial (M)	0.500	~0	0
change (M)	−x	+x	+x
equil (M)	0.500 − x	x	x

$$K_a = \frac{[H_3O^+][F^-]}{[HF]} = 3.5 \times 10^{-4} = \frac{x^2}{0.500 - x}$$

$x^2 + (3.5 \times 10^{-4})x - (1.75 \times 10^{-4}) = 0$

Solve for x using the quadratic formula.

$$x = \frac{-(3.5 \times 10^{-4}) \pm \sqrt{(3.5 \times 10^{-4})^2 - (4)(-1.75 \times 10^{-4})}}{2(1)} = \frac{(-3.5 \times 10^{-4}) \pm 0.0265}{2}$$

$x = -0.0134$ and 0.0131

Of the two solutions for x, only the positive value of x has physical meaning, because x is the $[H_3O^+]$.

$pH = -\log[H_3O^+] = -\log(0.0131) = 1.883 = 1.88$

(b) % dissociation $= \dfrac{0.0131 \text{ M}}{0.500} \times 100\% = 2.62\% = 2.6\%$

New % dissociation $= (3)(2.62\%) = 7.86\%$

Let X equal the concentration of HF dissociated and Y the new volume (in liters) that would triple the % dissociation.

$K_a = \dfrac{X^2}{[(0.0250/Y) - X]} = 3.5 \times 10^{-4}$

% dissociation $= \dfrac{X}{(0.0250/Y)} \times 100\% = 7.86\%$ and $\dfrac{X}{(0.0250/Y)} = 0.0786$

$X = 1.965 \times 10^{-3}/Y$

Substitute X into the K_a equation.

$\dfrac{(1.965 \times 10^{-3}/Y)^2}{(0.0250/Y) - (1.965 \times 10^{-3}/Y)} = 3.5 \times 10^{-4}$

$\dfrac{3.861 \times 10^{-6}/Y^2}{0.0230/Y} = 3.5 \times 10^{-4}$

$\dfrac{3.861 \times 10^{-6}/Y}{0.0230} = 3.5 \times 10^{-4}$

$\dfrac{3.861 \times 10^{-6}}{8.05 \times 10^{-6}} = Y = 0.48 \text{ L}$

The result in Problem 14.126 can't be used here because the concentration of HF that dissociates can't be neglected compared with the initial HF concentration.

14.140 (a) Rate $= k[OCl^-]^x[NH_3]^y[OH^-]^z$

From experiments 1 & 2, the $[OCl^-]$ doubles and the rate doubles, therefore $x = 1$.
From experiments 2 & 3, the $[NH_3]$ triples and the rate triples, therefore $y = 1$.
From experiments 3 & 4, the $[OH^-]$ goes up by a factor of 10 and the rate goes down by a factor of 10, therefore $z = -1$.

Rate $= k \dfrac{[OCl^-][NH_3]}{[OH^-]}$

$[H_3O^+] = 10^{-pH} = 10^{-12} = 1 \times 10^{-12} \text{ M}$

$[OH^-] = \dfrac{K_w}{[H_3O^+]} = \dfrac{1.0 \times 10^{-14}}{1 \times 10^{-12}} = 0.01 \text{ M}$

Using experiment 1: $k = \dfrac{(\text{Rate})[OH^-]}{[OCl^-][NH_3]} = \dfrac{(0.017 \text{ M/s})(0.01 \text{ M})}{(0.001 \text{ M})(0.01 \text{ M})} = 17 \text{ s}^{-1}$

(b) $K_1 = \dfrac{[HOCl][OH^-]}{[OCl^-]} = K_b(OCl^-) = \dfrac{K_w}{K_a(HOCl)} = \dfrac{1.0 \times 10^{-14}}{3.5 \times 10^{-8}} = 2.9 \times 10^{-7}$

For second step, Rate = k_2 [HOCl][NH$_3$]

Multiply the Rate by $\dfrac{K_1}{K_1} = \dfrac{K_1}{\left(\dfrac{[HOCl][OH^-]}{[OCl^-]}\right)}$

Rate $= K_1 k_2 \dfrac{[HOCl][NH_3]}{\left(\dfrac{[HOCl][OH^-]}{[OCl^-]}\right)} = K_1 k_2 \dfrac{[HOCl][NH_3][OCl^-]}{[HOCl][OH^-]} = K_1 k_2 \dfrac{[OCl^-][NH_3]}{[OH^-]}$

$K_1 k_2 = k = 17\ s^{-1}$

$k_2 = \dfrac{17\ s^{-1}}{2.9 \times 10^{-7}\ M} = 5.9 \times 10^7\ M^{-1}s^{-1}$

15

Applications of Aqueous Equilibria

15.1   (a) $HNO_2(aq) + OH^-(aq) \rightleftharpoons NO_2^-(aq) + H_2O(l)$; NO_2^- (basic anion), pH > 7.00

    (b) $H_3O^+(aq) + NH_3(aq) \rightleftharpoons NH_4^+(aq) + H_2O(l)$; NH_4^+ (acidic cation), pH < 7.00

    (c) $OH^-(aq) + H_3O^+(aq) \rightleftharpoons 2\ H_2O(l)$;   pH = 7.00

15.2   (a) $HF(aq) + OH^-(aq) \rightleftharpoons H_2O(l) + F^-(aq)$

$$K_n = \frac{K_a}{K_w} = \frac{3.5 \times 10^{-4}}{1.0 \times 10^{-14}} = 3.5 \times 10^{10}$$

    (b) $H_3O^+(aq) + OH^-(aq) \rightleftharpoons 2\ H_2O(l)$

$$K_n = \frac{1}{K_w} = \frac{1}{1.0 \times 10^{-14}} = 1.0 \times 10^{14}$$

    (c) $HF(aq) + NH_3(aq) \rightleftharpoons NH_4^+(aq) + F^-(aq)$

$$K_n = \frac{K_a K_b}{K_w} = \frac{(3.5 \times 10^{-4})(1.8 \times 10^{-5})}{1.0 \times 10^{-14}} = 6.3 \times 10^5$$

The tendency to proceed to completion is determined by the magnitude of K_n. The larger the value of K_n, the further does the reaction proceed to completion.

The tendency to proceed to completion is: reaction (c) < reaction (a) < reaction (b)

15.3

	$HCN(aq)$	$+\ H_2O(l)$	$\rightleftharpoons$	$H_3O^+(aq)$	$+\ CN^-(aq)$
initial (M)	0.025			~0	0.010
change (M)	−x			+x	+x
equil (M)	0.025 − x			x	0.010 + x

$$K_a = \frac{[H_3O^+][CN^-]}{[HCN]} = 4.9 \times 10^{-10} = \frac{x(0.010 + x)}{0.025 - x} \approx \frac{x(0.010)}{0.025}$$

Solve for x. $x = 1.23 \times 10^{-9}\ M = 1.2 \times 10^{-9}\ M = [H_3O^+]$

$pH = -\log[H_3O^+] = -\log(1.23 \times 10^{-9}) = 8.91$

$$[OH^-] = \frac{K_w}{[H_3O^+]} = \frac{1.0 \times 10^{-14}}{1.23 \times 10^{-9}} = 8.1 \times 10^{-6}\ M$$

$[Na^+] = [CN^-] = 0.010\ M$; $[HCN] = 0.025\ M$

$$\%\ \text{dissociation} = \frac{[HCN]_{diss}}{[HCN]_{initial}} \times 100\% = \frac{1.23 \times 10^{-9}\ M}{0.025\ M} \times 100\% = 4.9 \times 10^{-6}\ \%$$

15.4　From $NH_4Cl(s)$, $[NH_4^+]_{initial} = \dfrac{0.10 \text{ mol}}{0.500 \text{ L}} = 0.20$ M

$$NH_3(aq) + H_2O(l) \rightleftharpoons NH_4^+(aq) + OH^-(aq)$$

initial (M)	0.40	0.20	~0
change (M)	−x	+x	+x
equil (M)	0.40 − x	0.20 + x	x

$K_b = \dfrac{[NH_4^+][OH^-]}{[NH_3]} = 1.8 \times 10^{-5} = \dfrac{(0.20 + x)(x)}{(0.40 - x)} \approx \dfrac{(0.20)(x)}{(0.40)}$

Solve for x.　$x = [OH^-] = 3.6 \times 10^{-5}$ M

$[H_3O^+] = \dfrac{K_w}{[OH^-]} = \dfrac{1.0 \times 10^{-14}}{3.6 \times 10^{-5}} = 2.8 \times 10^{-10}$ M

$pH = -\log[H_3O^+] = -\log(2.8 \times 10^{-10}) = 9.55$

15.5　Each solution contains the same number of B molecules. The presence of BH^+ from BHCl lowers the percent dissociation of B. Solution (2) contains no BH^+, therefore it has the largest percent dissociation. BH^+ is the conjugate acid of B. Solution (1) has the largest amount of BH^+, and it would be the most acidic solution and have the lowest pH.

15.6　(a) (1) and (3). Both pictures show equal concentrations of HA and A^-.

　　　(b) (3). It contains a higher concentration of HA and A^-.

15.7　
$$HF(aq) + H_2O(l) \rightleftharpoons H_3O^+(aq) + F^-(aq)$$

initial (M)	0.25	~0	0.50
change (M)	−x	+x	+x
equil (M)	0.25 − x	x	0.50 + x

$K_a = \dfrac{[H_3O^+][F^-]}{[HF]} = 3.5 \times 10^{-4} = \dfrac{x(0.50 + x)}{0.25 - x} \approx \dfrac{x(0.50)}{0.25}$

Solve for x.　$x = 1.75 \times 10^{-4}$ M $= [H_3O^+]$

For the buffer, $pH = -\log[H_3O^+] = -\log(1.75 \times 10^{-4}) = 3.76$

(a) mol HF = 0.025 mol; mol F^- = 0.050 mol; vol = 0.100 L

$$\overset{100\%}{F^-(aq) + H_3O^+(aq) \rightarrow HF(aq) + H_2O(l)}$$

before (mol)	0.050	0.002	0.025
change (mol)	−0.002	−0.002	+0.002
after (mol)	0.048	0	0.027

$[H_3O^+] = K_a \dfrac{[HF]}{[F^-]} = (3.5 \times 10^{-4})\left(\dfrac{0.27}{0.48}\right) = 1.97 \times 10^{-4}$ M

$pH = -\log[H_3O^+] = -\log(1.97 \times 10^{-4}) = 3.71$

(b) mol HF = 0.025 mol; mol F⁻ = 0.050 mol; vol = 0.100 L

$$\overset{100\%}{HF(aq) \ + \ OH^-(aq) \ \rightarrow \ F^-(aq) \ + \ H_2O(l)}$$

before (mol)	0.025	0.004	0.050
change (mol)	–0.004	–0.004	+0.004
after (mol)	0.021	0	0.054

$$[H_3O^+] = K_a\frac{[HF]}{[F^-]} = (3.5 \times 10^{-4})\left(\frac{0.21}{0.54}\right) = 1.36 \times 10^{-4} \ M$$

$$pH = -\log[H_3O^+] = -\log(1.36 \times 10^{-4}) = 3.87$$

15.8
$$HF(aq) \ + \ H_2O(l) \ \rightleftharpoons \ H_3O^+(aq) \ + \ F^-(aq)$$

initial (M)	0.050	~0	0.100
change (M)	–x	+x	+x
equil (M)	0.050 – x	x	0.100 + x

$$K_a = \frac{[H_3O^+][F^-]}{[HF]} = 3.5 \times 10^{-4} = \frac{x(0.100 + x)}{0.050 - x} \approx \frac{x(0.100)}{0.050}$$

Solve for x. $x = [H_3O^+] = 1.75 \times 10^{-4} \ M$

$pH = -\log[H_3O^+] = -\log(1.75 \times 10^{-4}) = 3.76$

mol HF = 0.050 mol/L x 0.100 L = 0.0050 mol HF

mol F⁻ = 0.100 mol/L x 0.100 L = 0.0100 mol F⁻

mol HNO₃ = mol H₃O⁺ = 0.002 mol

$$\overset{100\%}{\text{Neutralization reaction:} \quad F^-(aq) \ + \ H_3O^+(aq) \ \rightarrow \ HF(aq) \ + \ H_2O(l)}$$

before reaction (mol)	0.0100	0.002	0.0050
change (mol)	–0.002	–0.002	+0.002
after reaction (mol)	0.008	0	0.007

$$[HF] = \frac{0.007 \ mol}{0.100 \ L} = 0.07 \ M; \qquad [F^-] = \frac{0.008 \ mol}{0.100 \ L} = 0.08 \ M$$

$$[H_3O^+] = K_a\frac{[HF]}{[F^-]} = (3.5 \times 10^{-4})\frac{(0.07)}{(0.08)} = 3 \times 10^{-4} \ M$$

$pH = -\log[H_3O^+] = -\log(3 \times 10^{-4}) = 3.5$

This solution has less buffering capacity than the solution in Problem 15.7 because it contains less HF and F⁻ per 100 mL. Note that the change in pH is greater than that in Problem 15.7.

15.9 When equal volumes of two solutions are mixed together, the concentration of each solution is cut in half.

$$pH = pK_a + \log\frac{[base]}{[acid]} = pK_a + \log\frac{[CO_3^{2-}]}{[HCO_3^-]}$$

For HCO₃⁻, $K_a = 5.6 \times 10^{-11}$, $pK_a = -\log K_a = -\log(5.6 \times 10^{-11}) = 10.25$

$$pH = 10.25 + \log\left(\frac{0.050}{0.10}\right) = 10.25 - 0.30 = 9.95$$

15.10 $pH = pK_a + \log \dfrac{[base]}{[acid]} = pK_a + \log \dfrac{[CO_3^{2-}]}{[HCO_3^-]}$

For HCO_3^-, $K_a = 5.6 \times 10^{-11}$, $pK_a = -\log K_a = -\log(5.6 \times 10^{-11}) = 10.25$

$10.40 = 10.25 + \log \dfrac{[CO_3^{2-}]}{[HCO_3^-]}$; $\log \dfrac{[CO_3^{2-}]}{[HCO_3^-]} = 10.40 - 10.25 = 0.15$

$\dfrac{[CO_3^{2-}]}{[HCO_3^-]} = 10^{0.15} = 1.4$

To obtain a buffer solution with pH 10.40, make the Na_2CO_3 concentration 1.4 times the concentration of $NaHCO_3$.

15.11 Look for an acid with pK_a near the required pH of 7.50.
$K_a = 10^{-pH} = 10^{-7.50} = 3.2 \times 10^{-8}$
Suggested buffer system: HOCl ($K_a = 3.5 \times 10^{-8}$) and NaOCl.

15.12 (a) serine is 66% dissociated at $pH = 9.15 + \log \left(\dfrac{66}{100 - 66} \right) = 9.44$

(b) serine is 5% dissociated at $pH = 9.15 + \log \left(\dfrac{5}{100 - 5} \right) = 7.87$

15.13 (a) mol HCl = mol H_3O^+ = 0.100 mol/L x 0.0400 L = 0.004 00 mol
mol NaOH = mol OH^- = 0.100 mol/L x 0.0350 L = 0.003 50 mol

Neutralization reaction:	$H_3O^+(aq)$	+ $OH^-(aq)$	→	$2 H_2O(l)$
before reaction (mol)	0.004 00	0.003 50		
change (mol)	−0.003 50	−0.003 50		
after reaction (mol)	0.000 50	0		

$[H_3O^+] = \dfrac{0.000\ 50\ mol}{(0.0400\ L + 0.0350\ L)} = 6.7 \times 10^{-3}$ M

$pH = -\log[H_3O^+] = -\log(6.7 \times 10^{-3}) = 2.17$

(b) mol HCl = mol H_3O^+ = 0.100 mol/L x 0.0400 L = 0.004 00 mol
mol NaOH = mol OH^- = 0.100 mol/L x 0.0450 L = 0.004 50 mol

Neutralization reaction:	$H_3O^+(aq)$	+ $OH^-(aq)$	→	$2 H_2O(l)$
before reaction (mol)	0.004 00	0.004 50		
change (mol)	−0.004 00	−0.004 00		
after reaction (mol)	0	0.000 50		

$[OH^-] = \dfrac{0.000\ 50\ mol}{(0.0400\ L + 0.0450\ L)} = 5.9 \times 10^{-3}$ M

$[H_3O^+] = \dfrac{K_w}{[OH^-]} = \dfrac{1.0 \times 10^{-14}}{5.9 \times 10^{-3}} = 1.7 \times 10^{-12}$ M

$pH = -\log[H_3O^+] = -\log(1.7 \times 10^{-12}) = 11.77$
The results obtained here are consistent with the pH data in Table 15.1.

15.14   (a) mol NaOH = mol OH^- = 0.100 mol/L x 0.0400 L = 0.004 00 mol
mol HCl = mol H_3O^+ = 0.0500 mol/L x 0.0600 L = 0.003 00 mol

Neutralization reaction:	$H_3O^+(aq)$ +	$OH^-(aq)$	→	$2 H_2O(l)$
before reaction (mol)	0.003 00	0.004 00		
change (mol)	−0.003 00	−0.003 00		
after reaction (mol)	0	0.001 00		

$$[OH^-] = \frac{0.001\ 00\ mol}{(0.0400\ L\ +\ 0.0600\ L)} = 1.0\ x\ 10^{-2}\ M$$

$$[H_3O^+] = \frac{K_w}{[OH^-]} = \frac{1.0\ x\ 10^{-14}}{1.0\ x\ 10^{-2}} = 1.0\ x\ 10^{-12}\ M$$

$$pH = -\log[H_3O^+] = -\log(1.0\ x\ 10^{-12}) = 12.00$$

(b) mol NaOH = mol OH^- = 0.100 mol/L x 0.0400 L = 0.004 00 mol
mol HCl = mol H_3O^+ = 0.0500 mol/L x 0.0802 L = 0.004 01 mol

Neutralization reaction:	$H_3O^+(aq)$ +	$OH^-(aq)$	→	$2 H_2O(l)$
before reaction (mol)	0.004 01	0.004 00		
change (mol)	−0.004 00	−0.004 00		
after reaction (mol)	0.000 01	0		

$$[H_3O^+] = \frac{0.000\ 01\ mol}{(0.0400\ L\ +\ 0.0802\ L)} = 8.3\ x\ 10^{-5}\ M$$

$$pH = -\log[H_3O^+] = -\log(8.3\ x\ 10^{-5}) = 4.08$$

(c) mol NaOH = mol OH^- = 0.100 mol/L x 0.0400 L = 0.004 00 mol
mol HCl = mol H_3O^+ = 0.0500 mol/L x 0.1000 L = 0.005 00 mol

Neutralization reaction:	$H_3O^+(aq)$ +	$OH^-(aq)$	→	$2 H_2O(l)$
before reaction (mol)	0.005 00	0.004 00		
change (mol)	−0.004 00	−0.004 00		
after reaction (mol)	0.001 00	0		

$$[H_3O^+] = \frac{0.001\ 00\ mol}{(0.0400\ L\ +\ 0.1000\ L)} = 7.1\ x\ 10^{-3}\ M$$

$$pH = -\log[H_3O^+] = -\log(7.1\ x\ 10^{-3}) = 2.15$$

15.15   (a) (3), only HA present   (b) (1), HA and A^- present
(c) (4), only A^- present   (d) (2), A^- and OH^- present

15.16   mol NaOH required = $\left(\dfrac{0.016\ mol\ HOCl}{L}\right)(0.100\ L)\left(\dfrac{1\ mol\ NaOH}{1\ mol\ HOCl}\right)$ = 0.0016 mol

vol NaOH required = $(0.0016\ mol)\left(\dfrac{1\ L}{0.0400\ mol}\right)$ = 0.040 L = 40 mL

40 mL of 0.0400 M NaOH are required to reach the equivalence point.

(a) mmol HOCl = 0.016 mmol/mL x 100.0 mL = 1.6 mmol
mmol NaOH = mmol OH^- = 0.0400 mmol/mL x 10.0 mL = 0.400 mmol

Neutralization reaction: $\quad HOCl(aq) + OH^-(aq) \rightarrow OCl^-(aq) + H_2O(l)$

before reaction (mmol)	1.6	0.400	0
change (mmol)	–0.400	–0.400	+0.400
after reaction (mmol)	1.2	0	0.400

$$[HOCl] = \frac{1.2 \text{ mmol}}{(100.0 \text{ mL} + 10.0 \text{ mL})} = 1.09 \times 10^{-2} \text{ M}$$

$$[OCl^-] = \frac{0.400 \text{ mmol}}{(100.0 \text{ mL} + 10.0 \text{ mL})} = 3.64 \times 10^{-3} \text{ M}$$

$$HOCl(aq) + H_2O(l) \rightleftharpoons H_3O^+(aq) + OCl^-(aq)$$

initial (M)	0.0109	~0	0.003 64
change (M)	–x	+x	+x
equil (M)	0.0109 – x	x	0.003 64 + x

$$K_a = \frac{[H_3O^+][OCl^-]}{[HOCl]} = 3.5 \times 10^{-8} = \frac{x(0.003\ 64 + x)}{0.0109 - x} \approx \frac{x(0.003\ 64)}{0.0109}$$

Solve for x. $x = [H_3O^+] = 1.05 \times 10^{-7}$ M
$pH = -\log[H_3O^+] = -\log(1.05 \times 10^{-7}) = 6.98$

(b) Halfway to the equivalence point, $[OCl^-] = [HOCl]$
$pH = pK_a = -\log K_a = -\log(3.5 \times 10^{-8}) = 7.46$

(c) At the equivalence point the solution contains the salt, NaOCl.
mol NaOCl = initial mol HOCl = 0.0016 mol = 1.6 mmol

$$[OCl^-] = \frac{1.6 \text{ mmol}}{(100.0 \text{ mL} + 40.0 \text{ mL})} = 1.1 \times 10^{-2} \text{ M}$$

For OCl^-, $K_b = \dfrac{K_w}{K_a \text{ for HOCl}} = \dfrac{1.0 \times 10^{-14}}{3.5 \times 10^{-8}} = 2.9 \times 10^{-7}$

$$OCl^-(aq) + H_2O(l) \rightleftharpoons HOCl(aq) + OH^-(aq)$$

initial (M)	0.011	0	~0
change (M)	–x	+x	+x
equil (M)	0.011 – x	x	x

$$K_b = \frac{[HOCl][OH^-]}{[OCl^-]} = 2.9 \times 10^{-7} = \frac{x^2}{0.011 - x} \approx \frac{x^2}{0.011}$$

Solve for x. $x = [OH^-] = 5.65 \times 10^{-5}$ M

$$[H_3O^+] = \frac{K_w}{[OH^-]} = \frac{1.0 \times 10^{-14}}{5.65 \times 10^{-5}} = 1.77 \times 10^{-10} = 1.8 \times 10^{-10} \text{ M}$$

$pH = -\log[H_3O^+] = -\log(1.77 \times 10^{-10}) = 9.75$

15.17 From Problem 15.16, pH = 9.75 at the equivalence point.
Use thymolphthalein (pH 9.4 – 10.6). Bromthymol blue is unacceptable because it changes color halfway to the equivalence point.

15.18   (a) mol NaOH required to reach first equivalence point

$$= \left(\frac{0.0800 \text{ mol } H_2SO_3}{L} \right) (0.0400 \text{ L}) \left(\frac{1 \text{ mol NaOH}}{1 \text{ mol } H_2SO_3} \right) = 0.003\ 20 \text{ mol}$$

vol NaOH required to reach first equivalence point

$$= (0.003\ 20 \text{ mol}) \left(\frac{1 \text{ L}}{0.160 \text{ mol}} \right) = 0.020 \text{ L} = 20.0 \text{ mL}$$

20.0 mL is enough NaOH solution to reach the first equivalence point for the titration of the diprotic acid, H_2SO_3.

For H_2SO_3,

$K_{a1} = 1.5 \times 10^{-2}$, $pK_{a1} = -\log K_{a1} = -\log(1.5 \times 10^{-2}) = 1.82$

$K_{a2} = 6.3 \times 10^{-8}$, $pK_{a2} = -\log K_{a2} = -\log(6.3 \times 10^{-8}) = 7.20$

At the first equivalence point, $pH = \dfrac{pK_{a1} + pK_{a2}}{2} = \dfrac{1.82 + 7.20}{2} = 4.51$

(b) mol NaOH required to reach second equivalence point

$$= \left(\frac{0.0800 \text{ mol } H_2SO_3}{L} \right) (0.0400 \text{ L}) \left(\frac{2 \text{ mol NaOH}}{1 \text{ mol } H_2SO_3} \right) = 0.006\ 40 \text{ mol}$$

vol NaOH required to reach second equivalence point

$$= (0.006\ 40 \text{ mol}) \left(\frac{1 \text{ L}}{0.160 \text{ mol}} \right) = 0.040 \text{ L} = 40.0 \text{ mL}$$

30.0 mL is enough NaOH solution to reach halfway to the second equivalent point. Halfway to the second equivalence point

$pH = pK_{a2} = -\log K_{a2} = -\log(6.3 \times 10^{-8}) = 7.20$

(c) mmol $HSO_3^- = 0.0800$ mmol/mL $\times$ 40.0 mL = 3.20 mmol

volume NaOH added after first equivalence point = 35.0 mL – 20.0 mL = 15.0 mL

mmol NaOH = mmol $OH^- = 0.160$ mmol/L $\times$ 15.0 mL = 2.40 mmol

Neutralization reaction:	$HSO_3^-(aq)$	+ $OH^-(aq)$	$\rightleftharpoons$ $SO_3^{2-}(aq)$	+ $H_2O(l)$
before reaction (mmol)	3.20	2.40	0	
change (mmol)	–2.40	–2.40	+2.40	
after reaction (mmol)	0.80	0	2.40	

$$[HSO_3^-] = \frac{0.80 \text{ mmol}}{(40.0 \text{ mL} + 35.0 \text{ mL})} = 0.0107 \text{ M}$$

$$[SO_3^{2-}] = \frac{2.40 \text{ mmol}}{(40.0 \text{ mL} + 35.0 \text{ mL})} = 0.0320 \text{ M}$$

	$HSO_3^-(aq)$	+ $H_2O(l)$	$\rightleftharpoons$ $H_3O^+(aq)$	+ $SO_3^{2-}(aq)$
initial (M)	0.0107		~0	0.0320
change (M)	–x		+x	+x
equil (M)	0.0107 – x		x	0.0320 + x

$$K_a = \frac{[H_3O^+][SO_3^{2-}]}{[HSO_3^-]} = 6.3 \times 10^{-8} = \frac{x(0.0320 + x)}{0.0107 - x} \approx \frac{x(0.0320)}{0.0107}$$

Solve for x. $x = [H_3O^+] = 2.1 \times 10^{-8}$ M

$pH = -\log[H_3O^+] = -\log(2.1 \times 10^{-8}) = 7.68$

15.19   Let H_2A^+ = valine cation

(a) mol NaOH required to reach first equivalence point

$$= \left(\frac{0.0250 \text{ mol } H_2A^+}{L}\right)(0.0400 \text{ L})\left(\frac{1 \text{ mol NaOH}}{1 \text{ mol } H_2A^+}\right) = 0.001\ 00 \text{ mol}$$

vol NaOH required to reach first equivalence point

$$= (0.001\ 00 \text{ mol})\left(\frac{1 \text{ L}}{0.100 \text{ mol}}\right) = 0.0100 \text{ L} = 10.0 \text{ mL}$$

10.0 mL is enough NaOH solution to reach the first equivalence point for the titration of the diprotic acid, H_2A^+.

For H_2A^+,

$K_{a1} = 4.8 \times 10^{-3}$, $pK_{a1} = -\log K_{a1} = -\log(4.8 \times 10^{-3}) = 2.32$

$K_{a2} = 2.4 \times 10^{-10}$, $pK_{a2} = -\log K_{a2} = -\log(2.4 \times 10^{-10}) = 9.62$

At the first equivalence point, $pH = \dfrac{pK_{a1} + pK_{a2}}{2} = \dfrac{2.32 + 9.62}{2} = 5.97$

(b) mol NaOH required to reach second equivalence point

$$= \left(\frac{0.0250 \text{ mol } H_2A^+}{L}\right)(0.0400 \text{ L})\left(\frac{2 \text{ mol NaOH}}{1 \text{ mol } H_2A^+}\right) = 0.002\ 00 \text{ mol}$$

vol NaOH required to reach second equivalence point

$$= (0.002\ 00 \text{ mol})\left(\frac{1 \text{ L}}{0.100 \text{ mol}}\right) = 0.0200 \text{ L} = 20.0 \text{ mL}$$

15.0 mL is enough NaOH solution to reach halfway to the second equivalent point. Halfway to the second equivalence point

$pH = pK_{a2} = -\log K_{a2} = -\log(2.4 \times 10^{-10}) = 9.62$

(c) 20.0 mL is enough NaOH to reach the second equivalence point.

At the second equivalence point

mmol A^- = (0.0250 mmol/mL)(40.0 mL) = 1.00 mmol A^-

solution volume = 40.0 mL + 20.0 mL = 60.0 mL

$$[A^-] = \frac{1.00 \text{ mmol}}{60.0 \text{ mL}} = 0.0167 \text{ M}$$

	$A^-(aq)$ +	$H_2O(l)$	⇌	$HA(aq)$ +	$OH^-(aq)$
initial (M)	0.0167			0	~0
change (M)	−x			+x	+x
equil (M)	0.0167 − x			x	x

$$K_b = \frac{K_w}{K_a \text{ for HA}} = \frac{K_w}{K_{a2}} = \frac{1.0 \text{ x } 10^{-14}}{2.4 \text{ x } 10^{-10}} = 4.17 \text{ x } 10^{-5}$$

$$K_b = \frac{[HA][OH^-]}{[A^-]} = 4.17 \text{ x } 10^{-5} = \frac{x^2}{0.0167 - x}$$

$x^2 + (4.17 \text{ x } 10^{-5})x - (6.964 \text{ x } 10^{-7}) = 0$

Use the quadratic formula to solve for x.

$$x = \frac{-(4.17 \text{ x } 10^{-5}) \pm \sqrt{(4.17 \text{ x } 10^{-5})^2 - (4)(1)(-6.964 \text{ x } 10^{-7})}}{2(1)} = \frac{(-4.17 \text{ x } 10^{-5}) \pm (1.67 \text{ x } 10^{-3})}{2}$$

$x = 8.14 \text{ x } 10^{-4}$ and $-8.56 \text{ x } 10^{-4}$

Of the two solutions for x, only the positive value has physical meaning because x is the $[OH^-]$.

$x = [OH^-] = 8.14 \text{ x } 10^{-4}$ M

$$[H_3O^+] = \frac{K_w}{[OH^-]} = \frac{1.0 \text{ x } 10^{-14}}{8.14 \text{ x } 10^{-4}} = 1.23 \text{ x } 10^{-11} \text{ M}$$

$pH = -\log[H_3O^+] = -\log(1.23 \text{ x } 10^{-11}) = 10.91$

15.20 (a) $K_{sp} = [Ag^+][Cl^-]$ (b) $K_{sp} = [Pb^{2+}][I^-]^2$
       (c) $K_{sp} = [Ca^{2+}]^3[PO_4^{3-}]^2$ (d) $K_{sp} = [Cr^{3+}][OH^-]^3$

15.21 $K_{sp} = [Ca^{2+}]^3[PO_4^{3-}]^2 = (2.01 \text{ x } 10^{-8})^3(1.6 \text{ x } 10^{-5})^2 = 2.1 \text{ x } 10^{-33}$

15.22 $[Ba^{2+}] = [SO_4^{2-}] = 1.05 \text{ x } 10^{-5}$ M; $K_{sp} = [Ba^{2+}][SO_4^{2-}] = (1.05 \text{ x } 10^{-5})^2 = 1.10 \text{ x } 10^{-10}$

15.23 (a) $AgCl(s) \rightleftharpoons Ag^+(aq) + Cl^-(aq)$

equil (M)   x   x

$K_{sp} = [Ag^+][Cl^-] = 1.8 \text{ x } 10^{-10} = (x)(x)$

molar solubility $= x = \sqrt{K_{sp}} = 1.3 \text{ x } 10^{-5}$ mol/L

AgCl, 143.32,   solubility $= \dfrac{\left(1.3 \text{ x } 10^{-5} \text{ mol x } \dfrac{143.32 \text{ g}}{1 \text{ mol}}\right)}{1 \text{ L}} = 0.0019$ g/L

(b) $Ag_2CrO_4(s) \rightleftharpoons 2 Ag^+(aq) + CrO_4^{2-}(aq)$

equil (M)   2x   x

$K_{sp} = [Ag^+]^2[CrO_4^{2-}] = 1.1 \text{ x } 10^{-12} = (2x)^2(x) = 4x^3$

molar solubility $= x = \sqrt[3]{\dfrac{1.1 \text{ x } 10^{-12}}{4}} = 6.5 \text{ x } 10^{-5}$ mol/L

Ag_2CrO_4, 331.73,   solubility $= \dfrac{\left(6.5 \text{ x } 10^{-5} \text{ mol x } \dfrac{331.73 \text{ g}}{1 \text{ mol}}\right)}{1 \text{ L}} = 0.022$ g/L

Ag_2CrO_4 has both the higher molar and gram solubility, despite its smaller value of K_{sp}.

15.24   Let the number of ions be proportional to its concentration.

For AgX, $K_{sp} = [Ag^+][X^-] \propto (4)(4) = 16$

For AgY, $K_{sp} = [Ag^+][Y^-] \propto (1)(9) = 9$

For AgZ, $K_{sp} = [Ag^+][Z^-] \propto (3)(6) = 18$

(a) AgZ   (b) AgY

15.25 $[Mg^{2+}]_0$ is from 0.10 M $MgCl_2$.

$$MgF_2(s) \rightleftharpoons Mg^{2+}(aq) + 2\,F^-(aq)$$

	Mg^{2+}	F^-
initial (M)	0.10	0
change (M)	+x	+2x
equil (M)	0.10 + x	2x

$K_{sp} = 7.4 \times 10^{-11} = [Mg^{2+}][F^-]^2 = (0.10 + x)(2x)^2 \approx (0.10)(4x^2)$

$x = 1.4 \times 10^{-5}$,   molar solubility $= x = 1.4 \times 10^{-5}$ M

15.26   Compounds that contain basic anions are more soluble in acidic solution than in pure water. AgCN, $Al(OH)_3$, and ZnS all contain basic anions.

15.27 $[Cu^{2+}] = (5.0 \times 10^{-3}$ mol$)/(0.500$ L$) = 0.010$ M

$$Cu^{2+}(aq) + 4\,NH_3(aq) \rightleftharpoons Cu(NH_3)_4^{2+}(aq)$$

	Cu^{2+}	NH_3	$Cu(NH_3)_4^{2+}$
before reaction (M)	0.010	0.40	0
assume 100 % reaction (M)	−0.010	− 4(0.010)	+0.010
after reaction (M)	0	0.36	0.010
assume small back reaction (M)	+x	+4x	−x
equil (M)	x	0.36 + 4x	0.010 − x

$$K_f = \frac{[Cu(NH_3)_4^{2+}]}{[Cu^{2+}][NH_3]^4} = 5.6 \times 10^{11} = \frac{(0.010 - x)}{(x)(0.36 + 4x)^4} \approx \frac{0.010}{x(0.36)^4}$$

Solve for x. $x = [Cu^{2+}] = 1.1 \times 10^{-12}$ M

15.28

$$AgBr(s) \rightleftharpoons Ag^+(aq) + Br^-(aq) \qquad\qquad K_{sp} = 5.4 \times 10^{-13}$$
$$\underline{Ag^+(aq) + 2\,S_2O_3^{2-} \rightarrow Ag(S_2O_3)_2^{3-}(aq)} \qquad\qquad K_f = 4.7 \times 10^{13}$$

dissolution $AgBr(s) + 2\,S_2O_3^{2-}(aq) \rightleftharpoons Ag(S_2O_3)_2^{3-}(aq) + Br^-(aq)$
reaction

$K = (K_{sp})(K_f) = (5.4 \times 10^{-13})(4.7 \times 10^{13}) = 25.4$

$$AgBr(s) + 2\,S_2O_3^{2-}(aq) \rightleftharpoons Ag(S_2O_3)_2^{3-}(aq) + Br^-(aq)$$

	$S_2O_3^{2-}$	$Ag(S_2O_3)_2^{3-}$	Br^-
initial (M)	0.10	0	0
change (M)	−2x	x	x
equil (M)	0.10 − 2x	x	x

$$K = \frac{[Ag(S_2O_3)_2^{3-}][Br^-]}{[S_2O_3^{2-}]^2} = 25.4 = \frac{x^2}{(0.10 - 2x)^2}$$

Take the square root of both sides and solve for x.

$$\sqrt{25.4} = \sqrt{\frac{x^2}{(0.10 - 2x)^2}}; \qquad 5.04 = \frac{x}{0.10 - 2x}; \qquad x = \text{molar solubility} = 0.045 \text{ mol/L}$$

15.29   On mixing equal volumes of two solutions, the concentrations of both solutions are cut in half.
For $BaCO_3$, $K_{sp} = 2.6 \times 10^{-9}$
(a) $IP = [Ba^{2+}][CO_3^{2-}] = (1.5 \times 10^{-3})(1.0 \times 10^{-3}) = 1.5 \times 10^{-6}$
     $IP > K_{sp}$; a precipitate of $BaCO_3$ will form.

(b) $IP = [Ba^{2+}][CO_3^{2-}] = (5.0 \times 10^{-6})(2.0 \times 10^{-5}) = 1.0 \times 10^{-10}$
     $IP < K_{sp}$; no precipitate will form.

15.30 $$pH = pK_a + \log \frac{[base]}{[acid]} = pK_a + \log \frac{[NH_3]}{[NH_4^+]}$$

For NH_4^+, $K_a = 5.6 \times 10^{-10}$, $pK_a = -\log K_a = -\log(5.6 \times 10^{-10}) = 9.25$

$$pH = 9.25 + \log \frac{(0.20)}{(0.20)} = 9.25; \quad [H_3O^+] = 10^{-pH} = 10^{-9.25} = 5.6 \times 10^{-10} \text{ M}$$

$$[OH^-] = \frac{K_w}{[H_3O^+]} = \frac{1.0 \times 10^{-14}}{5.6 \times 10^{-10}} = 1.8 \times 10^{-5} \text{ M}$$

$$[Fe^{2+}] = [Mn^{2+}] = \frac{(25 \text{ mL})(1.0 \times 10^{-3} \text{ M})}{250 \text{ mL}} = 1.0 \times 10^{-4} \text{ M}$$

For $Mn(OH)_2$, $K_{sp} = 2.1 \times 10^{-13}$
     $IP = [Mn^{2+}][OH^-]^2 = (1.0 \times 10^{-4})(1.8 \times 10^{-5})^2 = 3.2 \times 10^{-14}$
     $IP < K_{sp}$; no precipitate will form.

For $Fe(OH)_2$, $K_{sp} = 4.9 \times 10^{-17}$
     $IP = [Fe^{2+}][OH^-]^2 = (1.0 \times 10^{-4})(1.8 \times 10^{-5})^2 = 3.2 \times 10^{-14}$
     $IP > K_{sp}$; a precipitate of $Fe(OH)_2$ will form.

15.31 $MS(s) + 2 H_3O^+(aq) \rightleftharpoons M^{2+}(aq) + H_2S(aq) + 2 H_2O(l)$

$$K_{spa} = \frac{[M^{2+}][H_2S]}{[H_3O^+]^2}$$

For ZnS, $K_{spa} = 3 \times 10^{-2}$;   for CdS, $K_{spa} = 8 \times 10^{-7}$
$[Cd^{2+}] = [Zn^{2+}] = 0.005 \text{ M}$
Because the two cation concentrations are equal, Q_c is the same for both.

$$Q_c = \frac{[M^{2+}]_t[H_2S]_t}{[H_3O^+]_t^2} = \frac{(0.005)(0.10)}{(0.3)^2} = 6 \times 10^{-3}$$

$Q_c > K_{spa}$ for CdS; CdS will precipitate. $Q_c < K_{spa}$ for ZnS; Zn^{2+} will remain in solution.

15.32   1 ppm = 1 mg/L = 0.001g/L

$$[F^-] = \frac{\left(0.001 \text{ g F} \times \dfrac{1 \text{ mol F}}{19.00 \text{ g F}}\right)}{1.0 \text{ L}} = 5.3 \times 10^{-5} \text{ M}$$

15.33 $Ca_5(PO_4)_3(OH)(s) \rightleftharpoons 5\ Ca^{2+}(aq) + 3\ PO_4^{3-}(aq) + OH^-(aq)$ $K_1 = K_{sp} = 2.3 \times 10^{-59}$

$\underline{5\ Ca^{2+}(aq) + 3\ PO_4^{3-}(aq) + F^-(aq) \rightleftharpoons Ca_5(PO_4)_3(F)(s)}$ $K_2 = 1/K_{sp} = 1/3.2 \times 10^{-60}$

$Ca_5(PO_4)_3(OH)(s) + F^-(aq) \rightleftharpoons Ca_5(PO_4)_3(F)(s) + OH^-(aq)$ $K_3 = K_1 K_2$

$$K_3 = \frac{2.3 \times 10^{-59}}{3.2 \times 10^{-60}} = 7.2$$

Conceptual Problems

15.34   A buffer solution contains a conjugate acid-base pair in about equal concentrations.
(a) (1), (3), and (4)
(b) (4) because it has the highest buffer concentration.

15.36   (4); only A^- and water should be present

15.38   (a) (1) corresponds to (iii); (2) to (i); (3) to (ii); and (4) to (iv)
(b)

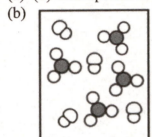

15.40   (2) is supersaturated; (3) is unsaturated; (4) is unsaturated

15.42   (a) The lower curve represents the titration of a strong acid; the upper curve represents the titration of a weak acid.
(b) pH = 7 for titration of the strong acid; pH = 10 for titration of the weak acid.
(c) Halfway to the equivalence point, the pH = $pK_a \sim 6.3$.

Section Problems
Neutralization Reactions (Section 15.1)

15.44   (a) $HI(aq) + NaOH(aq) \rightarrow H_2O(l) + NaI(aq)$
net ionic equation: $H_3O^+(aq) + OH^-(aq) \rightarrow 2\ H_2O(l)$
The solution at neutralization contains a neutral salt (NaI); pH = 7.00.
(b) $2\ HOCl(aq) + Ba(OH)_2(aq) \rightarrow 2\ H_2O(l) + Ba(OCl)_2(aq)$
net ionic equation: $HOCl(aq) + OH^-(aq) \rightarrow H_2O(l) + OCl^-(aq)$
The solution at neutralization contains a basic anion (OCl^-); pH > 7.00
(c) $HNO_3(aq) + C_6H_5NH_2(aq) \rightarrow C_6H_5NH_3NO_3(aq)$
net ionic equation: $H_3O^+(aq) + C_6H_5NH_2(aq) \rightarrow H_2O(l) + C_6H_5NH_3^+(aq)$
The solution at neutralization contains an acidic cation ($C_6H_5NH_3^+$); pH < 7.00.

(d) $C_6H_5CO_2H(aq) + KOH(aq) \rightarrow H_2O(l) + C_6H_5CO_2K(aq)$

net ionic equation: $C_6H_5CO_2H(aq) + OH^-(aq) \rightarrow H_2O(l) + C_6H_5CO_2^-(aq)$

The solution at neutralization contains a basic anion ($C_6H_5CO_2^-$); pH > 7.00.

15.46   (a) After mixing, the solution contains the basic salt, NaF; pH > 7.00

      (b) After mixing, the solution contains the neutral salt, NaCl; pH = 7.00

     Solution (a) has the higher pH.

15.48   Weak acid - weak base reaction $K_n = \dfrac{K_a K_b}{K_w} = \dfrac{(1.3 \times 10^{-10})(1.8 \times 10^{-9})}{1.0 \times 10^{-14}} = 2.3 \times 10^{-5}$

     K_n is small so the neutralization reaction does not proceed very far to completion.

15.50 $K_n = \dfrac{K_a K_b}{K_w} = 2.1 \times 10^{-4};$ $K_b = \dfrac{K_n K_w}{K_a} = \dfrac{(2.1 \times 10^{-4})(1.0 \times 10^{-14})}{(1.4 \times 10^{-4})} = 1.5 \times 10^{-14}$

The Common-Ion Effect (Section 15.2)

15.52   (a) $HF(aq) + H_2O(l) \rightleftharpoons H_3O^+(aq) + F^-(aq)$

     LiF is a source of F^- (reaction product). The equilibrium shifts toward reactants, and the $[H_3O^+]$ decreases. The pH increases.

     (b) Because HI is a strong acid, addition of KI, a neutral salt, does not change the pH.

     (c) $NH_3(aq) + H_2O(l) \rightleftharpoons NH_4^+(aq) + OH^-(aq)$

     NH_4Cl is a source of NH_4^+ (reaction product). The equilibrium shifts toward reactants, and the $[OH^-]$ decreases. The pH decreases.

15.54   For 0.25 M HF and 0.10 M NaF

	$HF(aq)$	$+ H_2O(l) \rightleftharpoons$	$H_3O^+(aq)$	$+ F^-(aq)$
initial (M)	0.25		~0	0.10
change (M)	−x		+x	+x
equil (M)	0.25 − x		x	0.10 + x

     $K_a = \dfrac{[H_3O^+][F^-]}{[HF]} = 3.5 \times 10^{-4} = \dfrac{x(0.10 + x)}{0.25 - x} \approx \dfrac{x(0.10)}{0.25}$

     Solve for x. $x = [H_3O^+] = 8.8 \times 10^{-4}$ M

     $pH = -\log[H_3O^+] = -\log(8.8 \times 10^{-4}) = 3.06$

15.56 $pH = 4.86;$ $[H_3O^+] = 10^{-pH} = 10^{-4.86} = 1.38 \times 10^{-5}$ M

     $HN_3(aq) + H_2O(l) \rightleftharpoons H_3O^+(aq) + N_3^-(aq)$

     $K_a = \dfrac{[H_3O^+][N_3^-]}{[HN_3]} = 1.9 \times 10^{-5} = \dfrac{(1.38 \times 10^{-5})[N_3^-]}{(0.016)}$

     $[N_3^-] = 0.022$ M

15.58   For 0.10 M HN_3:

$$HN_3(aq) + H_2O(l) \rightleftharpoons H_3O^+(aq) + N_3^-(aq)$$

initial (M)	0.10	~0	0
change (M)	−x	+x	+x
equil (M)	0.10 − x	x	x

$$K_a = \frac{[H_3O^+][N_3^-]}{[HN_3]} = 1.9 \times 10^{-5} = \frac{x^2}{0.10-x} \approx \frac{x^2}{0.10}$$

Solve for x. $x = 1.4 \times 10^{-3}$ M

$$\% \text{ dissociation} = \frac{[HN_3]_{diss}}{[HN_3]_{initial}} \times 100\% = \frac{1.4 \times 10^{-3} \text{ M}}{0.10 \text{ M}} \times 100\% = 1.4\%$$

For 0.10 M HN_3 in 0.10 M HCl:

$$HN_3(aq) + H_2O(l) \rightleftharpoons H_3O^+(aq) + N_3^-(aq)$$

initial (M)	0.10	0.10	0
change (M)	−x	+x	+x
equil (M)	0.10 − x	0.10 + x	x

$$K_a = \frac{[H_3O^+][N_3^-]}{[HN_3]} = 1.9 \times 10^{-5} = \frac{(0.10+x)(x)}{0.10-x} \approx \frac{(0.10)(x)}{0.10} = x$$

Solve for x. $x = 1.9 \times 10^{-5}$ M

$$\% \text{ dissociation} = \frac{[HN_3]_{diss}}{[HN_3]_{initial}} \times 100\% = \frac{1.9 \times 10^{-5} \text{ M}}{0.10 \text{ M}} \times 100\% = 0.019\%$$

The % dissociation is less because of the common ion (H_3O^+) effect.

Buffer Solutions (Sections 15.3 and 15.4)

15.60   Solutions (a), (c), and (d) are buffer solutions. Neutralization reactions for (c) and (d) result in solutions with equal concentrations of HF and F^-.

15.62   Both solutions buffer at the same pH because in both cases the $[NO_2^-]/[HNO_2] = 1$. Solution (a), however, has a higher concentration of both HNO_2 and NO_2^-, and therefore it has the greater buffer capacity.

15.64 $$pH = pK_a + \log\frac{[base]}{[acid]} = pK_a + \log\frac{[CN^-]}{[HCN]}$$

For HCN, $K_a = 4.9 \times 10^{-10}$, $pK_a = -\log K_a = -\log(4.9 \times 10^{-10}) = 9.31$

$$pH = 9.31 + \log\left(\frac{0.12}{0.20}\right) = 9.09$$

The pH of a buffer solution will not change on dilution because the acid and base concentrations will change by the same amount and their ratio will remain the same.

15.66

$$HCO_2H(aq) \ + \ H_2O(l) \ \rightleftharpoons \ H_3O^+(aq) \ + \ HCO_2^-(aq)$$

initial (M)	0.36	~0	0.30
change (M)	–x	+x	+x
equil (M)	0.36 – x	x	0.30 + x

$$K_a = \frac{[H_3O^+][HCO_2^-]}{[HCO_2H]} = 1.8 \times 10^{-4} = \frac{x(0.30 + x)}{0.36 - x} \approx \frac{x(0.30)}{0.36}$$

Solve for x. $x = 2.16 \times 10^{-4}$ M = $[H_3O^+]$

For the buffer, pH = $-\log[H_3O^+] = -\log(2.16 \times 10^{-4}) = 3.67$

mol HCO_2H = (0.36 mol/L)(0.250 L) = 0.090 mol HCO_2H

mol HCO_2^- = (0.30 mol/L)(0.250 L) = 0.075 mol HCO_2^-

(a)   100%

$$HCO_2H(aq) \ + \ OH^-(aq) \ \rightarrow \ HCO_2^-(aq) \ + \ H_2O(l)$$

before (mol)	0.090	0.0050	0.075
change (mol)	–0.0050	–0.0050	+0.0050
after (mol)	0.085	0	0.080

$$[H_3O^+] = K_a \frac{[HCO_2H]}{[HCO_2^-]} = (1.8 \times 10^{-4})\left(\frac{0.085}{0.080}\right) = 1.91 \times 10^{-4} \text{ M}$$

pH = $-\log[H_3O^+] = -\log(1.91 \times 10^{-4}) = 3.72$

(b)   100%

$$HCO_2^-(aq) \ + \ H_3O^+(aq) \ \rightarrow \ HCO_2H(aq) \ + \ H_2O(l)$$

before (mol)	0.075	0.0050	0.090
change (mol)	–0.0050	–0.0050	+0.0050
after (mol)	0.070	0	0.095

$$[H_3O^+] = K_a \frac{[HCO_2H]}{[HCO_2^-]} = (1.8 \times 10^{-4})\left(\frac{0.095}{0.070}\right) = 2.44 \times 10^{-4} \text{ M}$$

pH = $-\log[H_3O^+] = -\log(2.44 \times 10^{-4}) = 3.61$

15.68 $pH = pK_a + \log\dfrac{[\text{base}]}{[\text{acid}]} = pK_a + \log\dfrac{[HCO_2^-]}{[HCO_2H]}$

For HCO_2H, $K_a = 1.8 \times 10^{-4}$; $pK_a = -\log K_a = -\log(1.8 \times 10^{-4}) = 3.74$

$pH = 3.74 + \log\dfrac{(0.50)}{(0.25)} = 4.04$

15.70 $pH = pK_a + \log\dfrac{[\text{base}]}{[\text{acid}]} = pK_a + \log\dfrac{[NH_3]}{[NH_4^+]}$

For NH_4^+, $K_a = 5.6 \times 10^{-10}$; $pK_a = -\log K_a = -\log(5.6 \times 10^{-10}) = 9.25$

$9.80 = 9.25 + \log\dfrac{[NH_3]}{[NH_4^+]}$

$$0.550 = \log \frac{[NH_3]}{[NH_4^+]}$$

$$\frac{[NH_3]}{[NH_4^+]} = 10^{0.55} = 3.5$$

The volume of the 1.0 M NH_3 solution should be 3.5 times the volume of the 1.0 M NH_4Cl solution so that the mixture will buffer at pH 9.80.

15.72 H_3PO_4, $K_{a1} = 7.5 \times 10^{-3}$; $pK_{a1} = -\log K_{a1} = 2.12$

$H_2PO_4^-$, $K_{a2} = 6.2 \times 10^{-8}$; $pK_{a2} = -\log K_{a2} = 7.21$

HPO_4^{2-}, $K_{a3} = 4.8 \times 10^{-13}$; $pK_{a3} = -\log K_{a3} = 12.32$

The buffer system of choice for pH 7.00 is (b) $H_2PO_4^- - HPO_4^{2-}$ because the pK_a for $H_2PO_4^-$ (7.21) is closest to 7.00.

15.74 $pH = pK_a + \log \dfrac{[base]}{[acid]}$

$$9.46 = pK_a + \log \left(\frac{34.5}{100 - 34.5} \right)$$

$$pK_a = 9.46 - \log \left(\frac{34.5}{100 - 34.5} \right) = 9.74$$

$$K_a = 10^{-pK_a} = 10^{-9.74} = 1.8 \times 10^{-10}$$

pH Titration Curves (Sections 15.5–15.9)

15.76   (a) $(0.060 \text{ L})(0.150 \text{ mol/L})(1000 \text{ mmol/mol}) = 9.00 \text{ mmol } HNO_3$

(b) vol NaOH $= (9.00 \text{ mmol } HNO_3)\left(\dfrac{1 \text{ mmol NaOH}}{1 \text{ mmol } HNO_3} \right)\left(\dfrac{1 \text{ mL NaOH}}{0.450 \text{ mmol NaOH}} \right) = 20.0 \text{ mL NaOH}$

(c) At the equivalence point the solution contains the neutral salt $NaNO_3$. The pH is 7.00.

(d)

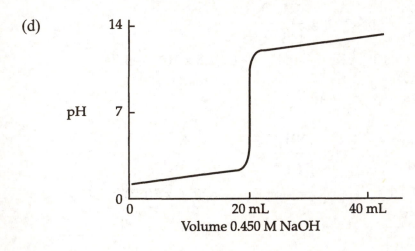

15.78   (0.0250 L)(0.125 mol/L)(1000 mmol/mol) = 3.125 mmol HCl

(a) (0.0030 L)(0.100 mol/L)(1000 mmol/mol) = 0.300 mmol NaOH

Neutralization reaction:	$H_3O^+(aq)$	+	$OH^-(aq)$	→	$H_2O(l)$
before reaction (mmol)	3.125		0.300		
change (mmol)	−0.300		−0.300		
after reaction (mmol)	2.825		0		

$$[H_3O^+] = \frac{2.825 \text{ mmol}}{(25.0 \text{ mL } + 3.0 \text{ mL})} = 0.1009 \text{ M}$$

$pH = -\log[H_3O^+] = -\log(0.1009) = 0.996$

(b) (0.020 L)(0.100 mol/L)(1000 mmol/mol) = 2.0 mmol NaOH

Neutralization reaction:	$H_3O^+(aq)$	+	$OH^-(aq)$	→	$H_2O(l)$
before reaction (mmol)	3.125		2.0		
change (mmol)	−2.0		−2.0		
after reaction (mmol)	1.125		0		

$$[H_3O^+] = \frac{1.125 \text{ mmol}}{(25.0 \text{ mL } + 20 \text{ mL})} = 0.025 \text{ M}$$

$pH = -\log[H_3O^+] = -\log(0.025) = 1.60$

(c) (0.065 L)(0.100 mol/L)(1000 mmol/mol) = 6.5 mmol NaOH

Neutralization reaction:	$H_3O^+(aq)$	+	$OH^-(aq)$	→	$H_2O(l)$
before reaction (mmol)	3.125		6.5		
change (mmol)	−3.125		−3.125		
after reaction (mmol)	0		3.375		

$$[OH^-] = \frac{3.375 \text{ mmol}}{(25.0 \text{ mL } + 65 \text{ mL})} = 0.0375 \text{ M}$$

$$[H_3O^+] = \frac{K_w}{[OH^-]} = \frac{1.0 \times 10^{-14}}{0.0375} = 2.7 \times 10^{-13} \text{ M}$$

$pH = -\log[H_3O^+] = -\log(2.7 \times 10^{-13}) = 12.57$

15.80   mmol HF = (40.0 mL)(0.250 mmol/mL) = 10.0 mmol

mmol NaOH required = mmol HF = 10.0 mmol

$$\text{mL NaOH required} = (10.0 \text{ mmol})\left(\frac{1.00 \text{ mL}}{0.200 \text{ mmol}}\right) = 50.0 \text{ mL}$$

50.0 mL of 0.200 M NaOH is required to reach the equivalence point.

For HF, $K_a = 3.5 \times 10^{-4}$; $pK_a = -\log K_a = -\log(3.5 \times 10^{-4}) = 3.46$

(a) mmol HF = 10.0 mmol

mmol NaOH = (0.200 mmol/mL)(10.0 mL) = 2.00 mmol

Neutralization reaction:	HF(aq)	+	$OH^-(aq)$	→	$F^-(aq)$	+	$H_2O(l)$
before reaction (mmol)	10.0		2.00		0		
change (mmol)	−2.00		−2.00		+2.00		
after reaction (mmol)	8.0		0		2.00		

$$[HF] = \frac{8.0 \text{ mmol}}{(40.0 \text{ mL } + 10.0 \text{ mL})} = 0.16 \text{ M}; \quad [F^-] = \frac{2.00 \text{ mmol}}{(40.0 \text{ mL } + 10.0 \text{ mL})} = 0.0400 \text{ M}$$

$$HF(aq) + H_2O(l) \rightleftharpoons H_3O^+(aq) + F^-(aq)$$

initial (M)	0.16	~0	0.0400
change (M)	−x	+x	+x
equil (M)	0.16 − x	x	0.0400 + x

$$K_a = \frac{[H_3O^+][F^-]}{[HF]} = 3.5 \times 10^{-4} = \frac{x(0.0400 + x)}{0.16 - x} \approx \frac{x(0.0400)}{0.16}$$

Solve for x. $x = [H_3O^+] = 1.4 \times 10^{-3}$ M

$pH = -\log[H_3O^+] = -\log(1.4 \times 10^{-3}) = 2.85$

(b) Halfway to the equivalence point,

$pH = pK_a = -\log K_a = -\log(3.5 \times 10^{-4}) = 3.46$

(c) At the equivalence point only the salt NaF is in solution.

$$[F^-] = \frac{10.0 \text{ mmol}}{(40.0 \text{ mL} + 50.0 \text{ mL})} = 0.111 \text{ M}$$

$$F^-(aq) + H_2O(l) \rightleftharpoons HF(aq) + OH^-(aq)$$

initial (M)	0.111	0	~0
change (M)	−x	+x	+x
equil (M)	0.111 − x	x	x

For F^-, $K_b = \dfrac{K_w}{K_a \text{ for HF}} = \dfrac{1.0 \times 10^{-14}}{3.5 \times 10^{-4}} = 2.9 \times 10^{-11}$

$$K_b = \frac{[HF][OH^-]}{[F^-]} = 2.9 \times 10^{-11} = \frac{x^2}{0.111 - x} \approx \frac{x^2}{0.111}$$

Solve for x. $x = [OH^-] = 1.8 \times 10^{-6}$ M

$$[H_3O^+] = \frac{K_w}{[OH^-]} = \frac{1.0 \times 10^{-14}}{1.8 \times 10^{-6}} = 5.6 \times 10^{-9} \text{ M}$$

$pH = -\log[H_3O^+] = -\log(5.6 \times 10^{-9}) = 8.25$

(d) mmol HF = 10.0 mmol

mol NaOH = (0.200 mmol/mL)(80.0 mL) = 16.0 mmol

Neutralization reaction: $\quad HF(aq) + OH^-(aq) \rightarrow F^-(aq) + H_2O(l)$

before reaction (mmol)	10.0	16.0	0
change (mmol)	−10.0	−10.0	+10.0
after reaction (mmol)	0	6.0	10.0

After the equivalence point, the pH of the solution is determined by the $[OH^-]$.

$$[OH^-] = \frac{6.0 \text{ mmol}}{(40.0 \text{ mL} + 80.0 \text{ mL})} = 5.0 \times 10^{-2} \text{ M}$$

$$[H_3O^+] = \frac{K_w}{[OH^-]} = \frac{1.0 \times 10^{-14}}{5.0 \times 10^{-2}} = 2.0 \times 10^{-13} \text{ M}$$

$pH = -\log[H_3O^+] = -\log(2.0 \times 10^{-13}) = 12.70$

15.82   For H_2A^+, $K_{a1} = 4.6 \times 10^{-3}$ and $K_{a2} = 2.0 \times 10^{-10}$

(a) (10.0 mL)(0.100 mmol/mL) = 1.00 mmol NaOH added = 1.00 mmol HA produced.

(50.0 mL)(0.100 mmol/mL) = 5.00 mmol H_2A^+

5.00 mmol H_2A^+ – 1.00 mmol NaOH = 4.00 mmol H_2A^+ after neutralization

$$[H_2A^+] = \frac{4.00 \text{ mmol}}{(50.0 \text{ mL} + 10.0 \text{ mL})} = 6.67 \times 10^{-2} \text{ M}$$

$$[HA] = \frac{1.00 \text{ mmol}}{(50.0 \text{ mL} + 10.0 \text{ mL})} = 1.67 \times 10^{-2} \text{ M}$$

$$pH = pK_{a1} + \log \frac{[HA]}{[H_2A^+]} = -\log(4.6 \times 10^{-3}) + \log \left(\frac{1.67 \times 10^{-2}}{6.67 \times 10^{-2}} \right) = 1.74$$

(b) Halfway to the first equivalence point, $pH = pK_{a1} = 2.34$

(c) At the first equivalence point, $pH = \dfrac{pK_{a1} + pK_{a2}}{2} = 6.02$

(d) Halfway between the first and second equivalence points, $pH = pK_{a2} = 9.70$

(e) At the second equivalence point only the basic salt, NaA, is in solution.

$$K_b = \frac{K_w}{K_a \text{ for HA}} = \frac{K_w}{K_{a2}} = \frac{1.0 \times 10^{-14}}{2.0 \times 10^{-10}} = 5.0 \times 10^{-5}$$

mmol A^- = (50.0 mL)(0.100 mmol/mL) = 5.00 mmol

$$[A^-] = \frac{5.0 \text{ mmol}}{(50.0 \text{ mL} + 100.0 \text{ mL})} = 3.3 \times 10^{-2} \text{ M}$$

	A^-(aq)	+ H_2O(l)	⇌	HA(aq)	+ OH^-(aq)
initial (M)	0.033			0	~0
change (M)	–x			+x	+x
equil (M)	0.033 – x			x	x

$$K_b = \frac{[HA][OH^-]}{[A^-]} = 5.0 \times 10^{-5} = \frac{(x)(x)}{0.033 - x} \approx \frac{x^2}{0.033}$$

Solve for x.

$$x = [OH^-] = \sqrt{(5.0 \times 10^{-5})(0.033)} = 1.3 \times 10^{-3} \text{ M}$$

$$[H_3O^+] = \frac{K_w}{[OH^-]} = \frac{1.0 \times 10^{-14}}{1.3 \times 10^{-3}} = 7.7 \times 10^{-12} \text{ M}$$

$$pH = -\log[H_3O^+] = -\log(7.7 \times 10^{-12}) = 11.11$$

15.84   (a) The strongest acid has the lowest pH at the equivalence point, C is the strongest acid.
(b) The weakest acid has the highest pH at the equivalence point, A is the weakest acid.

15.86   When equal volumes of acid and base react, all concentrations are cut in half.
(a) At the equivalence point, only the salt $NaNO_2$ is in solution.
$[NO_2^-] = 0.050$ M

$$\text{For } NO_2^-, K_b = \frac{K_w}{K_a \text{ for } HNO_2} = \frac{1.0 \times 10^{-14}}{4.5 \times 10^{-4}} = 2.2 \times 10^{-11}$$

	NO_2^-(aq)	+ H_2O(l)	⇌	HNO_2(aq)	+ OH^-(aq)
initial (M)	0.050			0	~0
change (M)	–x			+x	+x
equil (M)	0.050 – x			x	x

$$K_b = \frac{[HNO_2][OH^-]}{[NO_2^-]} = 2.2 \times 10^{-11} = \frac{(x)(x)}{0.050 - x} \approx \frac{x^2}{0.050}$$

Solve for x. $x = [OH^-] = 1.1 \times 10^{-6}$ M

$$[H_3O^+] = \frac{K_w}{[OH^-]} = \frac{1.0 \times 10^{-14}}{1.1 \times 10^{-6}} = 9.1 \times 10^{-9}\ M$$

pH $= -\log[H_3O^+] = -\log(9.1 \times 10^{-9}) = 8.04$

Phenol red would be a suitable indicator. (see Figure 15.4)

(b) The pH is 7.00 at the equivalence point for the titration of a strong acid (HI) with a strong base (NaOH).

Bromthymol blue or phenol red would be suitable indicators. (Any indicator that changes color in the pH range 4–10 is satisfactory for a strong acid–strong base titration.)

(c) At the equivalence point only the salt CH_3NH_3Cl is in solution.
$[CH_3NH_3^+] = 0.050$ M

For $CH_3NH_3^+$, $K_a = \dfrac{K_w}{K_b \text{ for } CH_3NH_2} = \dfrac{1.0 \times 10^{-14}}{3.7 \times 10^{-4}} = 2.7 \times 10^{-11}$

$$CH_3NH_3^+(aq) + H_2O(l) \rightleftharpoons H_3O^+(aq) + CH_3NH_2(aq)$$

initial (M)	0.050	~0	0
change (M)	–x	+x	+x
equil (M)	0.050 – x	x	x

$$K_a = \frac{[H_3O^+][CH_3NH_2]}{[CH_3NH_3^+]} = 2.7 \times 10^{-11} = \frac{(x)(x)}{0.050 - x} \approx \frac{x^2}{0.050}$$

Solve for x. $x = [H_3O^+] = 1.2 \times 10^{-6}$ M;   pH $= -\log[H_3O^+] = -\log(1.2 \times 10^{-6}) = 5.92$
Chlorphenol red would be a suitable indicator.

Solubility Equilibria (Sections 15.10 and 15.11)

15.88 (a) $Ag_2CO_3(s) \rightleftharpoons 2\ Ag^+(aq) + CO_3^{2-}(aq)$ $K_{sp} = [Ag^+]^2[CO_3^{2-}]$

(b) $PbCrO_4(s) \rightleftharpoons Pb^{2+}(aq) + CrO_4^{2-}(aq)$ $K_{sp} = [Pb^{2+}][CrO_4^{2-}]$

(c) $Al(OH)_3(s) \rightleftharpoons Al^{3+}(aq) + 3\ OH^-(aq)$ $K_{sp} = [Al^{3+}][OH^-]^3$

(d) $Hg_2Cl_2(s) \rightleftharpoons Hg_2^{2+}(aq) + 2\ Cl^-(aq)$ $K_{sp} = [Hg_2^{2+}][Cl^-]^2$

15.90 (a) $K_{sp} = [Pb^{2+}][I^-]^2 = (5.0 \times 10^{-3})(1.3 \times 10^{-3})^2 = 8.5 \times 10^{-9}$

(b) $[I^-] = \sqrt{\dfrac{K_{sp}}{[Pb^{2+}]}} = \sqrt{\dfrac{(8.5 \times 10^{-9})}{(2.5 \times 10^{-4})}} = 5.8 \times 10^{-3}\ M$

(c) $[Pb^{2+}] = \dfrac{K_{sp}}{[I^-]^2} = \dfrac{(8.5 \times 10^{-9})}{(2.5 \times 10^{-4})^2} = 0.14\ M$

15.92
$$Ag_2CO_3(s) \rightleftharpoons 2\,Ag^+(aq) + CO_3^{2-}(aq)$$
equil (M)   2x   x

$[Ag^+] = 2x = 2.56 \times 10^{-4}$ M; $[CO_3^{2-}] = x = (2.56 \times 10^{-4}$ M$)/2 = 1.28 \times 10^{-4}$ M

$K_{sp} = [Ag^+]^2[CO_3^{2-}] = (2.56 \times 10^{-4})^2(1.28 \times 10^{-4}) = 8.39 \times 10^{-12}$

15.94   (a)
$$SrF_2(s) \rightleftharpoons Sr^{2+}(aq) + 2\,F^-(aq)$$
equil (M)   x   2x

$[Sr^{2+}] = x = 1.03 \times 10^{-3}$ M; $[F^-] = 2x = 2(1.03 \times 10^{-3}$ M$) = 2.06 \times 10^{-3}$ M

$K_{sp} = [Sr^{2+}][F^-]^2 = (1.03 \times 10^{-3})(2.06 \times 10^{-3})^2 = 4.37 \times 10^{-9}$

(b)
$$CuI(s) \rightleftharpoons Cu^+(aq) + I^-(aq)$$
equil (M)   x   x

$[Cu^+] = [I^-] = x = 1.05 \times 10^{-6}$ M

$K_{sp} = [Cu^+][I^-] = (1.05 \times 10^{-6})^2 = 1.10 \times 10^{-12}$

(c) MgC_2O_4, 112.32

$$[Mg^{2+}] = \text{molarity of } MgC_2O_4 = \frac{\left(0.094\text{ g} \times \dfrac{1\text{ mol}}{112.32\text{ g}}\right)}{1\text{ L}} = 8.37 \times 10^{-4}\text{ M}$$

$$MgC_2O_4(s) \rightleftharpoons Mg^{2+}(aq) + C_2O_4^-(aq)$$
equil (M)   x   x

$[Mg^{2+}] = [C_2O_4^-] = x = 8.37 \times 10^{-4}$ M

$K_{sp} = [Mg^{2+}][C_2O_4^-] = (8.37 \times 10^{-4})^2 = 7.0 \times 10^{-7}$

(d) $Zn(CN)_2$, 117.41

$$[Zn^{2+}] = \text{molarity of } Zn(CN)_2 = \frac{\left(4.95 \times 10^{-4}\text{ g} \times \dfrac{1\text{ mol}}{117.41\text{ g}}\right)}{1\text{ L}} = 4.22 \times 10^{-6}\text{ M}$$

$$Zn(CN)_2(s) \rightleftharpoons Zn^{2+}(aq) + 2\,CN^-(aq)$$
equil (M)   x   2x

$[Zn^{2+}] = x = 4.22 \times 10^{-6}$ M; $[CN^-] = 2x = 2(4.22 \times 10^{-6}$ M$) = 8.44 \times 10^{-6}$ M

$K_{sp} = [Zn^{2+}][CN^-]^2 = (4.22 \times 10^{-6})(8.44 \times 10^{-6})^2 = 3.01 \times 10^{-16}$

15.96   (a)
$$BaCrO_4(s) \rightleftharpoons Ba^{2+}(aq) + CrO_4^{2-}(aq)$$
equil (M)   x   x

$K_{sp} = [Ba^{2+}][CrO_4^{2-}] = 1.2 \times 10^{-10} = (x)(x)$

molar solubility $= x = \sqrt{1.2 \times 10^{-10}} = 1.1 \times 10^{-5}$ M

(b)
$$Mg(OH)_2(s) \rightleftharpoons Mg^{2+}(aq) + 2\,OH^-(aq)$$
equil (M)   x   2x

$K_{sp} = [Mg^{2+}][OH^-]^2 = 5.6 \times 10^{-12} = x(2x)^2 = 4x^3$

$$\text{molar solubility} = x = \sqrt[3]{\frac{5.6 \times 10^{-12}}{4}} = 1.1 \times 10^{-4} \text{ M}$$

(c) $\qquad$ $Ag_2SO_3(s) \rightleftharpoons 2 Ag^+(aq) + SO_3^{2-}(aq)$

equil (M) $\qquad\qquad\qquad$ 2x $\qquad$ x

$K_{sp} = [Ag^+]^2[SO_3^{2-}] = 1.5 \times 10^{-14} = (2x)^2 x = 4x^3$

$$\text{molar solubility} = x = \sqrt[3]{\frac{1.5 \times 10^{-14}}{4}} = 1.6 \times 10^{-5} \text{ M}$$

Factors That Affect Solubility (Section 15.12)

15.98 $\quad Ag_2CO_3(s) \rightleftharpoons 2 Ag^+(aq) + CO_3^{2-}(aq)$

(a) $AgNO_3$, source of Ag^+; equilibrium shifts left

(b) HNO_3, source of H_3O^+, removes CO_3^{2-}; equilibrium shifts right

(c) Na_2CO_3, source of CO_3^{2-}; equilibrium shifts left

(d) NH_3, forms $Ag(NH_3)_2^+$, removes Ag^+; equilibrium shifts right

15.100 (a) $\qquad$ $PbCrO_4(s) \rightleftharpoons Pb^{2+}(aq) + CrO_4^{2-}(aq)$

equil (M) $\qquad\qquad\qquad$ x $\qquad$ x

$K_{sp} = [Pb^{2+}][CrO_4^{2-}] = 2.8 \times 10^{-13} = (x)(x)$

molar solubility $= x = \sqrt{2.8 \times 10^{-13}} = 5.3 \times 10^{-7} \text{ M}$

(b) $\qquad$ $PbCrO_4(s) \rightleftharpoons Pb^{2+}(aq) + CrO_4^{2-}(aq)$

initial(M) $\qquad\qquad\qquad$ 0 $\qquad$ 1.0×10^{-3}

equil (M) $\qquad\qquad\qquad$ x $\qquad$ $1.0 \times 10^{-3} + x$

$K_{sp} = [Pb^{2+}][CrO_4^{2-}] = 2.8 \times 10^{-13} = (x)(1.0 \times 10^{-3} + x) \approx (x)(1.0 \times 10^{-3})$

$$\text{molar solubility} = x = \frac{2.8 \times 10^{-13}}{1 \times 10^{-3}} = 2.8 \times 10^{-10} \text{ M}$$

15.102 (b), (c), and (d) are more soluble in acidic solution.

(a) $AgBr(s) \rightleftharpoons Ag^+(aq) + Br^-(aq)$

(b) $CaCO_3(s) + H_3O^+(aq) \rightleftharpoons Ca^{2+}(aq) + HCO_3^-(aq) + H_2O(l)$

(c) $Ni(OH)_2(s) + 2 H_3O^+(aq) \rightleftharpoons Ni^{2+}(aq) + 4 H_2O(l)$

(d) $Ca_3(PO_4)_2(s) + 2 H_3O^+(aq) \rightleftharpoons 3 Ca^{2+}(aq) + 2 HPO_4^{2-}(aq) + 2 H_2O(l)$

15.104 (a) Because $Zn(OH)_2$ contains a basic anion, it becomes more soluble as the acidity of the solution increases. $\qquad Zn(OH)_2(s) + 2 H_3O^+(aq) \rightleftharpoons Zn^{2+}(aq) + 4 H_2O(l)$

(b) Because $Zn(OH)_2$ forms the complex anion, $Zn(OH)_4^{2-}$, $Zn(OH)_2$ becomes more soluble in basic solution. $\qquad Zn(OH)_2(s) + 2 OH^-(aq) \rightleftharpoons Zn(OH)_4^{2-}(aq)$

(c) Because $Zn(OH)_2$ forms the complex anion, $Zn(CN)_4^{2-}$, $Zn(OH)_2$ becomes more soluble in the presence of CN^-.

$Zn(OH)_2(s) + 4 CN^-(aq) \rightleftharpoons Zn(CN)_4^{2-}(aq) + 2 OH^-(aq)$

15.106 On mixing equal volumes of two solutions, the concentrations of both solutions are cut in half.

	$Ag^+(aq)$	+	$2 CN^-(aq)$	$\rightleftharpoons$	$Ag(CN)_2^-(aq)$
before reaction (M)	0.0010		0.10		0
assume 100% reaction	−0.0010		−2(0.0010)		0.0010
after reaction (M)	0		0.098		0.0010
assume small back rxn	+x		+2x		−x
equil (M)	x		0.098 + 2x		0.0010 − x

$$K_f = 3.0 \times 10^{20} = \frac{[Ag(CN)_2^-]}{[Ag^+][CN^-]^2} = \frac{(0.0010 - x)}{x(0.098 + 2x)^2} \approx \frac{0.0010}{x(0.098)^2}$$

Solve for x. $x = [Ag^+] = 3.5 \times 10^{-22}$ M

15.108 (a)

$AgI(s) \rightleftharpoons Ag^+(aq) + I^-(aq)$	$K_{sp} = 8.5 \times 10^{-17}$
$\underline{Ag^+(aq) + 2 CN^-(aq) \rightarrow Ag(CN)_2^-(aq)}$	$K_f = 3.0 \times 10^{20}$

dissolution rxn $AgI(s) + 2 CN^-(aq) \rightleftharpoons Ag(CN)_2^-(aq) + I^-(aq)$

$K = (K_{sp})(K_f) = (8.5 \times 10^{-17})(3.0 \times 10^{20}) = 2.6 \times 10^4$

(b)

$Al(OH)_3(s) \rightleftharpoons Al^{3+}(aq) + 3\ OH^-(aq)$	$K_{sp} = 1.9 \times 10^{-33}$
$\underline{Al^{3+}(aq) + 4 OH^-(aq) \rightarrow Al(OH)_4^-(aq)}$	$K_f = 3 \times 10^{33}$

dissolution rxn $Al(OH)_3(s) + OH^-(aq) \rightleftharpoons Al(OH)_4^-(aq)$

$K = (K_{sp})(K_f) = (1.9 \times 10^{-33})(3 \times 10^{33}) = 6$

(c)

$Zn(OH)_2(s) \rightleftharpoons Zn^{2+}(aq) + 2\ OH^-(aq)$	$K_{sp} = 4.1 \times 10^{-17}$
$\underline{Zn^{2+}(aq) + 4 NH_3(aq) \rightarrow Zn(NH_3)_4^{2+}(aq)}$	$K_f = 7.8 \times 10^8$

dissolution rxn $Zn(OH)_2(s) + 4 NH_3(aq) \rightleftharpoons Zn(NH_3)_4^{2+} + 2\ OH^-(aq)$

$K = (K_{sp})(K_f) = (4.1 \times 10^{-17})(7.8 \times 10^8) = 3.2 \times 10^{-8}$

15.110 (a)

	$AgI(s) \rightleftharpoons$	$Ag^+(aq) +$	$I^-(aq)$
equil (M)		x	x

$K_{sp} = [Ag^+][I^-] = 8.5 \times 10^{-17} = (x)(x)$

molar solubility $= x = \sqrt{8.5 \times 10^{-17}} = 9.2 \times 10^{-9}$ M

(b)

	$AgI(s) + 2 CN^-(aq) \rightleftharpoons$	$Ag(CN)_2^-(aq) +$	$I^-(aq)$
initial (M)	0.10	0	0
change (M)	−2x	+x	+x
equil (M)	0.10 − 2x	x	x

$K = (K_{sp})(K_f) = (8.5 \times 10^{-17})(3.0 \times 10^{20}) = 2.6 \times 10^4$

$$K = 2.6 \times 10^4 = \frac{[Ag(CN)_2^-][I^-]}{[CN^-]^2} = \frac{x^2}{(0.10 - 2x)^2}$$

Take the square root of both sides and solve for x.

molar solubility $= x = 0.050$ M

Precipitation; Qualitative Analysis (Sections 15.13–15.15)

15.112 $BaSO_4$, $K_{sp} = 1.1 \times 10^{-10}$; $Fe(OH)_3$, $K_{sp} = 2.6 \times 10^{-39}$
Total volume = 80 mL + 20 mL = 100 mL

$$[Ba^{2+}] = \frac{(1.0 \times 10^{-5} \text{ M})(80 \text{ mL})}{(100 \text{ mL})} = 8.0 \times 10^{-6} \text{ M}$$

$$[OH^-] = 2[Ba^{2+}] = 2(8.0 \times 10^{-6}) = 1.6 \times 10^{-5} \text{ M}$$

$$[Fe^{3+}] = \frac{2(1.0 \times 10^{-5} \text{ M})(20 \text{ mL})}{(100 \text{ mL})} = 4.0 \times 10^{-6} \text{ M}$$

$$[SO_4^{2-}] = \frac{3(1.0 \times 10^{-5} \text{ M})(20 \text{ mL})}{(100 \text{ mL})} = 6.0 \times 10^{-6} \text{ M}$$

For $BaSO_4$, IP = $[Ba^{2+}]_t[SO_4^{2-}]_t = (8.0 \times 10^{-6})(6.0 \times 10^{-6}) = 4.8 \times 10^{-11}$
IP < K_{sp}; $BaSO_4$ will not precipitate.
For $Fe(OH)_3$, IP = $[Fe^{3+}]_t[OH^-]_t^3 = (4.0 \times 10^{-6})(1.6 \times 10^{-5})^3 = 1.6 \times 10^{-20}$
IP > K_{sp}; $Fe(OH)_3(s)$ will precipitate.

15.114 pH = 10.80; $[H_3O^+] = 10^{-pH} = 10^{-10.80} = 1.6 \times 10^{-11}$ M

$$[OH^-] = \frac{K_w}{[H_3O^+]} = \frac{1.0 \times 10^{-14}}{1.6 \times 10^{-11}} = 6.3 \times 10^{-4} \text{ M}$$

For $Mg(OH)_2$, $K_{sp} = 5.6 \times 10^{-12}$
IP = $[Mg^{2+}]_t[OH^-]_t^2 = (2.5 \times 10^{-4})(6.3 \times 10^{-4})^2 = 9.9 \times 10^{-11}$
IP > K_{sp}; $Mg(OH)_2(s)$ will precipitate

15.116 $K_{spa} = \dfrac{[M^{2+}][H_2S]}{[H_3O^+]^2}$; FeS, $K_{spa} = 6 \times 10^2$; SnS, $K_{spa} = 1 \times 10^{-5}$

Fe^{2+} and Sn^{2+} can be separated by bubbling H_2S through an acidic solution containing the two cations because their K_{spa} values are so different.
For FeS and SnS, $Q_c = \dfrac{(0.01)(0.10)}{(0.3)^2} = 1.1 \times 10^{-2}$

For FeS, $Q_c < K_{spa}$, and FeS will not precipitate.
For SnS, $Q_c > K_{spa}$, and SnS will precipitate.

15.118 FeS, $K_{spa} = \dfrac{[Fe^{2+}][H_2S]}{[H_3O^+]^2} = 6 \times 10^2$

(i) In 0.4 M HCl, $[H_3O^+] = 0.4$ M

$$Q_c = \frac{[Fe^{2+}]_t[H_2S]_t}{[H_3O^+]_t^2} = \frac{(0.10)(0.10)}{(0.4)^2} = 0.0625; \quad Q_c < K_{spa}; \text{ FeS will not precipitate}$$

(ii) pH = 8; $[H_3O^+] = 10^{-pH} = 10^{-8} = 1 \times 10^{-8}$ M

$$Q_c = \frac{[Fe^{2+}]_t[H_2S]_t}{[H_3O^+]_t^2} = \frac{(0.10)(0.10)}{(1 \times 10^{-8})^2} = 1 \times 10^{14}; \quad Q_c > K_{spa}; \text{ FeS(s) will precipitate}$$

15.120 (a) add Cl⁻ to precipitate AgCl
(b) add CO_3^{2-} to precipitate $CaCO_3$
(c) add H_2S to precipitate MnS
(d) add NH_3 and NH_4Cl to precipitate $Cr(OH)_3$
(Need buffer to control [OH⁻]; excess OH⁻ produces the soluble $Cr(OH)_4^-$.)

Chapter Problems

15.122 Prepare aqueous solutions of the three salts. Add a solution of $(NH_4)_2HPO_4$. If a white precipitate forms, the solution contains Mg^{2+}. Perform flame test on the other two solutions. A yellow flame test indicates Na^+. A violet flame test indicates K^+.

15.124 (a)

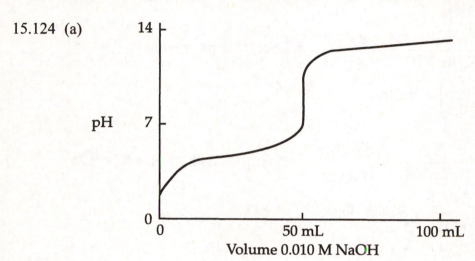

(b) mol NaOH required $= \left(\dfrac{0.010 \text{ mol HA}}{L} \right)(0.0500 \text{ L})\left(\dfrac{1 \text{ mol NaOH}}{1 \text{ mol HA}} \right) = 0.000\ 50 \text{ mol}$

vol NaOH required $= (0.000\ 50 \text{ mol})\left(\dfrac{1 \text{ L}}{0.010 \text{ mol}} \right) = 0.050 \text{ L} = 50 \text{ mL}$

(c) A basic salt is present at the equivalence point; pH > 7.00
(d) Halfway to the equivalence point, the pH = pK_a = 4.00

15.126 pH = 10.35; $[H_3O^+] = 10^{-pH} = 10^{-10.35} = 4.5 \times 10^{-11}$ M

$[OH^-] = \dfrac{K_w}{[H_3O^+]} = \dfrac{1.0 \times 10^{-14}}{4.5 \times 10^{-11}} = 2.2 \times 10^{-4}$ M

$[Mg^{2+}] = \dfrac{[OH^-]}{2} = \dfrac{2.2 \times 10^{-4}}{2} = 1.1 \times 10^{-4}$ M

$K_{sp} = [Mg^{2+}][OH^-]^2 = (1.1 \times 10^{-4})(2.2 \times 10^{-4})^2 = 5.3 \times 10^{-12}$

15.128 NaOH, 40.0; $20 \text{ g} \times \dfrac{1 \text{ mol}}{40.0 \text{ g}} = 0.50 \text{ mol NaOH}$

$(0.500 \text{ L})(1.5 \text{ mol/L}) = 0.75 \text{ mol NH}_4\text{Cl}$

$$NH_4^+(aq) \; + \; OH^-(aq) \; \rightleftharpoons \; NH_3(aq) \; + \; H_2O(l)$$

before reaction (mol)	0.75	0.50	0
change (mol)	−0.50	−0.50	+0.50
after reaction (mol)	0.25	0	0.50

This reaction produces a buffer solution.

$[NH_4^+] = 0.25 \text{ mol}/0.500 \text{ L} = 0.50 \text{ M}; \quad [NH_3] = 0.50 \text{ mol}/0.500 \text{ L} = 1.0 \text{ M}$

$$pH = pK_a + \log \frac{[\text{base}]}{[\text{acid}]} = pK_a + \log \frac{[NH_3]}{[NH_4^+]}$$

For NH_4^+, $K_a = \dfrac{K_w}{K_b \text{ for } NH_3} = \dfrac{1.0 \times 10^{-14}}{1.8 \times 10^{-5}} = 5.6 \times 10^{-10}$; $pK_a = -\log K_a = 9.25$

$pH = 9.25 + \log\left(\dfrac{1.0}{0.5}\right) = 9.55$

15.130 For NH_4^+, $K_a = \dfrac{K_w}{K_b \text{ for } NH_3} = \dfrac{1.0 \times 10^{-14}}{1.8 \times 10^{-5}} = 5.6 \times 10^{-10}$; $pK_a = -\log K_a = 9.25$

$pH = pK_a + \log \dfrac{[NH_3]}{[NH_4^+]} = 9.25 + \log \dfrac{(0.50)}{(0.30)} = 9.47$

$[H_3O^+] = 10^{-pH} = 10^{-9.47} = 3.4 \times 10^{-10} \text{ M}$

For MnS, $K_{spa} = \dfrac{[Mn^{2+}][H_2S]}{[H_3O^+]^2} = 3 \times 10^7$

molar solubility $= [Mn^{2+}] = \dfrac{K_{spa}[H_3O^+]^2}{[H_2S]} = \dfrac{(3 \times 10^7)(3.4 \times 10^{-10})^2}{(0.10)} = 3.5 \times 10^{-11} \text{ M}$

MnS, 87.00; solubility $= (3.5 \times 10^{-11} \text{ mol/L})(87.00 \text{ g/mol}) = 3 \times 10^{-9} \text{ g/L}$

15.132 $60.0 \text{ mL} = 0.0600 \text{ L}$

$\text{mol } H_3PO_4 = 0.0600 \text{ L} \times \dfrac{1.00 \text{ mol } H_3PO_4}{1.00 \text{ L}} = 0.0600 \text{ mol } H_3PO_4$

$\text{mol LiOH} = 1.00 \text{ L} \times \dfrac{0.100 \text{ mol LiOH}}{1.00 \text{ L}} = 0.100 \text{ mol LiOH}$

$$H_3PO_4(aq) \; + \; OH^-(aq) \; \rightarrow \; H_2PO_4^-(aq) \; + \; H_2O(l)$$

before reaction (mol)	0.0600	0.100	0
change (mol)	−0.0600	−0.0600	+0.0600
after reaction (mol)	0	0.040	0.0600

$$H_2PO_4^-(aq) + OH^-(aq) \rightarrow HPO_4^{2-}(aq) + H_2O(l)$$

before reaction (mol)	0.0600	0.040	0
change (mol)	−0.040	−0.040	+0.040
after reaction (mol)	0.020	0	0.040

The resulting solution is a buffer because it contains the conjugate acid-base pair, $H_2PO_4^-$ and HPO_4^{2-}, at acceptable buffer concentrations.
For $H_2PO_4^-$, $K_{a2} = 6.2 \times 10^{-8}$ and $pK_{a2} = -\log K_{a2} = -\log(6.2 \times 10^{-8}) = 7.21$

$$pH = pK_{a2} + \log\frac{[HPO_4^{2-}]}{[H_2PO_4^-]} = 7.21 + \log\frac{(0.040 \text{ mol}/1.06 \text{ L})}{(0.020 \text{ mol}/1.06 \text{ L})}$$

$$pH = 7.21 + \log\frac{(0.040)}{(0.020)} = 7.21 + 0.30 = 7.51$$

15.134 For CH_3CO_2H, $K_a = 1.8 \times 10^{-5}$ and $pK_a = -\log K_a = -\log(1.8 \times 10^{-5}) = 4.74$
The mixture will be a buffer solution containing the conjugate acid-base pair, CH_3CO_2H and $CH_3CO_2^-$, having a pH near the pK_a of CH_3CO_2H.

$$pH = pK_a + \log\frac{[CH_3CO_2^-]}{[CH_3CO_2H]}$$

$$4.85 = 4.74 + \log\frac{[CH_3CO_2^-]}{[CH_3CO_2H]}; \quad 4.85 - 4.74 = \log\frac{[CH_3CO_2^-]}{[CH_3CO_2H]}$$

$$0.11 = \log\frac{[CH_3CO_2^-]}{[CH_3CO_2H]}; \quad \frac{[CH_3CO_2^-]}{[CH_3CO_2H]} = 10^{0.11} = 1.3$$

In the Henderson-Hasselbalch equation, moles can be used in place of concentrations because both components are in the same volume so the volume terms cancel.
20.0 mL = 0.0200 L

Let X equal the volume of 0.10 M CH_3CO_2H and Y equal the volume of 0.15 M $CH_3CO_2^-$. Therefore, X + Y = 0.0200 L and

$$\frac{Y \times [CH_3CO_2^-]}{X \times [CH_3CO_2H]} = \frac{Y(0.15 \text{ mol/L})}{X(0.10 \text{ mol/L})} = 1.3$$

X = 0.0200 − Y

$$\frac{Y(0.15 \text{ mol/L})}{(0.020 - Y)(0.10 \text{ mol/L})} = 1.3$$

$$\frac{0.15Y}{0.0020 - 0.10Y} = 1.3$$

0.15Y = 1.3(0.0020 − 0.10Y)
0.15Y = 0.0026 − 0.13Y
0.15Y + 0.13Y = 0.0026
0.28Y = 0.0026

$Y = 0.0026/0.28 = 0.0093$ L

$X = 0.0200 - Y = 0.0200 - 0.0093 = 0.0107$ L

$X = 0.0107$ L $= 10.7$ mL and $Y = 0.0093$ L $= 9.3$ mL

You need to mix together 10.7 mL of 0.10 M CH_3CO_2H and 9.3 mL of 0.15 M $NaCH_3CO_2$ to prepare 20.0 mL of a solution with a pH of 4.85.

15.136   (a)   HCl is a strong acid. HCN is a weak acid with $K_a = 4.9 \times 10^{-10}$. Before the titration, the $[H_3O^+] = 0.100$ M. The HCN contributes an insignificant amount of additional H_3O^+, so the pH $= -\log[H_3O^+] = -\log(0.100) = 1.00$

      (b)   100.0 mL $= 0.1000$ L

$$\text{mol } H_3O^+ = 0.1000 \text{ L} \times \frac{0.100 \text{ mol HCl}}{1.00 \text{ L}} = 0.0100 \text{ mol } H_3O^+$$

add 75.0 mL of 0.100 M NaOH; 75.0 mL $= 0.0750$ L

$$\text{mol OH}^- = 0.0750 \text{ L} \times \frac{0.100 \text{ mol NaOH}}{1.00 \text{ L}} = 0.00750 \text{ mol OH}^-$$

	H_3O^+(aq) +	OH$^-$(aq) $\rightarrow$	2 H_2O(l)
before reaction (mol)	0.0100	0.0075	
change (mol)	−0.0075	−0.0075	
after reaction (mol)	0.0025	0	

$$[H_3O^+] = \frac{0.0025 \text{ mol } H_3O^+}{0.1000 \text{ L} + 0.0750 \text{ L}} = 0.0143 \text{ M}$$

pH $= -\log[H_3O^+] = -\log(0.0143) = 1.84$

(c)   100.0 mL of 0.100 M NaOH will completely neutralize all of the H_3O^+ from 100.0 mL of 0.100 M HCl. Only NaCl and HCN remain in the solution. NaCl is a neutral salt and does not affect the pH of the solution. [HCN] changes because of dilution. Because the solution volume is doubled, [HCN] is cut in half.

[HCN] $= 0.100$ M/2 $= 0.0500$ M

	HCN(aq) +	H_2O(l) $\rightleftharpoons$	H_3O^+(aq) +	CN$^-$(aq)
initial (M)	0.0500		~0	0
change (M)	−x		+x	+x
equil (M)	0.0500 − x		x	x

$$K_a = \frac{[H_3O^+][CN^-]}{\text{HCN}} = 4.9 \times 10^{-10} = \frac{x^2}{0.0500 - x} \approx \frac{x^2}{0.0500}$$

$[H_3O^+] = x = \sqrt{(0.0500)(4.9 \times 10^{-10})} = 4.95 \times 10^{-6}$ M

pH $= -\log[H_3O^+] = -\log(4.95 \times 10^{-6}) = 5.31$

(d)   Add an additional 25.0 mL of 0.100 M NaOH.

25.0 mL $= 0.0250$ L

$$\text{additional mol OH}^- = 0.0250 \text{ L} \times \frac{0.100 \text{ mol NaOH}}{1.00 \text{ L}} = 0.00250 \text{ mol OH}^-$$

$$\text{mol HCN} = 0.200 \text{ L} \times \frac{0.0500 \text{ mol HCN}}{1.00 \text{ L}} = 0.0100 \text{ mol HCN}$$

$$HCN(aq) \ + \ OH^-(aq) \ \rightarrow \ CN^-(aq) \ + \ H_2O(l)$$

	HCN(aq)	OH$^-$(aq)	CN$^-$(aq)	H$_2$O(l)
before reaction (mol)	0.0100	0.00250	0	
change (mol)	−0.00250	−0.00250	+0.00250	
after reaction (mol)	0.0075	0	0.00250	

The resulting solution is a buffer because it contains the conjugate acid-base pair, HCN and CN$^-$, at acceptable buffer concentrations.

For HCN, $K_a = 4.9 \times 10^{-10}$ and $pK_a = -\log K_a = -\log (4.9 \times 10^{-10}) = 9.31$

$$pH = pK_a \ + \ \log \frac{[CN^-]}{[HCN]} = 9.31 \ + \ \log \frac{(0.00250 \ mol/0.2250 \ L)}{(0.0075 \ mol/0.2250 \ L)}$$

$$pH = 9.31 \ + \ \log \frac{(0.00250)}{(0.0075)} = 9.31 - 0.48 = 8.83$$

15.138 (a) $$Zn(OH)_2(s) \ \rightleftharpoons \ Zn^{2+}(aq) \ + \ 2 \ OH^-(aq)$$

	Zn^{2+}	OH$^-$
initial (M)	0	~0
equil (M)	x	2x

$$K_{sp} = [Zn^{2+}][OH^-]^2 = 4.1 \times 10^{-17} = (x)(2x)^2 = 4x^3$$

$$\text{molar solubility} = x = \sqrt[3]{\frac{4.1 \times 10^{-17}}{4}} = 2.2 \times 10^{-6} \ M$$

(b) $[OH^-] = 2x = 2(2.2 \times 10^{-6} \ M) = 4.4 \times 10^{-6} \ M$

$$[H_3O^+] = \frac{1.0 \times 10^{-14}}{4.4 \times 10^{-6}} = 2.3 \times 10^{-9} \ M$$

$pH = -\log[H_3O^+] = -\log(2.3 \times 10^{-9}) = 8.64$

(c)

$Zn(OH)_2(s) \ \rightleftharpoons \ Zn^{2+}(aq) \ + \ 2 \ OH^-(aq)$		$K_{sp} = 4.1 \times 10^{-17}$
$Zn^{2+}(aq) \ + \ 4 \ OH^-(aq) \ \rightleftharpoons \ Zn(OH)_4^{2-}(aq)$		$K_f = 3 \times 10^{15}$
$Zn(OH)_2(s) \ + 2 \ OH^-(aq) \ \rightleftharpoons \ Zn(OH)_4^{2-}(aq)$		$K = K_{sp} \cdot K_f = 0.123$

	OH$^-$	Zn(OH)$_4^{2-}$
initial (M)	0.10	0
change (M)	−2x	+x
equil (M)	0.10 − 2x	x

$$K = \frac{[Zn(OH)_4^{2-}]}{[OH^-]^2} = 0.123 = \frac{x}{(0.10 - 2x)^2}$$

$0.492x^2 - 1.0492x + 0.00123 = 0$

Use the quadratic formula to solve for x.

$$x = \frac{-(-1.0492) \pm \sqrt{(-1.0492)^2 - (4)(0.492)(0.00123)}}{2(0.492)} = \frac{1.0492 \pm 1.0480}{0.984}$$

$x = 2.1$ and 1.2×10^{-3}

Of the two solutions for x, only 1.2×10^{-3} has physical meaning because the other solution leads to a negative [OH$^-$].

molar solubility of Zn(OH)$_4^{2-}$ in 0.10 M NaOH = x = 1.2×10^{-3} M

Multiconcept Problems

15.140 (a) $HA^-(aq) + H_2O(l) \rightleftharpoons H_3O^+(aq) + A^{2-}(aq)$ $K_{a2} = 10^{-10}$

   $HA^-(aq) + H_2O(l) \rightleftharpoons H_2A(aq) + OH^-(aq)$ $K_b = \dfrac{K_w}{K_{a1}} = 10^{-10}$

   $2\, HA^-(aq) \rightleftharpoons H_2A(aq) + A^{2-}(aq)$ $K = \dfrac{K_{a2}}{K_{a1}} = 10^{-6}$

   $2\, H_2O(l) \rightleftharpoons H_3O^+(aq) + OH^-(aq)$ $K_w = 1.0 \times 10^{-14}$

The principal reaction of the four is the one with the largest K, and that is the third reaction.

(b) $K_{a1} = \dfrac{[H_3O^+][HA^-]}{[H_2A]}$ and $K_{a2} = \dfrac{[H_3O^+][A^{2-}]}{[HA^-]}$

$[H_3O^+] = \dfrac{K_{a1}[H_2A]}{[HA^-]}$ and $[H_3O^+] = \dfrac{K_{a2}[HA^-]}{[A^{2-}]}$

$\dfrac{K_{a1}[H_2A]}{[HA^-]} \times \dfrac{K_{a2}[HA^-]}{[A^{2-}]} = [H_3O^+]^2;$ $\dfrac{K_{a1}K_{a2}[H_2A]}{[A^{2-}]} = [H_3O^+]^2$

Because the principal reaction is $2\, HA^-(aq) \rightleftharpoons H_2A(aq) + A^{2-}(aq)$, $[H_2A] = [A^{2-}]$.

$K_{a1}K_{a2} = [H_3O^+]^2$

$\log K_{a1} + \log K_{a2} = 2 \log [H_3O^+]$

$\dfrac{\log K_{a1} + \log K_{a2}}{2} = \log[H_3O^+];$ $\dfrac{-\log K_{a1} + (-\log K_{a2})}{2} = -\log[H_3O^+]$

$\dfrac{pK_{a1} + pK_{a2}}{2} = pH$

(c)

	$2\, HA^-(aq)$	$\rightleftharpoons$	$H_2A(aq)$	+	$A^{2-}(aq)$
initial (M)	1.0		0		0
change (M)	$-2x$		$+x$		$+x$
equil (M)	$1.0 - 2x$		x		x

$K = \dfrac{[H_2A][A^{2-}]}{[HA^-]^2} = 1 \times 10^{-6} = \dfrac{x^2}{(1.0 - 2x)^2}$

Take the square root of both sides and solve for x.

$x = [A^{2-}] = 1 \times 10^{-3}\ M$

mol $A^{2-} = (1 \times 10^{-3}\ mol/L)(0.0500\ L) = 5 \times 10^{-5}\ mol\ A^{2-}$

number of A^{2-} ions $= (5 \times 10^{-5}\ mol\ A^{2-})(6.022 \times 10^{23}\ ions/mol) = 3 \times 10^{19}\ A^{2-}$ ions

15.142 (a) The first equivalence point is reached when all the H_3O^+ from the HCl, and the H_3O^+ from the first ionization of H_3PO_4, is consumed.

At the first equivalence point pH $= \dfrac{pK_{a1} + pK_{a2}}{2} = 4.66$

$[H_3O^+] = 10^{-pH} = 10^{(-4.66)} = 2.2 \times 10^{-5}\ M$

(88.0 mL)(0.100 mmol/mL) = 8.80 mmol NaOH are used to get to the first equivalence point

(b) mmol (HCl + H_3PO_4) = mmol NaOH = 8.8 mmol

mmol H_3PO_4 = (126.4 mL – 88.0 mL)(0.100 mmol/mL) = 3.84 mmol

mmol HCl = (8.8 – 3.84) = 4.96 mmol

$$[HCl] = \frac{4.96 \text{ mmol}}{40.0 \text{ mL}} = 0.124 \text{ M}; \quad [H_3PO_4] = \frac{3.84 \text{ mmol}}{40.0 \text{ mL}} = 0.0960 \text{ M}$$

(c) 100% of the HCl is neutralized at the first equivalence point.

(d)

	$H_3PO_4(aq)$	$+$ $H_2O(l)$	$\rightleftharpoons$	$H_3O^+(aq)$	$+$ $H_2PO_4^-(aq)$
initial (M)	0.0960			0.124	0
change (M)	$-x$			$+x$	$+x$
equil (M)	$0.0960 - x$			$0.124 + x$	x

$$K_{a1} = \frac{[H_3O^+][H_2PO_4^-]}{[H_3PO_4]} = 7.5 \times 10^{-3} = \frac{(0.124 + x)(x)}{0.0960 - x}$$

$x^2 + 0.132x - (7.2 \times 10^{-4}) = 0$

Use the quadratic formula to solve for x.

$$x = \frac{-(0.132) \pm \sqrt{(0.132)^2 - 4(1)(-7.2 \times 10^{-4})}}{2(1)} = \frac{-0.132 \pm 0.142}{2}$$

$x = -0.137$ and 0.005

Of the two solutions for x, only the positive value of x has physical meaning because the other solution would give a negative $[H_3O^+]$.

$[H_3O^+] = 0.124 + x = 0.124 + 0.005 = 0.129 \text{ M}$

$pH = -\log[H_3O^+] = -\log(0.129) = 0.89$

(e)

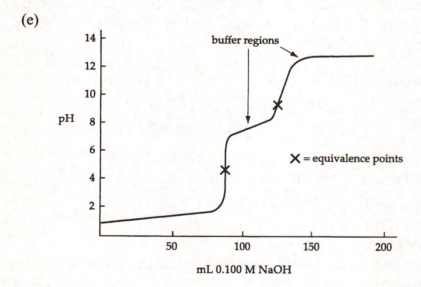

(f) Bromcresol green or methyl orange are suitable indicators for the first equivalence point. Thymolphthalein is a suitable indicator for the second equivalence point.

15.144 25 °C = 298 K

$$\Pi = 2MRT; \quad M = \frac{\Pi}{2RT} = \frac{\left(74.4 \text{ mm Hg} \times \dfrac{1.00 \text{ atm}}{760 \text{ mm Hg}}\right)}{(2)\left(0.082\ 06\ \dfrac{L \cdot atm}{K \cdot mol}\right)(298 \text{ K})} = 0.00200 \text{ M}$$

$[M^+] = [X^-] = 0.00200$ M

$K_{sp} = [M^+][X^-] = (0.00200)^2 = 4.00 \times 10^{-6}$

15.146 (a) species present initially:

NH_4^+	CO_3^{2-}	H_2O
acid	base	acid or base

$2H_2O(l) \rightleftharpoons H_3O^+(aq) + OH^-(aq)$

$NH_4^+(aq) + H_2O(l) \rightleftharpoons NH_3(aq) + H_3O^+(aq)$

$CO_3^{2-}(aq) + H_2O(l) \rightleftharpoons HCO_3^-(aq) + OH^-(aq)$

NH_3, $K_b = 1.8 \times 10^{-5}$

NH_4^+, $K_a = 5.6 \times 10^{-10}$

CO_3^{2-}, $K_b = 1.8 \times 10^{-4}$

HCO_3^-, $K_a = 5.6 \times 10^{-11}$

In the mixture, proton transfer takes place from the stronger acid to the stronger base, so the principal reaction is $NH_4^+(aq) + CO_3^{2-}(aq) \rightleftharpoons HCO_3^-(aq) + NH_3(aq)$

(b)

	$NH_4^+(aq) + OH^-(aq) \rightleftharpoons NH_3(aq) + H_2O(l)$	$K_1 = 1/K_b(NH_3)$
	$CO_3^{2-}(aq) + H_2O(l) \rightleftharpoons HCO_3^-(aq) + OH^-(aq)$	$K_2 = K_b(CO_3^{2-})$
	$NH_4^+(aq) + CO_3^{2-}(aq) \rightleftharpoons HCO_3^-(aq) + NH_3(aq)$	$K = K_1 \cdot K_2$

	NH_4^+	CO_3^{2-}	HCO_3^-	NH_3
initial (M)	0.16	0.080	0	0.16
change (M)	–x	–x	+x	+x
equil (M)	0.16 – x	0.080 – x	x	0.16 + x

$$K = \frac{[HCO_3^-][NH_3]}{[NH_4^+][CO_3^{2-}]} = \frac{1.8 \times 10^{-4}}{1.8 \times 10^{-5}} = 10 = \frac{x(0.16 + x)}{(0.16 - x)(0.080 - x)}$$

$9x^2 - 2.56x + 0.128 = 0$

Use the quadratic formula to solve for x.

$$x = \frac{-(-2.56) \pm \sqrt{(-2.56)^2 - (4)(9)(0.128)}}{2(9)} = \frac{2.56 \pm 1.395}{18}$$

x = 0.220 and 0.0647

Of the two solutions for x, only 0.00647 has physical meaning because 0.220 leads to negative concentrations.

$[NH_4^+] = 0.16 - x = 0.16 - 0.0647 = 0.0953$ M $= 0.095$ M

$[NH_3] = 0.16 + x = 0.16 + 0.0647 = 0.225$ M $= 0.23$ M

$[CO_3^{2-}] = 0.080 - x = 0.080 - 0.0647 = 0.0153$ M $= 0.015$ M

$[HCO_3^-] = x = 0.0647$ M $= 0.065$ M

The solution is a buffer containing two different sets of conjugate acid-base pairs.

Either pair can be used to calculate the pH.

For NH_4^+, $K_a = 5.6 \times 10^{-10}$ and $pK_a = 9.25$

$$pH = pK_a + \log \frac{[NH_3]}{[NH_4^+]} = 9.25 + \log \frac{(0.225)}{(0.0953)} = 9.62$$

$$[H_3O^+] = 10^{-pH} = 10^{-9.62} = 2.4 \times 10^{-10} \text{ M}$$

$$[OH^-] = \frac{1.0 \times 10^{-14}}{2.4 \times 10^{-10}} = 4.2 \times 10^{-5} \text{ M}$$

$$[H_2CO_3] = \frac{[HCO_3^-][H_3O^+]}{K_a} = \frac{(0.647)(2.4 \times 10^{-10})}{(4.3 \times 10^{-7})} = 3.6 \times 10^{-4} \text{ M}$$

(c) For MCO_3, IP = $[M^{2+}][CO_3^{2-}]$ = (0.010)(0.0153) = 1.5×10^{-4}

$K_{sp}(CaCO_3) = 5.0 \times 10^{-9}$, $10^3 \, K_{sp} = 5.0 \times 10^{-6}$

$K_{sp}(BaCO_3) = 2.6 \times 10^{-9}$, $10^3 \, K_{sp} = 2.6 \times 10^{-6}$

$K_{sp}(MgCO_3) = 6.8 \times 10^{-6}$, $10^3 \, K_{sp} = 6.8 \times 10^{-3}$

IP $> 10^3 \, K_{sp}$ for $CaCO_3$ and $BaCO_3$, but IP $< 10^3 \, K_{sp}$ for $MgCO_3$ so the $[CO_3^{2-}]$ is large enough to give observable precipitation of $CaCO_3$ and $BaCO_3$, but not $MgCO_3$.

(d) For $M(OH)_2$, IP = $[M^{2+}][OH^-]^2$ = (0.010)(4.17 $\times 10^{-5}$)^2 = 1.7×10^{-11}

$K_{sp}(Ca(OH)_2) = 4.7 \times 10^{-6}$, $10^3 \, K_{sp} = 4.7 \times 10^{-3}$

$K_{sp}(Ba(OH)_2) = 5.0 \times 10^{-3}$, $10^3 \, K_{sp} = 5.0$

$K_{sp}(Mg(OH)_2) = 5.6 \times 10^{-12}$, $10^3 \, K_{sp} = 5.6 \times 10^{-9}$

IP $< 10^3 \, K_{sp}$ for all three $M(OH)_2$. None precipitate.

(e)
	$CO_3^{2-}(aq)$	+	$H_2O(l)$	$\rightleftharpoons$	$HCO_3^-(aq)$	+	$OH^-(aq)$
initial (M)	0.08				0		~0
change (M)	−x				+x		+x
equil (M)	0.08 − x				x		x

$$K_b = \frac{[HCO_3^-][OH^-]}{[CO_3^{2-}]} = 1.8 \times 10^{-4} = \frac{x^2}{(0.08 - x)}$$

$x^2 + (1.8 \times 10^{-4})x - (1.44 \times 10^{-5}) = 0$

Use the quadratic formula to solve for x.

$$x = \frac{-(1.8 \times 10^{-4}) \pm \sqrt{(1.8 \times 10^{-4})^2 - (4)(1)(-1.44 \times 10^{-5})}}{2(1)} = \frac{-(1.8 \times 10^{-4}) \pm 7.59 \times 10^{-3}}{2}$$

x = 0.0037 and −0.0039

Of the two solutions for x, only 0.0037 has physical meaning because −0.0039 leads to negative concentrations.

$[OH^-] = x = 3.7 \times 10^{-3}$ M

For MCO_3, IP = $[M^{2+}][CO_3^{2-}]$ = (0.010)(0.08) = 8.0×10^{-4}

For $M(OH)_2$, IP = $[M^{2+}][OH^-]^2$ = (0.010)(3.7 $\times 10^{-3}$)^2 = 1.4×10^{-7}

Comparing IP's here and $10^3 \, K_{sp}$'s in (c) and (d) above, Ca^{2+} and Ba^{2+} cannot be separated from Mg^{2+} using 0.08 M Na_2CO_3. Na_2CO_3 is more basic than $(NH_4)_2CO_3$ and $Mg(OH)_2$ would precipitate along with $CaCO_3$ and $BaCO_3$.

15.148 $Pb(CH_3CO_2)_2$, 325.29; PbS, 239.27

(a) mass PbS = $(2 \text{ mL})(1 \text{ g/mL})(0.003) \times \dfrac{1 \text{ mol } Pb(CH_3CO_2)_2}{325.29 \text{ g } Pb(CH_3CO_2)_2} \times$

$\dfrac{1 \text{ mol PbS}}{1 \text{ mol } Pb(CH_3CO_2)_2} \times \dfrac{239.27 \text{ g PbS}}{1 \text{ mol PbS}} \times (30/100) = 0.0013 \text{ g}$

$= 1.3 \text{ mg PbS per dye application}$

(b) $[H_3O^+] = 10^{-pH} = 10^{-5.50} = 3.16 \times 10^{-6} \text{ M}$

$$PbS(s) + 2 H_3O^+(aq) \rightleftarrows Pb^{2+}(aq) + H_2S(aq) + 2 H_2O(l)$$

initial (M)	3.16×10^{-6}	0	0
change (M)	$-2x$	$+x$	$+x$
equil (M)	$3.16 \times 10^{-6} - 2x$	x	x

$K_{spa} = \dfrac{[Pb^{2+}][H_2S]}{[H_3O^+]^2} = \dfrac{x^2}{(3.16 \times 10^{-6} - 2x)^2} \approx \dfrac{x^2}{(3.16 \times 10^{-6})^2} = 3 \times 10^{-7}$

$x^2 = (3.16 \times 10^{-6})^2(3 \times 10^{-7}) = 3.0 \times 10^{-18}$

$x = 1.7 \times 10^{-9} \text{ M} = [Pb^{2+}]$ for a saturated solution.

mass of PbS dissolved per washing =

$(3 \text{ gal})(3.7854 \text{ L/1 gal})(1.7 \times 10^{-9} \text{ mol/L}) \times \dfrac{239.27 \text{ g PbS}}{1 \text{ mol PbS}} = 4.6 \times 10^{-6} \text{ g PbS/washing}$

Number of washings required to remove 50% of the PbS from one application =

$\dfrac{(0.0013 \text{ g PbS})(50/100)}{(4.6 \times 10^{-6} \text{ g PbS/washing})} = 1.4 \times 10^2 \text{ washings}$

(c) The number of washings does not look reasonable. It seems too high considering that frequent dye application is recommended. If the PbS is located mainly on the surface of the hair, as is believed to be the case, solid particles of PbS can be lost by abrasion during shampooing.

16

Thermodynamics: Entropy, Free Energy, and Equilibrium

16.1   (a) spontaneous;   (b), (c), and (d) nonspontaneous

16.2   (a) $H_2O(g) \rightarrow H_2O(l)$
A liquid has less randomness than a gas. Therefore, ΔS is negative.
(b) $I_2(g) \rightarrow 2\ I(g)$
ΔS is positive because the reaction increases the number of gaseous particles from 1 mol to 2 mol.
(c) $CaCO_3(s) \rightarrow CaO(s) + CO_2(g)$
ΔS is positive because the reaction increases the number of gaseous molecules.
(d) $Ag^+(aq) + Br^-(aq) \rightarrow AgBr(s)$
A solid has less randomness than +1 and −1 charged ions in an aqueous solution. Therefore, ΔS is negative.
(e) Deposition of frost on a cold morning, $H_2O(g) \rightarrow H_2O(s)$
Deposition is the formation of a solid from a gas. A solid has less randomness than a gas. Therefore, ΔS is negative.

16.3   (a) $A_2 + AB_3 \rightarrow 3\ AB$
(b) ΔS is positive because the reaction increases the number of gaseous molecules.

16.4   (a) disordered N_2O (more randomness)
(b) quartz glass (amorphous solid, more randomness)
(c) 1 mole N_2 at STP (larger volume, more randomness)
(d) 1 mole N_2 at 273 K and 0.25 atm (larger volume, more randomness)

16.5 $CaCO_3(s) \rightarrow CaO(s) + CO_2(g)$
$\Delta S^\circ = [S^\circ(CaO) + S^\circ(CO_2)] - S^\circ(CaCO_3)$
$\Delta S^\circ = [(1\ mol)(38.1\ J/(K \cdot mol)) + (1\ mol)(213.6\ J/(K \cdot mol))]$
$$- (1\ mol)(91.7\ J/(K \cdot mol)) = +160.0\ J/K$$

16.6   From Problem 16.5, $\Delta S_{sys} = \Delta S^\circ = 160.0\ J/K$
$CaCO_3(s) \rightarrow CaO(s) + CO_2(g)$
$\Delta H^\circ = [\Delta H^\circ_f(CaO) + \Delta H^\circ_f(CO_2)] - \Delta H^\circ_f(CaCO_3)$
$\Delta H^\circ = [(1\ mol)(-634.9\ kJ/mol) + (1\ mol)(-393.5\ kJ/mol)]$
$$- (1\ mol)(-1207.6\ kJ/mol) = +179.2\ kJ$$

$$\Delta S_{surr} = \frac{-\Delta H^\circ}{T} = \frac{-179,200\ J}{298\ K} = -601\ J/K$$

$\Delta S_{total} = \Delta S_{sys} + \Delta S_{surr} = 160.0\ J/K + (-601\ J/K) = -441\ J/K$
Because ΔS_{total} is negative, the reaction is not spontaneous under standard-state conditions at 25 °C.

16.7   (a) $\Delta G = \Delta H - T\Delta S = 55.3 \text{ kJ} - (298 \text{ K})(0.1757 \text{ kJ/K}) = +2.9 \text{ kJ}$
Because $\Delta G > 0$, the reaction is nonspontaneous at 25 °C (298 K)
(b) Set $\Delta G = 0$ and solve for T.
$0 = \Delta H - T\Delta S$
$$T = \frac{\Delta H}{\Delta S} = \frac{55.3 \text{ kJ}}{0.1757 \text{ kJ/K}} = 315 \text{ K} = 42 \text{ °C}$$

16.8   (a) $\Delta G = \Delta H - T\Delta S = 59.11 \text{ kJ/mol} - (598 \text{ K})[0.0939 \text{ kJ/(K} \cdot \text{mol)}] = +3.0 \text{ kJ/mol}$
Because $\Delta G > 0$, Hg does not boil at 325 °C and 1 atm.
(b) The boiling point (phase change) is associated with an equilibrium. Set $\Delta G = 0$ and solve for T, the boiling point.
$0 = \Delta H_{vap} - T\Delta S_{vap}$
$$T_{bp} = \frac{\Delta H_{vap}}{\Delta S_{vap}} = \frac{59.11 \text{ kJ/mol}}{0.0939 \text{ kJ/(K} \cdot \text{mol)}} = 629 \text{ K} = 356 \text{ °C}$$

16.9 $\Delta H < 0$ (reaction involves bond making - exothermic)
$\Delta S < 0$ (the reaction has less randomness in going from reactants (2 atoms) to
            products (1 molecule)
$\Delta G < 0$ (the reaction is spontaneous)

16.10   From Problems 16.5 and 16.6: $\Delta H° = 179.2 \text{ kJ}$ and $\Delta S° = 160.0 \text{ J/K} = 0.1600 \text{ kJ/K}$
(a) $\Delta G° = \Delta H° - T\Delta S° = 179.2 \text{ kJ} - (298 \text{ K})(0.1600 \text{ kJ/K}) = +131.5 \text{ kJ}$
(b) Because $\Delta G > 0$, the reaction is nonspontaneous at 25 °C (298 K).
(c) Set $\Delta G = 0$ and solve for T, the temperature above which the reaction becomes spontaneous.
$$0 = \Delta H - T\Delta S; \qquad T = \frac{\Delta H}{\Delta S} = \frac{179.2 \text{ kJ}}{0.1600 \text{ kJ/K}} = 1120 \text{ K} = 847 \text{ °C}$$

16.11 $2 \text{ AB}_2 \rightarrow \text{A}_2 + 2 \text{ B}_2$
(a) $\Delta S°$ is positive because the reaction increases the number of molecules.
(b) $\Delta H°$ is positive because the reaction is endothermic.
$\Delta G° = \Delta H° - T\Delta S°$
For the reaction to be spontaneous, $\Delta G°$ must be negative. This will only occur at high temperature where $T\Delta S°$ is greater than $\Delta H°$.

16.12   (a) $\text{CaC}_2(s) + 2 \text{ H}_2\text{O}(l) \rightarrow \text{C}_2\text{H}_2(g) + \text{Ca(OH)}_2(s)$
$\Delta G° = [\Delta G°_f(\text{C}_2\text{H}_2) + \Delta G°_f(\text{Ca(OH)}_2)] - [\Delta G°_f(\text{CaC}_2) + 2 \Delta G°_f(\text{H}_2\text{O})]$
$\Delta G° = [(1 \text{ mol})(209.9 \text{ kJ/mol}) + (1 \text{ mol})(-897.5 \text{ kJ/mol})]$
        $- [(1 \text{ mol})(-64.8 \text{ kJ/mol}) + (2 \text{ mol})(-237.2 \text{ kJ/mol})] = -148.4 \text{ kJ}$
This reaction can be used for the synthesis of C_2H_2 because $\Delta G < 0$.
(b) It is not possible to synthesize acetylene from solid graphite and gaseous H_2 at 25 °C and 1 atm because $\Delta G°_f(\text{C}_2\text{H}_2) > 0$.

16.13 $C(s) + 2 H_2(g) \rightarrow C_2H_4(g)$

$$Q_p = \frac{P_{C_2H_4}}{(P_{H_2})^2} = \frac{(0.10)}{(100)^2} = 1.0 \times 10^{-5}$$

$\Delta G = \Delta G° + RT \ln Q_p$

$\Delta G = 68.1 \text{ kJ/mol} + [8.314 \times 10^{-3} \text{ kJ/(K} \cdot \text{mol)}](298 \text{ K})\ln(1.0 \times 10^{-5}) = +39.6 \text{ kJ/mol}$

Because $\Delta G > 0$, the reaction is spontaneous in the reverse direction.

16.14 $\Delta G = \Delta G° + RT \ln Q$ and $\Delta G° = 15 \text{ kJ}$

For $A_2(g) + B_2(g) \rightleftharpoons 2 AB(g)$, $Q_p = \dfrac{(P_{AB})^2}{(P_{A_2})(P_{B_2})}$

Let the number of molecules be proportional to the partial pressure.

(1) $Q_p = 1.0$ (2) $Q_p = 0.0667$ (3) $Q_p = 18$

(a) Reaction (3) has the largest ΔG because Q_p is the largest. Reaction (2) has the smallest ΔG because Q_p is the smallest.

(b) $\Delta G = \Delta G° = 15 \text{ kJ}$ because $Q_p = 1$ and $\ln (1) = 0$.

16.15   From Problem 16.10, $\Delta G° = +131.5 \text{ kJ}$

$\Delta G° = -RT \ln K_p$

$$\ln K_p = \frac{-\Delta G°}{RT} = \frac{-131.5 \text{ kJ/mol}}{[8.314 \times 10^{-3} \text{ kJ/(K} \cdot \text{mol)}](298 \text{ K})} = -53.1$$

$K_p = e^{-53.1} = 9 \times 10^{-24}$

16.16 $H_2O(l) \rightleftharpoons H_2O(g)$

$K_p = P_{H_2O}$; K_p is equal to the vapor pressure for H_2O.

$\Delta G° = \Delta G°_f(H_2O(g)) - \Delta G°_f(H_2O(l))$

$\Delta G° = (1 \text{ mol})(-228.6 \text{ kJ/mol}) - (1 \text{ mol})(-237.2 \text{ kJ/mol}) = +8.6 \text{ kJ}$

$\Delta G° = -RT \ln K_p$

$$\ln K_p = \frac{-\Delta G°}{RT} = \frac{-8.6 \text{ kJ/mol}}{[8.314 \times 10^{-3} \text{ kJ/(K} \cdot \text{mol)}](298 \text{ K})} = -3.5$$

$K_p = P_{H_2O} = e^{-3.5} = 0.03 \text{ atm}$

16.17 $\Delta G° = -RT \ln K = -[8.314 \times 10^{-3} \text{ kJ/(K} \cdot \text{mol)}](298 \text{ K}) \ln (1.0 \times 10^{-14}) = 80 \text{ kJ/mol}$

16.18   The growth of a human adult from a single cell does not violate the second law of thermodynamics. The energy a human obtains from glucose is used to build and organize complex molecules, resulting in a decrease in entropy for the human. At the same time, however, the entropy of the surroundings increases as the human releases small, simple waste products such as CO_2 and H_2O. Furthermore, heat is released by the human, further increasing the entropy of the surroundings. Thus, an organism pays for its decrease in entropy by increasing the entropy of the rest of the universe.

16.19 You would expect to see violations of the second law if you watched a movie run backwards. Consider an action-adventure movie with a lot of explosions. An explosion is a spontaneous process that increases the entropy of the universe. You would see an explosion go backwards if you run the the movie backwards but this is impossible because it would decrease the entropy of the universe.

Conceptual Problems

16.20 (a)

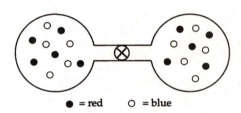

● = red   ○ = blue

(b) $\Delta H = 0$ (no heat is gained or lost in the mixing of ideal gases)
   $\Delta S > 0$ (the mixture of the two gases has more randomness)
   $\Delta G < 0$ (the mixing of the two gases is spontaneous)
(c) For an isolated system, $\Delta S_{surr} = 0$ and $\Delta S_{sys} = \Delta S_{Total} > 0$ for the spontaneous process.
(d) $\Delta G > 0$ and the process is nonspontaneous.

16.22 $\Delta H < 0$ (heat is lost during condensation)
   $\Delta S < 0$ (liquid has less randomness than vapor)
   $\Delta G < 0$ (the reaction is spontaneous)

16.24 (a) $2\ A_2\ +\ B_2\ \rightarrow\ 2\ A_2B$
   (b) $\Delta H < 0$ (because ΔS is negative, ΔH must also be negative in order for ΔG to be negative)
      $\Delta S < 0$ (the mixture becomes less random going from reactants (3 molecules) to products (2 molecules))
      $\Delta G < 0$ (the reaction is spontaneous)

16.26 (a) $\Delta H° > 0$ (reaction involves bond breaking - endothermic)
      $\Delta S° > 0$ (2 A's have more randomness than A_2)
   (b) $\Delta S°$ is for the complete conversion of 1 mole of A_2 in its standard state to 2 moles of A in its standard state.
   (c) There is not enough information to say anything about the sign of $\Delta G°$. $\Delta G°$ decreases (becomes less positive or more negative) as the temperature increases.
   (d) K_p increases as the temperature increases. As the temperature increases there will be more A and less A_2.
   (e) $\Delta G = 0$ at equilibrium.

16.28 $\Delta G° = -RT \ln K$ where $K = \dfrac{[X]}{[A]}$ or $\dfrac{[Y]}{[A]}$ or $\dfrac{[Z]}{[A]}$

   Let the number of molecules be proportional to the concentration.

(1) $K = 1$, $\ln K = 0$, and $\Delta G° = 0$.
(2) $K > 1$, $\ln K$ is positive, and $\Delta G°$ is negative.
(3) $K < 1$, $\ln K$ is negative, and $\Delta G°$ is positive.

Section Problems
Spontaneous Processes (Section 16.1)

16.30 (a) and (d) nonspontaneous; (b) and (c) spontaneous

16.32 (b) and (d) spontaneous (because of the large positive K_p's)

Entropy (Sections 16.2–16.4)

16.34 Molecular randomness is called entropy. For the following reaction, the entropy increases: $H_2O(s) \rightarrow H_2O(l)$ at 25 °C.

16.36 (a) + (solid → gas)
  (b) − (liquid → solid)
  (c) − (aqueous ions → solid)
  (d) + ($CO_2(aq) \rightarrow CO_2(g)$)

16.38 (a) − (liquid → solid)
  (b) − (decrease in number of O_2 molecules)
  (c) + (gas has more randomness in larger volume)
  (d) − (aqueous ions → solid)

16.40 $S = k \ln W$, $k = 1.38 \times 10^{-23}$ J/K
  (a) $S = (1.38 \times 10^{-23}$ J/K$) \ln (4^{12}) = 2.30 \times 10^{-22}$ J/K
  (b) $S = (1.38 \times 10^{-23}$ J/K$) \ln (4^{120}) = 2.30 \times 10^{-21}$ J/K
  (c) $S = (1.38 \times 10^{-23}$ J/K$) \ln (4^{6.02 \times 10^{23}}) = (1.38 \times 10^{-23}$ J/K$)(6.022 \times 10^{23})\ln 4 = 11.5$ J/K
  If all C–D bonds point in the same direction, $S = 0$.

16.42 $S = k \ln W$, $k = 1.38 \times 10^{-23}$ J/K
  $W = 1000^{100}$; $S = (1.38 \times 10^{-23}$ J/K$) \ln (1000^{100}) = 9.53 \times 10^{-21}$ J/K

16.44 $S = k \ln W$
  $S_i = k \ln W_i = k \ln (1.00 \times 10^6)^{1000}$ and $S_f = k \ln W_f = k \ln (1.00 \times 10^7)^{1000}$

$$\frac{S_f}{S_i} = \frac{k \ln (1.00 \times 10^7)^{1000}}{k \ln (1.00 \times 10^6)^{1000}} = \frac{k (1000) \ln (1.00 \times 10^7)}{k (1000) \ln (1.00 \times 10^6)} = \frac{\ln(1.00 \times 10^7)}{\ln(1.00 \times 10^6)} = 1.17$$

and

  $S_i = k \ln W_i = k \ln (1.00 \times 10^{16})^{1000}$ and $S_f = k \ln W_f = k \ln (1.00 \times 10^{17})^{1000}$

$$\frac{S_f}{S_i} = \frac{k \ln (1.00 \times 10^{17})^{1000}}{k \ln (1.00 \times 10^{16})^{1000}} = \frac{k (1000) \ln (1.00 \times 10^{17})}{k (1000) \ln (1.00 \times 10^{16})} = \frac{\ln(1.00 \times 10^{17})}{\ln(1.00 \times 10^{16})} = 1.06$$

16.46 (a) H_2 at 25 °C in 50 L (larger volume)

(b) O_2 at 25 °C, 1 atm (larger volume)

(c) H_2 at 100 °C, 1 atm (larger volume and higher T)

(d) CO_2 at 100 °C, 0.1 atm (larger volume and higher T)

16.48 $\Delta S = nR \ln\left(\dfrac{V_f}{V_i}\right) = (0.050 \text{ mol})(8.314 \text{ J/K} \cdot \text{mol}) \ln\left(\dfrac{3.5 \text{ L}}{2.5 \text{ L}}\right) = 0.14 \text{ J/K}$

Standard Molar Entropies and Standard Entropies of Reaction (Section 16.5)

16.50 (a) $C_2H_6(g)$; more atoms/molecule

(b) $CO_2(g)$; more atoms/molecule

(c) $I_2(g)$; gas has more randomness than the solid

(d) $CH_3OH(g)$; gas has more randomness than the liquid

16.52 $2 \text{ CO}(g) + O_2(g) \rightarrow 2 \text{ CO}_2(g)$

$\Delta S° = [2 \text{ S}°(CO_2)] - [2 \text{ S}°(CO) + \text{S}°(O_2)]$

$\Delta S° = [(2 \text{ mol})(213.6 \text{ J/(K} \cdot \text{mol))}]$

$\qquad\qquad - [(2 \text{ mol})(197.6 \text{ J/(K} \cdot \text{mol})) + (1 \text{ mol})(205.0 \text{ J/(K} \cdot \text{mol}))] = -173.0 \text{ J/K}$

16.54 (a) $2 \text{ H}_2O_2(l) \rightarrow 2 \text{ H}_2O(l) + O_2(g)$

$\Delta S° = [2 \text{ S}°(H_2O(l)) + \text{S}°(O_2)] - 2 \text{ S}°(H_2O_2)$

$\Delta S° = [(2 \text{ mol})(69.9 \text{ J/(K} \cdot \text{mol})) + (1 \text{ mol})(205.0 \text{ J/(K} \cdot \text{mol}))]$

$\qquad\qquad - (2 \text{ mol})(110 \text{ J/(K} \cdot \text{mol})) = +125 \text{ J/K}$ (+, because moles of gas increase)

(b) $2 \text{ Na}(s) + Cl_2(g) \rightarrow 2 \text{ NaCl}(s)$

$\Delta S° = 2 \text{ S}°(NaCl) - [2 \text{ S}°(Na) + \text{S}°(Cl_2)]$

$\Delta S° = (2 \text{ mol})(72.1 \text{ J/(K} \cdot \text{mol})) - [(2 \text{ mol})(51.2 \text{ J/(K} \cdot \text{mol})) + (1 \text{ mol})(223.0 \text{ J/(K} \cdot \text{mol}))]$

$\Delta S° = -181.2 \text{ J/K}$ (−, because moles of gas decrease)

(c) $2 \text{ O}_3(g) \rightarrow 3 \text{ O}_2(g)$

$\Delta S° = 3 \text{ S}°(O_2) - 2 \text{ S}°(O_3)$

$\Delta S° = (3 \text{ mol})(205.0 \text{ J/(K} \cdot \text{mol})) - (2 \text{ mol})(238.8 \text{ J/(K} \cdot \text{mol}))$

$\Delta S° = +137.4 \text{ J/K}$ (+, because moles of gas increase)

(d) $4 \text{ Al}(s) + 3 \text{ O}_2(g) \rightarrow 2 \text{ Al}_2O_3(s)$

$\Delta S° = 2 \text{ S}°(Al_2O_3) - [4 \text{ S}°(Al) + 3 \text{ S}°(O_2)]$

$\Delta S° = (2 \text{ mol})(50.9 \text{ J/(K} \cdot \text{mol})) - [(4 \text{ mol})(28.3 \text{ J/(K} \cdot \text{mol})) + (3 \text{ mol})(205.0 \text{ J/(K} \cdot \text{mol}))]$

$\Delta S° = -626.4 \text{ J/K}$ (−, because moles of gas decrease)

Entropy and the Second Law of Thermodynamics (Section 16.6)

16.56 In any spontaneous process, the total entropy of a system and its surroundings always increases.

16.58 $\Delta S_{surr} = \dfrac{-\Delta H}{T}$; the temperature (T) is always positive.

(a) For an exothermic reaction, ΔH is negative and ΔS_{surr} is positive.

(b) For an endothermic reaction, ΔH is positive and ΔS_{surr} is negative.

16.60 HgO(s) + Zn(s) → ZnO(s) + Hg(l)

(a) $\Delta S_{surr} = \dfrac{-\Delta H^\circ}{T} = \dfrac{-(-259.7 \times 10^3 \text{ J})}{298 \text{ K}} = +871.5 \text{ J/K}$

$\Delta S_{total} = \Delta S^\circ + \Delta S_{surr} = +7.8 \text{ J/K} + 871.5 \text{ J/K} = +879.3 \text{ J/K}$

The reaction is spontaneous because ΔS_{total} is > 0.

(b) Because $\Delta S^\circ > 0$ and $\Delta H^\circ < 0$, there is no temperature at which the reaction is not spontaneous.

16.62 3 O$_2$(g) → 2 O$_3$(g)

$\Delta H^\circ = 2 \Delta H^\circ_f(O_3) = (2 \text{ mol})(143 \text{ kJ/mol}) = 286 \text{ kJ} = 286 \times 10^3 \text{ J}$

$\Delta S^\circ = [2 \text{ S}^\circ(O_3) - 3 \text{ S}^\circ(O_2)]$

$\Delta S^\circ = [(2 \text{ mol})(238.8 \text{ J/(K} \cdot \text{mol})) - (3 \text{ mol})(205.0 \text{ J/(K} \cdot \text{mol}))]$

$\Delta S^\circ = -137.4 \text{ J/K}$

$\Delta S_{total} = \Delta S^\circ + \Delta S_{surr} = \Delta S^\circ + \dfrac{-\Delta H^\circ}{T} = -137.4 \text{ J/K} + \dfrac{-(286 \times 10^3 \text{ J})}{298 \text{ K}} = -1097 \text{ J/K}$

Because $\Delta S_{total} < 0$, the reaction is not spontaneous under standard-state conditions at 25°C.

16.64 (a) 2 HgO(s) → 2 Hg(l) + O$_2$(g)

$\Delta H^\circ = 0 - 2 \Delta H^\circ_f(HgO)$

$\Delta H^\circ = -(2 \text{ mol})(-90.8 \text{ kJ/mol}) = +181.6 \text{ kJ} = +181.6 \times 10^3 \text{ J}$

$\Delta S^\circ = [2 \text{ S}^\circ(Hg) + S^\circ(O_2)] - 2 \text{ S}^\circ(HgO)$

$\Delta S^\circ = [(2 \text{ mol})(76.0 \text{ J/(K} \cdot \text{mol})) + (1 \text{ mol})(205.0 \text{ J/(K} \cdot \text{mol}))] - (2 \text{ mol})(70.3 \text{ J/(K} \cdot \text{mol}))$

$\Delta S^\circ = \Delta S_{sys} = +216.4 \text{ J/K}$

$\Delta S_{surr} = \dfrac{-\Delta H^\circ}{T} = \dfrac{-(181.6 \times 10^3 \text{ J})}{298 \text{ K}} = -609.4 \text{ J/K}$

$\Delta S_{total} = \Delta S^\circ + \Delta S_{surr} = 216.4 \text{ J/K} + (-609.4 \text{ J/K}) = -393.0 \text{ J/K}$

Because $\Delta S_{total} < 0$, the reaction is not spontaneous under standard-state conditions at 25°C.

(b) $\Delta S_{total} = \Delta S^\circ + \Delta S_{surr} = \Delta S^\circ + \dfrac{-\Delta H^\circ}{T}$

To find the temperature where the reaction becomes spontaneous, set $\Delta S_{total} = 0$ and solve for T.

$0 = \Delta S^\circ + \dfrac{-\Delta H^\circ}{T} = +216.4 \text{ J/K} + \dfrac{-(181.6 \times 10^3 \text{ J})}{T}$

$T = \dfrac{-(181.6 \times 10^3 \text{ J})}{-216.4 \text{ J/K}} = 839.2 \text{ K}$

16.66 (a) $\Delta S_{surr} = \dfrac{-\Delta H_{vap}}{T} = \dfrac{-30,700 \text{ J/mol}}{343 \text{ K}} = -89.5 \text{ J/(K} \cdot \text{mol})$

$\Delta S_{total} = \Delta S_{vap} + \Delta S_{surr} = 87.0 \text{ J/(K} \cdot \text{mol}) + (-89.5 \text{ J/(K} \cdot \text{mol})) = -2.5 \text{ J/(K} \cdot \text{mol})$

(b) $\Delta S_{surr} = \dfrac{-\Delta H_{vap}}{T} = \dfrac{-30,700 \text{ J/mol}}{353 \text{ K}} = -87.0 \text{ J/(K} \cdot \text{mol})$

$\Delta S_{total} = \Delta S_{vap} + \Delta S_{surr} = 87.0 \text{ J/(K} \cdot \text{mol}) + (-87.0 \text{ J/(K} \cdot \text{mol})) = 0$

(c) $\Delta S_{surr} = \dfrac{-\Delta H_{vap}}{T} = \dfrac{-30,700 \text{ J/mol}}{363 \text{ K}} = -84.6 \text{ J/(K} \cdot \text{mol)}$

$\Delta S_{total} = \Delta S_{vap} + \Delta S_{surr} = 87.0 \text{ J/(K} \cdot \text{mol)} + (-84.6 \text{ J/(K} \cdot \text{mol)}) = +2.4 \text{ J/(K} \cdot \text{mol)}$

Benzene does not boil at 70 °C (343 K) because ΔS_{total} is negative.

The normal boiling point for benzene is 80 °C (353 K), where $\Delta S_{total} = 0$.

Free Energy (Section 16.7)

16.68

ΔH	ΔS	$\Delta G = \Delta H - T\Delta S$	Reaction Spontaneity
–	+	–	Spontaneous at all temperatures
–	–	– or +	Spontaneous at low temperatures where $\|\Delta H\| > \|T\Delta S\|$ Nonspontaneous at high temperatures where $\|\Delta H\| < \|T\Delta S\|$
+	–	+	Nonspontaneous at all temperatures
+	+	– or +	Spontaneous at high temperatures where $T\Delta S > \Delta H$ Nonspontaneous at low temperature where $T\Delta S < \Delta H$

16.70 (a) 0 °C   (temperature is below mp); $\Delta H > 0$, $\Delta S > 0$, $\Delta G > 0$

(b) 15 °C (temperature is above mp); $\Delta H > 0$, $\Delta S > 0$, $\Delta G < 0$

16.72 $\Delta H_{vap} = 30.7 \text{ kJ/mol}$

$\Delta S_{vap} = 87.0 \text{ J/(K} \cdot \text{mol)} = 87.0 \times 10^{-3} \text{ kJ/(K} \cdot \text{mol)}$

$\Delta G_{vap} = \Delta H_{vap} - T\Delta S_{vap}$

(a) $\Delta G_{vap} = 30.7 \text{ kJ/mol} - (343 \text{ K})(87.0 \times 10^{-3} \text{ kJ/(K} \cdot \text{mol)}) = +0.9 \text{ kJ/mol}$

At 70 °C (343 K), benzene does not boil because ΔG_{vap} is positive.

(b) $\Delta G_{vap} = 30.7 \text{ kJ/mol} - (353 \text{ K})(87.0 \times 10^{-3} \text{ kJ/(K} \cdot \text{mol)}) = 0$

80 °C (353 K) is the boiling point for benzene because $\Delta G_{vap} = 0$

(c) $\Delta G_{vap} = 30.7 \text{ kJ/mol} - (363 \text{ K})(87.0 \times 10^{-3} \text{ kJ/(K} \cdot \text{mol)}) = -0.9 \text{ kJ/mol}$

At 90 °C (363 K), benzene boils because ΔG_{vap} is negative.

16.74 At the melting point (phase change), $\Delta G_{fusion} = 0$.

$\Delta G_{fusion} = \Delta H_{fusion} - T\Delta S_{fusion}$

$0 = \Delta H_{fusion} - T\Delta S_{fusion}; \quad T = \dfrac{\Delta H_{fusion}}{\Delta S_{fusion}} = \dfrac{18.02 \text{ kJ/mol}}{45.56 \times 10^{-3} \text{ kJ/(K} \cdot \text{mol)}} = 395.5 \text{ K} = 122.4 \text{ °C}$

Standard Free-Energy Changes and Standard Free Energies of Formation (Sections 16.8 and 16.9)

16.76 (a) $\Delta G°$ is the change in free energy that occurs when reactants in their standard states are converted to products in their standard states.

(b) $\Delta G°_f$ is the free-energy change for formation of one mole of a substance in its standard state from the most stable form of the constituent elements in their standard states.

16.78　(a) $N_2(g) + 2 O_2(g) \rightarrow 2 NO_2(g)$

$\Delta H^\circ = 2 \Delta H^\circ_f(NO_2) = (2\ mol)(33.2\ kJ/mol) = 66.4\ kJ$

$\Delta S^\circ = 2 S^\circ(NO_2) - [S^\circ(N_2) + 2 S^\circ(O_2)]$

$\Delta S^\circ = (2\ mol)(240.0\ J/(K \cdot mol)) - [(1\ mol)(191.5\ J/(K \cdot mol)) + (2\ mol)(205.0\ J/(K \cdot mol))]$

$\Delta S^\circ = -121.5\ J/K = -121.5 \times 10^{-3}\ kJ/K$

$\Delta G^\circ = \Delta H^\circ - T\Delta S^\circ = 66.4\ kJ - (298\ K)(-121.5 \times 10^{-3}\ kJ/K) = +102.6\ kJ$

Because ΔG° is positive, the reaction is nonspontaneous under standard-state conditions at 25 °C.

(b) $2 KClO_3(s) \rightarrow 2 KCl(s) + 3 O_2(g)$

$\Delta H^\circ = 2 \Delta H^\circ_f(KCl) - 2 \Delta H^\circ_f(KClO_3)$

$\Delta H^\circ = (2\ mol)(-436.5\ kJ/mol) - (2\ mol)(-397.7\ kJ/mol) = -77.6\ kJ$

$\Delta S^\circ = [2 S^\circ(KCl) + 3 S^\circ(O_2)] - 2 S^\circ(KClO_3)$

$\Delta S^\circ = [(2\ mol)(82.6\ J/(K \cdot mol)) + (3\ mol)(205.0\ J/(K \cdot mol))] - (2\ mol)(143.1\ J/(K \cdot mol))$

$\Delta S^\circ = 494.0\ J/K = 494.0 \times 10^{-3}\ kJ/K$

$\Delta G^\circ = \Delta H^\circ - T\Delta S^\circ = -77.6\ kJ - (298\ K)(494.0 \times 10^{-3}\ kJ/K) = -224.8\ kJ$

Because ΔG° is negative, the reaction is spontaneous under standard-state conditions at 25 °C.

(c) $CH_3CH_2OH(l) + O_2(g) \rightarrow CH_3CO_2H(l) + H_2O(l)$

$\Delta H^\circ = [\Delta H^\circ_f(CH_3CO_2H) + \Delta H^\circ_f(H_2O)] - \Delta H^\circ_f(CH_3CH_2OH)$

$\Delta H^\circ = [(1\ mol)(-484.5\ kJ/mol) + (1\ mol)(-285.8\ kJ/mol)] - (1\ mol)(-277.7\ kJ/mol) = -492.6\ kJ$

$\Delta S^\circ = [S^\circ(CH_3CO_2H) + S^\circ(H_2O)] - [S^\circ(CH_3CH_2OH) + S^\circ(O_2)]$

$\Delta S^\circ = [(1\ mol)(160\ J/(K \cdot mol)) + (1\ mol)(69.9\ J/(K \cdot mol))]$

$\qquad - [(1\ mol)(161\ J/(K \cdot mol)) + (1\ mol)(205.0\ J/(K \cdot mol))]$

$\Delta S^\circ = -136.1\ J/K = -136.1 \times 10^{-3}\ kJ/K$

$\Delta G^\circ = \Delta H^\circ - T\Delta S^\circ = -492.6\ kJ - (298\ K)(-136.1 \times 10^{-3}\ kJ/K) = -452.0\ kJ$

Because ΔG° is negative, the reaction is spontaneous under standard-state conditions at 25 °C.

16.80　(a) $N_2(g) + 2 O_2(g) \rightarrow 2 NO_2(g)$

$\Delta G^\circ = 2 \Delta G^\circ_f(NO_2) = (2\ mol)(51.3\ kJ/mol) = +102.6\ kJ$

(b) $2 KClO_3(s) \rightarrow 2 KCl(s) + 3 O_2(g)$

$\Delta G^\circ = 2 \Delta G^\circ_f(KCl) - 2 \Delta G^\circ_f(KClO_3)$

$\Delta G^\circ = (2\ mol)(-408.5\ kJ/mol) - (2\ mol)(-296.3\ kJ/mol) = -224.4\ kJ$

(c) $CH_3CH_2OH(l) + O_2(g) \rightarrow CH_3CO_2H(l) + H_2O(l)$

$\Delta G^\circ = [\Delta G^\circ_f(CH_3CO_2H) + \Delta G^\circ_f(H_2O)] - \Delta G^\circ_f(CH_3CH_2OH)$

$\Delta G^\circ = [(1\ mol)(-390\ kJ/mol) + (1\ mol)(-237.2\ kJ/mol)] - (1\ mol)(-174.9\ kJ/mol) = -452\ kJ$

16.82　A compound is thermodynamically stable with respect to its constituent elements at 25 °C if ΔG°_f is negative.

	ΔG°_f (kJ/mol)	Stable
(a) $BaCO_3(s)$	−1134.4	yes
(b) $HBr(g)$	−53.4	yes
(c) $N_2O(g)$	+104.2	no
(d) $C_2H_4(g)$	+68.1	no

16.84 $C_2H_4(g) + Cl_2(g) \rightarrow CH_2ClCH_2Cl(l)$
$\Delta G^{\circ} = \Delta G^{\circ}_f(CH_2ClCH_2Cl) - \Delta G^{\circ}_f(C_2H_4)$
$\Delta G^{\circ} = (1\ mol)(-79.6\ kJ/mol) - (1\ mol)(68.1\ kJ/mol) = -147.7\ kJ$
Because $\Delta G^{\circ} < 0$, dichloroethane can be synthesized from gaseous C_2H_4 and Cl_2, each at 25 °C and 1 atm pressure.

16.86 $CH_2=CH_2(g) + H_2O(l) \rightarrow CH_3CH_2OH(l)$
$\Delta H^{\circ} = \Delta H^{\circ}_f(CH_3CH_2OH) - [\Delta H^{\circ}_f(CH_2=CH_2) + \Delta H^{\circ}_f(H_2O)]$
$\Delta H^{\circ} = (1\ mol)(-277.7\ kJ/mol) - [(1\ mol)(52.3\ kJ/mol) + (1\ mol)(-285.8\ kJ/mol)]$
$\Delta H^{\circ} = -44.2\ kJ$
$\Delta S^{\circ} = S^{\circ}(CH_3CH_2OH) - [S^{\circ}(CH_2=CH_2) + S^{\circ}(H_2O)]$
$\Delta S^{\circ} = (1\ mol)(161\ J/(K \cdot mol)) - [(1\ mol)(219.5\ J/(K \cdot mol)) + (1\ mol)(69.9\ J/(K \cdot mol))]$
$\Delta S^{\circ} = -128\ J/(K \cdot mol) = -128 \times 10^{-3}\ kJ/(K \cdot mol)$
$\Delta G^{\circ} = \Delta H^{\circ} - T\Delta S^{\circ} = -44.2\ kJ - (298\ K)(-128 \times 10^{-3}\ kJ/K) = -6.1\ kJ$
Because ΔG° is negative, the reaction is spontaneous under standard-state conditions at 25 °C. The reaction becomes nonspontaneous at high temperatures because ΔS° is negative. To find the crossover temperature, set $\Delta G = 0$ and solve for T.

$$T = \frac{\Delta H^{\circ}}{\Delta S^{\circ}} = \frac{-44,200\ J}{-128\ J/K} = 345\ K = 72\ °C$$

The reaction becomes nonspontaneous at 72 °C.

16.88 $3\ C_2H_2(g) \rightarrow C_6H_6(l)$
$\Delta G^{\circ} = \Delta G^{\circ}_f(C_6H_6) - 3\ \Delta G^{\circ}_f(C_2H_2)$
$\Delta G^{\circ} = (1\ mol)(124.5\ kJ/mol) - (3\ mol)(209.9\ kJ/mol) = -505.2\ kJ$
Because ΔG° is negative, the reaction is possible. Look for a catalyst.
Because ΔG°_f for benzene is positive (+124.5 kJ/mol), the synthesis of benzene from graphite and gaseous H_2 at 25 °C and 1 atm pressure is not possible.

Free Energy, Composition, and Chemical Equilibrium (Sections 16.10 and 16.11)

16.90 $\Delta G = \Delta G^{\circ} + RT \ln Q$

16.92 $2\ NO(g) + Cl_2(g) \rightarrow 2\ NOCl(g)$
$\Delta G^{\circ} = 2\ \Delta G^{\circ}_f(NOCl) - 2\ \Delta G^{\circ}_f(NO)$
$\Delta G^{\circ} = (2\ mol)(66.1\ kJ/mol) - (2\ mol)(87.6\ kJ/mol) = -43.0\ kJ$

$$\Delta G = \Delta G^{\circ} + RT \ln\left[\frac{(P_{NOCl})^2}{(P_{NO})^2(P_{Cl_2})}\right]$$

$$\Delta G = (-43.0\ kJ/mol) + [8.314 \times 10^{-3}\ kJ/(K \cdot mol)](298\ K)\ln\left[\frac{(2.00)^2}{(1.00 \times 10^{-3})^2(1.00 \times 10^{-3})}\right]$$

$\Delta G = +11.8\ kJ/mol$
The reaction is spontaneous in the reverse direction.

16.94 $\Delta G = \Delta G^\circ + RT \ln \left[\dfrac{(P_{SO_3})^2}{(P_{SO_2})^2(P_{O_2})} \right]$

(a) $\Delta G = (-141.8 \text{ kJ/mol}) + [8.314 \times 10^{-3} \text{ kJ/(K} \cdot \text{mol)}](298 \text{ K}) \ln \left[\dfrac{(1.0)^2}{(100)^2(100)} \right] = -176.0 \text{ kJ/mol}$

(b) $\Delta G = (-141.8 \text{ kJ/mol}) + [8.314 \times 10^{-3} \text{ kJ/(K} \cdot \text{mol)}](298 \text{ K}) \ln \left[\dfrac{(10)^2}{(2.0)^2(1.0)} \right] = -133.8 \text{ kJ/mol}$

(c) $Q = 1$, $\ln Q = 0$, $\Delta G = \Delta G^\circ = -141.8 \text{ kJ/mol}$

16.96 $\Delta G^\circ = -RT \ln K$
   (a) If $K > 1$, ΔG° is negative.   (b) If $K = 1$, $\Delta G^\circ = 0$.   (c) If $K < 1$, ΔG° is positive.

16.98 $\Delta G^\circ = -RT \ln K_p = -141.8 \text{ kJ}$

$\ln K_p = \dfrac{-\Delta G^\circ}{RT} = \dfrac{-(-141.8 \text{ kJ/mol})}{[8.314 \times 10^{-3} \text{ kJ/(K} \cdot \text{mol)}](298 \text{ K})} = 57.23$

$K_p = e^{57.23} = 7.2 \times 10^{24}$

16.100 $C_2H_5OH(l) \rightleftharpoons C_2H_5OH(g)$
   $\Delta G^\circ = \Delta G^\circ_f(C_2H_5OH(g)) - \Delta G^\circ_f(C_2H_5OH(l))$
   $\Delta G^\circ = (1 \text{ mol})(-167.9 \text{ kJ/mol}) - (1 \text{ mol})(-174.9 \text{ kJ/mol}) = +7.0 \text{ kJ}$
   $\Delta G^\circ = -RT \ln K$

$\ln K = \dfrac{-\Delta G^\circ}{RT} = \dfrac{-(7.0 \text{ kJ/mol})}{[8.314 \times 10^{-3} \text{ kJ/(K} \cdot \text{mol)}](298 \text{ K})} = -2.83$

$K = e^{-2.83} = 0.059$; $K = K_p = P_{C_2H_5OH} = 0.059 \text{ atm}$

16.102 $Br_2(l) \rightleftharpoons Br_2(g)$
   $\Delta G^\circ = \Delta G^\circ_f(Br_2(g)) = 3.14 \text{ kJ/mol}$
   $\Delta G^\circ = -RT \ln K$

$\ln K = \dfrac{-\Delta G^\circ}{RT} = \dfrac{-(3.14 \text{ kJ/mol})}{[8.314 \times 10^{-3} \text{ kJ/(K} \cdot \text{mol)}](298 \text{ K})} = -1.267$

$K = e^{-1.267} = 0.282$; $K = K_p = P_{Br_2} = 0.282 \text{ atm} = 0.28 \text{ atm}$

16.104 $2 \text{ CH}_2=\text{CH}_2(g) + O_2(g) \rightarrow 2 \text{ C}_2H_4O(g)$
   $\Delta G^\circ = 2 \Delta G^\circ_f(C_2H_4O) - 2 \Delta G^\circ_f(CH_2=CH_2)$
   $\Delta G^\circ = (2 \text{ mol})(-13.1 \text{ kJ/mol}) - (2 \text{ mol})(68.1 \text{ kJ/mol}) = -162.4 \text{ kJ}$
   $\Delta G^\circ = -RT \ln K$

$\ln K = \dfrac{-\Delta G^\circ}{RT} = \dfrac{-(-162.4 \text{ kJ/mol})}{[8.314 \times 10^{-3} \text{ kJ/(K} \cdot \text{mol)}](298 \text{ K})} = 65.55$

$K = K_p = e^{65.55} = 2.9 \times 10^{28}$

Chapter Problems

16.106 C_3H_8, 44.10; 20 °C = 293 K

$$\text{mol } C_3H_8 = 1.32 \text{ g} \times \frac{1 \text{ mol } C_3H_8}{44.10 \text{ g}} = 0.0300 \text{ mol } C_3H_8$$

$$V = \frac{nRT}{P} = \frac{(0.0300 \text{ mol})\left(0.082\ 06 \dfrac{L \cdot atm}{K \cdot mol}\right)(293 \text{ K})}{0.100 \text{ atm}} = 7.21 \text{ L}$$

Compress 7.21 L by a factor of 5 (7.21/5) to 1.44 L

$$\Delta S = nR \ln\left(\frac{V_f}{V_i}\right) = (0.0300 \text{ mol})(8.314 \text{ J/K} \cdot \text{mol})\ln\left(\frac{1.44 \text{ L}}{7.21 \text{ L}}\right) = -0.402 \text{ J/K}$$

16.108 (a) Spontaneous does not mean fast, just possible.
(b) For a spontaneous reaction $\Delta S_{total} > 0$. ΔS_{sys} can be positive or negative.
(c) An endothermic reaction can be spontaneous if $\Delta S_{sys} > 0$.
(d) True, because the sign of ΔG changes when the direction of a reaction is reversed.

16.110

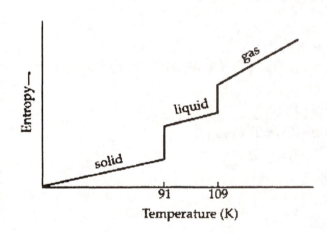

16.112 At the normal boiling point, $\Delta G = 0$.

$$\Delta G_{vap} = \Delta H_{vap} - T\Delta S_{vap}; \qquad T = \frac{\Delta H_{vap}}{\Delta S_{vap}} = \frac{38,600 \text{ J}}{110 \text{ J/K}} = 351 \text{ K} = 78 \text{ °C}$$

16.114 $\Delta G = \Delta H - T\Delta S$
(a) ΔH must be positive (endothermic) and greater than $T\Delta S$ in order for ΔG to be positive (nonspontaneous reaction).
(b) Set $\Delta G = 0$ and solve for ΔH.
$\Delta G = 0 = \Delta H - T\Delta S = \Delta H - (323 \text{ K})(104 \text{ J/K}) = \Delta H - (33592 \text{ J}) = \Delta H - (33.6 \text{ kJ})$
$\Delta H = 33.6 \text{ kJ}$
ΔH must be greater than 33.6 kJ.

16.116 (a) $2 Mg(s) + O_2(g) \rightarrow 2 MgO(s)$

$\Delta H^\circ = 2 \Delta H^\circ_f(MgO) = (2 \text{ mol})(-601.7 \text{ kJ/mol}) = -1203.4 \text{ kJ}$

$\Delta S^\circ = 2 S^\circ(MgO) - [2 S^\circ(Mg) + S^\circ(O_2)]$

$\Delta S^\circ = (2 \text{ mol})(26.9 \text{ J/(K} \cdot \text{mol)}) - [(2 \text{ mol})(32.7 \text{ J/(K} \cdot \text{mol)}) + (1 \text{ mol})(205.0 \text{ J/(K} \cdot \text{mol)})]$

$\Delta S^\circ = -216.6 \text{ J/K} = -216.6 \times 10^{-3} \text{ kJ/K}$

$\Delta G^\circ = \Delta H^\circ - T\Delta S^\circ = -1203.4 \text{ kJ} - (298 \text{ K})(-216.6 \times 10^{-3} \text{ kJ/K}) = -1138.8 \text{ kJ}$

Because ΔG° is negative, the reaction is spontaneous at 25 °C. ΔG° becomes less negative as the temperature is raised.

(b) $MgCO_3(s) \rightarrow MgO(s) + CO_2(g)$

$\Delta H^\circ = [\Delta H^\circ_f(MgO) + \Delta H^\circ_f(CO_2)] - \Delta H^\circ_f(MgCO_3)$

$\Delta H^\circ = [(1 \text{ mol})(-601.1 \text{ kJ/mol}) + (1 \text{ mol})(-393.5 \text{ kJ/mol})] - (1 \text{ mol})(-1096 \text{ kJ/mol}) = +101 \text{ kJ}$

$\Delta S^\circ = [S^\circ(MgO) + S^\circ(CO_2)] - S^\circ(MgCO_3)$

$\Delta S^\circ = [(1 \text{ mol})(26.9 \text{ J/(K} \cdot \text{mol)}) + (1 \text{ mol})(213.6 \text{ J/(K} \cdot \text{mol)})] - (1 \text{ mol})(65.7 \text{ J/(K} \cdot \text{mol)})$

$\Delta S^\circ = 174.8 \text{ J/K} = 174.8 \times 10^{-3} \text{ kJ/K}$

$\Delta G^\circ = \Delta H^\circ - T\Delta S^\circ = 101 \text{ kJ} - (298 \text{ K})(174.8 \times 10^{-3} \text{ kJ/K}) = +49 \text{ kJ}$

Because ΔG° is positive, the reaction is not spontaneous at 25 °C. ΔG° becomes less positive as the temperature is raised.

(c) $Fe_2O_3(s) + 2 Al(s) \rightarrow Al_2O_3(s) + 2 Fe(s)$

$\Delta H^\circ = \Delta H^\circ_f(Al_2O_3) - \Delta H^\circ_f(Fe_2O_3)$

$\Delta H^\circ = (1 \text{ mol})(-1676 \text{ kJ/mol}) - (1 \text{ mol})(-824.2 \text{ kJ/mol}) = -852 \text{ kJ}$

$\Delta S^\circ = [S^\circ(Al_2O_3) + 2 S^\circ(Fe)] - [S^\circ(Fe_2O_3) + 2 S^\circ(Al)]$

$\Delta S^\circ = [(1 \text{ mol})(50.9 \text{ J/(K} \cdot \text{mol)}) + (2 \text{ mol})(27.3 \text{ J/(K} \cdot \text{mol)})]$
$$ - [(1 \text{ mol})(87.4 \text{ J/(K} \cdot \text{mol)}) + (2 \text{ mol})(28.3 \text{ J/(K} \cdot \text{mol)})]$$

$\Delta S^\circ = -38.5 \text{ J/K} = -38.5 \times 10^{-3} \text{ kJ/K}$

$\Delta G^\circ = \Delta H^\circ - T\Delta S^\circ = -852 \text{ kJ} - (298 \text{ K})(-38.5 \times 10^{-3} \text{ kJ/K}) = -841 \text{ kJ}$

Because ΔG° is negative, the reaction is spontaneous at 25 °C. ΔG° becomes less negative as the temperature is raised.

(d) $2 NaHCO_3(s) \rightarrow Na_2CO_3(s) + CO_2(g) + H_2O(g)$

$\Delta H^\circ = [\Delta H^\circ_f(Na_2CO_3) + \Delta H^\circ_f(CO_2) + \Delta H^\circ_f(H_2O)] - 2 \Delta H^\circ_f(NaHCO_3)$

$\Delta H^\circ = [(1 \text{ mol})(-1130.7 \text{ kJ/mol}) + (1 \text{ mol})(-393.5 \text{ kJ/mol})$
$$ + (1 \text{ mol})(-241.8 \text{ kJ/mol})] - (2 \text{ mol})(-950.8 \text{ kJ/mol}) = +135.6 \text{ kJ}$$

$\Delta S^\circ = [S^\circ(Na_2CO_3) + S^\circ(CO_2) + S^\circ(H_2O)] - 2 S^\circ(NaHCO_3)$

$\Delta S^\circ = [(1 \text{ mol})(135.0 \text{ J/(K} \cdot \text{mol)}) + (1 \text{ mol})(213.6 \text{ J/(K} \cdot \text{mol)})$
$$ + (1 \text{ mol})(188.7 \text{ J/(K} \cdot \text{mol)})] - (2 \text{ mol})(102 \text{ J/(K} \cdot \text{mol)})$$

$\Delta S^\circ = +333 \text{ J/K} = +333 \times 10^{-3} \text{ kJ/K}$

$\Delta G^\circ = \Delta H^\circ - T\Delta S^\circ = +135.6 \text{ kJ} - (298 \text{ K})(+333 \times 10^{-3} \text{ kJ/K}) = +36.4 \text{ kJ}$

Because ΔG° is positive, the reaction is not spontaneous at 25 °C. ΔG° becomes less positive as the temperature is raised.

16.118 $NH_4HS(s) \rightarrow NH_3(g) + H_2S(g)$

$P_{total} = P_{NH_3} + P_{H_2S} = 0.658$ atm, and $P_{NH_3} = P_{H_2S} = 0.658$ atm/2 = 0.329 atm

$\Delta G = \Delta G^\circ + RT \ln[(P_{NH_3})(P_{H_2S})]$

At equilibrium, $\Delta G = 0$

$\Delta G^\circ = - RT \ln[(P_{NH_3})(P_{H_2S})]$

315

$\Delta G° = - [8.314 \times 10^{-3} \text{ kJ/(K} \cdot \text{mol)}](298 \text{ K}) \ln[(0.329)(0.329)] = +5.51 \text{ kJ/mol}$

$\Delta G° = [\Delta G°_f(NH_3) + \Delta G°_f(H_2S)] - \Delta G°_f(NH_4HS)$

$5.51 \text{ kJ} = [(1 \text{ mol})(-16.5 \text{ kJ/mol}) + (1 \text{ mol})(-33.6 \text{ kJ/mol})] - \Delta G°_f(NH_4HS)$

$\Delta G°_f(NH_4HS) = [(1 \text{ mol})(-16.5 \text{ kJ/mol}) + (1 \text{ mol})(-33.6 \text{ kJ/mol})] - 5.51 \text{ kJ}$

$\Delta G°_f(NH_4HS) = -55.6 \text{ kJ/mol}$

16.120 (a) $6 \text{ C(s)} + 3 \text{ H}_2\text{(g)} \rightarrow \text{C}_6\text{H}_6\text{(l)}$

$\Delta S°_f = S°(C_6H_6) - [6 S°(C) + 3 S°(H_2)]$

$\Delta S°_f = (1 \text{ mol})(173.4 \text{ J/(K} \cdot \text{mol))} - [(6 \text{ mol})(5.7 \text{ J/(K} \cdot \text{mol))} + (3 \text{ mol})(130.6 \text{ J/(K} \cdot \text{mol))}]$

$\Delta S°_f = -253 \text{ J/K} = -253 \text{ J/(K} \cdot \text{mol)}$

$\Delta G°_f = \Delta H°_f - T\Delta S°_f$

$\Delta S°_f = \dfrac{\Delta H°_f - \Delta G°_f}{T} = \dfrac{49.0 \text{ kJ/mol} - 124.5 \text{ kJ/mol}}{298 \text{ K}} = -0.2533 \text{ kJ/(K} \cdot \text{mol)}$

$\Delta S°_f = -253.3 \text{ J/(K} \cdot \text{mol)}$

Both calculations lead to the same value of $\Delta S°_f$.

(b) $\text{Ca(s)} + \text{S(s)} + 2 \text{ O}_2\text{(g)} \rightarrow \text{CaSO}_4\text{(s)}$

$\Delta S°_f = S°(CaSO_4) - [S°(Ca) + S°(S) + 2 S°(O_2)]$

$\Delta S°_f = (1 \text{ mol})(107 \text{ J/(K} \cdot \text{mol))}$
$\quad - [(1 \text{ mol})(41.4 \text{ J/(K} \cdot \text{mol))} + (1 \text{ mol})(31.8 \text{ J/(K} \cdot \text{mol))} + (2 \text{ mol})(205.0 \text{ J/(K} \cdot \text{mol))}]$

$\Delta S°_f = -376 \text{ J/K} = -376 \text{ J/(K} \cdot \text{mol)}$

$\Delta G°_f = \Delta H°_f - T\Delta S°_f$

$\Delta S°_f = \dfrac{\Delta H°_f - \Delta G°_f}{T} = \dfrac{-1434.1 \text{ kJ/mol} - (-1321.9 \text{ kJ/mol})}{298 \text{ K}} = -0.377 \text{ kJ/(K} \cdot \text{mol)}$

$\Delta S°_f = -377 \text{ J/(K} \cdot \text{mol)}$

Both calculations lead to the same value of $\Delta S°_f$.

(c) $2 \text{ C(s)} + 3 \text{ H}_2\text{(g)} + 1/2 \text{ O}_2\text{(g)} \rightarrow \text{C}_2\text{H}_5\text{OH(l)}$

$\Delta S°_f = S°(C_2H_5OH) - [S°(C) + S°(H_2) + 1/2 S°(O_2)]$

$\Delta S°_f = (1 \text{ mol})(161 \text{ J/(K} \cdot \text{mol))}$
$- [(2 \text{ mol})(5.7 \text{ J/(K} \cdot \text{mol))} + (3 \text{ mol})(130.6 \text{ J/(K} \cdot \text{mol))} + (0.5 \text{ mol})(205.0 \text{ J/(K} \cdot \text{mol))}]$

$\Delta S°_f = -345 \text{ J/K} = -345 \text{ J/(K} \cdot \text{mol)}$

$\Delta G°_f = \Delta H°_f - T\Delta S°_f$

$\Delta S°_f = \dfrac{\Delta H°_f - \Delta G°_f}{T} = \dfrac{-277.7 \text{ kJ/mol} - (-174.9 \text{ kJ/mol})}{298 \text{ K}} = -0.345 \text{ kJ/(K} \cdot \text{mol)}$

$\Delta S°_f = -345 \text{ J/(K} \cdot \text{mol)}$

Both calculations lead to the same value of $\Delta S°_f$.

16.122 $\Delta G° = -RT \ln K_b$

At 20 °C: $\Delta G° = -[8.314 \times 10^{-3} \text{ kJ/(K} \cdot \text{mol)}](293 \text{ K}) \ln(1.710 \times 10^{-5}) = +26.74 \text{ kJ/mol}$

At 50 °C: $\Delta G° = -[8.314 \times 10^{-3} \text{ kJ/(K} \cdot \text{mol)}](323 \text{ K}) \ln(1.892 \times 10^{-5}) = +29.20 \text{ kJ/mol}$

$\Delta G° = \Delta H° - T\Delta S°$

$26.74 = \Delta H° - 293\Delta S°$

$29.20 = \Delta H° - 323\Delta S°$ Solve these two equations simultaneously for $\Delta H°$ and $\Delta S°$.

$26.74 + 293\Delta S° = \Delta H°$
$29.20 + 323\Delta S° = \Delta H°$ Set these two equations equal to each other.

$26.74 + 293\Delta S° = 29.20 + 323\Delta S°$
$26.74 - 29.20 = 323\Delta S° - 293 \Delta S°$
$-2.46 = 30\Delta S°$
$\Delta S° = -2.46/30 = -0.0820 = -0.0820 \text{ kJ/K} = -82.0 \text{ J/K}$
$26.74 + 293\Delta S° = 26.74 + 293(-0.0820) = \Delta H° = +2.71 \text{ kJ}$

16.124   (a) $\Delta H° = [\Delta H°_f(Ag^+(aq)) + \Delta H°_f(Br^-(aq))] - \Delta H°_f(AgBr(s))$
$\Delta H° = [(1 \text{ mol})(105.6 \text{ kJ/mol}) + (1 \text{ mol})(-121.5 \text{ kJ/mol})] - (1 \text{ mol})(-100.4 \text{ kJ/mol}) = +84.5 \text{ kJ}$
$\Delta S° = [S°(Ag^+(aq)) + S°(Br^-(aq))] - S°(AgBr(s))$
$\Delta S° = [(1 \text{ mol})(72.7 \text{ J/(K} \cdot \text{mol})) + (1 \text{ mol})(82.4 \text{ J/(K} \cdot \text{mol}))]$
$$- (1 \text{ mol})(107.1 \text{ J/(K} \cdot \text{mol})) = +48.0 \text{ J}$$
$\Delta G° = \Delta H° - T\Delta S° = 84.5 \text{ kJ} - (298 \text{ K})(48.0 \times 10^{-3} \text{ kJ/K}) = +70.2 \text{ kJ}$
(b) $\Delta G° = -RT \ln K_{sp}$

$\ln K_{sp} = \dfrac{-\Delta G°}{RT} = \dfrac{-70.2 \text{ kJ/mol}}{[8.314 \times 10^{-3} \text{ kJ/(K} \cdot \text{mol})](298 \text{ K})} = -28.3$

$K_{sp} = e^{-28.3} = 5 \times 10^{-13}$
(c) $Q = [Ag^+][Br^-] = (1.00 \times 10^{-5})(1.00 \times 10^{-5}) = 1.00 \times 10^{-10}$
$\Delta G = \Delta G° + RT\ln Q$
$\Delta G = 70.2 \text{ kJ/mol} + [8.314 \times 10^{-3} \text{ kJ/(K} \cdot \text{mol})](298 \text{ K}) \ln(1.00 \times 10^{-10}) = 13.2 \text{ kJ/mol}$
A positive value of ΔG means that the forward reaction is nonspontaneous under these conditions. The reverse reaction is therefore spontaneous, which is consistent with the fact that $Q > K_{sp}$.

16.126 $Br_2(l) \rightleftharpoons Br_2(g)$
$\Delta S° = S°(Br_2(g)) - S°(Br_2(l))$
$\Delta S° = (1 \text{ mol})(245.4 \text{ J/(K} \cdot \text{mol})) - (1 \text{ mol})(152.2 \text{ J/(K} \cdot \text{mol})) = 93.2 \text{ J/K} = 93.2 \times 10^{-3} \text{ kJ/K}$
$\Delta G = \Delta H° - T\Delta S°$
At the boiling point, $\Delta G = 0$.
$0 = \Delta H° - T_{bp}\Delta S°$

$T_{bp} = \dfrac{\Delta H°}{\Delta S°}$

$\Delta H° = T_{bp} \Delta S° = (332 \text{ K})(93.2 \times 10^{-3} \text{ kJ/K}) = 30.9 \text{ kJ}$

$K_p = P_{Br_2} = \left(227 \text{ mm Hg} \times \dfrac{1 \text{ atm}}{760 \text{ mm Hg}} \right) = 0.299 \text{ atm}$

$\Delta G° = -RT \ln K_p$ and $\Delta G° = \Delta H° - T\Delta S°$ (set equations equal to each other)
$\Delta H° - T\Delta S° = -RT \ln K_p$ (rearrange)

$\ln K_p = \dfrac{-\Delta H°}{R} \dfrac{1}{T} + \dfrac{\Delta S°}{R}$ (solve for T)

$$T = \frac{\left(\dfrac{-\Delta H^\circ}{R}\right)}{\left(\ln K_p - \dfrac{\Delta S^\circ}{R}\right)} = \frac{\left(\dfrac{-30.9 \text{ kJ/mol}}{8.314 \times 10^{-3} \text{ kJ/(K} \cdot \text{mol)}}\right)}{\left(\ln(0.299) - \dfrac{93.2 \times 10^{-3} \text{ kJ/(K} \cdot \text{mol)}}{8.314 \times 10^{-3} \text{ kJ/(K} \cdot \text{mol)}}\right)} = 299 \text{ K} = 26\ ^\circ\text{C}$$

$Br_2(l)$ has a vapor pressure of 227 mm Hg at 26 °C.

16.128 $\Delta H^\circ = [2\ \Delta H^\circ_f(Cl^-(aq))] - [2\ \Delta H^\circ_f(Br^-(aq))]$

$\Delta H^\circ = [(2 \text{ mol})(-167.2 \text{ kJ/mol})] - [(2 \text{ mol})(-121.5 \text{ kJ/mol})] = -91.4 \text{ kJ}$

$\Delta S^\circ = [S^\circ(Br_2(l)) + 2\ S^\circ(Cl^-(aq))] - [2\ S^\circ(Br^-(aq)) + S^\circ(Cl_2(g))]$

$\Delta S^\circ = [(1 \text{ mol})(152.2 \text{ J/(K} \cdot \text{mol))} + (2 \text{ mol})(56.5 \text{ J/(K} \cdot \text{mol))}]$
$\qquad - [(2 \text{ mol})(82.4 \text{ J/(K} \cdot \text{mol))} + (1 \text{ mol})(223.0 \text{ J/(K} \cdot \text{mol))}] = -122.6 \text{ J/K}$

80 °C = 80 + 273 = 353 K

$\Delta G^\circ = \Delta H^\circ - T\Delta S^\circ = -91.4 \text{ kJ} - (353 \text{ K})(-122.6 \times 10^{-3} \text{ kJ/K}) = -48.1 \text{ kJ}$

$\Delta G^\circ = -RT \ln K$

$\ln K = \dfrac{-\Delta G^\circ}{RT} = \dfrac{-(-48.1 \text{ kJmol})}{[8.314 \times 10^{-3} \text{ kJ/(K} \cdot \text{mol)}](353 \text{ K})} = 16.4$

$K = e^{16.4} = 1.3 \times 10^7$

16.130   35 °C = 35 + 273 = 308 K

$\Delta G^\circ = \Delta H^\circ - T\Delta S^\circ = -352 \text{ kJ} - (308 \text{ K})(-899 \times 10^{-3} \text{ kJ/K}) = -75.1 \text{ kJ}$

$\Delta G^\circ = -RT \ln K_p$

$\ln K_p = \dfrac{-\Delta G^\circ}{RT} = \dfrac{-(-75.1 \text{ kJ/mol})}{[8.314 \times 10^{-3} \text{ kJ/(K} \cdot \text{mol)}](308 \text{ K})} = 29.33$

$K_p = e^{29.33} = 5.5 \times 10^{12}$

$K_p = \dfrac{1}{(P_{H_2O})^6} = 5.5 \times 10^{12}$

$P_{H_2O} = \sqrt[6]{\dfrac{1}{5.5 \times 10^{12}}} = 0.0075 \text{ atm}$

$P_{H_2O} = 0.0075 \text{ atm} \times \dfrac{760 \text{ mm Hg}}{1 \text{ atm}} = 5.7 \text{ mm Hg}$

16.132 $N_2O_4(g) \rightleftharpoons 2\ NO_2(g)$

$\Delta H^\circ = 2\ \Delta H^\circ_f(NO_2) - \Delta H^\circ_f(N_2O_4) = (2 \text{ mol})(33.2 \text{ kJ}) - (1 \text{ mol})(11.1 \text{ kJ}) = 55.3 \text{ kJ}$

$\Delta S^\circ = 2\ S^\circ(NO_2) - S^\circ(N_2O_4) = (2 \text{ mol})(240.0 \text{ J/(K} \cdot \text{mol))} - (1 \text{ mol})(304.3 \text{ J/(K} \cdot \text{mol))}$

$\Delta S^\circ = 175.7 \text{ J/K} = 175.7 \times 10^{-3} \text{ kJ/K}$

$\Delta G^\circ = \Delta H^\circ - T\Delta S^\circ$ and $\Delta G^\circ = -RT \ln K_p$

Set these two equations equal to each other and solve for T.

$\Delta H^\circ - T\Delta S^\circ = -RT \ln K_p$

$\Delta H^\circ = T\Delta S^\circ - RT \ln K_p = T(\Delta S^\circ - R \ln K_p)$

$T = \dfrac{\Delta H^\circ}{\Delta S^\circ - R \ln K_p}$

(a) $P_{N_2O_4} + P_{NO_2} = 1.00$ atm and $P_{NO_2} = 2 P_{N_2O_4}$

$P_{N_2O_4} + 2 P_{N_2O_4} = 3 P_{N_2O_4} = 1.00$ atm

$P_{N_2O_4} = 1.00$ atm$/3 = 0.333$ atm

$P_{NO_2} = 1.00$ atm $- P_{N_2O_4} = 1.00 - 0.333 = 0.667$ atm

$$K_p = \frac{(P_{NO_2})^2}{P_{N_2O_4}} = \frac{(0.667)^2}{(0.333)} = 1.34$$

$$T = \frac{\Delta H^\circ}{\Delta S^\circ - R \ln K_p}$$

$$T = \frac{55.3 \text{ kJ/mol}}{[175.7 \times 10^{-3} \text{ kJ/(K} \cdot \text{mol)}] - [8.314 \times 10^{-3} \text{ kJ/(K} \cdot \text{mol)}] \ln(1.34)} = 319 \text{ K}$$

$T = 319$ K $= 319 - 273 = 46$ °C

(b) $P_{N_2O_4} + P_{NO_2} = 1.00$ atm and $P_{NO_2} = P_{N_2O_4}$ so $P_{NO_2} = P_{N_2O_4} = 0.50$ atm

$$K_p = \frac{(P_{NO_2})^2}{P_{N_2O_4}} = \frac{(0.500)^2}{(0.500)} = 0.500$$

$$T = \frac{\Delta H^\circ}{\Delta S^\circ - R \ln K_p}$$

$$T = \frac{55.3 \text{ kJ/mol}}{[175.7 \times 10^{-3} \text{ kJ/(K} \cdot \text{mol)}] - [8.314 \times 10^{-3} \text{ kJ/(K} \cdot \text{mol)}] \ln(0.500)} = 305 \text{ K}$$

$T = 305$ K $= 305 - 273 = 32$ °C

Multiconcept Problems

16.134 $N_2(g) + 3 H_2(g) \rightleftharpoons 2 NH_3(g)$

$\Delta H^\circ = 2 \Delta H^\circ_f(NH_3) - [\Delta H^\circ_f(N_2) + 3 \Delta H^\circ_f(H_2)] = (2 \text{ mol})(-46.1 \text{ kJ}) - [0] = -92.2 \text{ kJ}$

$\Delta S^\circ = 2 S^\circ(NH_3) - [S^\circ(N_2) + 3 S^\circ(H_2)]$

$\Delta S^\circ = (2 \text{ mol})(192.3 \text{ J/(K} \cdot \text{mol)})$

$\qquad - [(1 \text{ mol})(191.5 \text{ J/(K} \cdot \text{mol)}) + (3 \text{ mol})(130.6 \text{ J/(K} \cdot \text{mol)})] = -198.7 \text{ J/K}$

$\Delta G^\circ = \Delta H^\circ - T \Delta S^\circ = -92.2 \text{ kJ} - (673 \text{ K})(-198.7 \times 10^{-3} \text{ kJ/K}) = 41.5 \text{ kJ}$

$\Delta G^\circ = -RT \ln K_p$

$$\ln K_p = \frac{-\Delta G^\circ}{RT} = \frac{-41.5 \text{ kJ/mol}}{[8.314 \times 10^{-3} \text{ kJ/(K} \cdot \text{mol)}](673 \text{ K})} = -7.42$$

$K_p = e^{-7.42} = 6.0 \times 10^{-4}$

Because $K_p = K_c(RT)^{\Delta n}$, $K_c = K_p(RT)^{-\Delta n}$

$K_c = K_p(RT)^2 = (6.0 \times 10^{-4})[(0.082\ 06)(673)]^2 = 1.83$

N_2, 28.01; H_2, 2.016

Initial concentrations:

$$[N_2] = \frac{(14.0\ g)\left(\dfrac{1\ mol}{28.01\ g}\right)}{5.00\ L} = 0.100\ M \quad \text{and} \quad [H_2] = \frac{(3.024\ g)\left(\dfrac{1\ mol}{2.016\ g}\right)}{5.00\ L} = 0.300\ M$$

$$N_2(g) \ + \ 3\ H_2(g) \ \rightleftharpoons \ 2\ NH_3(g)$$

initial (M)	0.100	0.300	0
change (M)	–x	–3x	+2x
equil (M)	0.100 – x	0.300 – 3x	2x

$$K_c = \frac{[NH_3]^2}{[N_2][H_2]^3} = \frac{(2x)^2}{(0.100-x)(0.300-3x)^3} = \frac{4x^2}{27(0.100-x)^4} = 1.83$$

$$\left(\frac{x}{(0.100-x)^2}\right)^2 = \frac{(27)(1.83)}{4} = 12.35; \qquad \frac{x}{(0.100-x)^2} = \sqrt{12.35} = 3.514$$

$$3.514x^2 - 1.703x + 0.03514 = 0$$

Use the quadratic formula to solve for x.

$$x = \frac{-(-1.703) \pm \sqrt{(-1.703)^2 - (4)(3.514)(0.03514)}}{2(3.514)} = \frac{1.703 \pm 1.551}{7.028}$$

x = 0.463 and 0.0216

Of the two solutions for x, only 0.0216 has physical meaning because 0.463 would lead to negative concentrations of N_2 and H_2.

$[N_2] = 0.100 - x = 0.100 - 0.0216 = 0.078\ M$

$[H_2] = 0.300 - 3x = 0.300 - 3(0.0216) = 0.235\ M$

$[NH_3] = 2x = 2(0.0216) = 0.043\ M$

16.136 $Pb(s) \ + \ PbO_2(s) \ + \ 2\ H^+(aq) \ + \ 2\ HSO_4^-(aq) \ \rightarrow \ 2\ PbSO_4(s) \ + \ 2\ H_2O(l)$

(a) $\Delta G° = [2\ \Delta G°_f(PbSO_4) + 2\ \Delta G°_f(H_2O)] - [\Delta G°_f(PbO_2) + 2\ \Delta G°_f(HSO_4^-)]$

$\Delta G° = (2\ mol)(-813.2\ kJ/mol) + (2\ mol)(-237.2\ kJ/mol)]$
$\qquad\qquad - [(1\ mol)(-217.4\ kJ/mol) + (2\ mol)(-756.0\ kJ/mol)] = -371.4\ kJ$

(b) $°C = 5/9(°F - 32) = 5/9(10 - 32) = -12.2\ °C; \quad -12.2\ °C = 261\ K$

$\Delta H° = [2\ \Delta H°_f(PbSO_4) + 2\ \Delta H°_f(H_2O)] - [\Delta H°_f(PbO_2) + 2\ \Delta H°_f(HSO_4^-)]$

$\Delta H° = [(2\ mol)(-919.9\ kJ/mol) + (2\ mol)(-285.8\ kJ/mol)]$
$\qquad\qquad - [(1\ mol)(-277\ kJ/mol) + (2\ mol)(-887.3\ kJ/mol)] = -359.8\ kJ$

$\Delta S° = [2\ S°(PbSO_4) + 2\ S°(H_2O)] - [S°(Pb) + S°(PbO_2) + 2\ S°(H^+) + 2\ S°(HSO_4^-)]$

$\Delta S° = [(2\ mol)(148.6\ J/(K \cdot mol)) + (2\ mol)(69.9\ J/(K \cdot mol))]$
$\qquad\qquad - [(1\ mol)(64.8\ J/(K \cdot mol)) + (1\ mol)(68.6\ J/(K \cdot mol))$
$\qquad\qquad\qquad + (2\ mol)(132\ J/(K \cdot mol))] = 39.6\ J/K = 39.6 \times 10^{-3}\ kJ/K$

$\Delta G° = \Delta H° - T\Delta S° = -359.8\ kJ - (261\ K)(39.6 \times 10^{-3}\ kJ/K) = -370.1\ kJ$ at 261 K

$$HSO_4^-(aq) \ + \ H_2O(l) \ \rightleftharpoons \ H_3O^+(aq) \ + \ SO_4^{2-}(aq)$$

initial (M)	0.100		0.100	0
change (M)	–x		+x	+x
equil (M)	0.100 – x		0.100 + x	x

$$K_{a2} = \frac{[H_3O^+][SO_4^{2-}]}{[HSO_4^-]} = 1.2 \times 10^{-2} = \frac{(0.100 + x)x}{0.100 - x}$$

$x^2 + 0.112x - (1.2 \times 10^{-3}) = 0$

Use the quadratic formula to solve for x.

$$x = \frac{-(0.112) \pm \sqrt{(0.112)^2 - (4)(1)(-1.2 \times 10^{-3})}}{2(1)} = \frac{-0.112 \pm 0.132}{2}$$

$x = -0.122$ and 0.010

Of the two solutions for x, only 0.010 has physical meaning because −0.122 would lead to negative concentrations of H_3O^+ and SO_4^{2-}.

$[H^+] = 0.100 + x = 0.100 + 0.010 = 0.110$ M

$[HSO_4^-] = 0.100 - x = 0.100 - 0.010 = 0.090$ M

$$\Delta G = \Delta G° + RT \ln \frac{1}{[H^+]^2[HSO_4^-]^2}$$

$$\Delta G = (-370.1 \text{ kJ/mol}) + [8.314 \times 10^{-3} \text{ kJ/(K} \cdot \text{mol)}](261 \text{ K}) \ln \frac{1}{(0.110)^2(0.090)^2}$$

$\Delta G = -350.1$ kJ/mol

16.138 $PV = nRT$

$$n_{NH_3} = \frac{PV}{RT} = \frac{\left(744 \text{ mm Hg} \times \frac{1.00 \text{ atm}}{760 \text{ mm Hg}}\right)(1.00 \text{ L})}{\left(0.082\ 06 \frac{L \cdot atm}{K \cdot mol}\right)(298.1 \text{ K})} = 0.0400 \text{ mol NH}_3$$

500.0 mL = 0.5000 L

$[NH_3]$ = 0.0400 mol/0.5000 L = 0.0800 M

$NH_3(aq) + H_2O(l) \rightleftharpoons NH_4^+(aq) + OH^-(aq)$

$\Delta H° = [\Delta H°_f(NH_4^+) + \Delta H°_f(OH^-)] - [\Delta H°_f(NH_3) + \Delta H°_f(H_2O)]$

$\Delta H° = [(1 \text{ mol})(-132.5 \text{ kJ/mol}) + (1 \text{ mol})(-230.0 \text{ kJ/mol})]$
$\qquad - [(1 \text{ mol})(-80.3 \text{ kJ/mol}) + (1 \text{ mol})(-285.8 \text{ kJ/mol})] = +3.6 \text{ kJ}$

$\Delta S° = [S°(NH_4^+) + S°(OH^-)] - [S°(NH_3) + S°(H_2O)]$

$\Delta S° = [(1 \text{ mol})(113 \text{ J/(K} \cdot \text{mol)}) + (1 \text{ mol})(-10.8 \text{ J/(K} \cdot \text{mol)})]$
$\qquad - [(1 \text{ mol})(111 \text{ J/(K} \cdot \text{mol)}) + (1 \text{ mol})(69.9 \text{ J/(K} \cdot \text{mol)})] = -78.7 \text{ J/K}$

T = 2.0 °C = 2.0 + 273.1 = 275.1 K

$\Delta G° = \Delta H° - T\Delta S° = 3.6 \text{ kJ} - (275.1 \text{ K})(-78.7 \times 10^{-3} \text{ kJ/K}) = 25.3 \text{ kJ}$

$\Delta G° = -RT \ln K_b$

$$\ln K_b = \frac{-\Delta G°}{RT} = \frac{-25.3 \text{ kJ/mol}}{[8.314 \times 10^{-3} \text{ kJ/(K} \cdot \text{mol)}](275.1 \text{ K})} = -11.06$$

$K_b = e^{-11.06} = 1.6 \times 10^{-5}$

	$NH_3(aq)$	+ $H_2O(l)$	$\rightleftharpoons$ $NH_4^+(aq)$	+ $OH^-(aq)$
initial (M)	0.0800		0	~0
change (M)	−x		+x	+x
equil (M)	0.0800 − x		x	x

at 2 °C, $K_b = \dfrac{[NH_4^+][OH^-]}{[NH_3]} = 1.6 \times 10^{-5} = \dfrac{x^2}{0.0800 - x} \approx \dfrac{x^2}{0.0800}$

$x^2 = (1.6 \times 10^{-5})(0.0800)$

$x = [OH^-] = \sqrt{(1.6 \times 10^{-5})(0.0800)} = 1.13 \times 10^{-3}$ M

$[H_3O^+] = \dfrac{1.0 \times 10^{-14}}{1.13 \times 10^{-3}} = 8.85 \times 10^{-12}$ M

$pH = -\log[H_3O^+] = -\log(8.85 \times 10^{-12}) = 11.05$

16.140 (a) $N_2O_4(g) \rightleftharpoons 2\, NO_2(g)$

$\Delta H° = 2\,\Delta H°_f(NO_2) - \Delta H°_f(N_2O_4) = (2\text{ mol})(33.2\text{ kJ/mol}) - (1\text{ mol})(11.1\text{ kJ/mol}) = 55.3$ kJ

$\Delta S° = 2\,S°(NO_2) - S°(N_2O_4) = (2\text{ mol})(240.0\text{ J/(K} \cdot \text{mol)}) - (1\text{ mol})(304.3\text{ J/(K} \cdot \text{mol)})$

$\Delta S° = 175.7$ J/K $= 175.7 \times 10^{-3}$ kJ/K

$\Delta G° = \Delta H° - T\Delta S° = 55.3\text{ kJ} - (373\text{ K})(175.7 \times 10^{-3}\text{ kJ/K}) = -10.2$ kJ

$K_p = \dfrac{(P_{NO_2})^2}{P_{N_2O_4}}$

$\Delta G° = -RT \ln K_p$

$\ln K_p = \dfrac{-\Delta G°}{RT} = \dfrac{-(-10.2\text{ kJ/mol})}{[8.314 \times 10^{-3}\text{ kJ/(K} \cdot \text{mol)}](373\text{ K})} = 3.29$

$K_p = e^{3.29} = 27$

	$N_2O_4(g)$	$\rightleftharpoons$	$2\, NO_2(g)$
initial (atm)	1.00		1.00
change (atm)	$-x$		$+2x$
equil (atm)	$1.00 - x$		$1.00 + 2x$

$K_p = \dfrac{(P_{NO_2})^2}{P_{N_2O_4}} = 27 = \dfrac{(1.00 + 2x)^2}{(1.00 - x)}$

$4x^2 + 31x - 26 = 0$

Use the quadratic formula to solve for x.

$x = \dfrac{-(31) \pm \sqrt{(31)^2 - (4)(4)(-26)}}{2(4)} = \dfrac{-31 \pm 37.1}{8}$

$x = 0.76$ and -8.5

Of the two solutions for x, only 0.76 has physical meaning because -8.5 would lead to a negative partial pressure for NO_2.

$P_{N_2O_4} = 1.00 - x = 1.00 - 0.76 = 0.24$ atm; $P_{NO_2} = 1.00 + 2x = 1.00 + 2(0.76) = 2.52$ atm

(b) One resonance structure is shown here.

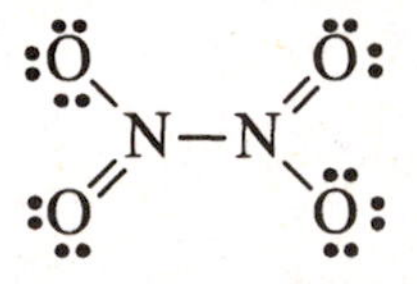

Each N is sp^2 hybridized. There is a trigonal planar geometry about each N.

17

Electrochemistry

17.1 $2 \text{ Ag}^+(aq) + \text{Ni}(s) \rightarrow 2 \text{ Ag}(s) + \text{Ni}^{2+}(aq)$

There is a Ni anode in an aqueous solution of Ni^{2+}, and a Ag cathode in an aqueous solution of Ag^+. A salt bridge connects the anode and cathode compartment. The electrodes are connected through an external circuit.

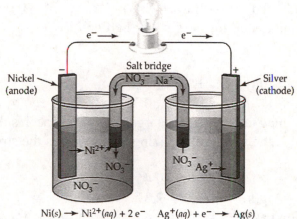

$$\text{Ni}(s) \rightarrow \text{Ni}^{2+}(aq) + 2 \text{ e}^- \qquad \text{Ag}^+(aq) + \text{e}^- \rightarrow \text{Ag}(s)$$

17.2 $\text{Pb}(s) + \text{Br}_2(l) \rightarrow \text{Pb}^{2+}(aq) + 2 \text{ Br}^-(aq)$

There is a Pb anode in an aqueous solution of Pb^{2+}. The cathode is a Pt wire that dips into a pool of liquid Br_2 and an aqueous solution that is saturated with Br_2. A salt bridge connects the anode and cathode compartment. The electrodes are connected through an external circuit.

17.3 $\text{Fe}(s)\,|\,\text{Fe}^{3+}(aq)\,\|\,\text{Cr}_2\text{O}_7^{2-}(aq),\, \text{Cr}^{3+}(aq)\,|\,\text{Pt}(s)$

17.4 (a) and (b)

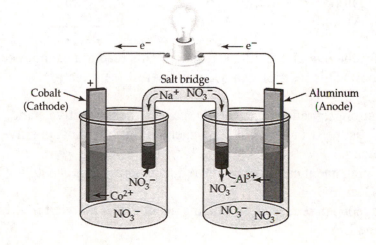

    (c) $2 \text{ Al(s)} + 3 \text{ Co}^{2+}(aq) \rightarrow 2 \text{ Al}^{3+}(aq) + 3 \text{ Co(s)}$

    (d) $\text{Al(s)} | \text{Al}^{3+}(aq) \| \text{Co}^{2+}(aq) | \text{Co(s)}$

17.5 $\text{Cr}_2\text{O}_7^{2-}(aq) + 3 \text{ Sn}^{2+}(aq) + 14 \text{ H}^+(aq) \rightarrow 2 \text{ Cr}^{3+}(aq) + 3 \text{ Sn}^{4+}(aq) + 7 \text{ H}_2\text{O(l)}$

    $n = 6 \text{ mol e}^-$

$$\Delta G^\circ = -nFE^\circ = -(6 \text{ mol e}^-)\left(\frac{96{,}500 \text{ C}}{1 \text{ mol e}^-}\right)(1.21 \text{ V})\left(\frac{1 \text{ J}}{1 \text{ C} \cdot \text{V}}\right) = -700{,}590 \text{ J} = -701 \text{ kJ}$$

17.6   oxidation: $\text{Al(s)} \rightarrow \text{Al}^{3+}(aq) + 3 \text{ e}^-$ $E^\circ = 1.66 \text{ V}$

          reduction: $\underline{\text{Cr}^{3+}(aq) + 3 \text{ e}^- \rightarrow \text{Cr(s)}}$ $\underline{E^\circ = \text{?}}$

          overall $\text{Al(s)} + \text{Cr}^{3+}(aq) \rightarrow \text{Al}^{3+}(aq) + \text{Cr(s)}$ $E^\circ = 0.92 \text{ V}$

    The standard reduction potential for the Cr^{3+}/Cr half cell is:

    $E^\circ = 0.92 - 1.66 = -0.74 \text{ V}$

17.7   (a) $\text{Cl}_2(g) + 2 \text{ e}^- \rightarrow 2 \text{ Cl}^-(aq)$ $E^\circ = 1.36 \text{ V}$

            $\text{Ag}^+(aq) + \text{e}^- \rightarrow \text{Ag(s)}$ $E^\circ = 0.80 \text{ V}$

    Cl_2 has the greater tendency to be reduced (larger E°). The species that has the greater tendency to be reduced is the stronger oxidizing agent. Cl_2 is the stronger oxidizing agent.

    (b) $\text{Fe}^{2+}(aq) + 2 \text{ e}^- \rightarrow \text{Fe(s)}$ $E^\circ = -0.45 \text{ V}$

            $\text{Mg}^{2+}(aq) + 2 \text{ e}^- \rightarrow \text{Mg(s)}$ $E^\circ = -2.37 \text{ V}$

    The second half-reaction has the lesser tendency to occur in the forward direction (more negative E°) and the greater tendency to occur in the reverse direction. Therefore, Mg is the stronger reducing agent.

17.8   (a) $2 \text{ Fe}^{3+}(aq) + 2 \text{ I}^-(aq) \rightarrow 2 \text{ Fe}^{2+}(aq) + \text{I}_2(s)$

          reduction: $\text{Fe}^{3+}(aq) + \text{e}^- \rightarrow \text{Fe}^{2+}(aq)$ $E^\circ = 0.77 \text{ V}$

          oxidation: $2 \text{ I}^-(aq) \rightarrow \text{I}_2(s) + 2 \text{ e}^-$ $\underline{E^\circ = -0.54 \text{ V}}$

                                      overall $E^\circ = 0.23 \text{ V}$

    Because E° for the overall reaction is positive, this reaction can occur under standard-state conditions.

    (b) $3 \text{ Ni(s)} + 2 \text{ Al}^{3+}(aq) \rightarrow 3 \text{ Ni}^{2+}(aq) + 2 \text{ Al(s)}$

          oxidation: $\text{Ni(s)} \rightarrow \text{Ni}^{2+}(aq) + 2 \text{ e}^-$ $E^\circ = 0.26 \text{ V}$

          reduction: $\text{Al}^{3+}(aq) + 3 \text{ e}^- \rightarrow \text{Al(s)}$ $\underline{E^\circ = -1.66 \text{ V}}$

                                        overall $E^\circ = -1.40 \text{ V}$

    Because E° for the overall reaction is negative, this reaction cannot occur under standard-state conditions. This reaction can occur in the reverse direction.

17.9   (a) D is the strongest reducing agent. D^+ has the most negative standard reduction potential. A^{3+} is the strongest oxidizing agent. It has the most positive standard reduction potential.

    (b) An oxidizing agent can oxidize any reducing agent that is below it in the table. B^{2+} can oxidize C and D.

    A reducing agent can reduce any oxidizing agent that is above it in the table. C can reduce A^{3+} and B^{2+}.

(c) Use the two half-reactions that have the most positive and the most negative standard reduction potentials, respectively.

$$
\begin{array}{ll}
A^{3+} + 2\,e^- \rightarrow A^+ & 1.47\ V \\
\underline{2 \times (D \rightarrow D^+ + e^-)} & \underline{1.38\ V} \\
A^{3+} + 2\,D \rightarrow A^+ + 2\,D^+ & 2.85\ V
\end{array}
$$

17.10 $Cu(s) + 2\,Fe^{3+}(aq) \rightarrow Cu^{2+}(aq) + 2\,Fe^{2+}(aq)$

$E^\circ = E^\circ_{Cu \rightarrow Cu^{2+}} + E^\circ_{Fe^{3+} \rightarrow Fe^{2+}} = -0.34\ V + 0.77\ V = 0.43\ V; \qquad n = 2\ mol\ e^-$

$E = E^\circ - \dfrac{0.0592\ V}{n} \log \dfrac{[Cu^{2+}][Fe^{2+}]^2}{[Fe^{3+}]^2} = 0.43\ V - \dfrac{(0.0592\ V)}{2} \log \dfrac{(0.25)(0.20)^2}{(1.0 \times 10^{-4})^2} = 0.25\ V$

17.11 $5\,[Cu(s) \rightarrow Cu^{2+}(aq) + 2\,e^-]$ (oxidation half reaction)
$2\,[5\,e^- + 8\,H^+(aq) + MnO_4^-(aq) \rightarrow Mn^{2+}(aq) + 4\,H_2O(l)]$ (reduction half reaction)

$5\,Cu(s) + 16\,H^+(aq) + 2\,MnO_4^-(aq) \rightarrow 5\,Cu^{2+}(aq) + 2\,Mn^{2+}(aq) + 8\,H_2O(l)$

$\Delta E = -\dfrac{0.0592\ V}{n} \log \dfrac{[Cu^{2+}]^5[Mn^{2+}]^2}{[MnO_4^-]^2[H^+]^{16}}$

(a) The anode compartment contains Cu^{2+}.

$\Delta E = -\dfrac{0.0592\ V}{10} \log \dfrac{(0.01)^5(1)^2}{(1)^2(1)^{16}} = +0.059\ V$

(b) The cathode compartment contains Mn^{2+}, MnO_4^-, and H^+.

$\Delta E = -\dfrac{0.0592\ V}{10} \log \dfrac{(1)^5(0.01)^2}{(0.01)^2(0.01)^{16}} = -0.19\ V$

17.12 $H_2(g) + Pb^{2+}(aq) \rightarrow 2\,H^+(aq) + Pb(s)$

$E^\circ = E^\circ_{H_2 \rightarrow H^+} + E^\circ_{Pb^{2+} \rightarrow Pb} = 0\ V + (-0.13\ V) = -0.13\ V; \qquad n = 2\ mol\ e^-$

$E = E^\circ - \dfrac{0.0592\ V}{n} \log \dfrac{[H_3O^+]^2}{[Pb^{2+}](P_{H_2})}$

$0.28\ V = -0.13\ V - \dfrac{(0.0592\ V)}{2} \log \dfrac{[H_3O^+]^2}{(1)(1)} = -0.13\ V - (0.0592\ V) \log [H_3O^+]$

$pH = -\log[H_3O^+]$ therefore $0.28\ V = -0.13\ V + (0.0592\ V)\ pH$

$pH = \dfrac{(0.28\ V + 0.13\ V)}{0.0592\ V} = 6.9$

17.13 $4\,Fe^{2+}(aq) + O_2(g) + 4\,H^+(aq) \rightarrow 4\,Fe^{3+}(aq) + 2\,H_2O(l)$

$E^\circ = E^\circ_{Fe^{2+} \rightarrow Fe^{3+}} + E^\circ_{O_2 \rightarrow H_2O} = -0.77\ V + 1.23\ V = 0.46\ V; \qquad n = 4\ mol\ e^-$

$E^\circ = \dfrac{0.0592\ V}{n} \log K; \quad \log K = \dfrac{nE^\circ}{0.0592\ V} = \dfrac{(4)(0.46\ V)}{0.0592\ V} = 31; \quad K = 10^{31}$ at 25 °C

17.14 $E° = \dfrac{0.0592\,V}{n}\,\log K = \dfrac{0.0592\,V}{2}\,\log(1.8 \times 10^{-5}) = -0.140\ V$

17.15   (a) $Zn(s) + 2\ MnO_2(s) + 2\ NH_4{}^+(aq) \rightarrow Zn^{2+}(aq) + Mn_2O_3(s) + 2\ NH_3(aq) + H_2O(l)$
      (b) $Zn(s) + 2\ MnO_2(s) \rightarrow ZnO(s) + Mn_2O_3(s)$
      (c) $Cd(s) + 2\ NiO(OH)(s) + 2\ H_2O(l) \rightarrow Cd(OH)_2(s) + 2\ Ni(OH)_2(s)$
      (d) $x\ Li(s) + MnO_2(s) \rightarrow Li_xMnO_2(s)$
      (e) $Li_xC_6(s) + Li_{1-x}CoO_2(s) \rightarrow 6\ C(s) + LiCoO_2(s)$

17.16   A fuel cell and a battery are both galvanic cells that convert chemical energy into electrical energy utilizing a spontaneous redox reaction. A fuel cell differs from an ordinary battery in that the reactants are not contained within the cell but instead are continuously supplied from an external reservoir.

17.17   (a)   anode reaction $2\ H_2(g) + 4\ OH^-(aq) \rightarrow 4\ H_2O(l) + 4\ e^-$ $E° = 0.83\ V$
                cathode reaction $\underline{O_2(g) + 2\ H_2O(l) + 4\ e^- \rightarrow 4\ OH^-(aq)}$ $\underline{E° = 0.40\ V}$
                overall reaction $2\ H_2(g) + O_2(g) \rightarrow 2\ H_2O(l)$ $E° = 1.23\ V$

      (b)   anode reaction $2\ H_2(g) \rightarrow 4\ H^+(aq) + 4\ e^-$ $E° = 0.00\ V$
                cathode reaction $\underline{O_2(g) + 4\ H^+(aq) + 4\ e^- \rightarrow 2\ H_2O(l)}$ $\underline{E° = 1.23\ V}$
                overall reaction $2\ H_2(g) + O_2(g) \rightarrow 2\ H_2O(l)$ $E° = 1.23\ V$

17.18   (a) $[Mg(s) \rightarrow Mg^{2+}(aq) + 2\ e^-]\ x\ 2$
          $\underline{O_2(g) + 4\ H^+(aq) + 4\ e^- \rightarrow 2\ H_2O(l)}$
          $2\ Mg(s) + O_2(g) + 4\ H^+(aq) \rightarrow 2\ Mg^{2+}(aq) + 2\ H_2O(l)$

      (b) $[Fe(s) \rightarrow Fe^{2+}(aq) + 2\ e^-]\ x\ 4$
          $[O_2(g) + 4\ H^+(aq) + 4\ e^- \rightarrow 2\ H_2O(l)]\ x\ 2$
          $4\ Fe^{2+}(aq) + O_2(g) + 4\ H^+(aq) \rightarrow 4\ Fe^{3+}(aq) + 2\ H_2O(l)$
          $\underline{[2\ Fe^{3+}(aq) + 4\ H_2O(l) \rightarrow Fe_2O_3 \cdot H_2O(s) + 6\ H^+(aq)]\ x\ 2}$
          $4\ Fe(s) + 3\ O_2(g) + 2\ H_2O(l) \rightarrow 2\ Fe_2O_3 \cdot H_2O(s)$

17.19   (a)

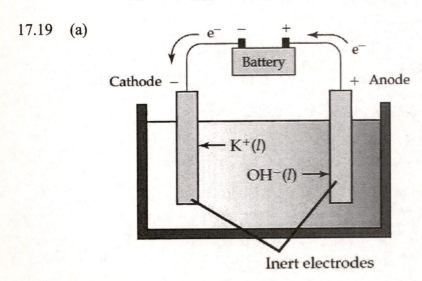

Inert electrodes

(b)   anode reaction $4 \, OH^-(l) \rightarrow O_2(g) + 2 \, H_2O(l) + 4 \, e^-$
cathode reaction $\underline{4 \, K^+(l) + 4 \, e^- \rightarrow 4 \, K(l)}$
overall reaction $4 \, K^+(l) + 4 \, OH^-(l) \rightarrow 4 \, K(l) + O_2(g) + 2 \, H_2O(l)$

17.20   (a)   anode reaction $2 \, Cl^-(aq) \rightarrow Cl_2(g) + 2 \, e^-$
cathode reaction $\underline{2 \, H_2O(l) + 2 \, e^- \rightarrow H_2(g) + 2 \, OH^-(aq)}$
overall reaction $2 \, Cl^-(aq) + 2 \, H_2O(l) \rightarrow Cl_2(g) + H_2(g) + 2 \, OH^-(aq)$

(b)   anode reaction $2 \, H_2O(l) \rightarrow O_2(g) + 4 \, H^+(aq) + 4 \, e^-$
cathode reaction $\underline{2 \, Cu^{2+}(aq) + 4 \, e^- \rightarrow 2 \, Cu(s)}$
overall reaction $2 \, Cu^{2+}(aq) + 2 \, H_2O(l) \rightarrow 2 \, Cu(s) + O_2(g) + 4 \, H^+(aq)$

(c)   anode reaction $2 \, H_2O(l) \rightarrow O_2(g) + 4 \, H^+(aq) + 4 \, e^-$
cathode reaction $\underline{4 \, H_2O(l) + 4 \, e^- \rightarrow 2 \, H_2(g) + 4 \, OH^-(aq)}$
overall reaction $2 \, H_2O(l) \rightarrow 2 \, H_2(g) + O_2(g)$

17.21

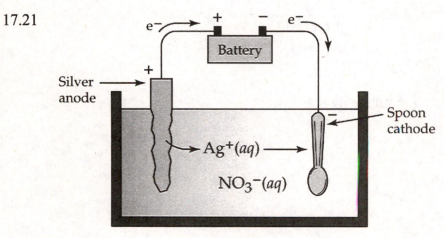

anode reaction $Ag(s) \rightarrow Ag^+(aq) + e^-$
cathode reaction $Ag^+(aq) + e^- \rightarrow Ag(s)$
The overall reaction is transfer of silver metal from the silver anode to the spoon.

17.22 $Charge = \left(1.00 \times 10^5 \, \dfrac{C}{s}\right)(8.00 \, h)\left(\dfrac{60 \, min}{h}\right)\left(\dfrac{60 \, s}{min}\right) = 2.88 \times 10^9 \, C$

$Moles \, of \, e^- = (2.88 \times 10^9 \, C)\left(\dfrac{1 \, mol \, e^-}{96,500 \, C}\right) = 2.98 \times 10^4 \, mol \, e^-$

cathode reaction: $Al^{3+} + 3 \, e^- \rightarrow Al$

$mass \, Al = (2.98 \times 10^4 \, mol \, e^-) \times \dfrac{1 \, mol \, Al}{3 \, mol \, e^-} \times \dfrac{26.98 \, g \, Al}{1 \, mol \, Al} \times \dfrac{1 \, kg}{1000 \, g} = 268 \, kg \, Al$

17.23 $3.00 \, g \, Ag \times \dfrac{1 \, mol \, Ag}{107.9 \, g \, Ag} = 0.0278 \, mol \, Ag$

cathode reaction: $Ag^+(aq) + e^- \rightarrow Ag(s)$

$$\text{Charge} = (0.0278 \text{ mol Ag})\left(\frac{1 \text{ mol e}^-}{1 \text{ mol Ag}}\right)\left(\frac{96,500 \text{ C}}{1 \text{ mol e}^-}\right) = 2682.7 \text{ C}$$

$$\text{Time} = \frac{C}{A} = \left(\frac{2682.7 \text{ C}}{0.100 \text{ C/s}} \times \frac{1 \text{ h}}{3600 \text{ s}}\right) = 7.45 \text{ h}$$

17.24 anode reaction $\text{Ti(s)} + 2 \text{ H}_2\text{O}(l) \rightarrow \text{TiO}_2(s) + 4 \text{ H}^+(aq) + 4 \text{ e}^-$ $E° = 1.066 \text{ V}$

cathode reaction $\underline{4 \text{ H}^+(aq) + 4 \text{ e}^- \rightarrow 2 \text{ H}_2(g)}$ $\underline{E° = 0.000 \text{ V}}$

overall reaction $\text{Ti(s)} + 2 \text{ H}_2\text{O}(l) \rightarrow \text{TiO}_2(s) + \quad 2 \text{ H}_2(g)$ $E° = 1.066 \text{ V}$

17.25 $\text{volume} = \left(0.0100 \text{ mm} \times \dfrac{1 \text{ cm}}{10 \text{ mm}}\right)(10.0 \text{ cm})^2 = 0.100 \text{ cm}^3$

$$\text{mol Al}_2\text{O}_3 = (0.100 \text{ cm}^3)(3.97 \text{ g/cm}^3)\frac{1 \text{ mol Al}_2\text{O}_3}{102.0 \text{ g Al}_2\text{O}_3} = 3.892 \times 10^{-3} \text{ mol Al}_2\text{O}_3$$

$$\text{mole e}^- = 3.892 \times 10^{-3} \text{ mol Al}_2\text{O}_3 \times \frac{6 \text{ mol e}^-}{1 \text{ mol Al}_2\text{O}_3} = 0.02335 \text{ mol e}^-$$

$$\text{coulombs} = 0.02335 \text{ mol e}^- \times \frac{96,500 \text{ C}}{1 \text{ mol e}^-} = 2253 \text{ C}$$

$$\text{time} = \frac{C}{A} = \frac{2253 \text{ C}}{0.600 \text{ C/s}} \times \frac{1 \text{ min}}{60 \text{ s}} = 62.6 \text{ min}$$

Conceptual Problems

17.26 (a) – (d)

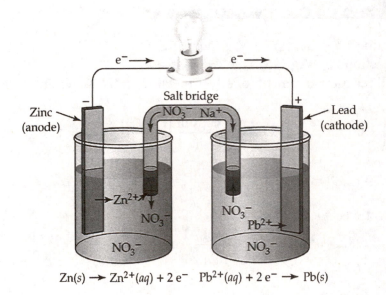

$\text{Zn(s)} \rightarrow \text{Zn}^{2+}(aq) + 2 \text{ e}^- \quad \text{Pb}^{2+}(aq) + 2 \text{ e}^- \rightarrow \text{Pb(s)}$

(e) anode reaction $\text{Zn(s)} \rightarrow \text{Zn}^{2+}(aq) + 2 \text{ e}^-$

cathode reaction $\underline{\text{Pb}^{2+}(aq) + 2 \text{ e}^- \rightarrow \text{Pb(s)}}$

overall reaction $\text{Zn(s)} + \text{Pb}^{2+}(aq) \rightarrow \text{Zn}^{2+}(aq) + \text{Pb(s)}$

17.28   (a) The three cell reactions are the same except for cation concentrations.

anode reaction	$Cu(s) \rightarrow Cu^{2+}(aq) + 2\,e^-$	$E° = -0.34$ V
cathode reaction	$2\,Fe^{3+}(aq) + 2\,e^- \rightarrow 2\,Fe^{2+}(aq)$	$E° = 0.77$ V
overall reaction	$Cu(s) + 2\,Fe^{3+}(aq) \rightarrow Cu^{2+}(aq) + 2\,Fe^{2+}(aq)$	$E° = 0.43$ V

(b)

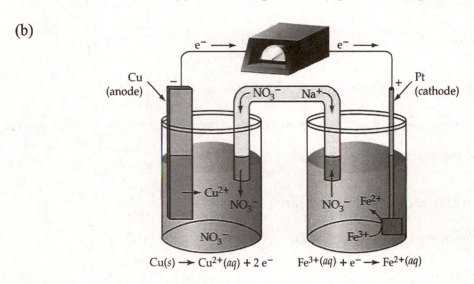

$$Cu(s) \rightarrow Cu^{2+}(aq) + 2\,e^- \qquad Fe^{3+}(aq) + e^- \rightarrow Fe^{2+}(aq)$$

(c) $E = E° - \dfrac{0.0592\,V}{n} \log \dfrac{[Cu^{2+}][Fe^{2+}]^2}{[Fe^{2+}]^2}$; $n = 2$ mol e^-

(1) $E = E° = 0.43$ V because all cation concentrations are 1 M.

(2) $E = E° - \dfrac{0.0592\,V}{2} \log \dfrac{(1)(5)^2}{(1)^2} = 0.39$ V

(3) $E = E° - \dfrac{0.0592\,V}{2} \log \dfrac{(0.1)(0.1)^2}{(0.1)^2} = 0.46$ V

Cell (3) has the largest potential, while cell (2) has the smallest as calculated from the Nernst equation.

17.30   (a) This is an electrolytic cell that has a battery connected between two inert electrodes.

(b)

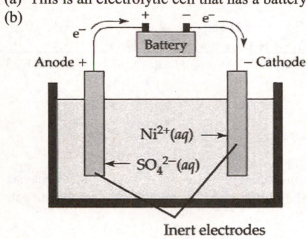

Inert electrodes

(c) anode reaction $2 H_2O(l) \rightarrow O_2(g) + 4 H^+(aq) + 4 e^-$
  cathode reaction $\underline{Ni^{2+}(aq) + 2 e^- \rightarrow Ni(s)}$
  overall reaction $2 Ni^{2+}(aq) + 2 H_2O(l) \rightarrow 2 Ni(s) + O_2(g) + 4 H^+(aq)$

17.32 $Zn(s) + Cu^{2+}(aq) \rightarrow Zn^{2+}(aq) + Cu(s); \quad E = E^\circ - \dfrac{0.0592 \text{ V}}{2} \log \dfrac{[Zn^{2+}]}{[Cu^{2+}]}$

(a) E increases because increasing $[Cu^{2+}]$ decreases $\log \dfrac{[Zn^{2+}]}{[Cu^{2+}]}$.

(b) E will decrease because addition of H_2SO_4 increases the volume which decreases $[Cu^{2+}]$ and increases $\log \dfrac{[Zn^{2+}]}{[Cu^{2+}]}$.

(c) E decreases because increasing $[Zn^{2+}]$ increases $\log \dfrac{[Zn^{2+}]}{[Cu^{2+}]}$.

(d) Because there is no change in $[Zn^{2+}]$, there is no change in E.

17.34 (a), (b) & (c)

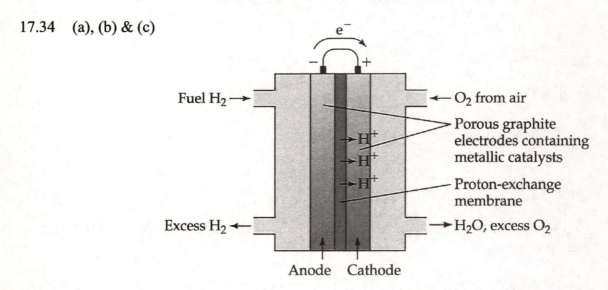

(d) anode reaction $2 H_2(g) \rightarrow 4 H^+(aq) + 4 e^-$ $E^\circ = 0.00$ V
  cathode reaction $\underline{O_2(g) + 4 H^+(aq) + 4 e^- \rightarrow 2 H_2O(l)}$ $\underline{E^\circ = 1.23 \text{ V}}$
  overall reaction $2 H_2(g) + O_2(g) \rightarrow 2 H_2O(l)$ $E^\circ = 1.23$ V

Section Problems
Galvanic Cells (Sections 17.1 and 17.2)

17.36 The cathode of a galvanic cell is considered to be the positive electrode because electrons flow through the external circuit toward the positive electrode (the cathode).

17.38   (a) Cd(s) + Sn²⁺(aq) → Cd²⁺(aq) + Sn(s)

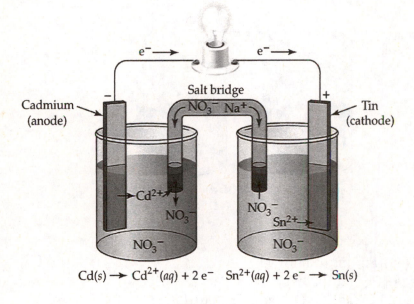

Cd(s) → Cd²⁺(aq) + 2 e⁻   Sn²⁺(aq) + 2 e⁻ → Sn(s)

(b) 2 Al(s) + 3 Cd²⁺(aq) → 2 Al³⁺(aq) + 3 Cd(s)

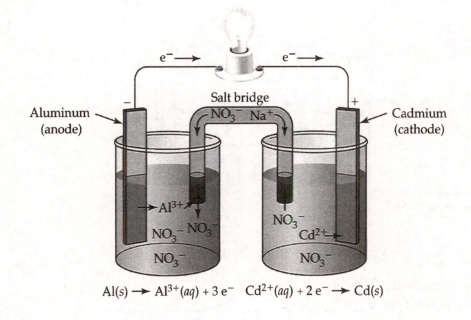

Al(s) → Al³⁺(aq) + 3 e⁻   Cd²⁺(aq) + 2 e⁻ → Cd(s)

(c) $6\ Fe^{2+}(aq) + Cr_2O_7^{2-}(aq) + 14\ H^+(aq) \rightarrow 6\ Fe^{3+}(aq) + 2\ Cr^{3+}(aq) + 7\ H_2O(l)$

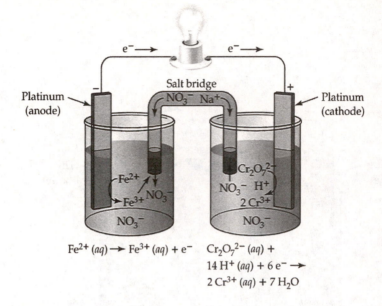

17.40 (a) $Cd(s)\,\big|\,Cd^{2+}(aq)\,\big\|\,Sn^{2+}(aq)\,\big|\,Sn(s)$
       (b) $Al(s)\,\big|\,Al^{3+}(aq)\,\big\|\,Cd^{2+}(aq)\,\big|\,Cd(s)$
       (c) $Pt(s)\,\big|\,Fe^{2+}(aq),\ Fe^{3+}(aq)\,\big\|\,Cr_2O_7^{2-}(aq),\ Cr^{3+}(aq)\,\big|\,Pt(s)$

17.42 $2\ Br^-(aq) + Cl_2(g) \rightarrow Br_2(l) + 2\ Cl^-(aq)$
       Inert electrodes are required because none of the reactants or products is an electrical conductor.

17.44 (a)

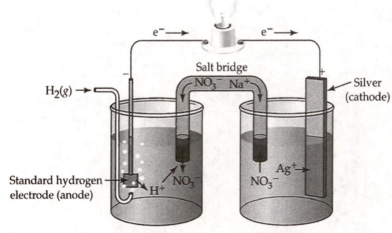

(b) anode reaction $H_2(g) \rightarrow 2\ H^+(aq) + 2\ e^-$
     cathode reaction $\underline{2\ Ag^+(aq) + 2\ e^- \rightarrow 2\ Ag(s)}$
     overall reaction $H_2(g) + 2\ Ag^+(aq) \rightarrow 2\ H^+(aq) + 2\ Ag(s)$

(c) $Pt(s)\,\big|\,H_2(g)\,\big|\,H^+(aq)\,\big\|\,Ag^+(aq)\,\big|\,Ag(s)$

17.46 (a) anode reaction $Co(s) \rightarrow Co^{2+}(aq) + 2\ e^-$
            cathode reaction $\underline{Cu^{2+}(aq) + 2\ e^- \rightarrow Cu(s)}$
            overall reaction $Co(s) + Cu^{2+}(aq) \rightarrow Co^{2+}(aq) + Cu(s)$

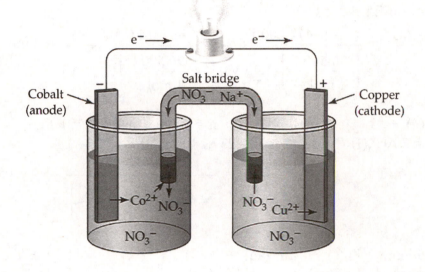

(b) anode reaction $2\ Fe(s) \rightarrow 2\ Fe^{2+}(aq) + 4\ e^-$
        cathode reaction $\underline{O_2(g) + 4\ H^+(aq) + 4\ e^- \rightarrow 2\ H_2O(l)}$
        overall reaction $2\ Fe(s) + O_2(g) + 4\ H^+(aq) \rightarrow 2\ Fe^{2+}(aq) + 2\ H_2O(l)$

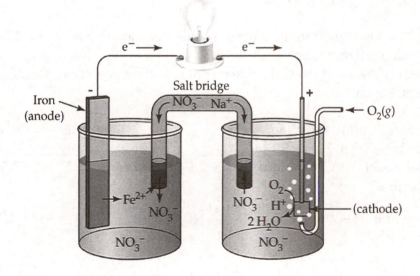

Cell Potentials and Free-Energy Changes; Standard Reduction Potentials (Sections 17.3–17.5)

17.48 E is the standard cell potential ($E°$) when all reactants and products are in their standard
states which are solutes at 1 M concentrations, gases at a partial pressure of 1 atm, solids
and liquids in pure form, all at 25 °C.

17.50 $Zn(s) + Ag_2O(s) \rightarrow ZnO(s) + 2\,Ag(s);$ $n = 2 \text{ mol e}^-$

$$\Delta G = -nFE = -(2 \text{ mol e}^-)\left(\frac{96,500 \text{ C}}{1 \text{ mol e}^-}\right)(1.60 \text{ V})\left(\frac{1 \text{ J}}{1 \text{ C} \cdot \text{V}}\right) = -308,800 \text{ J} = -309 \text{ kJ}$$

17.52 $\Delta G^\circ = -nFE^\circ;$ $1 \text{ J} = \text{C} \cdot \text{V}$

$$n = \frac{-\Delta G^\circ}{FE^\circ} = \frac{-(-414,000 \text{ J})}{\left(\dfrac{96,500 \text{ C}}{1 \text{ mol e}^-}\right)(1.43 \text{ V})} = \frac{-(-414,000 \text{ C} \cdot \text{V})}{\left(\dfrac{96,500 \text{ C}}{1 \text{ mol e}^-}\right)(1.43 \text{ V})} = 3 \text{ mol e}^-$$

17.54 $2\,H_2(g) + O_2(g) \rightarrow 2\,H_2O(l);$ $n = 4 \text{ mol e}^-$ and $1 \text{ V} = 1 \text{ J/C}$
$\Delta G^\circ = 2\,\Delta G^\circ_f(H_2O(l)) = (2 \text{ mol})(-237.2 \text{ kJ/mol}) = -474.4 \text{ kJ}$

$$\Delta G^\circ = -nFE^\circ \qquad E^\circ = \frac{-\Delta G^\circ}{nF} = \frac{-(-474,400 \text{ J})}{(4 \text{ mol e}^-)\left(\dfrac{96,500 \text{ C}}{1 \text{ mol e}^-}\right)} = +1.23 \text{ J/C} = +1.23 \text{ V}$$

17.56
oxidation:	$Zn(s) \rightarrow Zn^{2+}(aq) + 2\,e^-$	$E^\circ = 0.76 \text{ V}$
reduction:	$\underline{Eu^{3+}(aq) + e^- \rightarrow Eu^{2+}(aq)}$	$\underline{E^\circ = ?}$
overall	$Zn(s) + 2\,Eu^{3+}(aq) \rightarrow Zn^{2+}(aq) + 2\,Eu^{2+}(aq)$	$E^\circ = 0.40 \text{ V}$

The standard reduction potential for the Eu^{3+}/Eu^{2+} half cell is:
$E^\circ = 0.40 - 0.76 = -0.36 \text{ V}$

17.58 $Sn^{4+}(aq) < Br_2(aq) < MnO_4^-$

17.60 $Cr_2O_7^{2-}(aq)$ is highest in the table of standard reduction potentials, therefore it is the strongest oxidizing agent.
$Fe^{2+}(aq)$ is lowest in the table of standard reduction potentials, therefore it is the weakest oxidizing agent.

17.62
oxidation:	$Co(s) \rightarrow Co^{2+}(aq) + 2\,e^-$	$E^\circ = 0.28 \text{ V}$
reduction:	$\underline{I_2(s) + 2\,e^- \rightarrow 2\,I^-(aq)}$	$\underline{E^\circ = 0.54 \text{ V}}$
overall	$I_2(s) + Co(s) \rightarrow Co^{2+}(aq) + 2\,I^-(aq)$	$E^\circ = 0.82 \text{ V}$

$n = 2 \text{ mol e}^-$

$$\Delta G^\circ = -nFE^\circ = -(2 \text{ mol e}^-)\left(\frac{96,500 \text{ C}}{1 \text{ mol e}^-}\right)(0.82 \text{ V})\left(\frac{1 \text{ J}}{1 \text{ C} \cdot \text{V}}\right) = -158,260 \text{ J} = -1.6 \times 10^2 \text{ kJ}$$

17.64   (a) $Cd(s) + Sn^{2+}(aq) \rightarrow Cd^{2+}(aq) + Sn(s)$

oxidation:	$Cd(s) \rightarrow Cd^{2+}(aq) + 2\,e^-$	$E^\circ = 0.40 \text{ V}$
reduction:	$Sn^{2+}(aq) + 2\,e^- \rightarrow Sn(s)$	$\underline{E^\circ = -0.14 \text{ V}}$
		overall $E^\circ = 0.26 \text{ V}$

$n = 2 \text{ mol e}^-$

$$\Delta G^\circ = -nFE^\circ = -(2 \text{ mol e}^-)\left(\frac{96,500 \text{ C}}{1 \text{ mol e}^-}\right)(0.26 \text{ V})\left(\frac{1 \text{ J}}{1 \text{ C} \cdot \text{V}}\right) = -50,180 \text{ J} = -50 \text{ kJ}$$

(b) $2\,Al(s) + 3\,Cd^{2+}(aq) \rightarrow 2\,Al^{3+}(aq) + 3\,Cd(s)$

oxidation: $2\,Al(s) \rightarrow 2\,Al^{3+}(aq) + 6\,e^-$ $E° = 1.66\ V$

reduction: $3\,Cd^{2+}(aq) + 6\,e^- \rightarrow 3\,Cd(s)$ $\underline{E° = -0.40\ V}$

overall $E° = 1.26\ V$

$n = 6\ mol\ e^-$

$$\Delta G° = -nFE° = -(6\ mol\ e^-)\left(\frac{96{,}500\ C}{1\ mol\ e^-}\right)(1.26\ V)\left(\frac{1\ J}{1\ C \cdot V}\right) = -729{,}540\ J = -730\ kJ$$

(c) $6\,Fe^{2+}(aq) + Cr_2O_7^{2-}(aq) + 14\,H^+(aq) \rightarrow 6\,Fe^{3+}(aq) + 2\,Cr^{3+}(aq) + 7\,H_2O(l)$

oxidation: $6\,Fe^{2+}(aq) \rightarrow 6\,Fe^{3+}(aq) + 6\,e^-$ $E° = -0.77\ V$

reduction: $Cr_2O_7^{2-}(aq) + 14\,H^+(aq) + 6\,e^- \rightarrow 2\,Cr^{3+}(aq) + 7\,H_2O(l)$ $\underline{E° = 1.36\ V}$

overall $E° = 0.59\ V$

$n = 6\ mol\ e^-$

$$\Delta G° = -nFE° = -(6\ mol\ e^-)\left(\frac{96{,}500\ C}{1\ mol\ e^-}\right)(0.59\ V)\left(\frac{1\ J}{1\ C \cdot V}\right) = -341{,}610\ J = -342\ kJ$$

17.66   (a) $2\,Fe^{2+}(aq) + Pb^{2+}(aq) \rightarrow 2\,Fe^{3+}(aq) + Pb(s)$

oxidation: $2\,Fe^{2+}(aq) \rightarrow 2\,Fe^{3+}(aq) + 2\,e^-$ $E° = -0.77\ V$

reduction: $Pb^{2+}(aq) + 2\,e^- \rightarrow Pb(s)$ $\underline{E° = -0.13\ V}$

overall $E° = -0.90\ V$

Because the overall $E°$ is negative, this reaction is nonspontaneous.

(b) $Mg(s) + Ni^{2+}(aq) \rightarrow Mg^{2+}(aq) + Ni(s)$

oxidation: $Mg(s) \rightarrow Mg^{2+}(aq) + 2\,e^-$ $E° = 2.37\ V$

reduction: $Ni^{2+}(aq) + 2\,e^- \rightarrow Ni(s)$ $\underline{E° = -0.26\ V}$

overall $E° = 2.11\ V$

Because the overall $E°$ is positive, this reaction is spontaneous.

17.68   (a) oxidation: $Sn^{2+}(aq) \rightarrow Sn^{4+}(aq) + 2\,e^-$ $E° = -0.15\ V$

reduction: $Br_2(aq) + 2\,e^- \rightarrow 2\,Br^-(aq)$ $\underline{E° = 1.09\ V}$

overall $E° = +0.94\ V$

Because the overall $E°$ is positive, $Sn^{2+}(aq)$ can be oxidized by $Br_2(aq)$.

(b) oxidation: $Sn^{2+}(aq) \rightarrow Sn^{4+}(aq) + 2\,e^-$ $E° = -0.15\ V$

reduction: $Ni^{2+}(aq) + 2\,e^- \rightarrow Ni(s)$ $\underline{E° = -0.26\ V}$

overall $E° = -0.41\ V$

Because the overall $E°$ is negative, $Ni^{2+}(aq)$ cannot be reduced by $Sn^{2+}(aq)$.

(c) oxidation: $2\,Ag(s) \rightarrow 2\,Ag^+(aq) + 2\,e^-$ $E° = -0.80\ V$

reduction: $Pb^{2+}(aq) + 2\,e^- \rightarrow Pb(s)$ $\underline{E° = -0.13\ V}$

overall $E° = -0.93\ V$

Because the overall $E°$ is negative, $Ag(s)$ cannot be oxidized by $Pb^{2+}(aq)$.

(d) oxidation: $H_2SO_3(aq) + H_2O(l) \rightarrow SO_4^{2-}(aq) + 4\,H^+(aq) + 2\,e^-$ $E° = -0.17\ V$

reduction: $I_2(s) + 2\,e^- \rightarrow 2\,I^-(aq)$ $\underline{E° = 0.54\ V}$

overall $E° = +0.37\ V$

Because the overall $E°$ is positive, $I_2(s)$ can be reduced by H_2SO_3.

17.70 (a) oxidation: $2\ Cr^{3+}(aq) + 7\ H_2O(l) \rightarrow Cr_2O_7^{2-}(aq) + 14\ H^+(aq) + 6\ e^-$ $E° = -1.36\ V$
reduction: $O_2(g) + 4\ H^+(aq) + 4\ e^- \rightarrow 2\ H_2O(l)$ $\underline{E° = \ \ 1.23\ V}$
overall $E° = -0.13\ V$

There is no reaction because the overall E° is negative.

(b) oxidation: $Pb(s) \rightarrow Pb^{2+}(aq) + 2\ e^-$ $E° = \ \ 0.13\ V$
reduction: $2\ Ag^+(aq) + 2\ e^- \rightarrow 2\ Ag(s)$ $\underline{E° = \ \ 0.80\ V}$
overall $E° = +0.93\ V$

$Pb(s) + 2\ Ag^+(aq) \rightarrow Pb^{2+}(aq) + 2\ Ag(s)$
The reaction is spontaneous because the overall E° is positive.

(c) oxidation: $H_2C_2O_4(aq) \rightarrow 2\ CO_2(g) + 2\ H^+(aq) + 2\ e^-$ $E° = \ \ 0.49\ V$
reduction: $Cl_2(g) + 2\ e^- \rightarrow 2\ Cl^-(aq)$ $\underline{E° = \ \ 1.36\ V}$
overall $E° = +1.85\ V$

$Cl_2(g) + H_2C_2O_4(aq) \rightarrow 2\ Cl^-(aq) + 2\ CO_2(g) + 2\ H^+(aq)$
The reaction is spontaneous because the overall E° is positive.

(d) oxidation: $Ni(s) \rightarrow Ni^{2+}(aq) + 2\ e^-$ $E° = \ \ 0.26\ V$
reduction: $2\ HClO(aq) + 2\ H^+(aq) + 2\ e^- \rightarrow Cl_2(g) + H_2O(l)$ $\underline{E° = \ \ 1.61\ V}$
overall $E° = +1.87\ V$

$Ni(s) + 2\ HClO(aq) + 2\ H^+(aq) \rightarrow Ni^{2+}(aq) + Cl_2(g) + H_2O(l)$
The reaction is spontaneous because the overall E° is positive.

The Nernst Equation (Sections 17.6 and 17.7)

17.72 $2\ Ag^+(aq) + Sn(s) \rightarrow 2\ Ag(s) + Sn^{2+}(aq)$
oxidation: $Sn(s) \rightarrow Sn^{2+}(aq) + 2\ e^-$ $E° = 0.14\ V$
reduction: $2\ Ag^+(aq) + 2\ e^- \rightarrow 2\ Ag(s)$ $\underline{E° = 0.80\ V}$
overall $E° = 0.94\ V$

$$E = E° - \frac{0.0592\ V}{n}\ \log\frac{[Sn^{2+}]}{[Ag^+]^2} = 0.94\ V - \frac{(0.0592\ V)}{2}\ \log\frac{(0.020)}{(0.010)^2} = 0.87\ V$$

17.74 $Pb(s) + Cu^{2+}(aq) \rightarrow Pb^{2+}(aq) + Cu(s)$
oxidation: $Pb(s) \rightarrow Pb^{2+}(aq) + 2\ e^-$ $E° = 0.13\ V$
reduction: $Cu^{2+}(aq) + 2\ e^- \rightarrow Cu(s)$ $\underline{E° = 0.34\ V}$
overall $E° = 0.47\ V$

$$E = E° - \frac{0.0592\ V}{n}\ \log\frac{[Pb^{2+}]}{[Cu^{2+}]} = 0.47\ V - \frac{(0.0592\ V)}{2}\ \log\frac{1.0}{(1.0 \times 10^{-4})} = 0.35\ V$$

When $E = 0$, $0 = E° - \dfrac{0.0592\ V}{n}\ \log\dfrac{[Pb^{2+}]}{[Cu^{2+}]} = 0.47\ V - \dfrac{(0.0592\ V)}{2}\ \log\dfrac{1.0}{[Cu^{2+}]}$

$0 = 0.47\ V + \dfrac{(0.0592\ V)}{2}\ \log\ [Cu^{2+}]$

$\log\ [Cu^{2+}] = (-0.47\ V)\left(\dfrac{2}{0.0592\ V}\right) = -15.88;$ $[Cu^{2+}] = 10^{-15.88} = 1 \times 10^{-16}\ M$

17.76 $Zn(s) + Cu^{2+}(aq) \rightarrow Zn^{2+}(aq) + Cu(s)$

oxidation: $Zn(s) \rightarrow Zn^{2+}(aq) + 2 e^-$ $E° = 0.76$ V

reduction: $Cu^{2+}(aq) + 2 e^- \rightarrow Cu(s)$ $E° = 0.34$ V

overall $E° = 1.10$ V

$$E = 1.07 \text{ V} = E° - \frac{0.0592 \text{ V}}{n} \log \frac{[Zn^{2+}]}{[Cu^{2+}]} = 1.10 \text{ V} - \frac{(0.0592 \text{ V})}{2} \log\left(\frac{[Zn^{2+}]}{[Cu^{2+}]}\right)$$

$$1.07 \text{ V} - 1.10 \text{ V} = - \frac{(0.0592 \text{ V})}{2} \log\left(\frac{[Zn^{2+}]}{[Cu^{2+}]}\right)$$

$$\frac{0.03 \text{ V}}{\frac{(0.0592 \text{ V})}{2}} = \log\left(\frac{[Zn^{2+}]}{[Cu^{2+}]}\right) = 1; \qquad \frac{[Zn^{2+}]}{[Cu^{2+}]} = 10^1 = 10$$

17.78   (a) $E = E° - \dfrac{0.0592 \text{ V}}{n} \log [I^-]^2 = 0.54 \text{ V} - \dfrac{(0.0592 \text{ V})}{2} \log (0.020)^2 = 0.64$ V

(b) $E = E° - \dfrac{0.0592 \text{ V}}{n} \log \dfrac{[Fe^{2+}]}{[Fe^{3+}]} = 0.77 \text{ V} - \dfrac{(0.0592 \text{ V})}{1} \log\left(\dfrac{0.10}{0.10}\right) = 0.77$ V

(c) $E = E° - \dfrac{0.0592 \text{ V}}{n} \log \dfrac{[Sn^{4+}]}{[Sn^{2+}]} = -0.15 \text{ V} - \dfrac{(0.0592 \text{ V})}{2} \log\left(\dfrac{0.40}{0.0010}\right) = -0.23$ V

(d) $E = E° - \dfrac{0.0592 \text{ V}}{n} \log \dfrac{[Cr_2O_7^{2-}][H^+]^{14}}{[Cr^{3+}]^2} = -1.36 \text{ V} - \dfrac{(0.0592 \text{ V})}{6} \log\left(\dfrac{(1.0)(0.010)^{14}}{1.0}\right)$

$E = -1.36 \text{ V} - \dfrac{(0.0592 \text{ V})}{6} (14) \log (0.010) = -1.08$ V

17.80 $H_2(g) + Ni^{2+}(aq) \rightarrow 2 H^+(aq) + Ni(s)$

$E° = E°_{H_2 \rightarrow H^+} + E°_{Ni^{2+} \rightarrow Ni} = 0 \text{ V} + (-0.26 \text{ V}) = -0.26$ V

$$E = E° - \frac{0.0592 \text{ V}}{n} \log \frac{[H_3O^+]^2}{[Ni^{2+}](P_{H_2})}$$

$$0.27 \text{ V} = -0.26 \text{ V} - \frac{(0.0592 \text{ V})}{2} \log \frac{[H_3O^+]^2}{(1)(1)}$$

$0.27 \text{ V} = -0.26 \text{ V} - (0.0592 \text{ V}) \log [H_3O^+]$

$pH = - \log [H_3O^+]$ therefore $0.27 \text{ V} = -0.26 \text{ V} + (0.0592 \text{ V}) \, pH$

$pH = \dfrac{(0.27 \text{ V} + 0.26 \text{ V})}{0.0592 \text{ V}} = 9.0$

Standard Cell Potentials and Equilibrium Constants (Section 17.8)

17.82 $\Delta G° = -nFE°$

Because n and F are always positive, $\Delta G°$ is negative when E° is positive because of the negative sign in the equation.

$$E° = \frac{0.0592 \text{ V}}{n} \log K; \quad \log K = \frac{nE°}{0.0592 \text{ V}}; \quad K = 10^{\frac{nE°}{0.0592}}$$

If E° is positive, the exponent is positive (because n is positive), and K is greater than 1.

17.84 $Ni(s) + 2 Ag^+(aq) \rightarrow Ni^{2+}(aq) + 2 Ag(s)$
oxidation: $Ni(s) \rightarrow Ni^{2+}(aq) + 2 e^-$ E° = 0.26 V
reduction: $2 Ag^+(aq) + 2 e^- \rightarrow 2 Ag(s)$ $\underline{E° = 0.80 \text{ V}}$
                                        overall E° = 1.06 V

$$E° = \frac{0.0592 \text{ V}}{n} \log K; \quad \log K = \frac{nE°}{0.0592 \text{ V}} = \frac{(2)(1.06 \text{ V})}{0.0592 \text{ V}} = 35.8; \quad K = 10^{35.8} = 6 \times 10^{35}$$

17.86   E° and n are from Problem 17.64.

$$E° = \frac{0.0592 \text{ V}}{n} \log K; \quad \log K = \frac{nE°}{0.0592 \text{ V}}$$

(a) $Cd(s) + Sn^{2+}(aq) \rightarrow Cd^{2+}(aq) + Sn(s)$;   E° = 0.26 V and n = 2 mol e⁻

$$\log K = \frac{(2)(0.26 \text{ V})}{0.0592 \text{ V}} = 8.8; \quad K = 10^{8.8} = 6 \times 10^8$$

(b) $2 Al(s) + 3 Cd^{2+}(aq) \rightarrow 2 Al^{3+}(aq) + 3 Cd(s)$;   E° = 1.26 V and n = 6 mol e⁻

$$\log K = \frac{(6)(1.26 \text{ V})}{0.0592 \text{ V}} = 128; \quad K = 10^{128}$$

(c) $6 Fe^{2+}(aq) + Cr_2O_7^{2-}(aq) + 14 H^+(aq) \rightarrow 6 Fe^{3+}(aq) + 2 Cr^{3+}(aq) + 7 H_2O(l)$
E° = 0.59 V and n = 6 mol e⁻

$$\log K = \frac{(6)(0.59 \text{ V})}{0.0592 \text{ V}} = 60; \quad K = 10^{60}$$

17.88 $Hg_2^{2+}(aq) \rightarrow Hg(l) + Hg^{2+}(aq)$
oxidation: $\frac{1}{2}[Hg_2^{2+}(aq) \rightarrow 2 Hg^{2+}(aq) + 2 e^-]$ E° = -0.92 V
reduction: $\frac{1}{2}[Hg_2^{2+}(aq) + 2 e^- \rightarrow 2 Hg(l)]$ $\underline{E° = 0.80 \text{ V}}$
                                        overall E° = -0.12 V

$$E° = \frac{0.0592 \text{ V}}{n} \log K$$

$$\log K = \frac{nE°}{0.0592 \text{ V}} = \frac{(1)(-0.12 \text{ V})}{0.0592 \text{ V}} = -2.027; \quad K = 10^{-2.027} = 9 \times 10^{-3}$$

Batteries; Fuel Cells; Corrosion (Sections 17.9–17.11)

17.90　(a)

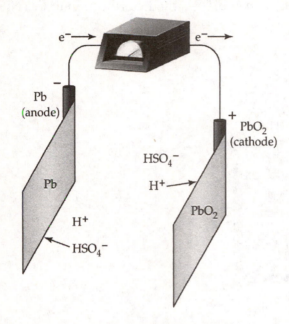

(b)　　anode:　Pb(s) + HSO$_4^-$(aq) → PbSO$_4$(s) + H$^+$(aq) + 2 e$^-$　　　　　E° = 0.296 V
　　　　cathode: PbO$_2$(s) + 3 H$^+$(aq) + HSO$_4^-$(aq) + 2 e$^-$ → PbSO$_4$(s) + 2 H$_2$O(l)　E° = 1.628 V
　overall:　Pb(s) + PbO$_2$(s) + 2 H$^+$(aq) + 2 HSO$_4^-$(aq) → 2 PbSO$_4$(s) + 2 H$_2$O(l)　E° = 1.924 V

(c)　$E° = \dfrac{0.0592\ V}{n} \log K$;　$\log K = \dfrac{nE°}{0.0592\ V} = \dfrac{(2)(1.924\ V)}{0.0592\ V} = 65.0$;　$K = 1 \times 10^{65}$

(d)　When the cell reaction reaches equilibrium the cell voltage = 0.

17.92　anode:　　　2 H$_2$(g) + 4 OH$^-$(aq) → 4 H$_2$O(l) + 4 e$^-$　　　　E° = 0.83 V
　　　　cathode:　　O$_2$(g) + 2 H$_2$O(l) + 4 e$^-$ → 4 OH$^-$(aq)　　　　　E° = 0.40 V
　　　　overall:　　H$_2$(g) + O$_2$(g) → 2 H$_2$O(l)　　　　　　　　　　E° = 1.23 V

n = 4 mol e$^-$ and 1 J = 1 C x 1 V

$$\Delta G° = -nFE° = -(4\ mol\ e^-)\left(\dfrac{96{,}500\ C}{1\ mol\ e^-}\right)(1.23\ V)\left(\dfrac{1\ J}{1\ C \cdot V}\right) = -474{,}780\ J = -475\ kJ$$

$$E° = \dfrac{0.0592\ V}{n} \log K;\quad \log K = \dfrac{nE°}{0.0592\ V} = \dfrac{(4)(1.23\ V)}{0.0592\ V} = 83.1;\quad K = 10^{83.1} = 1 \times 10^{83}$$

$$E = E° - \dfrac{0.0592\ V}{n} \log \dfrac{1}{(P_{H_2})^2(P_{O_2})} = 1.23\ V - \dfrac{0.0592\ V}{4} \log \dfrac{1}{(25)^2(25)} = 1.29\ V$$

17.94　Rust is a hydrated form of iron(III) oxide (Fe$_2$O$_3 \cdot$ H$_2$O). Rust forms from the oxidation of Fe in the presence of O$_2$ and H$_2$O. Rust can be prevented by coating Fe with Zn (galvanizing).

17.96   Cathodic protection is the attachment of a more easily oxidized metal to the metal you want to protect. This forces the metal you want to protect to be the cathode, hence the name, cathodic protection. Zn and Al can offer cathodic protection to Fe (Ni and Sn cannot).

Electrolysis (Sections 17.12–17.14)

17.98   (a)

   (b) anode: $2 Cl^-(l) \rightarrow Cl_2(g) + 2 e^-$
   cathode: $\underline{Mg^{2+}(l) + 2 e^- \rightarrow Mg(l)}$
   overall: $Mg^{2+}(l) + 2 Cl^-(l) \rightarrow Mg(l) + Cl_2(g)$

17.100   Possible anode reactions:
       $2 Cl^-(aq) \rightarrow Cl_2(g) + 2 e^-$
       $2 H_2O(l) \rightarrow O_2(g) + 4 H^+(aq) + 4 e^-$

   Possible cathode reactions:
       $2 H_2O(l) + 2 e^- \rightarrow H_2(g) + 2 OH^-(aq)$
       $Mg^{2+}(aq) + 2 e^- \rightarrow Mg(s)$

   Actual reactions:
   anode: $2 Cl^-(aq) \rightarrow Cl_2(g) + 2 e^-$
   cathode: $2 H_2O(l) + 2 e^- \rightarrow H_2(g) + 2 OH^-(aq)$

   This anode reaction takes place instead of $2 H_2O(l) \rightarrow O_2(g) + 4 H^+(aq) + 4 e^-$ because of a high overvoltage for formation of gaseous O_2.
   This cathode reaction takes place instead of $Mg^{2+}(aq) + 2 e^- \rightarrow Mg(s)$ because H_2O is easier to reduce than Mg^{2+}.

17.102   (a) NaBr
   anode: $2 Br^-(aq) \rightarrow Br_2(l) + 2 e^-$
   cathode: $\underline{2 H_2O(l) + 2 e^- \rightarrow H_2(g) + 2 OH^-(aq)}$
   overall: $2 H_2O(l) + 2 Br^-(aq) \rightarrow Br_2(l) + H_2(g) + 2 OH^-(aq)$

   (b) $CuCl_2$
   anode: $2 Cl^-(aq) \rightarrow Cl_2(g) + 2 e^-$
   cathode: $\underline{Cu^{2+}(aq) + 2 e^- \rightarrow Cu(s)}$
   overall: $Cu^{2+}(aq) + 2 Cl^-(aq) \rightarrow Cu(s) + Cl_2(g)$

    (c) LiOH

anode:	$4\ OH^-(aq) \rightarrow O_2(g) + 2\ H_2O(l) + 4\ e^-$
cathode:	$\underline{4\ H_2O(l) + 4\ e^- \rightarrow 2\ H_2(g) + 4\ OH^-(aq)}$
overall:	$2\ H_2O(l) \rightarrow O_2(g) + 2\ H_2(g)$

17.104 $Ag^+(aq) + e^- \rightarrow Ag(s);\quad 1\ A = 1\ C/s$

$$\text{mass Ag} = 2.40\ \frac{C}{s} \times 20.0\ \text{min} \times \frac{60\ s}{1\ \text{min}} \times \frac{1\ \text{mol}\ e^-}{96,500\ C} \times \frac{1\ \text{mol Ag}}{1\ \text{mol}\ e^-} \times \frac{107.87\ \text{g Ag}}{1\ \text{mol Ag}} = 3.22\ g$$

17.106 $2\ Na^+(l) + 2\ Cl^-(l) \rightarrow 2\ Na(l) + Cl_2(g)$

$Na^+(l) + e^- \rightarrow Na(l);\quad 1\ A = 1\ C/s;\quad 1.00 \times 10^3\ kg = 1.00 \times 10^6\ g$

$$\text{Charge} = 1.00 \times 10^6\ \text{g Na} \times \frac{1\ \text{mol Na}}{22.99\ \text{g Na}} \times \frac{1\ \text{mol}\ e^-}{1\ \text{mol Na}} \times \frac{96,500\ C}{1\ \text{mol}\ e^-} = 4.20 \times 10^9\ C$$

$$\text{Time} = \frac{4.20 \times 10^9\ C}{30,000\ C/s} \times \frac{1\ h}{3600\ s} = 38.9\ h$$

$$1.00 \times 10^6\ \text{g Na} \times \frac{1\ \text{mol Na}}{22.99\ \text{g Na}} \times \frac{1\ \text{mol}\ Cl_2}{2\ \text{mol Na}} = 21,748.6\ \text{mol}\ Cl_2$$

$PV = nRT$

$$V = \frac{nRT}{P} = \frac{(21,748.6\ \text{mol})\left(0.082\ 06\ \dfrac{L \cdot atm}{K \cdot mol}\right)(273.15\ K)}{1.00\ atm} = 4.87 \times 10^5\ L\ Cl_2$$

17.108 $M^{2+} + 2\ e^- \rightarrow M;\quad 20.0\ A = 20.0\ C/s$

$$\text{mol}\ e^- = 20.0\ \frac{C}{s} \times 325\ \text{min} \times \frac{60\ s}{\text{min}} \times \frac{1\ \text{mol}\ e^-}{96,500\ C} = 4.04\ \text{mol}\ e^-$$

$$4.04\ \text{mol}\ e^- \times \frac{1\ \text{mol M}}{2\ \text{mol}\ e^-} = 2.02\ \text{mol M}$$

$$\text{molar mass} = \frac{111\ \text{g M}}{2.02\ \text{mol M}} = 54.9\ \text{g/mol},\quad M^{2+} = Mn^{2+}$$

Chapter Problems

17.110 (a) $2\ MnO_4^-(aq) + 16\ H^+(aq) + 5\ Sn^{2+}(aq) \rightarrow 2\ Mn^{2+}(aq) + 5\ Sn^{4+}(aq) + 8\ H_2O(l)$
      (b) MnO_4^- is the oxidizing agent; Sn^{2+} is the reducing agent.
      (c) $E° = 1.51\ V + (-0.15\ V) = 1.36\ V$

17.112 (a) Ag^+ is the strongest oxidizing agent because Ag^+ has the most positive standard reduction potential.
      Pb is the strongest reducing agent because Pb^{2+} has the most negative standard reduction potential.

(b)

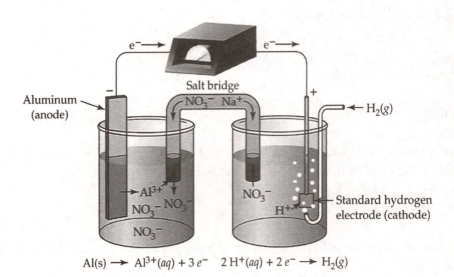

$Pb(s) \longrightarrow Pb^{2+}(aq) + 2\,e^-$ $Ag^+(aq) + e^- \longrightarrow Ag(s)$

(c) $Pb(s) + 2\,Ag^+(aq) \rightarrow Pb^{2+}(aq) + 2\,Ag(s);$ $n = 2 \text{ mol } e^-$

$E° = E°_{ox} + E°_{red} = 0.13 \text{ V} + 0.80 \text{ V} = 0.93 \text{ V}$

$$\Delta G° = -nFE° = -(2 \text{ mol } e^-)\left(\frac{96{,}500 \text{ C}}{1 \text{ mol } e^-}\right)(0.93 \text{ V})\left(\frac{1 \text{ J}}{1 \text{ C} \cdot \text{V}}\right) = -179{,}490 \text{ J} = -180 \text{ kJ}$$

$$E° = \frac{0.0592 \text{ V}}{n} \log K; \quad \log K = \frac{nE°}{0.0592 \text{ V}} = \frac{(2)(0.93 \text{ V})}{0.0592 \text{ V}} = 31; \quad K = 10^{31}$$

(d) $E = E° - \dfrac{0.0592 \text{ V}}{n} \log \dfrac{[Pb^{2+}]}{[Ag^+]^2} = 0.93 \text{ V} - \dfrac{0.0592 \text{ V}}{2} \log\left(\dfrac{0.01}{(0.01)^2}\right) = 0.87 \text{ V}$

17.114 (a)

$Al(s) \longrightarrow Al^{3+}(aq) + 3\,e^-$ $2\,H^+(aq) + 2\,e^- \longrightarrow H_2(g)$

(b) $2\,Al(s) + 6\,H^+(aq) \rightarrow 2\,Al^{3+}(aq) + 3\,H_2(g)$

$E° = E°_{ox} + E°_{red} = 1.66 \text{ V} + 0.00 \text{ V} = 1.66 \text{ V}$

(c)

$$E = E° - \frac{0.0592 \text{ V}}{n} \log \frac{[Al^{3+}]^2(P_{H_2})^3}{[H^+]^6} = 1.66 \text{ V} - \frac{(0.0592 \text{ V})}{6} \log\left(\frac{(0.10)^2(10.0)^3}{(0.10)^6}\right) = 1.59 \text{ V}$$

(d) $\Delta G° = -nFE° = -(6 \text{ mol } e^-)\left(\frac{96,500 \text{ C}}{1 \text{ mol } e^-}\right)(1.66 \text{ V})\left(\frac{1 \text{ J}}{1 \text{ C} \cdot \text{V}}\right) = -961,140 \text{ J} = -961 \text{ kJ}$

$$E° = \frac{0.0592 \text{ V}}{n} \log K; \quad \log K = \frac{nE°}{0.0592 \text{ V}} = \frac{(6)(1.66 \text{ V})}{0.0592 \text{ V}} = 168; \quad K = 10^{168}$$

(e) mass Al $= 10.0 \dfrac{C}{s} \times 25.0 \text{ min} \times \dfrac{60 \text{ s}}{1 \text{ min}} \times \dfrac{1 \text{ mol } e^-}{96,500 \text{ C}} \times \dfrac{1 \text{ mol Al}}{3 \text{ mol } e^-} \times \dfrac{26.98 \text{ g Al}}{1 \text{ mol Al}} = 1.40 \text{ g}$

17.116 (a)

$5 \times [2 \text{ Cl}^-(aq) \rightarrow \text{Cl}_2(g) + 2 \text{ e}^-]$		$E° = -1.36 \text{ V}$
$2 \times [\text{MnO}_4^-(aq) + 8 \text{ H}^+(aq) + 5 \text{ e}^- \rightarrow \text{Mn}^{2+}(aq) + 4 \text{ H}_2\text{O}(l)]$		$E° = 1.51 \text{ V}$
$10 \text{ Cl}^-(aq) + 2 \text{ MnO}_4^-(aq) + 16 \text{ H}^+(aq) \rightarrow 5 \text{ Cl}_2(g) + 2 \text{ Mn}^{2+}(aq) + 8 \text{ H}_2\text{O}(l)$		$E° = 0.15 \text{ V}$

(b) $E° = 0.15 \text{ V}$

$$\Delta G° = -nFE° = -(10 \text{ mol } e^-)\left(\frac{96,500 \text{ C}}{1 \text{ mol } e^-}\right)(0.15 \text{ V})\left(\frac{1 \text{ J}}{1 \text{ C} \cdot \text{V}}\right) = -144,750 \text{ J} = -1.4 \times 10^2 \text{ kJ}$$

(c) $KMnO_4$, 158.0

$$179 \text{ g KMnO}_4 \times \frac{1 \text{ mol KMnO}_4}{158.0 \text{ g KMnO}_4} \times \frac{5 \text{ mol Cl}_2}{2 \text{ mol KMnO}_4} = 2.83 \text{ mol Cl}_2$$

$$PV = nRT; \quad V = \frac{nRT}{P} = \frac{(2.83 \text{ mol})\left(0.082\ 06 \dfrac{L \cdot atm}{K \cdot mol}\right)(298 \text{ K})}{1.0 \text{ atm}} = 69 \text{ L}$$

17.118 (a) oxidizing agents: PbO_2, H^+, $Cr_2O_7^{2-}$;   reducing agents: Al, Fe, Ag

(b) PbO_2 is the strongest oxidizing agent. H^+ is the weakest oxidizing agent.

(c) Al is the strongest reducing agent.   Ag is the weakest reducing agent.

(d) oxidized by Cu^{2+}: Fe and Al;   reduced by H_2O_2: PbO_2 and $Cr_2O_7^{2-}$

17.120 (a) $3 \text{ CH}_3\text{CH}_2\text{OH}(aq) + 2 \text{ Cr}_2\text{O}_7^{2-}(aq) + 16 \text{ H}^+(aq) \rightarrow$

$$3 \text{ CH}_3\text{CO}_2\text{H}(aq) + 4 \text{ Cr}^{3+}(aq) + 11 \text{ H}_2\text{O}(l)$$

oxidation:
$3 \text{ CH}_3\text{CH}_2\text{OH}(aq) + 3 \text{ H}_2\text{O}(l) \rightarrow 3 \text{ CH}_3\text{CO}_2\text{H}(aq) + 12 \text{ H}^+(aq) + 12 \text{ e}^-$ $E° = -0.058 \text{V}$

reduction:

$2 \text{ Cr}_2\text{O}_7^{2-}(aq) + 28 \text{ H}^+(aq) + 12 \text{ e}^- + \rightarrow 4 \text{ Cr}^{3+}(aq) + 14 \text{ H}_2\text{O}(l)$	$E° = 1.36 \text{ V}$
	overall $E° = 1.30 \text{ V}$

(b) $E = E° - \dfrac{0.0592 \text{ V}}{n} \log \dfrac{[CH_3CO_2H]^3[Cr^{3+}]^4}{[CH_3CH_2OH]^3[Cr_2O_7^{2-}]^2[H^+]^{16}}$

$pH = 4.00$, $[H^+] = 0.000\ 10$ M

$$E = 1.30 \text{ V} - \frac{(0.0592 \text{ V})}{12} \log\left(\frac{(1.0)^3(1.0)^4}{(1.0)^3(1.0)^2(0.000\ 10)^{16}}\right)$$

$$E = 1.30 \text{ V} - \frac{(0.0592 \text{ V})}{12} \log \frac{1}{(0.000\ 10)^{16}} = 0.98 \text{ V}$$

17.122　anode:　　　　$Ag(s) + Cl^-(aq) \rightarrow AgCl(s) + e^-$
　　　　cathode:　　　$\underline{Ag^+(aq) + e^- \rightarrow Ag(s)}$
　　　　overall:　　　 $Ag^+(aq) + Cl^-(aq) \rightarrow AgCl(s) \qquad E° = 0.578 \text{ V}$

For: $AgCl(s) \rightleftharpoons Ag^+(aq) + Cl^-(aq) \qquad E° = -0.578 \text{ V}$

$$E° = \frac{0.0592 \text{ V}}{n} \log K; \quad \log K = \frac{nE°}{0.0592 \text{ V}} = \frac{(1)(-0.578 \text{ V})}{0.0592 \text{ V}} = -9.76$$

$$K = K_{sp} = 10^{-9.76} = 1.7 \times 10^{-10}$$

17.124　$4 Fe^{2+}(aq) + O_2(g) + 4 H^+(aq) \rightarrow 4 Fe^{3+}(aq) + 2 H_2O(l)$
　　　　oxidation:　$4 Fe^{2+}(aq) \rightarrow 4 Fe^{3+}(aq) + 4 e^- \qquad\qquad E° = -0.77 \text{ V}$
　　　　reduction:　$\underline{O_2(g) + 4 H^+(aq) + 4 e^- \rightarrow 2 H_2O(l) \qquad\quad E° = 1.23 \text{ V}}$
　　　　　　　　　　　　　　　　　　　　　　　　　overall $E° = 0.46 \text{ V}$

$$P_{O_2} = 160 \text{ mm Hg} \times \frac{1.00 \text{ atm}}{760 \text{ mm Hg}} = 0.211 \text{ atm}$$

$$E = E° - \frac{0.0592 \text{ V}}{n} \log \frac{[Fe^{3+}]^4}{[Fe^{2+}]^4[H^+]^4(P_{O_2})}$$

$$E = 0.46 \text{ V} - \frac{0.0592 \text{ V}}{4} \log \frac{(1 \times 10^{-7})^4}{(1 \times 10^{-7})^4(1 \times 10^{-7})^4(0.211)}$$

$E = 0.46 \text{ V} - 0.42 \text{ V} = 0.04 \text{ V}$
Because E is positive, the reaction is spontaneous.

17.126　First calculate $E°$ for the galvanic cell in order to determine $E°_1$.
　　　　anode:　　　$5 [2 Hg(l) + 2 Br^-(aq) \rightarrow Hg_2Br_2(s) + 2 e^-] \qquad\qquad E°_1 = \ ?$
　　　　cathode:　　$\underline{2 [MnO_4^-(aq) + 8 H^+(aq) + 5 e^- \rightarrow Mn^{2+}(aq) + 4 H_2O(l)]} \quad E°_2 = 1.51 \text{ V}$
　　　　overall:　　$2 MnO_4^-(aq) + 10 Hg(l) + 10 Br^-(aq) + 16 H^+(aq) \rightarrow$
　　　　　　　　　　　　　　　　　　$2 Mn^{2+}(aq) + 5 Hg_2Br_2(s) + 8 H_2O(l)$

$n = 10 \text{ mol } e^-$

$$E = E° - \frac{0.0592 \text{ V}}{n} \log \frac{[Mn^{2+}]^2}{[Br^-]^{10}[MnO_4^-]^2[H^+]^{16}}$$

$$1.214 \text{ V} = E° - \frac{(0.0592 \text{ V})}{10} \log\left(\frac{(0.10)^2}{(0.10)^{10}(0.10)^2(0.10)^{16}}\right)$$

$$1.214 \text{ V} = E° - \frac{(0.0592 \text{ V})}{10} \log \frac{1}{(0.10)^{26}} = E° - 0.154 \text{ V}$$

$E° = 1.214 + 0.154 = 1.368 \text{ V}$
$E°_1 + E°_2 = 1.368 \text{ V}; \quad E°_1 + 1.51 \text{ V} = 1.368 \text{ V}; \quad E°_1 = 1.368 \text{ V} - 1.51 \text{ V} = -0.142 \text{ V}$

oxidation: $2 \, Hg(l) \rightarrow Hg_2^{2+}(aq) + 2 \, e^-$ $E° = -0.80 \, V$ (Appendix D)

reduction: $\underline{Hg_2Br_2(s) + 2 \, e^- \rightarrow 2 \, Hg(l) + 2 \, Br^-(aq)}$ $E° = +0.142 \, V$ (from $E°_1$)

overall: $Hg_2Br_2(s) \rightarrow Hg_2^{2+}(aq) + 2 \, Br^-(aq)$ $E° = -0.658 \, V$

$$E° = \frac{0.0592 \, V}{n} \log K; \quad \log K = \frac{nE°}{0.0592 \, V} = \frac{(2)(-0.658 \, V)}{0.0592 \, V} = -22.2$$

$$K = K_{sp} = 10^{-22.2} = 6 \times 10^{-23}$$

17.128 (a) anode: $4[Al(s) \rightarrow Al^{3+}(aq) + 3 \, e^-]$ $E° = 1.66 \, V$

           cathode: $\underline{3[O_2(g) + 4 \, H^+(aq) + 4 \, e^- \rightarrow 2 \, H_2O(l)]}$ $E° = 1.23 \, V$

           overall: $4 \, Al(s) + 3 \, O_2(g) + 12 \, H^+(aq) \rightarrow 4 \, Al^{3+}(aq) + 6 \, H_2O(l)$ $E° = 2.89 \, V$

(b) & (c) $E = E° - \dfrac{2.303 \, RT}{nF} \log \dfrac{[Al^{3+}]^4}{(P_{O_2})^3 [H^+]^{12}}$

$$E = 2.89 \, V - \frac{(2.303)\left(8.314 \, \dfrac{J}{K \cdot mol}\right)(310 \, K)}{(12 \, mol \, e^-)(96{,}500 \, C/mol \, e^-)} \log\left(\frac{(1.0 \times 10^{-9})^4}{(0.20)^3 (1.0 \times 10^{-7})^{12}}\right)$$

$$E = 2.89 \, V - 0.257 \, V = 2.63 \, V$$

Multiconcept Problems

17.130 (a) $4 \, CH_2{=}CHCN + 2 \, H_2O \rightarrow 2 \, NC(CH_2)_4CN + O_2$

(b) $mol \, e^- = 3000 \, C/s \times 10.0 \, h \times \dfrac{3600 \, s}{1 \, h} \times \dfrac{1 \, mol \, e^-}{96{,}500 \, C} = 1119.2 \, mol \, e^-$

mass adiponitrile =

$1119.2 \, mol \, e^- \times \dfrac{1 \, mol \, adiponitrile}{2 \, mol \, e^-} \times \dfrac{108.14 \, g \, adiponitrile}{1 \, mol \, adiponitrile} \times \dfrac{1.0 \, kg}{1000 \, g} = 60.5 \, kg$

(c) $1119.2 \, mol \, e^- \times \dfrac{1 \, mol \, O_2}{4 \, mol \, e^-} = 279.8 \, mol \, O_2$

$$PV = nRT; \quad V = \frac{nRT}{P} = \frac{(279.8 \, mol)\left(0.082\,06 \, \dfrac{L \cdot atm}{K \cdot mol}\right)(298 \, K)}{\left(740 \, mm \, Hg \times \dfrac{1 \, atm}{760 \, mm \, Hg}\right)} = 7030 \, L \, O_2$$

17.132 $Ba(s) + Cl_2(g) \rightarrow BaCl_2(s)$ $\Delta G°_f = -806.7 \, kJ/mol$

       $\underline{BaCl_2(s) \rightarrow Ba^{2+}(aq) + 2 \, Cl^-(aq)}$ $\underline{\Delta G°_1 = -16.7 \, kJ/mol}$

       $Ba(s) + Cl_2(g) \rightarrow Ba^{2+}(aq) + 2 \, Cl^-(aq)$ $\Delta G°_2 = \Delta G°_f + \Delta G°_1 = -823.4 \, kJ/mol$

$$E° = \frac{-\Delta G°}{nF} = \frac{-(-823{,}400 \, J)}{(2 \, mol \, e^-)\left(\dfrac{96{,}500 \, C}{1 \, mol \, e^-}\right)} = +4.266 \, J/C = +4.266 \, V$$

$$Ba(s) + Cl_2(g) \rightarrow Ba^{2+}(aq) + 2 Cl^-(aq) \qquad E° = +4.266 \text{ V}$$
$$\underline{2 Cl^-(aq) \rightarrow Cl_2(g) + 2 e^- \qquad\qquad\quad E° = -1.36 \text{ V}}$$
$$Ba(s) \rightarrow Ba^{2+}(aq) + 2 e^- \qquad\qquad\quad E° = 2.91 \text{ V}$$

$$Ba^{2+}(aq) + 2 e^- \rightarrow Ba(s) \qquad\qquad\quad E° = -2.91 \text{ V}$$

17.134 (a) $Cr_2O_7^{2-}(aq) + 6 Fe^{2+}(aq) + 14 H^+(aq) \rightarrow 2 Cr^{3+}(aq) + 6 Fe^{3+}(aq) + 7 H_2O(l)$
(b) The two half reactions are:
oxidation: $\quad Fe^{2+}(aq) \rightarrow Fe^{3+}(aq) + e^- \qquad\qquad\qquad\qquad E° = -0.77 \text{ V}$
reduction: $\quad Cr_2O_7^{2-}(aq) + 14 H^+(aq) + 6 e^- \rightarrow 2 Cr^{3+}(aq) + 7 H_2O(l) \quad E° = 1.36 \text{ V}$
At the equivalence point the potential is given by either of the following expressions:

$$\text{(1)} \quad E = 1.36 \text{ V} - \frac{0.0592 \text{ V}}{6} \log \frac{[Cr^{3+}]^2}{[Cr_2O_7^{2-}][H^+]^{14}}$$

$$\text{(2)} \quad E = 0.77 \text{ V} - \frac{0.0592 \text{ V}}{1} \log \frac{[Fe^{2+}]}{[Fe^{3+}]}$$

where E is the same in both because equilibrium is reached and the solution can have only one potential. Multiplying (1) by 6, adding it to (2), and using some stoichiometric relationships at the equivalence point will simplify the log term.

$$7E = [(6 \times 1.36 \text{ V}) + 0.77 \text{ V}] - (0.0592 \text{ V}) \log \frac{[Fe^{2+}][Cr^{3+}]^2}{[Fe^{3+}][Cr_2O_7^{2-}][H^+]^{14}}$$

At the equivalence point, $[Fe^{2+}] = 6[Cr_2O_7^{2-}]$ and $[Fe^{3+}] = 3[Cr^{3+}]$. Substitute these equalities into the previous equation.

$$7E = [(6 \times 1.36 \text{ V}) + 0.77 \text{ V}] - (0.0592 \text{ V}) \log \frac{6[Cr_2O_7^{2-}][Cr^{3+}]^2}{3[Cr^{3+}][Cr_2O_7^{2-}][H^+]^{14}}$$

Cancel identical terms.

$$7E = [(6 \times 1.36 \text{ V}) + 0.77 \text{ V}] - (0.0592 \text{ V}) \log \frac{6[Cr^{3+}]}{3[H^+]^{14}}$$

mol Fe^{2+} = (0.120 L)(0.100 mol/L) = 0.0120 mol Fe^{2+}

$$\text{mol } Cr_2O_7^{2-} = 0.0120 \text{ mol } Fe^{2+} \times \frac{1 \text{ mol } Cr_2O_7^{2-}}{6 \text{ mol } Fe^{2+}} = 0.002\,00 \text{ mol } Cr_2O_7^{2-}$$

$$\text{volume } Cr_2O_7^{2-} = 0.002\,00 \text{ mol} \times \frac{1 \text{ L}}{0.120 \text{ mol}} = 0.0167 \text{ L}$$

At the equivalence point assume mol Fe^{3+} = initial mol Fe^{2+} = 0.0120 mol
Total volume at the equivalence point is 0.120 L + 0.0167 L = 0.1367 L

$$[Fe^{3+}] = \frac{0.0120 \text{ mol}}{0.1367 \text{ L}} = 0.0878 \text{ M}; \quad [Cr^{3+}] = [Fe^{3+}]/3 = (0.0878 \text{ M})/3 = 0.0293 \text{ M}$$

$[H^+] = 10^{-pH} = 10^{-2.00} = 0.010 \text{ M}$

$$7E = [(6 \times 1.36 \text{ V}) + 0.77 \text{ V}] - (0.0592 \text{ V}) \log \frac{6(0.0293)}{3(0.010)^{14}} = 8.93 - 1.585 = 7.345 \text{ V}$$

$$E = \frac{7.345 \text{ V}}{7} = 1.05 \text{ V at the equivalence point.}$$

17.136 (a) $Zn(s) + 2 Ag^+(aq) + H_2O(l) \rightarrow ZnO(s) + 2 Ag(s) + 2 H^+(aq)$

$\Delta H^\circ_{rxn} = \Delta H^\circ_f(ZnO) - [2\ \Delta H^\circ_f(Ag^+) + \Delta H^\circ_f(H_2O)]$

$\Delta H^\circ_{rxn} = [(1\ mol)(-350.5\ kJ/mol)] - [(2\ mol)(105.6\ kJ/mol) + (1\ mol)(-285.8\ kJ/mol)]$

$\Delta H^\circ_{rxn} = -275.9\ kJ$

$\Delta S^\circ = [S^\circ(ZnO) + 2\ S^\circ(Ag)] - [S^\circ(Zn) + 2\ S^\circ(Ag^+) + S^\circ(H_2O)]$

$\Delta S^\circ = [(1\ mol)(43.7\ J/(K \cdot mol)) + (2\ mol)(42.6\ J/(K \cdot mol))]$

$\qquad - [(1\ mol)(41.6\ J/(K \cdot mol)) + (2\ mol)(72.7\ J/(K \cdot mol)) + (1\ mol)(69.9\ J/(K \cdot mol))]$

$\Delta S^\circ = -128.0\ J/K$

$\Delta G^\circ = \Delta H^\circ - T\Delta S^\circ = -275.9\ kJ - (298\ K)(-128.0 \times 10^{-3}\ kJ/K) = -237.8\ kJ$

(b) 1 V = 1 J/C

$$\Delta G^\circ = -nFE^\circ \quad E^\circ = \frac{-\Delta G^\circ}{nF} = \frac{-(-237.8 \times 10^3\ J)}{(2\ mol\ e^-)\left(\dfrac{96,500\ C}{1\ mol\ e^-}\right)} = 1.232\ J/C = 1.232\ V$$

$$E^\circ = \frac{0.0592\ V}{n} \log K; \quad \log K = \frac{nE^\circ}{0.0592\ V} = \frac{(2)(1.232\ V)}{0.0592\ V} = 41.62$$

$K = 10^{41.62} = 4 \times 10^{41}$

(c) $E = E^\circ - \dfrac{0.0592\ V}{n} \log \dfrac{[H^+]^2}{[Ag^+]^2}$

The addition of NH_3 to the cathode compartment would result in the formation of the $Ag(NH_3)_2{}^+$ complex ion which results in a decrease in Ag^+ concentration. The log term in the Nernst equation becomes larger and the cell voltage decreases.

On mixing equal volumes of two solutions, the concentrations of both solutions are cut in half.

	$Ag^+(aq)$	+	$2 NH_3(aq)$	$\rightleftarrows$	$Ag(NH_3)_2{}^+(aq)$
before reaction (M)	0.0500		2.00		0
assume 100% reaction	−0.0500		−2(0.0500)		+0.0500
after reaction (M)	0		1.90		0.0500
assume small back rxn	+x		+2x		−x
equil (M)	x		1.90 + 2x		0.0500 − x

$$K_f = 1.7 \times 10^7 = \frac{[Ag(NH_3)_2^+]}{[Ag^+][NH_3]^2} = \frac{(0.0500 - x)}{(x)(1.90 + 2x)^2} \approx \frac{0.0500}{(x)(1.90)^2}$$

Solve for x. $x = [Ag^+] = 8.15 \times 10^{-10}\ M$

$$E = E^\circ - \frac{0.0592\ V}{n} \log \frac{[H^+]^2}{[Ag^+]^2} = 1.232\ V - \frac{0.0592\ V}{2} \log \frac{(1.00\ M)^2}{(8.15 \times 10^{-10}\ M)^2} = 0.694\ V$$

(d) Calculate new initial concentrations because of dilution to 110.0 mL.

$$M_i \times V_i = M_f \times V_f; \quad M_f = [Cl^-] = \frac{M_i \times V_i}{V_f} = \frac{0.200\ M\ \times 10.0\ mL}{110.0\ mL} = 0.0182\ M$$

$$M_i \times V_i = M_f \times V_f; \quad M_f = [Ag^+] = \frac{M_i \times V_i}{V_f} = \frac{0.0500\ M\ \times 100.0\ mL}{110.0\ mL} = 0.0455\ M$$

$$M_i \times V_i = M_f \times V_f; \quad M_f = [NH_3] = \frac{M_i \times V_i}{V_f} = \frac{2.00 \text{ M} \times 100.0 \text{ mL}}{110.0 \text{ mL}} = 1.82 \text{ M}$$

Now calculate the $[Ag^+]$ as a result of the following equilibrium:

	$Ag^+(aq)$	$+$	$2\,NH_3(aq)$	$\rightleftharpoons$	$Ag(NH_3)_2^+(aq)$
before reaction (M)	0.0455		1.82		0
assume 100% reaction	−0.0455		−2(0.0455)		+0.0455
after reaction (M)	0		1.73		0.0455
assume small back rxn	+x		+2x		−x
equil (M)	x		1.73 + 2x		0.0455 − x

$$K_f = 1.7 \times 10^7 = \frac{[Ag(NH_3)_2^+]}{[Ag^+][NH_3]^2} = \frac{(0.0455 - x)}{(x)(1.73 + 2x)^2} \approx \frac{0.0455}{(x)(1.73)^2}$$

Solve for x. $x = [Ag^+] = 8.94 \times 10^{-10} \text{ M}$
For AgCl, $K_{sp} = 1.8 \times 10^{-10}$
$IP = [Ag^+][Cl^-] = (8.94 \times 10^{-10} \text{ M})(0.0182 \text{ M}) = 1.6 \times 10^{-11}$
$IP < K_{sp}$, AgCl will not precipitate.
Now calculate new initial concentrations because of dilution to 120.0 mL.

$$M_i \times V_i = M_f \times V_f; \quad M_f = [Br^-] = \frac{M_i \times V_i}{V_f} = \frac{0.200 \text{ M} \times 10.0 \text{ mL}}{120.0 \text{ mL}} = 0.0167 \text{ M}$$

$$M_i \times V_i = M_f \times V_f; \quad M_f = [Ag^+] = \frac{M_i \times V_i}{V_f} = \frac{0.0500 \text{ M} \times 100.0 \text{ mL}}{120.0 \text{ mL}} = 0.0417 \text{ M}$$

$$M_i \times V_i = M_f \times V_f; \quad M_f = [NH_3] = \frac{M_i \times V_i}{V_f} = \frac{2.00 \text{ M} \times 100.0 \text{ mL}}{120.0 \text{ mL}} = 1.67 \text{ M}$$

Now calculate the $[Ag^+]$ as a result of the following equilibrium:

	$Ag^+(aq)$	$+$	$2\,NH_3(aq)$	$\rightleftharpoons$	$Ag(NH_3)_2^+(aq)$
before reaction (M)	0.0417		1.67		0
assume 100% reaction	−0.0417		−2(0.0417)		+0.0417
after reaction (M)	0		1.59		0.0417
assume small back rxn	+x		+2x		−x
equil (M)	x		1.59 + 2x		0.0417 − x

$$K_f = 1.7 \times 10^7 = \frac{[Ag(NH_3)_2^+]}{[Ag^+][NH_3]^2} = \frac{(0.0417 - x)}{(x)(1.59 + 2x)^2} \approx \frac{0.0417}{(x)(1.59)^2}$$

Solve for x. $x = [Ag^+] = 9.70 \times 10^{-10} \text{ M}$
For AgBr, $K_{sp} = 5.4 \times 10^{-13}$
$IP = [Ag^+][Br^-] = (9.70 \times 10^{-10} \text{ M})(0.0167 \text{ M}) = 1.6 \times 10^{-11}$
$IP > K_{sp}$, AgBr will precipitate.

17.138 (a) Oxidation half reaction: $2\,[C_4H_{10}(g) + 13\,O^{2-}(s) \rightarrow 4\,CO_2(g) + 5\,H_2O(l) + 26\,e^-]$
Reduction half reaction: $\underline{13\,[O_2(g) + 4\,e^- \rightarrow 2\,O^{2-}(s)]}$

Cell reaction: $2\,C_4H_{10}(g) + 13\,O_2(g) \rightarrow 8\,CO_2(g) + 10\,H_2O(l)$

(b) $\Delta H^\circ = [8 \Delta H^\circ_f(CO_2) + 10 \Delta H^\circ_f(H_2O)] - [2 \Delta H^\circ_f(C_4H_{10})]$

$\Delta H^\circ = [(8 \text{ mol})(-393.5 \text{ kJ/mol}) + (10 \text{ mol})(-285.8 \text{ kJ/mol})]$

$- [(2 \text{ mol})(-126 \text{ kJ/mol})] = -5754 \text{ kJ}$

$\Delta S^\circ = [8 S^\circ(CO_2) + 10 S^\circ(H_2O)] - [2 S^\circ(C_4H_{10}) + 13 S^\circ(O_2)]$

$\Delta S^\circ = [(8 \text{ mol})(213.6 \text{ J/(K} \cdot \text{mol)}) + (10 \text{ mol})(69.9 \text{ J/(K} \cdot \text{mol)})]$

$- [(2 \text{ mol})(310 \text{ J/(K} \cdot \text{mol)}) + (13 \text{ mol})(205 \text{ J/(K} \cdot \text{mol)})] = -877.2 \text{ J/K}$

$\Delta G^\circ = \Delta H^\circ - T\Delta S^\circ = -5754 \text{ kJ} - (298 \text{ K})(-877.2 \times 10^{-3} \text{ kJ/K}) = -5493 \text{ kJ}$

$1 \text{ V} = 1 \text{ J/C}$

$\Delta G^\circ = - nFE^\circ; \quad E^\circ = -\dfrac{\Delta G^\circ}{nF} = -\dfrac{-5493 \times 10^3 \text{ J}}{(52)(96,500 \text{ C})} = 1.09 \text{ J/C} = 1.09 \text{ V}$

$\Delta G^\circ = -RT \ln K$

$\ln K = \dfrac{-\Delta G^\circ}{RT} = \dfrac{-(-5493 \text{ kJ})}{(8.314 \times 10^{-3} \text{ kJ/K})(298 \text{ K})} = 2217$

$K = e^{2217} = 7 \times 10^{962}$

On raising the temperature, both K and E° will decrease because the reaction is exothermic ($\Delta H^\circ < 0$).

(c) C_4H_{10}, 58.12; 10.5 A = 10.5 C/s

mass C_4H_{10} = 10.5 C/s x 8 hr x $\dfrac{60 \text{ min}}{1 \text{ hr}}$ x $\dfrac{60 \text{ s}}{1 \text{ min}}$ x $\dfrac{1 \text{ mol e}^-}{96,500 \text{ C}}$ x $\dfrac{2 \text{ mol } C_4H_{10}}{52 \text{ mol e}^-}$ x

$\dfrac{58.12 \text{ g } C_4H_{10}}{1 \text{ mol } C_4H_{10}} = 7.00 \text{ g } C_4H_{10}$

$n = 7.00 \text{ g } C_4H_{10} \text{ x } \dfrac{1 \text{ mol } C_4H_{10}}{58.12 \text{ g } C_4H_{10}} = 0.120 \text{ mol } C_4H_{10}$

20 °C = 20 + 273 = 293 K

$PV = nRT \qquad V = \dfrac{nRT}{P} = \dfrac{(0.120 \text{ mol})\left(0.082\ 06 \dfrac{L \cdot atm}{K \cdot mol}\right)(293 \text{ K})}{\left(815 \text{ mm Hg x } \dfrac{1.00 \text{ atm}}{760 \text{ mm Hg}}\right)} = 2.69 \text{ L}$

17.140 (a) $4 [Au(s) + 2 CN^-(aq) \rightarrow Au(CN)_2^-(aq) + e^-]$ (oxidation half reaction)

$O_2(g) \rightarrow 2 H_2O(l)$

$O_2(g) + 4 H^+(aq) \rightarrow 2 H_2O(l)$

$4 e^- + O_2(g) + 4 H^+(aq) \rightarrow 2 H_2O(l)$ (reduction half reaction)

Combine the two half reactions.

$4 Au(s) + 8 CN^-(aq) + O_2(g) + 4 H^+(aq) \rightarrow 4 Au(CN)_2^-(aq) + 2 H_2O(l)$

$4 Au(s) + 8 CN^-(aq) + O_2(g) + 4 H^+(aq) + 4 OH^-(aq)$

$\rightarrow 4 Au(CN)_2^-(aq) + 2 H_2O(l) + 4 OH^-(aq)$

$4 Au(s) + 8 CN^-(aq) + O_2(g) + 4 H_2O(l)$

$\rightarrow 4 Au(CN)_2^-(aq) + 2 H_2O(l) + 4 OH^-(aq)$

$4 Au(s) + 8 CN^-(aq) + O_2(g) + 2 H_2O(l) \rightarrow 4 Au(CN)_2^-(aq) + 4 OH^-(aq)$

(b) Add the following five reactions together. ΔG° is calculated below each reaction.

$4\,[\text{Au}^+(aq) + 2\,\text{CN}^-(aq) \rightarrow \text{Au(CN)}_2^-(aq)]$ $K = (K_f)^4$

$\Delta G^\circ = -RT\ln K = -(8.314 \times 10^{-3}\,\text{kJ/K})(298\,\text{K})\ln(6.2 \times 10^{38})^4 = -885.2\,\text{kJ}$

$O_2(g) + 4\,\text{H}^+(aq) + 4\,e^- \rightarrow 2\,\text{H}_2\text{O}(l)$ $E^\circ = 1.229\,\text{V}$

$\Delta G^\circ = -nFE^\circ = -(4\,\text{mol}\,e^-)\left(\dfrac{96{,}500\,\text{C}}{1\,\text{mol}\,e^-}\right)(1.229\,\text{V})\left(\dfrac{1\,\text{J}}{1\,\text{C}\cdot\text{V}}\right) = -474{,}394\,\text{J} = -474.4\,\text{kJ}$

$4\,[\text{H}_2\text{O}(l) \rightleftharpoons \text{H}^+(aq) + \text{OH}^-(aq)]$ $K = (K_w)^4$

$\Delta G^\circ = -RT\ln K = -(8.314 \times 10^{-3}\,\text{kJ/K})(298\,\text{K})\ln(1.0 \times 10^{-14})^4 = +319.5\,\text{kJ}$

$4\,[\text{Au}(s) \rightarrow \text{Au}^{3+}(aq) + 3\,e^-]$ $E^\circ = -1.498\,\text{V}$

$\Delta G^\circ = -nFE^\circ = -(12\,\text{mol}\,e^-)\left(\dfrac{96{,}500\,\text{C}}{1\,\text{mol}\,e^-}\right)(-1.498\,\text{V})\left(\dfrac{1\,\text{J}}{1\,\text{C}\cdot\text{V}}\right) = +1{,}734{,}684\,\text{J} = +1{,}734.7\,\text{kJ}$

$4\,[\text{Au}^{3+}(aq) + 2\,e^- \rightarrow \text{Au}^+(aq)]$ $E^\circ = 1.401\,\text{V}$

$\Delta G^\circ = -nFE^\circ = -(8\,\text{mol}\,e^-)\left(\dfrac{96{,}500\,\text{C}}{1\,\text{mol}\,e^-}\right)(1.401\,\text{V})\left(\dfrac{1\,\text{J}}{1\,\text{C}\cdot\text{V}}\right) = -1{,}081{,}572\,\text{J} = -1{,}081.6\,\text{kJ}$

Overall reaction:

$4\,\text{Au}(s) + 8\,\text{CN}^-(aq) + O_2(g) + 2\,\text{H}_2\text{O}(l) \rightarrow 4\,\text{Au(CN)}_2^-(aq) + 4\,\text{OH}^-(aq)$

$\Delta G^\circ = -885.2\,\text{kJ} - 474.4\,\text{kJ} + 319.5\,\text{kJ} + 1{,}734.7\,\text{kJ} - 1{,}081.6\,\text{kJ} = -387.0\,\text{kJ}$

Hydrogen, Oxygen, and Water

18.1 $PV = nRT; \quad PV = \dfrac{g}{\text{molar mass}} RT$

$d_{H_2} = \dfrac{g}{V} = \dfrac{P(\text{molar mass})}{RT} = \dfrac{(1.00 \text{ atm})(2.016 \text{ g/mol})}{\left(0.08206 \dfrac{L \cdot atm}{K \cdot mol}\right)(298 \text{ K})} = 0.0824 \text{ g/L}$

$1 \text{ L} = 1000 \text{ mL} = 1000 \text{ cm}^3$

$d_{H_2} = 0.0824 \text{ g}/1000 \text{ cm}^3 = 8.24 \times 10^{-5} \text{ g/cm}^3$

$\dfrac{d_{air}}{d_{H_2}} = \dfrac{1.185 \times 10^{-3} \text{ g/cm}^3}{8.24 \times 10^{-5} \text{ g/cm}^3} = 14.4; \quad$ Air is 14 times more dense than H_2.

18.2 For every 100.0 g, there are:

61.4 g O, 22.9 g C, 10.0 g H, 2.6 g N, and 3.1 g other

$10.0 \text{ g H} \times \dfrac{1 \text{ mol H}}{1.008 \text{ g H}} = 9.921 \text{ mol H}$

Assume the sample contains 9.921 mol D.

$\text{mass D} = 9.921 \text{ mol D} \times \dfrac{2.0141 \text{ g D}}{1 \text{ mol D}} = 20.0 \text{ g D}$

(a) Total mass if all H is D is:

$61.4 \text{ g O} + 22.9 \text{ C} + 20.0 \text{ g D} + 2.6 \text{ g N} + 3.1 \text{ g other} = 110.0 \text{ g}$

$\text{mass \% D} = \dfrac{20.0 \text{ g D}}{110.0 \text{ g}} \times 100\% = 18.2\% \text{ D}$

(b) Because the mass % of H in the human body is 10.0%, the mass of H in a 150-pound person is (150 lbs x 0.100) 15.0 pounds. ^{2}H weighs twice as much as ^{1}H. If all the ^{1}H were replaced by ^{2}H, the mass of ^{2}H would be 30.0 pounds and the 150-pound person would have gained 15.0 pounds.

18.3

(a) $H_2O(g) + C(s) \xrightarrow{1000\,°C} CO(g) + H_2(g)$

(b) $C_3H_8(g) + 3 H_2O(g) \rightarrow 7 H_2(g) + 3 CO(g)$

18.4 (a) SiH_4, covalent (b) KH, ionic (c) H_2Se, covalent

18.5 (a) $SrH_2(s) + 2 H_2O(l) \rightarrow 2 H_2(g) + Sr^{2+}(aq) + 2 OH^-(aq)$

(b) $KH(s) + H_2O(l) \rightarrow H_2(g) + K^+(aq) + OH^-(aq)$

18.6 (a) (1) ZrH_x, interstitial   (2) PH_3, covalent   (3) HBr, covalent   (4) LiH, ionic

(b) (1) and (4) are likely to be solids at 25 °C. (2) and (3) are likely to be gases at 25 °C. Covalent hydrides, like (2) and (3), form discrete molecules and have only relatively weak intermolecular forces, resulting in gases. (4) is an ionic metal hydride with strong ion-ion forces holding the 3-dimensional lattice together in the solid state. (1) is an interstitial hydride with the metal atoms in a solid crystal lattice and H's occupying holes.

(c) $LiH(s) + H_2O(l) \rightarrow H_2(g) + Li^+(aq) + OH^-(aq)$

18.7 Assume 12.0 g of Pd with a volume of 1.0 cm^3.

$V_{H_2} = 935\ cm^3 = 935\ mL = 0.935\ L$

$$PV = nRT; \quad n_{H_2} = \frac{PV}{RT} = \frac{(1.00\ atm)(0.935\ L)}{\left(0.082\ 06\ \dfrac{L \cdot atm}{K \cdot mol}\right)(273\ K)} = 0.0417\ mol\ H_2$$

$n_H = 2\,n_{H_2} = 0.0834\ mol\ H$

$12.0\ g\ Pd \times \dfrac{1\ mol\ Pd}{106.42\ g\ Pd} = 0.113\ mol\ Pd$

$Pd_{0.113}H_{0.0834}$

$Pd_{0.113/0.113}H_{0.0834/0.113}$

$PdH_{0.74}$

g H = (0.0834 mol H)(1.008 g/mol) = 0.0841 g H

$d_H = 0.0841\ g/cm^3; \quad M_H = \dfrac{0.0834\ mol}{0.001\ L} = 83.4\ M$

18.8 A is Li; B is Ga; C is C
(a) Li_2O, Ga_2O_3, CO_2
(b) Li_2O is the most ionic. CO_2 is the most covalent.
(c) CO_2 is the most acidic. Li_2O is the most basic.
(d) Ga_2O_3 is amphoteric and can react with both $H^+(aq)$ and $OH^-(aq)$.

18.9 (a) $Li_2O(s) + H_2O(l) \rightarrow 2\ Li^+(aq) + 2\ OH^-(aq)$
(b) $SO_3(l) + H_2O(l) \rightarrow H^+(aq) + HSO_4^-(aq)$
(c) $Cr_2O_3(s) + 6\ H^+(aq) \rightarrow 2\ Cr^{3+}(aq) + 3\ H_2O(l)$
(d) $Cr_2O_3(s) + 2\ OH^-(aq) + 3\ H_2O(l) \rightarrow 2\ Cr(OH)_4^-(aq)$

18.10 (a) Rb_2O_2 Rb +1, O –1, peroxide   (b) CaO   Ca +2, O –2, oxide
(c) CsO_2 Cs +1, O –1/2, superoxide   (d) SrO_2 Sr +2, O –1, peroxide
(e) CO_2 C +4, O –2, oxide

18.11 (a) $Rb_2O_2(s) + H_2O(l) \rightarrow 2\ Rb^+(aq) + HO_2^-(aq) + OH^-(aq)$
(b) $CaO(s) + H_2O(l) \rightarrow Ca^{2+}(aq) + 2\ OH^-(aq)$
(c) $2\ CsO_2(s) + H_2O(l) \rightarrow O_2(g) + 2\ Cs^+(aq) + HO_2^-(aq) + OH^-(aq)$
(d) $SrO_2(s) + H_2O(l) \rightarrow Sr^{2+}(aq) + HO_2^-(aq) + OH^-(aq)$
(e) $CO_2(g) + H_2O(l) \rightleftharpoons H^+(aq) + HCO_3^-(aq)$

18.12

$$\begin{array}{ll} \sigma^*_{2p} & \underline{\quad} \\ \pi^*_{2p} & \underline{\uparrow\downarrow}\ \underline{\uparrow} \\ \pi_{2p} & \underline{\uparrow\downarrow}\ \underline{\uparrow\downarrow} \\ \sigma_{2p} & \underline{\uparrow\downarrow} \\ \sigma^*_{2s} & \underline{\uparrow\downarrow} \\ \sigma_{2s} & \underline{\uparrow\downarrow} \\ & O_2^- \end{array}$$

O_2^- is paramagnetic with one unpaired electron.

$$O_2^- \text{ bond order} = \frac{\left(\begin{array}{c}\text{number of}\\\text{bonding electrons}\end{array}\right) - \left(\begin{array}{c}\text{number of}\\\text{antibonding electrons}\end{array}\right)}{2} = \frac{8-5}{2} = 1.5$$

18.13   H—Ö—Ö—H The electron dot structure indicates a single bond (see text Table 18.2) which is consistent with an O–O bond length of 148 pm.

18.14 $PbS(s) + 4\ H_2O_2(aq) \rightarrow PbSO_4(s) + 4\ H_2O(l)$

18.15   mass of H_2O = 5.62 g – 3.10 g = 2.52 g H_2O

$$2.52\ \text{g}\ H_2O \times \frac{1\ \text{mol}\ H_2O}{18.02\ \text{g}\ H_2O} = 0.140\ \text{mol}\ H_2O$$

$$3.10\ \text{g}\ NiSO_4 \times \frac{1\ \text{mol}\ NiSO_4}{154.8\ \text{g}\ NiSO_4} = 0.0200\ \text{mol}\ NiSO_4$$

$$\text{number of}\ H_2O\text{'s in hydrate} = \frac{n_{H_2O}}{n_{NiSO_4}} = \frac{0.140\ \text{mol}}{0.0200\ \text{mol}} = 7$$

Hydrate formula is $NiSO_4 \cdot 7\ H_2O$

18.16   Hydrogen can be stored as a solid in the form of solid interstitial hydrides or in the recently discovered tube-shaped molecules called carbon nanotubes.

18.17 $H_2(g) + 1/2\ O_2(g) \rightarrow H_2O(g)$ $\Delta H° = -242$ kJ

$$\text{mol}\ H_2 = 1.45 \times 10^6\ \text{L} \times \frac{0.088\ \text{kg}}{1\ \text{L}} \times \frac{1000\ \text{g}}{1\ \text{kg}} \times \frac{1\ \text{mol}\ H_2}{2.016\ \text{g}\ H_2} = 6.33 \times 10^7\ \text{mol}\ H_2$$

$$q = 6.33 \times 10^7\ \text{mol}\ H_2 \times \frac{242\ \text{kJ}}{1\ \text{mol}\ H_2} = 1.5 \times 10^{10}\ \text{kJ}$$

$$\text{mass}\ O_2 = 6.33 \times 10^7\ \text{mol}\ H_2 \times \frac{0.5\ \text{mol}\ O_2}{1\ \text{mol}\ H_2} \times \frac{32.00\ \text{g}\ O_2}{1\ \text{mol}\ O_2} \times \frac{1\ \text{kg}}{1000\ \text{g}} = 1.0 \times 10^6\ \text{kg}\ O_2$$

Conceptual Problems

18.18   (a) (1) covalent   (2) ionic   (3) covalent   (4) interstitial
          (b) (1) H, +1; other element, –3   (2) H, –1; other element, +1
                (3) H, +1; other element, –2

18.20   (a) Because of the unpaired electron, the compound is a superoxide. The chemical formula is KO_2.

(b) Because of the unpaired electron, the compound is paramagnetic and attracted by a magnetic field.

(c) O_2 has a bond order of 2; O_2^- has a bond order of 1.5. The O–O bond length in O_2^- is longer and the bond energy is smaller than in O_2.

(d) The solution is basic because of the following reaction that produces OH^-:

$$2\ KO_2(s) + H_2O(l) \rightarrow O_2(g) + 2\ K^+(aq) + HO_2^-(aq) + OH^-(aq)$$

18.22   (a) (1) −2, +2; (2) −2, +1; (3) −2, +5

(b) (1) three-dimensional; (2) molecular; (3) molecular

(c) (1) solid; (2) gas or liquid; (3) gas or liquid

(d) (2) hydrogen; (3) nitrogen

18.24   (a) The ionic hydride (4) has the highest melting point.

(b) (1), (2), and (3) are covalent hydrides. (1) and (2) can hydrogen bond, (3) cannot. Consequently, (3) has the lowest boiling point.

(c) (1), water, and (4), the ionic hydride react together to form $H_2(g)$.

18.26   (a) A, KH; B, MgH_2; C, H_2O; D, HCl

(b) HCl

(c) $KH(s) + H_2O(l) \rightarrow H_2(g) + K^+(aq) + OH^-(aq)$

      $MgH_2(s) + 2\ H_2O(l) \rightarrow 2\ H_2(g) + Mg^{2+}(aq) + 2\ OH^-(aq)$

(d) HCl reacts with water to give an acidic solution. KH and MgH_2 react with water to give a basic solution.

Section Problems
Chemistry of Hydrogen (Sections 18.1–18.4)

18.28   There are 18 kinds of H_2O.

$H_2^{16}O$	$H_2^{17}O$	$H_2^{18}O$
$D_2^{16}O$	$D_2^{17}O$	$D_2^{18}O$
$T_2^{16}O$	$T_2^{17}O$	$T_2^{18}O$
$HD^{16}O$	$HD^{17}O$	$HD^{18}O$
$HT^{16}O$	$HT^{17}O$	$HT^{18}O$
$DT^{16}O$	$DT^{17}O$	$DT^{18}O$

18.30   (a) $Zn(s) + 2\ H^+(aq) \rightarrow H_2(g) + Zn^{2+}(aq)$

(b) at 1000 °C, $H_2O(g) + C(s) \rightarrow CO(g) + H_2(g)$

(c) at 1100 °C with a Ni catalyst, $H_2O(g) + CH_4(g) \rightarrow CO(g) + 3\ H_2(g)$

(d) There are a number of possibilities. (b) and (c) above are two; electrolysis is another: $2\ H_2O(l) \rightarrow 2\ H_2(g) + O_2(g)$

18.32   The steam-hydrocarbon reforming process is the most important industirial preparation of hydrogen.

$$CH_4(g) + H_2O(g) \xrightarrow[\text{Ni catalyst}]{1100\ °C} CO(g) + 3\ H_2(g)$$

$$CO(g) + H_2O(g) \xrightarrow{400\ °C} CO_2(g) + H_2(g)$$

$$CO_2(g) + 2\ OH^-(aq) \rightarrow CO_3^{2-}(aq) + H_2O(l)$$

18.34   (a) LiH, 7.95; CaH_2, 42.09

$$LiH(s) + H_2O(l) \rightarrow H_2(g) + Li^+(aq) + OH^-(aq)$$

$$CaH_2(s) + 2\ H_2O(l) \rightarrow 2\ H_2(g) + Ca^{2+}(aq) + 2\ OH^-(aq)$$

You obtain $\dfrac{1\ \text{mol } H_2}{7.95\ \text{g LiH}} = 0.126$ mol H_2/g LiH

and $\dfrac{2\ \text{mol } H_2}{42.09\ \text{g CaH}_2} = 0.0475$ mol H_2/g CaH_2;   Therefore, LiH gives more H_2.

(b) 25 °C = 298 K

$$PV = nRT; \quad n_{H_2} = \frac{PV}{RT} = \frac{(150\ \text{atm})(100\ \text{L})}{\left(0.082\ 06\ \dfrac{\text{L} \cdot \text{atm}}{\text{K} \cdot \text{mol}}\right)(298\text{K})} = 613.4\ \text{mol } H_2$$

$$\text{mass CaH}_2 = 613.4\ \text{mol } H_2 \times \frac{1\ \text{mol CaH}_2}{2\ \text{mol } H_2} \times \frac{42.09\ \text{g CaH}_2}{1\ \text{mol CaH}_2} \times \frac{1\ \text{kg}}{1000\ \text{g}} = 12.9\ \text{kg CaH}_2$$

18.36   (a) MgH_2, H^- (b) PH_3, covalent   (c) KH, H^- (d) HBr, covalent

18.38 H_2S covalent hydride, gas, weak acid in H_2O
NaH   ionic hydride, solid (salt like), reacts with H_2O to produce H_2
PdH_x metallic (interstitial) hydride, solid, stores hydrogen

18.40   (a) CH_4, covalent bonding   (b) NaH, ionic bonding

18.42   (a) H—S̈e—H , bent   (b) H—Äs—H, trigonal pyramidal
                                                              |
                                                              H

(c)   H   , tetrahedral
                      |
            H—Si—H
                      |
                      H

18.44   A nonstoichiometric compound is a compound whose atomic composition cannot be expressed as a ratio of small whole numbers. An example is PdH_x. The lack of stoichiometry results from the hydrogen occupying holes in the solid state structure.

18.46 $BaH_2(s) + 2 H_2O(l) \rightarrow 2 H_2(g) + Ba^{2+}(aq) + 2 OH^-(aq)$
BaH_2, 139.3; 20 °C = 293 K

$$PV = nRT; \qquad n_{H_2} = \frac{PV}{RT} = \frac{(0.975 \text{ atm})(4.36 \text{ L})}{\left(0.082\ 06\ \dfrac{\text{L} \cdot \text{atm}}{\text{K} \cdot \text{mol}}\right)(293 \text{ K})} = 0.177 \text{ mol } H_2$$

$$0.177 \text{ mol } H_2 \times \frac{1 \text{ mol } BaH_2}{2 \text{ mol } H_2} \times \frac{139.3 \text{ g } BaH_2}{1 \text{ mol } BaH_2} = 12.3 \text{ g } BaH_2$$

18.48   (a) TiH_2, 49.88; Assume 1.0 cm³ of TiH_2 which has a mass of 3.75 g.

$$3.75 \text{ g } TiH_2 \times \frac{1 \text{ mol } TiH_2}{49.88 \text{ g } TiH_2} = 0.0752 \text{ mol } TiH_2$$

$$0.0752 \text{ mol } TiH_2 \times \frac{2 \text{ mol H}}{1 \text{ mol } TiH_2} = 0.150 \text{ mol H}$$

$$0.150 \text{ mol H} \times \frac{1.008 \text{ g H}}{1 \text{ mol H}} = 0.151 \text{ g H}$$

$d_H = 0.15$ g/cm³; the density of H in TiH_2 is about 2.1 times the density of liquid H_2.

(b)

$$PV = nRT; \quad V = \frac{nRT}{P} = \frac{\left(0.15 \text{ g} \times \dfrac{1 \text{ mol}}{2.016 \text{ g}}\right)\left(0.082\ 06\ \dfrac{\text{L} \cdot \text{atm}}{\text{K} \cdot \text{mol}}\right)(273 \text{ K})}{1.00 \text{ atm}} = 1.7 \text{ L } H_2$$

1.7 L = 1.7 x 10³ mL = 1.7 x 10³ cm³

Chemistry of Oxygen (Sections 18.5–18.11)

18.50   (a) O_2 is obtained in industry by the fractional distillation of liquid air.
(b) In the laboratory, O_2 is prepared by the thermal decomposition of $KClO_3(s)$.

$$2 KClO_3(s) \xrightarrow[\text{MnO}_2]{\text{heat}} 2 KCl(s) + 3 O_2(g)$$

18.52 $2 KClO_3(s) \rightarrow 2 KCl(s) + 3 O_2(g)$
$KClO_3$, 122.6; 25 °C = 298 K

$$\text{mol } O_2 = 0.200 \text{ g } KClO_3 \times \frac{1 \text{ mol } KClO_3}{122.6 \text{ g } KClO_3} \times \frac{3 \text{ mol } O_2}{2 \text{ mol } KClO_3} = 2.45 \times 10^{-3} \text{ mol } O_2$$

$$PV = nRT; \quad V = \frac{nRT}{P} = \frac{(2.45 \times 10^{-3} \text{ mol})\left(0.082\ 06\ \dfrac{\text{L} \cdot \text{atm}}{\text{K} \cdot \text{mol}}\right)(298 \text{ K})}{1.00 \text{ atm}} = 0.0599 \text{ L}$$

$$V = 0.0599 \text{ L} \times \frac{1000 \text{ mL}}{1 \text{ L}} = 59.9 \text{ mL } O_2$$

18.54 $$2\,H_2O_2(aq) \xrightarrow{\text{catalyst}} 2\,H_2O(l) + O_2(g)$$

H_2O_2, 34.01; 25 °C = 298 K

$$\text{mol } O_2 = 20.4 \text{ g } H_2O_2 \times \frac{1 \text{ mol } H_2O_2}{34.01 \text{ g } H_2O_2} \times \frac{1 \text{ mol } O_2}{2 \text{ mol } H_2O_2} = 0.300 \text{ mol } O_2$$

$$PV = nRT; \quad V = \frac{nRT}{P} = \frac{(0.300 \text{ mol})\left(0.082\,06\,\dfrac{L \cdot atm}{K \cdot mol}\right)(298 \text{ K})}{1.00 \text{ atm}} = 7.34 \text{ L } O_2$$

18.56   (a) $4\,Li(s) + O_2(g) \rightarrow 2\,Li_2O(s)$ (b) $P_4(s) + 5\,O_2(g) \rightarrow P_4O_{10}(s)$
(c) $4\,Al(s) + 3\,O_2(g) \rightarrow 2\,Al_2O_3(s)$ (d) $Si(s) + O_2(g) \rightarrow SiO_2(s)$

18.58 $:\ddot{O}::\ddot{O}:$ The electron dot structure shows an O=O double bond. It also shows all electrons paired. This is not consistent with the fact that O_2 is paramagnetic.

18.60   An element that forms an acidic oxide is more likely to form a covalent hydride. C and N are examples.

18.62 $Li_2O < BeO < B_2O_3 < CO_2 < N_2O_5$ (see Figure 18.6)

18.64 $N_2O_5 < Al_2O_3 < K_2O < Cs_2O$ (see Figure 18.6)

18.66   (a) CrO_3 (higher Cr oxidation state)   (b) N_2O_5 (higher N oxidation state)
(c) SO_3 (higher S oxidation state)

18.68   (a) $Cl_2O_7(l) + H_2O(l) \rightarrow 2\,H^+(aq) + 2\,ClO_4^-(aq)$
(b) $K_2O(s) + H_2O(l) \rightarrow 2\,K^+(aq) + 2\,OH^-(aq)$
(c) $SO_3(l) + H_2O(l) \rightarrow H^+(aq) + HSO_4^-(aq)$

18.70   (a) $ZnO(s) + 2\,H^+(aq) \rightarrow Zn^{2+}(aq) + H_2O(l)$
(b) $ZnO(s) + 2\,OH^-(aq) + H_2O(l) \rightarrow Zn(OH)_4^{2-}(aq)$

18.72   A peroxide has oxygen in the −1 oxidation state, for example, H_2O_2. A superoxide has oxygen in the −1/2 oxidation state, for example, KO_2.

18.74   (a) BaO_2 (b) CaO (c) CsO_2 (d) Li_2O (e) Na_2O_2

18.76

	O_2	O_2^-	O_2^{2-}
σ^*_{2p}	— —	— —	— —
π^*_{2p}	↑ ↑	↑↓ ↑	↑↓ ↑↓
π_{2p}	↑↓ ↑↓	↑↓ ↑↓	↑↓ ↑↓
σ_{2p}	↑↓	↑↓	↑↓
	Bond order = 2	Bond order = 1.5	Bond order = 1

(a) The O–O bond length increases because the bond order decreases. The bond order decreases because of the increased occupancy of antibonding orbitals.

(b) O_2^- has 1 unpaired electron and is paramagnetic. O_2^{2-} has no unpaired electrons and is diamagnetic.

18.78 (a) $H_2O_2(aq) + 2 H^+(aq) + 2 I^-(aq) \rightarrow I_2(aq) + 2 H_2O(l)$
(b) $3 H_2O_2(aq) + 8 H^+(aq) + Cr_2O_7^{2-}(aq) \rightarrow 2 Cr^{3+}(aq) + 3 O_2(g) + 7 H_2O(l)$

18.80

Ozone has two resonance structures, consistent with two equivalent O–O bond lengths.

18.82 $3 O_2(g) \xrightarrow{\text{electric discharge}} 2 O_3(g)$

Water and Hydrates (Sections 18.12 and 18.13)

18.84 Convert mi^3 to cm^3; $(1.0 \text{ mi}^3)\left(\dfrac{1609 \text{ m}}{1 \text{ mi}}\right)^3\left(\dfrac{100 \text{ cm}}{1 \text{ m}}\right)^3 = 4.2 \times 10^{15} \text{ cm}^3$

mass of sea water = volume x density = $(4.2 \times 10^{15} \text{ cm}^3)(1.025 \text{ g/cm}^3) = 4.3 \times 10^{15} \text{ g}$

mass of salts = $(0.035)(4.3 \times 10^{15} \text{ g})\left(\dfrac{1 \text{ kg}}{1000 \text{ g}}\right) = 1.5 \times 10^{11} \text{ kg}$

18.86 Water that contains appreciable concentrations of doubly charged cations such as Ca^{2+} and Mg^{2+} is called hard water. They form the unwanted metal carbonate precipitates, known as boiler scale, that deposit in boilers, hot water heaters, and teakettles. When hard water containing HCO_3^- anions is heated, the equilibrium for the decomposition of HCO_3^- to CO_2 and CO_3^{2-} shifts to the right as CO_2 gas escapes from the solution. The resulting CO_3^{2-} ions then combine with cations, such as Ca^{2+} and Mg^{2+}, to form insoluble metal carbonates.

18.88 $AlCl_3 \cdot 6 H_2O$

18.90 $CaSO_4 \cdot \frac{1}{2} H_2O$, 145.15; H_2O, 18.02
Assume one mole of $CaSO_4 \cdot \frac{1}{2} H_2O$

mass % $H_2O = \dfrac{\text{mass } H_2O}{\text{mass hydrate}} \times 100\% = \dfrac{\frac{1}{2}(18.02 \text{ g})}{145.15 \text{ g}} \times 100\% = 6.21\%$

18.92 $CaSO_4 \cdot \frac{1}{2} H_2O$, 145.15; H_2O, 18.02

   mass of H_2O lost = 3.44 g – 2.90 g = 0.54 g H_2O

   2.90 g $CaSO_4 \cdot \frac{1}{2} H_2O$ x $\dfrac{1 \text{ mol}}{145.15 \text{ g}}$ = 0.020 mol $CaSO_4 \cdot \frac{1}{2} H_2O$

   0.54 g H_2O x $\dfrac{1 \text{ mol}}{18.02 \text{ g}}$ = 0.030 mol

   number of H_2O's lost = $\dfrac{0.030 \text{ mol}}{0.020 \text{ mol}}$ = 1.5 H_2O per $CaSO_4 \cdot \frac{1}{2} H_2O$ formed

   The mineral gypsum is $CaSO_4 \cdot 2 H_2O$; x = 2

Chapter Problems

18.94   React H_2O with a reducing agent to produce H_2. Ca or Al could be used.

18.96   Butadiene, C_4H_6, 54.09; 2.7 kg = 2700 g

   moles H_2 = 2700 g C_4H_6 x $\dfrac{1 \text{ mol } C_4H_6}{54.09 \text{ g } C_4H_6}$ x $\dfrac{2 \text{ mol } H_2}{1 \text{ mol } C_4H_6}$ = 99.8 mol H_2

   At STP, P = 1.00 atm and T = 273 K

   $PV = nRT;\quad V = \dfrac{nRT}{P} = \dfrac{(99.8 \text{ mol})\left(0.082\ 06\ \dfrac{\text{L} \cdot \text{atm}}{\text{K} \cdot \text{mol}}\right)(273 \text{ K})}{1.00 \text{ atm}}$ = 2.2 x 10^3 L of H_2

18.98   (a) B_2O_3, diboron trioxide   (b) H_2O_2, hydrogen peroxide
   (c) SrH_2, strontium hydride   (d) CsO_2, cesium superoxide
   (e) $HClO_4$, perchloric acid   (f) BaO_2, barium peroxide

18.100 (a) 6; $^{16}O_2$, $^{17}O_2$, $^{18}O_2$, $^{16}O^{17}O$, $^{16}O^{18}O$, $^{17}O^{18}O$
   (b) 18; $^{16}O_3$ $^{16}O_2{}^{17}O$ $^{18}O_3$
   $^{17}O_2{}^{16}O$ $^{17}O_3$ $^{16}O_2{}^{18}O$
   $^{18}O_2{}^{16}O$ $^{18}O_2{}^{17}O$ $^{17}O_2{}^{18}O$
   $^{16}O^{17}O^{16}O$ $^{17}O^{18}O^{17}O$ $^{16}O^{17}O^{18}O$
   $^{16}O^{18}O^{16}O$ $^{17}O^{16}O^{17}O$ $^{18}O^{16}O^{18}O$
   $^{17}O^{18}O^{16}O$ $^{18}O^{16}O^{17}O$ $^{18}O^{17}O^{18}O$

18.102 (a) $2 H_2(g) + O_2(g) \rightarrow 2 H_2O(l)$
   (b) $O_3(g) + 2 I^-(aq) + H_2O(l) \rightarrow O_2(g) + I_2(aq) + 2 OH^-(aq)$
   (c) $H_2O_2(aq) + 2 H^+(aq) + 2 Br^-(aq) \rightarrow 2 H_2O(l) + Br_2(aq)$
   (d) $2 Na(l) + H_2(g) \rightarrow 2 NaH(s)$
   (e) $2 Na(s) + 2 H_2O(l) \rightarrow H_2(g) + 2 Na^+(aq) + 2 OH^-(aq)$

18.104 (a) $CO(g) + 2 H_2(g) \rightarrow CH_3OH(l)$
   $\Delta H° = \Delta H°_f(CH_3OH) - \Delta H°_f(CO)$
   $\Delta H° = (1 \text{ mol})(-239.2 \text{ kJ/mol}) - (1 \text{ mol})(-110.5 \text{ kJ/mol}) = -128.7 \text{ kJ}$

(b) $CO(g) + H_2O(g) \rightarrow CO_2(g) + H_2(g)$

$\Delta H° = \Delta H°_f(CO_2) - [\Delta H°_f(CO) + \Delta H°_f(H_2O)]$

$\Delta H° = (1 \text{ mol})(-393.5 \text{ kJ/mol}) - [(1 \text{ mol})(-110.5 \text{ kJ/mol})$
$+ (1 \text{ mol})(-241.8 \text{ kJ/mol})] = -41.2 \text{ kJ}$

(c) $2 KClO_3(s) \rightarrow 2 KCl(s) + 3 O_2(g)$

$\Delta H° = 2 \Delta H°_f(KCl) - 2 \Delta H°_f(KClO_3)$

$\Delta H° = (2 \text{ mol})(-436.5 \text{ kJ/mol}) - (2 \text{ mol})(-397.7 \text{ kJ/mol}) = -77.6 \text{ kJ}$

(d) $6 CO_2(g) + 6 H_2O(l) \rightarrow 6 O_2(g) + C_6H_{12}O_6(s)$

$\Delta H° = \Delta H°_f(C_6H_{12}O_6) - [6 \Delta H°_f(CO_2) + 6 \Delta H°_f(H_2O)]$

$\Delta H° = (1 \text{ mol})(-1273.3 \text{ kJ/mol}) - [(6 \text{ mol})(-393.5 \text{ kJ/mol})$
$+ (6 \text{ mol})(-285.8 \text{ kJ/mol})] = 2802.5 \text{ kJ}$

18.106 C_6H_{12}, 84.16

$$555 \text{ g C}_6\text{H}_{12} \times \frac{1 \text{ mol C}_6\text{H}_{12}}{84.16 \text{ g C}_6\text{H}_{12}} \times \frac{1 \text{ mol O}_2}{2 \text{ mol C}_6\text{H}_{12}} = 3.30 \text{ mol O}_2$$

$$PV = nRT; \quad V = \frac{nRT}{P} = \frac{(3.30 \text{ mol})\left(0.082\ 06\ \dfrac{\text{L} \cdot \text{atm}}{\text{K} \cdot \text{mol}}\right)(298.15 \text{ K})}{1.00 \text{ atm}} = 80.7 \text{ L}$$

18.108 Assume a 100.0 g sample. From the percent composition data, a 100.0 g sample contains 10.04 g C, 76.90 g O, and 13.06 g H.

$$10.04 \text{ g C} \times \frac{1 \text{ mol C}}{12.01 \text{ g C}} = 0.8360 \text{ mol C}$$

$$76.90 \text{ g O} \times \frac{1 \text{ mol O}}{16.00 \text{ g O}} = 4.806 \text{ mol O}$$

$(CH_4)_x(H_2O)_{23}$

$$\frac{23}{4.806} = \frac{x}{0.8360}$$

$$x = \frac{(23)(0.8360)}{4.806} = 4$$

18.110 $CaSO_4$, 136.14

Assume a 100.0 g sample. From the percent composition data, a 100.0 g sample contains 55.76 g O, and 2.34 g H.

$$2.34 \text{ g H} \times \frac{1 \text{ mol H}}{1.008 \text{ g H}} = 2.32 \text{ mol H}$$

$$2.32 \text{ mol H} \times \frac{1 \text{ mol O}}{2 \text{ mol H}} = 1.16 \text{ mol O from water}$$

$$1.16 \text{ mol O} \times \frac{16.00 \text{ g O}}{1 \text{ mol O}} = 18.6 \text{ g O from water}$$

mass $CaSO_4$ = 100.0 g – 2.34 g H – 18.6 g O = 79.06 g $CaSO_4$

$$79.06 \text{ g CaSO}_4 \times \frac{1 \text{ mol CaSO}_4}{136.14 \text{ g CaSO}_4} = 0.581 \text{ mol CaSO}_4$$

$(\text{CaSO}_4)_{0.581} (\text{H}_2\text{O})_{1.16}$
$(\text{CaSO}_4)_{0.581 \,/\, 0.581} (\text{H}_2\text{O})_{1.16 \,/\, 0.581}$
$(\text{CaSO}_4) (\text{H}_2\text{O})_2, \ x = 2$
$\text{CaSO}_4 \cdot 2 \text{ H}_2\text{O}$

18.112 Mass % O = 58.19% and mass % Na = 100.00 – 58.19 = 41.81%

Assume a 100.0 g sample. From the percent composition data, a 100.0 g sample contains 58.19 g O, and 41.81 g Na.

$$58.19 \text{ g O} \times \frac{1 \text{ mol O}}{16.00 \text{ g O}} = 3.637 \text{ mol O}; \quad 41.81 \text{ g Na} \times \frac{1 \text{ mol Na}}{22.99 \text{ g Na}} = 1.819 \text{ mol Na}$$

$\text{Na}_{1.819}\text{O}_{3.637}$
$\text{Na}_{1.819 \,/\, 1.819}\text{O}_{3.637 \,/\, 1.819}$
NaO_2 is a superoxide.

18.114 $n_{\text{H}_2} = \dfrac{PV}{RT} = \dfrac{(4.192 \text{ atm})(0.500 \text{ L})}{\left(0.082\ 06 \ \dfrac{\text{L} \cdot \text{atm}}{\text{K} \cdot \text{mol}}\right)(448 \text{ K})} = 0.0570 \text{ mol H}_2$

$$0.0570 \text{ mol H}_2 \times \frac{2 \text{ mol H}}{1 \text{ mol H}_2} = 0.114 \text{ mol H}$$

$$0.114 \text{ mol H} \times \frac{1.0079 \text{ g H}}{1 \text{ mol H}} = 0.115 \text{ g H}$$

mass Ge = 3.22 g – 0.115 g = 3.105 g Ge

$$3.105 \text{ g Ge} \times \frac{1 \text{ mol Ge}}{72.61 \text{ g Ge}} = 0.0428 \text{ mol Ge}$$

$\text{Ge}_{0.0428}\text{H}_{0.114}$
$\text{Ge}_{0.0428 \,/\, 0.0428}\text{H}_{0.114 \,/\, 0.0428}$
$\text{GeH}_{2.66}$, multiply each subscript by 3.
$\text{Ge}_{(1 \times 3)}\text{H}_{(2.66 \times 3)}$
Ge_3H_8

Multiconcept Problems

18.116 $2 \text{ O}_3(g) \ \rightarrow \ 3 \text{ O}_2(g) \qquad \Delta H° = -258 \text{ kJ}$
PV = nRT

$$n_{\text{O}_3} = \frac{PV}{RT} = \frac{\left(63.6 \text{ mm Hg} \times \dfrac{1.00 \text{ atm}}{760 \text{ mm Hg}}\right)(1.000 \text{ L})}{\left(0.082\ 06 \ \dfrac{\text{L} \cdot \text{atm}}{\text{K} \cdot \text{mol}}\right)(293 \text{ K})} = 3.48 \times 10^{-3} \text{ mol O}_3$$

$$q = 3.48 \times 10^{-3} \text{ mol O}_3 \times \frac{285 \text{ kJ}}{2 \text{ mol O}_3} \times \frac{1000 \text{ J}}{1 \text{ kJ}} = 496 \text{ J are liberated.}$$

18.118 $MH_2(s) + 2 HCl(aq) \rightarrow 2 H_2(g) + M^{2+}(aq) + 2 Cl^-(aq)$

$PV = nRT$

$$n_{H_2} = \frac{PV}{RT} = \frac{\left(750 \text{ mm Hg} \times \dfrac{1.00 \text{ atm}}{760 \text{ mm Hg}}\right)(1.000 \text{ L})}{\left(0.082\ 06 \dfrac{\text{L} \cdot \text{atm}}{\text{K} \cdot \text{mol}}\right)(293 \text{ K})} = 0.0410 \text{ mol H}_2$$

$$0.0410 \text{ mol H}_2 \times \frac{1 \text{ mol MH}_2}{2 \text{ mol H}_2} = 0.0205 \text{ mol MH}_2$$

$$\text{molar mass} = \frac{1.84 \text{ g}}{0.0205 \text{ mol}} = 89.8 \text{ g/mol}$$

mass M = 89.8 – mass of 2 H = 89.8 – 2(1.008) = 87.7 g/mol;   M = Sr; SrH_2

18.120 PH_3, 34.00

(a) $PH_3(g) + 2 O_2(g) \rightarrow H_3PO_4(s)$

(b) $0.646 \text{ g PH}_3 \times \dfrac{1 \text{ mol PH}_3}{34.00 \text{ g PH}_3} \times \dfrac{1 \text{ mol H}_3PO_4}{1 \text{ mol PH}_3} = 0.0190 \text{ mol H}_3PO_4$

$$[H_3PO_4] = \frac{0.0190 \text{ mol H}_3PO_4}{0.250 \text{ L}} = 0.0760 \text{ M}$$

For the dissociation of the first proton, the following equilibrium must be considered:

	$H_3PO_4(aq)$ + $H_2O(l)$	$\rightleftharpoons$	$H_3O^+(aq)$ +	$H_2PO_4^-(aq)$
initial (M)	0.0760		~0	0
change (M)	–x		+x	+x
equil (M)	0.0760 – x		x	x

$$K_{a1} = \frac{[H_3O^+][H_2PO_4^-]}{[H_3PO_4]} = 7.5 \times 10^{-3} = \frac{x^2}{0.0760 - x}$$

$x^2 + (7.5 \times 10^{-3})x - (5.7 \times 10^{-4}) = 0$

Solve for x using the quadratic formula.

$$x = \frac{-(7.5 \times 10^{-3}) \pm \sqrt{(7.5 \times 10^{-3})^2 - (4)(-5.7 \times 10^{-4})}}{2(1)} = \frac{(-7.5 \times 10^{-3}) \pm 0.0483}{2}$$

x = 0.0204 and –0.0279

Of the two solutions for x, only the positive value of x has physical meaning, because x is the $[H_3O^+]$.

x = 0.0204 M = $[H_3O^+]$

The second and third proton dissociation produce an insignificant amount of additional H_3O^+.

pH = –log$[H_3O^+]$ = –log(0.0204) = 1.69

18.122 Anode (oxidation): $2 H_2O(l) \rightarrow O_2(g) + 4 H^+(aq) + 4 e^-$

Cathode (reduction): $\underline{4 H_2O(l) + 4 e^- \rightarrow 2 H_2(g) + 4 OH^-(aq)}$

Overall cell reaction: $6 H_2O(l) \rightarrow 2 H_2(g) + O_2(g) + 4 H^+(aq) + 4 OH^-(aq)$

$2 H_2O(l) \rightarrow 2 H_2(g) + O_2(g)$

5.00 A = 5.00 C/s

$$C = 5.00 \text{ C/s} \times 10 \text{ min} \times \frac{60 \text{ s}}{1 \text{ min}} = 3000 \text{ C}$$

$$\text{mole } e^- = 3000 \text{ C} \times \frac{1 \text{ mol } e^-}{96{,}500 \text{ C}} = 0.031 \text{ 09 mole } e^-$$

$$0.031 \text{ 09 mole } e^- \times \frac{2 \text{ mol H}_2}{4 \text{ mol } e^-} = 0.015 \text{ 55 mol H}_2$$

$$0.031 \text{ 09 mole } e^- \times \frac{1 \text{ mol O}_2}{4 \text{ mol } e^-} = 0.007 \text{ 773 mol O}_2$$

$$n_{H_2 \text{ gas}} = \frac{PV}{RT} = \frac{\left(740 \text{ mm Hg} \times \dfrac{1.00 \text{ atm}}{760 \text{ mm Hg}}\right)(0.3525 \text{ L})}{\left(0.082 \text{ 06} \dfrac{L \cdot atm}{K \cdot mol}\right)(298 \text{ K})} = 0.014 \text{ 04 mol H}_2$$

$$n_{H_2 \text{ total}} - n_{H_2 \text{ gas}} = 0.015 \text{ 55 mol} - 0.014 \text{ 04 mol} = 0.001 \text{ 51 mol H}_2 \text{ in solution}$$

$$k = \frac{M}{P} = \frac{(0.001 \text{ 51 mol/2.0 L})}{\left(740 \text{ mm Hg} \times \dfrac{1.00 \text{ atm}}{760 \text{ mm Hg}}\right)} = 7.8 \times 10^{-4} \text{ mol/(L} \cdot atm)$$

18.124 N_2O_5, 108.01

$$N_2O_5(g) + H_2O(l) \rightarrow 2 \text{ HNO}_3(aq)$$
$$2 \text{ HNO}_3(aq) + Zn(s) \rightarrow H_2(g) + Zn(NO_3)_2(aq)$$

$$5.4 \text{ g N}_2O_5 \times \frac{1 \text{ mol N}_2O_5}{108.01 \text{ g N}_2O_5} \times \frac{2 \text{ mol HNO}_3}{1 \text{ mol N}_2O_5} \times \frac{1 \text{ mol H}_2}{2 \text{ mol HNO}_3} = 0.050 \text{ mol H}_2$$

$$PV = nRT; \quad P_{H_2} = \frac{nRT}{V} = \frac{(0.050 \text{ mol})\left(0.082 \text{ 06} \dfrac{L \cdot atm}{K \cdot mol}\right)(298 \text{ K})}{0.500 \text{ L}} = 2.45 \text{ atm}$$

$$P_{H_2} = 2.45 \text{ atm} \times \frac{760 \text{ mm Hg}}{1.00 \text{ atm}} = 1862 \text{ mm Hg}$$

(a) In H there is 0.0156 atom % D.

To get P_{HD} multiply P_{H_2} by the atom % D and then by 2 because H_2 is diatomic.

$$P_{H_2} = (1862 \text{ mm Hg})(0.000156)(2) = 0.58 \text{ mm Hg}$$

(b) $$n_{HD} = \frac{PV}{RT} = \frac{\left(0.58 \text{ mm Hg} \times \dfrac{1.00 \text{ atm}}{760 \text{ mm Hg}}\right)(0.5000 \text{ L})}{\left(0.082 \text{ 06} \dfrac{L \cdot atm}{K \cdot mol}\right)(298 \text{ K})} = 1.56 \times 10^{-5} \text{ mol HD}$$

$$1.56 \times 10^{-5} \text{ mol HD} \times \frac{6.022 \times 10^{23} \text{ HD molecules}}{1 \text{ mol HD}} = 9.4 \times 10^{18} \text{ HD molecules}$$

(c) $P_{D_2} = (1862 \text{ mm Hg})(0.000\ 156)^2 = 4.53 \times 10^{-5} \text{ mm Hg}$

$$n_{D_2} = \frac{PV}{RT} = \frac{\left(4.53 \times 10^{-5} \text{ mm Hg} \times \dfrac{1.00 \text{ atm}}{760 \text{ mm Hg}}\right)(0.5000 \text{ L})}{\left(0.082\ 06 \dfrac{\text{L} \cdot \text{atm}}{\text{K} \cdot \text{mol}}\right)(298 \text{ K})} = 1.22 \times 10^{-9} \text{ mol D}_2$$

$$1.22 \times 10^{-9} \text{ mol D}_2 \times \frac{6.022 \times 10^{23} \text{ D}_2 \text{ molecules}}{1 \text{ mol D}_2} = 7.3 \times 10^{14} \text{ D}_2 \text{ molecules}$$

18.126
$$\ln P_2 = \ln P_1 + \frac{\Delta H_{vap}}{R}\left(\frac{1}{T_1} - \frac{1}{T_2}\right)$$

$$\Delta H_{vap} = \frac{(\ln P_2 - \ln P_1)(R)}{\left(\dfrac{1}{T_1} - \dfrac{1}{T_2}\right)}$$

$P_1 = 75.0 \text{ mm Hg};$ $T_1 = 89\ {}^\circ\text{C} = 89 + 273 = 362 \text{ K}$
$P_2 = 319.2 \text{ mm Hg};$ $T_2 = 125\ {}^\circ\text{C} = 125 + 273 = 398 \text{ K}$

$$\Delta H_{vap} = \frac{[\ln(319.2) - \ln(75.0)](8.314 \times 10^{-3} \text{ kJ/(K} \cdot \text{mol)})}{\left(\dfrac{1}{362 \text{ K}} - \dfrac{1}{398 \text{ K}}\right)} = 48.2 \text{ kJ/mol}$$

Now calculate the normal boiling point.
$P_1 = 75.0 \text{ mm Hg};$ $T_1 = 89\ {}^\circ\text{C} = 89 + 273 = 362 \text{ K}$
$P_2 = 760 \text{ mm Hg};$ $T_2 = ?$

$$\ln P_2 = \ln P_1 + \frac{\Delta H_{vap}}{R}\left(\frac{1}{T_1} - \frac{1}{T_2}\right)$$

$$(\ln P_2 - \ln P_1)\left(\frac{R}{\Delta H_{vap}}\right) = \frac{1}{T_1} - \frac{1}{T_2}$$

Solve for T_2 (the boiling point for H_2O_2 at 760 mm Hg).

$$\frac{1}{T_1} - (\ln P_2 - \ln P_1)\left(\frac{R}{\Delta H_{vap}}\right) = \frac{1}{T_2}$$

$$\frac{1}{362 \text{ K}} - [\ln(760) - \ln(75.0)]\left(\frac{8.314 \times 10^{-3} \text{ kJ/(K} \cdot \text{mol)}}{48.2 \text{ kJ/mol}}\right) = \frac{1}{T_2} = 0.002\ 363/\text{K}$$

$T_2 = 1/0.002\ 363/\text{K} = 423 \text{ K} = 423 - 273 = 150\ {}^\circ\text{C}$ (boiling point for H_2O_2)
The calculated boiling point is the same as the one listed in section 18.11.

19 The Main-Group Elements

19.1 (a) B is above Al in group 3A, and therefore B is more nonmetallic than Al.
(b) Ge and Br are in the same row of the periodic table, but Br (group 7A) is to the right of Ge (group 4A). Therefore, Br is more nonmetallic.
(c) Se (group 6A) is more nonmetallic than In because it is above and to the right of In (group 3A).
(d) Cl (group 7A) is more nonmetallic than Te because it is above and to the right of Te (group 6A).

19.2 (a) HNO_3 $\qquad\qquad\qquad\qquad\qquad\qquad$ H_3PO_4

Nitrogen can form very strong $p\pi$ - $p\pi$ bonds. Phosphorus forms weaker $p\pi$ - $p\pi$ bonds, so it tends to form more single bonds.
(b) The larger S atom can accommodate six bond pairs in its valence shell, but the smaller O atom is limited to two bond pairs and two lone pairs.

19.3 Carbon forms strong π bonds with oxygen. Silicon does not form strong π bonds with oxygen, and what results are chains of alternating silicon and oxygen singly bonded to each other.

19.4 An ethane-like structure is unlikely for diborane because it would require 14 valence electrons and diborane only has 12. The result is two three-center, two-electron bonds between the borons and the bridging hydrogen atoms.

19.5 $Hb–O_2 + CO \rightleftharpoons Hb–CO + O_2$
Mild cases of carbon monoxide poisoning can be treated with O_2. Le Châtelier's principle says that adding a product (O_2) will cause the reaction to proceed in the reverse direction, back to $Hb–O_2$.

19.6 (a) $Si_8O_{24}{}^{16-}$ $\qquad\qquad$ (b) $Si_2O_5{}^{2-}$

19.7   Formal charge = $\left(\begin{array}{c}\text{Number of}\\\text{valence electrons}\\\text{in free atom}\end{array}\right) - \dfrac{1}{2}\left(\begin{array}{c}\text{Number of}\\\text{bonding}\\\text{electrons}\end{array}\right) - \left(\begin{array}{c}\text{Number of}\\\text{nonbonding}\\\text{electrons}\end{array}\right)$

Nitrous oxide, N_2O

0 +1 –1   –1 +1 0   –2 +1 +1

:N≡N–Ö: ⟷ :N̈=N=Ö: ⟷ :N̈–N≡O:

The first structure is most important because it has a –1 formal charge on an electronegative oxygen.

Nitric oxide, NO

0   0   –1 +1   paramagnetic

:N̈=Ö: ⟷ ·N̈=Ö:

The first structure is more important because it has no formal charges.

Nitrous acid, HNO_2

0   0   0   +1   0   –1

H–Ö–N̈=Ö: ⟷ H–Ö=N–Ö:

The first structure is more important because it has no formal charges.

Nitrogen dioxide, NO_2

–1 +1   0   0   +1 –1   paramagnetic

:Ö–N̈=Ö: ⟷ :Ö=N̈–Ö:

Both structures are of equal importance.

Nitric acid, HNO_3

0 +1   0   0 +1 –1   +1 +1 –1

H–Ö–N̈=Ö: ⟷ H–Ö–N–Ö: ⟷ H–Ö=N–Ö:

      | ‖ |

    :Ö: Ö: :Ö:

     –1   0   –1

The first two structures are of equal importance. Both are more important than structure three because it has a +1 formal charge on an electronegative oxygen.

19.8   (a) SO_3^{2-}, HSO_3^-, SO_4^{2-}, HSO_4^- (b) HSO_4^- (c) SO_3^{2-} (d) HSO_4^-

19.9   (a)   H–S̈–H , bent.

       (b)   :Ö–S̈=Ö: ⟷ :Ö=S̈–Ö: , bent, S is sp^2 hybridized.

       (c)   :Ö: :Ö: :Ö

                | | ‖

        :Ö–S̈=Ö: ⟷ :Ö=S̈–Ö: ⟷ :Ö–S̈–Ö:

        trigonal planar, S is sp^2 hybridized.

19.10   Because copper is always conducting, a laser printer drum coated with copper would not hold any charge so no document image would adhere to the drum for printing.

Conceptual Problems

19.12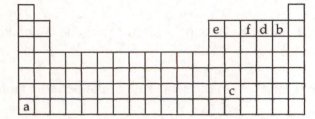

19.14 (a) PF_5 and SF_6 (b) CH_4 and NH_4^+ (c) CO and NO_2 (d) P_4O_{10}

19.16 (a)  is OF_2;  is NF_3; is CF_4 and SiF_4;

 is PF_5; 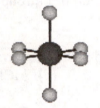 is SF_6

(b) The small N atom is limited to three nearest neighbors in NF_3, whereas the larger P atom can accommodate five nearest neighbors in PF_5. N uses its three unpaired electrons in bonding to three F atoms and has one lone pair, whereas P uses all five valence electrons in bonding to five F atoms. Both C and Si have four valence electrons and use sp^3 hybrid orbitals to bond to four F atoms.

19.18 (a) H_2O, CH_4, HF, B_2H_6, NH_3

(b)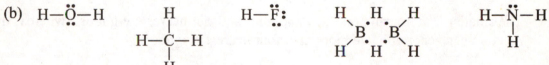

There is a problem in drawing an electron-dot structure for B_2H_6 because this molecule is electron deficient and has two three-center, two-electron bonds.

Section Problems
General Properties and Periodic Trends (Sections 19.1 and 19.2)

19.20 (a) Cl (group 7A) is to the right of S (group 6A) in the same row of the periodic table. Cl has the higher ionization energy.
(b) Si is above Ge in group 4A. Si has the higher ionization energy.
(c) O (group 6A) is above and to the right of In (group 3A) in the periodic table. O has the higher ionization energy.

19.22   (a) Al is below B in group 3A. Al has the larger atomic radius.
(b) P (group 5A) is to the left of S (group 6A) in the same row of the periodic table. P has the larger atomic radius.
(c) Pb (group 4A) is below and to the left of Br (group 7A) in the periodic table. Pb has the larger atomic radius.

19.24   (a) I (group 7A) is to the right of Te (group 6A) in the same row of the periodic table. I has the higher electronegativity.
(b) N is above P in group 5A. N has the higher electronegativity.
(c) F (group 7A) is above and to the right of In (group 3A) in the periodic table. F has the higher electronegativity.

19.26   (a) Sn is below Si in group 4A. Sn has more metallic character.
(b) Ge (group 4A) is to the left of Se (group 6A) in the same row of the periodic table. Ge has more metallic character.
(c) Bi (group 5A) is below and to the left of I (group 7A) in the periodic table. Bi has more metallic character.

19.28   In each case the more ionic compound is the one formed between a metal and nonmetal.
(a) CaH_2 (b) Ga_2O_3 (c) KCl   (d) $AlCl_3$

19.30   Molecular   (a) B_2H_6 (c) SO_3 (d) $GeCl_4$
Extended three-dimensional structure   (b) $KAlSi_3O_8$

19.32   Nonmetal and semimetal oxides are acidic. Metal oxides are basic or amphoteric.
(a) P_4O_{10} (b) B_2O_3 (c) SO_2
(d) N_2O_3 N_2O_3 is in the same group but above As_2O_3 in the periodic table.

19.34   (a) Sn   (b) Cl   (c) Sn   (d) Se   (e) B

19.36   The smaller B atom can bond to a maximum of four nearest neighbors, whereas the larger Al atom can accommodate more than four nearest neighbors.

19.38   In O_2 a π bond is formed by 2p orbitals on each O. S does not form strong π bonds with its 3p orbitals, which leads to the S_8 ring structure with single bonds.

The Group 3A Elements (Sections 19.3–19.5)

19.40   +3 for B, Al, Ga and In;   +1 for Tl

19.42   Boron is a hard semiconductor with a high melting point. Boron forms only molecular compounds and does not form an aqueous B^{3+} ion. $B(OH)_3$ is an acid.

19.44   (a) An electron deficient molecule is a molecule that doesn't have enough electrons to form a two-center, two-electron bond between each pair of bonded atoms. B_2H_6 is an electron deficient molecule.

(b) A three-center, two-electron bond has three atoms bonded together using just two electrons. The B-H-B bridging bond in B_2H_6 is a three-center, two-electron bond.

19.46   (a) Al   (b) Tl   (c) B   (d) B

The Group 4A Elements (Sections 19.6–19.8)

19.48   (a) Pb   (b) C   (c) Si   (d) C

19.50   (a) $GeBr_4$, tetrahedral; Ge is sp^3 hybridized.
          (b) CO_2, linear; C is sp hybridized.
          (c) CO_3^{2-}, trigonal planar; C is sp^2 hybridized.
          (d) $SnCl_3^-$, trigonal pyramidal; Sn is sp^3 hybridized.

19.52   Diamond is a very hard, high melting solid. It is an electrical insulator.
          Diamond has a covalent network structure in which each C atom uses sp^3 hybrid orbitals to form a tetrahedral array of σ bonds. The interlocking, three-dimensional network of strong bonds makes diamond the hardest known substance with the highest melting point for an element. Because the valence electrons are localized in the σ bonds, diamond is an electrical insulator.

19.54   Graphene is a two-dimensional array of hexagonally arranged carbon atoms just one atom thick, essentially one layer of graphite. Graphene is extremely strong and flexible, and is a superb conductor of electricity.

19.56   (a) carbon tetrachloride, CCl_4 (b) carbon monoxide, CO   (c) methane, CH_4

19.58   Some uses for CO_2 are:
          (1) To provide the "bite" in soft drinks; $CO_2(aq) + H_2O(l) \rightleftharpoons H_2CO_3(aq)$
          (2) CO_2 fire extinguishers; CO_2 is nonflammable and 1.5 times more dense than air.
          (3) Refrigerant; dry ice, sublimes at −78 °C.

19.60 $SiO_2(l) + 2\ C(s) \rightarrow Si(l) + 2\ CO(g)$
          (sand)

          Purification of silicon for semiconductor devices:
          $Si(s) + 2\ Cl_2(g) \rightarrow SiCl_4(l)$; $SiCl_4$ is purified by distillation.
          $SiCl_4(g) + 2\ H_2(g) \xrightarrow{heat} Si(s) + 4\ HCl(g)$; Si is purified by zone refining.

19.62   (a) SiO_4^{4-} (b) $Si_4O_{13}^{10-}$

          The charge on the anion is equal to the number of terminal O atoms.

369

19.64   (a) $Si_3O_{10}^{8-}$

(b) The charge on the anion is 8–. Because the Ca^{2+} to Cu^{2+} ratio is 1:1, there must be 2 Ca^{2+} and 2 Cu^{2+} ions in the formula for the mineral. There are also 2 waters. The formula of the mineral is: $Ca_2Cu_2Si_3O_{10} \cdot 2\,H_2O$

The Group 5A Elements (Sections 19.9–19.11)

19.66   (a) P   (b) Sb and Bi   (c) N   (d) Bi

19.68   (a) N_2O, +1   (b) N_2H_4, –2   (c) Ca_3P_2, –3
(d) H_3PO_3, +3   (e) H_3AsO_4, +5

19.70   :N≡N:

N_2 is unreactive because of the large amount of energy necessary to break the N≡N triple bond.

19.72   (a) NO_2^-, bent
(b) PH_3, trigonal pyramidal
(c) PF_5, trigonal bipyramidal
(d) PCl_4^+, tetrahedral

19.74   White phosphorus consists of tetrahedral P_4 molecules with 60° bond angles.

Red phosphorus is polymeric.
White phosphorus is reactive due to the considerable strain in the P_4 molecule.

19.76   (a) The structure for phosphorous acid is

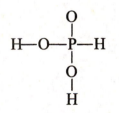

Only the two hydrogens bonded to oxygen are acidic.
(b) Nitrogen forms strong π bonds, and in N_2 the nitrogen atoms are triple bonded to each other. Phosphorus does not form strong pπ – pπ bonds, and so the P atoms are single bonded to each other in P_4.

19.78   (a) $2 NO(g) + O_2(g) \rightarrow 2 NO_2(g)$
       (b) $4 HNO_3(aq) \rightarrow 4 NO_2(aq) + O_2(g) + 2 H_2O(l)$
       (c) $3 Ag(s) + 4 H^+(aq) + NO_3^-(aq) \rightarrow 3 Ag^+(aq) + NO(g) + 2 H_2O(l)$
       (d) $N_2H_4(aq) + 2 I_2(aq) \rightarrow N_2(g) + 4 H^+(aq) + 4 I^-(aq)$

The Group 6A Elements (Sections 19.12 and 19.13)

19.80   (a) O   (b) Te   (c) Po   (d) O

19.82   (a) rhombic sulfur – yellow crystalline solid (mp 113 °C) that contains crown-shaped S_8 rings.
      (b) monoclinic sulfur – an allotrope of sulfur in which the S_8 rings pack differently in the crystal.
      (c) plastic sulfur – when sulfur is cooled rapidly, the sulfur forms disordered, tangled chains, yielding an amorphous, rubbery material called plastic sulfur.
      (d) Liquid sulfur between 160 and 195 °C becomes dark reddish-brown and very viscous forming long polymer chains (S_n, n > 200,000).

19.84   (a) hydrogen sulfide, H_2S (b) sulfur dioxide, SO_2 (c) sulfur trioxide, SO_3
       lead(II) sulfide, PbS   sulfurous acid, H_2SO_3 sulfur hexafluoride, SF_6

19.86   Sulfuric acid (H_2SO_4) is manufactured by the contact process, a three-step reaction sequence in which (1) sulfur burns in air to give SO_2, (2) SO_2 is oxidized to SO_3 in the presence of a vanadium(V) oxide catalyst, and (3) SO_3 reacts with water to give H_2SO_4.
      (1) $S(s) + O_2(g) \rightarrow SO_2(g)$

                       heat
      (2) $2 SO_2(g) + O_2(g) \rightarrow 2 SO_3(g)$
               V_2O_5 catalyst

      (3) $SO_3(g) + H_2O$ (in conc H_2SO_4) $\rightarrow H_2SO_4(l)$

19.88   (a) $Zn(s) + 2 H_3O^+(aq) \rightarrow Zn^{2+}(aq) + H_2(g) + 2 H_2O(l)$
       (b) $BaSO_3(s) + 2 H_3O^+(aq) \rightarrow H_2SO_3(aq) + Ba^{2+}(aq) + 2 H_2O(l)$
       (c) $Cu(s) + 2 H_2SO_4(l) \rightarrow Cu^{2+}(aq) + SO_4^{2-}(aq) + SO_2(g) + 2 H_2O(l)$
       (d) $H_2S(aq) + I_2(aq) \rightarrow S(s) + 2 H^+(aq) + 2 I^-(aq)$

19.90   (a) Acid strength increases as the number of O atoms increases.
      (b) In comparison with S, O is much too electronegative to form compounds of O in the +4 oxidation state. Also, an S atom is large enough to accommodate four bond pairs and a lone pair in its valence shell, but an O atom is too small to do so.
      (c) Each S is sp^3 hybridized with two lone pairs of electrons. The bond angles are therefore 109.5°. A planar ring would require bond angles of 135°.

Halogen Oxoacids and Oxoacid Salts (Section 19.14)

19.92   (a) $HBrO_3$, +5   (b) HIO, +1   (c) $NaClO_2$, +3   (d) $NaIO_4$, +7

19.94   (a) iodic acid   (b) chlorous acid   (c) sodium hypobromite
        (d) lithium perchlorate

19.96   (a) HIO_3 :Ö—Ï—Ö—H   trigonal pyramidal
                            :Ö:

        (b) ClO_2^- [:Ö—Cl—Ö:]$^-$ bent

        (c) HOCl   H—Ö—Cl: bent

        (d) IO_6^{5-} [octahedral lewis structure]$^{5-}$ octahedral

19.98   Oxygen atoms are highly electronegative. Increasing the number of oxygen atoms
        increases the polarity of the O–H bond and increases the acid strength.

19.100 (a) $Br_2(l) + 2\ OH^-(aq) \rightarrow OBr^-(aq) + Br^-(aq) + H_2O(l)$
        (b) $Cl_2(g) + H_2O(l) \rightarrow HOCl(aq) + H^+(aq) + Cl^-(aq)$
        (c) $3\ Cl_2(g) + 6\ OH^-(aq) \rightarrow ClO_3^-(aq) + 5\ Cl^-(aq) + 3\ H_2O(l)$

Chapter Problems

19.102 $Mg(s) + 2\ H_2SO_4(l) \rightarrow Mg^{2+}(aq) + SO_4^{2-}(aq) + SO_2(g) + 2\ H_2O(l)$

19.104 Hedenbergite contains single-stranded silicate chains with an $Si_2O_6^{4-}$ repeating unit, as in
        diopside.

19.106 $N_2H_4(aq) + O_2(aq) \rightarrow N_2(aq) + 2\ H_2O(l)$

19.108 (a) LiCl is an ionic compound. PCl_3 is a covalent molecular compound.
        The ionic compound, LiCl, has the higher melting point.
        (b) Carbon forms strong π bonds with oxygen, and CO_2 is a covalent molecular
        compound with a low melting point. Silicon prefers to form single bonds with oxygen.
        SiO_2 is three dimensional extended structure with alternating silicon and oxygen singly
        bonded to each other. SiO_2 is a high melting solid.
        (c) Nitrogen forms strong π bonds with oxygen and NO_2 is a covalent molecular
        compound with a low melting point. Phosphorus prefers to form single bonds with
        oxygen. P_4O_{10} is a larger covalent molecular compound than NO_2, with a higher melting
        point.

19.110 A portion of the extended graphene is shown here.

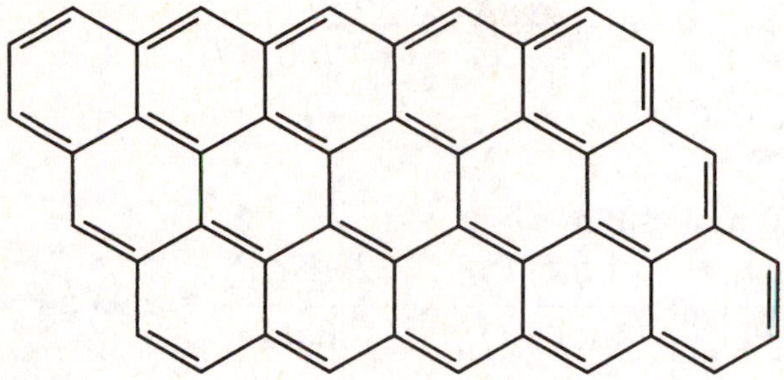

19.112 Earth's crust: O, Si, Al, Fe;   Human body: O, C, H, N

19.114 C, Si, Ge and Sn have allotropes with the diamond structure.
Sn and Pb have metallic allotropes.
C (nonmetal), Si (semimetal), Ge (semimetal), Sn (semimetal and metal), Pb (metal)

19.116 (a) $H_3PO_4(aq) + H_2O(l) \rightleftharpoons H_3O^+(aq) + H_2PO_4^-(aq)$
H_3PO_4 is a Brønsted-Lowry acid.
(b) $B(OH)_3(aq) + 2\ H_2O(l) \rightleftharpoons B(OH)_4^-(aq) + H_3O^+(aq)$
$B(OH)_3$ is a Lewis acid.

19.118 (a) In diamond each C is covalently bonded to four additional C atoms in a rigid three-dimensional network solid. Graphite is a two-dimensional covalent network solid of carbon sheets that can slide over each other. Both are high melting because melting requires the breaking of C–C bonds.
(b) Chlorine does not form perhalic acids of the type H_5XO_6 because its smaller size favors a tetrahedral structure over an octahedral one.

19.120 Cl—S—S—Cl

19.122 NH_3, $K_b = 1.8 \times 10^{-5}$; N_2H_4, $K_b = 8.9 \times 10^{-7}$; NH_2OH, $K_b = 9.1 \times 10^{-9}$
The strongest base, NH_3, will react to the greatest extent with HNO_2.

19.124 H   H   H
           \Al   \B/
          /   \ /   \
        H   H   H

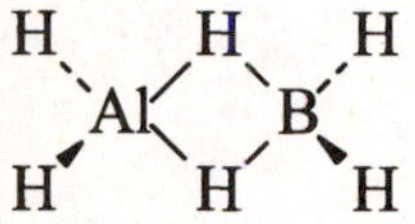

19.126 When there is excess oxygen, the reaction is
$P_4(s) + 5\ O_2(g) \rightarrow P_4O_{10}(s)$
$P_4O_{10}(s) + 6\ H_2O(l) \rightarrow 4\ H_3PO_4(aq)$
Let $Y = [H_3PO_4]$

	$H_3PO_4(aq)$	$+ H_2O(l) \rightarrow$	$H_3O^+(aq)$	$+ H_2PO_4^{2-}(aq)$
initial (M)	Y		0	~0
change (M)	−x		+x	+x
equil (M)	Y − x		x	x

$x = [H_3O^+] = 10^{-pH} = 10^{-1.93} = 0.0117$ M

$$K_a = \frac{[H_3O^+][H_2PO_4^{2-}]}{[H_3PO_4]} = 7.5 \times 10^{-3} = \frac{x^2}{(Y-x)} = \frac{(0.0117)^2}{(Y-0.0117)}$$

Solve for Y. $Y = [H_3PO_4] = 0.030$ M

(0.030 mol/L)(1.00 L) = 0.030 mol H_3PO_4

$$0.030 \text{ mol } H_3PO_4 \times \frac{1 \text{ mol } P_4O_{10}}{4 \text{ mol } H_3PO_4} \times \frac{1 \text{ mol } P_4}{1 \text{ mol } P_4O_{10}} \times \frac{123.9 \text{ g } P_4}{1 \text{ mol } P_4} = 0.93 \text{ g } P_4 \text{ burned}$$

19.128 The angle required by P_4 is 60°. The strain would not be reduced by using sp^3 hybrid orbitals because their angle is ~109°.

Multiconcept Problems

19.130 (a) $\cdot\ddot{N}{=}\ddot{O}:$ $\left[:\ddot{O}{-}\ddot{O}\cdot\right]^- \longleftrightarrow \left[\cdot\ddot{O}{-}\ddot{O}:\right]^- \quad \left[:\ddot{O}{=}\ddot{N}{-}\ddot{O}{-}\ddot{O}:\right]^-$

The O–N–O bond angle should be ~120°.

(b)

$$\begin{array}{ll}
\sigma^*_{2p} & \underline{} \\[4pt]
\pi^*_{2p} & \underline{\uparrow}\ \underline{} \\[4pt]
\sigma_{2p} & \underline{\uparrow\downarrow} \\[4pt]
\pi_{2p} & \underline{\uparrow\downarrow}\ \underline{\uparrow\downarrow} \\[4pt]
\sigma^*_{2s} & \underline{\uparrow\downarrow} \\[4pt]
\sigma_{2s} & \underline{\uparrow\downarrow}
\end{array}$$

The bond order is 2½ with one unpaired electron.

19.132

$$\begin{array}{lll}
2\ In^+(aq) + 2\ e^- \rightarrow 2\ In(s) & E° = -0.14 \text{ V} \\
\underline{In^+(aq) \rightarrow In^{3+}(aq) + 2\ e^-} & \underline{E° = 0.44 \text{ V}} \\
3\ In^+(aq) \rightarrow In^{3+}(aq) + 2\ In(s) & E° = 0.30 \text{ V}
\end{array}$$

$1\ J = 1\ V \cdot C$

$\Delta G° = -nFE° = -(2)(96,500\ C)(0.30\ V) = -5.8 \times 10^4\ J = -58$ kJ

Because $\Delta G° < 0$, the disproportionation of In^+ is spontaneous.

$$\begin{array}{lll}
2\ Tl^+(aq) + 2\ e^- \rightarrow 2\ Tl(s) & E° = -0.34 \text{ V} \\
\underline{Tl^+(aq) \rightarrow Tl^{3+}(aq) + 2\ e^-} & \underline{E° = -1.25 \text{ V}} \\
3\ Tl^+(aq) \rightarrow Tl^{3+}(aq) + 2\ Tl(s) & E° = -1.59 \text{ V}
\end{array}$$

$\Delta G° = -nFE° = -(2)(96,500\ C)(-1.59\ V) = +3.07 \times 10^5\ J = +307$ kJ

Because $\Delta G° > 0$, the disproportionation of Tl^+ is nonspontaneous.

19.134 NH_4NO_3, 80.04; $(NH_4)_2HPO_4$, 132.06

(a) % N in $NH_4NO_3 = \dfrac{2 \times (14.007 \text{ amu N})}{80.04 \text{ amu } NH_4NO_3} \times 100\% = 35.0\%$ N

$$\% \text{ N in } (NH_4)_2HPO_4 = \frac{2 \times (14.007 \text{ amu N})}{132.06 \text{ amu } (NH_4)_2HPO_4} \times 100\% = 21.2\% \text{ N}$$

Let x equal the fraction of NH_4NO_3 and $(1 - x)$ equal the fraction of $(NH_4)_2HPO_4$ in the mixture.

$$0.3381 = x(0.350) + (1 - x)(0.212) = 0.138x + 0.212$$

$$0.1261 = 0.138x \qquad x = \frac{0.1261}{0.138} = 0.9138$$

$$(1 - x) = (1 - 0.9138) = 0.0862$$

There is 8.62 % $(NH_4)_2HPO_4$ and 91.38% NH_4NO_3 in the mixture.

(b) sample mass = 0.965 g

mass $NH_4NO_3 = (0.965 \text{ g})(0.9138) = 0.8818 \text{ g}$

mass $(NH_4)_2HPO_4 = 0.965 \text{ g} - 0.8818 \text{ g} = 0.0832 \text{ g}$

$$\text{mol } NH_4NO_3 = 0.8818 \text{ g} \times \frac{1 \text{ mol } NH_4NO_3}{80.04 \text{ g } NH_4NO_3} = 0.0110 \text{ mol } NH_4NO_3$$

$$\text{mol } (NH_4)_2HPO_4 = 0.0832 \text{ g} \times \frac{1 \text{ mol } (NH_4)_2HPO_4}{132.06 \text{ g } (NH_4)_2HPO_4} = 6.30 \times 10^{-4} \text{ mol } (NH_4)_2HPO_4$$

$$[NH_4^+] = \frac{0.0110 \text{ mol} + 2(6.30 \times 10^{-4} \text{ mol})}{0.0500 \text{ L}} = 0.245 \text{ M}$$

$$[HPO_4^{2-}] = \frac{6.30 \times 10^{-4} \text{ mol}}{0.0500 \text{ L}} = 0.0126 \text{ M}$$

The equilibrium constant for the transfer of a proton from the cation of a salt to the anion of a salt is equal to $\frac{K_a K_b}{K_w}$.

$$K_a(NH_4^+) = 5.56 \times 10^{-10} \text{ and } K_b(HPO_4^{2-}) = 1.61 \times 10^{-7}$$

$$K = \frac{K_a K_b}{K_w} = \frac{(5.56 \times 10^{-10})(1.61 \times 10^{-7})}{1.0 \times 10^{-14}} = 8.95 \times 10^{-3}$$

	$NH_4^+(aq)$	$+$	$HPO_4^{2-}(aq)$	$\rightleftarrows$	$H_2PO_4^-(aq)$	$+$	$NH_3(aq)$
initial (M)	0.245		0.0126		0		0
change (M)	–x		–x		+x		+x
equil (M)	0.245 – x		0.0126 – x		x		x

$$K = \frac{[H_2PO_4^-][NH_3]}{[NH_4^+][HPO_4^{2-}]} = 8.95 \times 10^{-3} = \frac{x^2}{(0.245 - x)(0.0126 - x)}$$

$$0.991x^2 + 0.002\,31x - (2.76 \times 10^{-5}) = 0$$

Use the quadratic formula to solve for x.

$$x = \frac{(-0.002\,31) \pm \sqrt{(0.002\,31)^2 - (4)(0.991)(-2.76 \times 10^{-5})}}{2(0.991)} = \frac{(-0.002\,31) \pm 0.010\,71}{2(0.991)}$$

$$x = 0.004\,24 \text{ and } -0.006\,67$$

Of the two solutions for x, only the positive value of x has physical meaning because x is equal to both the $[H_2PO_4^-]$ and $[NH_3]$.

$$K_a(NH_4^+) = \frac{[H_3O^+][NH_3]}{[NH_4^+]} = 5.56 \times 10^{-10} = \frac{[H_3O^+](0.004\,24)}{(0.245 - 0.004\,24)}$$

$$[H_3O^+] = \frac{(5.56 \times 10^{-10})(0.245 - 0.004\,24)}{0.004\,24} = 3.157 \times 10^{-8}\ M$$

$$pH = -\log[H_3O^+] = -\log(3.157 \times 10^{-8}) = 7.50$$

19.136 $N_2O_4(g) \rightleftarrows 2\,NO_2(g)$

$$P_{Total} = 753\ mm\ Hg \times \frac{1.00\ atm}{760\ mm\ Hg} = 0.991\ atm$$

$$P_{Total} = P_{N_2O_4} + P_{NO_2} = 0.991\ atm$$

$$P_{N_2O_4} = 0.991\ atm - P_{NO_2}$$

$$K_p = \frac{(P_{NO_2})^2}{P_{N_2O_4}} = 0.113; \qquad K_p = \frac{(P_{NO_2})^2}{(0.991\ atm - P_{NO_2})} = 0.113$$

$$(P_{NO_2})^2 + 0.113(P_{NO_2}) - 0.112 = 0$$

Use the quadratic formula to solve for P_{NO_2}.

$$P_{NO_2} = \frac{-(0.113) \pm \sqrt{(0.113)^2 - (4)(1)(-0.112)}}{2(1)} = \frac{-0.113 \pm 0.679}{2}$$

$$P_{NO_2} = -0.396\ and\ 0.283$$

Of the two solutions for P_{NO_2}, only 0.283 has physical meaning because NO_2 can't have

a negative partial pressure.

$P_{NO_2} = 0.283\ atm$ and $P_{N_2O_4} = 0.991\ atm - P_{NO_2} = 0.991 - 0.283 = 0.708\ atm$

	$N_2O_4(g)$	$\rightarrow$	$2\,NO_2(g)$
before reaction (atm)	0.708		0.283
change (atm)	−0.708		+2(0.708)
after reaction (atm)	0		0.283 + 2(0.708) = 1.70 atm

$PV = nRT$; 25 °C = 298 K

$$n_{NO_2} = \frac{PV}{RT} = \frac{(1.70\ atm)(0.5000\ L)}{\left(0.082\,06\ \dfrac{L \cdot atm}{K \cdot mol}\right)(298\ K)} = 0.0348\ mol\ NO_2$$

(a)	$2\,NO_2(aq)$	$+\ 2\,H_2O(l)$	$\rightarrow$	$HNO_2(aq)$	$+\ H_3O^+(aq)$	$+\ NO_3^-(aq)$
before reaction (mol)	0.0348			0	~0	0
change (mol)	−0.0348			+0.0348/2	+0.0348/2	+0.0348/2
after reaction (mol)	0			0.0174	0.0174	0.0174

(b) $[HNO_2] = [H_3O^+] = 0.0174 \text{ mol}/0.250 \text{ L} = 0.0696 \text{ M}$

$$HNO_2(aq) \ + \ H_2O(l) \ \rightleftharpoons \ H_3O^+(aq) \ + \ NO_2^-(aq)$$

initial (M)	0.0696	0.0696	0
change (M)	−x	+x	+x
equil (M)	0.0696 − x	0.0696 + x	x

$$K_a = \frac{[H_3O^+][NO_2^-]}{[HNO_2]} = 4.5 \times 10^{-4} = \frac{(0.0696 \ + \ x)x}{(0.0696 \ - \ x)}$$

$x^2 + 0.070\,05 - (3.132 \times 10^{-5}) = 0$

Use the quadratic formula to solve for x.

$$x = \frac{-(0.070\,05) \pm \sqrt{(0.070\,05)^2 - (4)(1)(-3.132 \times 10^{-5})}}{2(1)} = \frac{-0.070\,05 \pm 0.070\,94}{2}$$

$x = -0.0705$ and 4.45×10^{-4}

Of the two solutions for x only 4.45×10^{-4} has physical meaning because −0.0705 leads to negative concentrations.

$[NO_2^-] = x = 4.45 \times 10^{-4} \text{ M} = 4.4 \times 10^{-4} \text{ M}$

$[H_3O^+] = 0.0696 + x = 0.0696 \ + \ 4.45 \times 10^{-4} = 0.0700 \text{ M}$

$pH = -\log[H_3O^+] = -\log(0.0700) = 1.15$

(c) Total solution molarity = $[NO_3^-] + [NO_2^-] + [H_3O^+] + [HNO_2]$

$= 0.0696 \text{ M} + (4.45 \times 10^{-4}) + 0.0700 \text{ M} + 0.0692 \text{ M} = 0.2092 \text{ M}$

$$\Pi = MRT = (0.2092 \text{ M})\left(0.082\,06 \ \frac{L \cdot atm}{K \cdot mol}\right)(298 \text{ K}) = 5.12 \text{ atm}$$

(d) mol H_3O^+ = (0.0700 mol/L)(0.250 L) = 0.0175 mol H_3O^+

mol HNO_2 = (0.0692 mol/L)(0.250 L) = 0.0173 mol HNO_2

Total mol of acid to neutralize = 0.0175 $+$ 0.0173 $=$ 0.0348 mol

$$\text{mass CaO} = 0.0348 \text{ mol acid} \times \frac{1 \text{ mol CaO}}{2 \text{ mol acid}} \times \frac{56.08 \text{ g CaO}}{1 \text{ mol CaO}} = 0.976 \text{ g CaO}$$

20 Transition Elements and Coordination Chemistry

20.1   (a) V, [Ar] $3d^3 4s^2$ (b) Co^{2+}, [Ar] $3d^7$
(c) Mn^{4+} in MnO_2, [Ar] $3d^3$ (d) Cu^{2+} in $CuCl_4^{2-}$, [Ar] $3d^9$

20.2

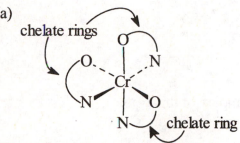

(a) Mn   (b) Ni^{2+}

(c) Ag   (d) Mo^{3+}

20.3 Z_{eff} increases from left to right across the first transition series.
(a) The transition metal with the lowest Z_{eff} (Ti) should be the strongest reducing agent because it is easier for Ti to lose its valence electrons. The transition metal with the highest Z_{eff} (Zn) should be the weakest reducing agent because it is more difficult for Zn to lose its valence electrons.
(b) The oxoanion with the highest Z_{eff} (FeO_4^{2-}) should be the strongest oxidizing agent because of the greater attraction for electrons. The oxoanion with the lowest Z_{eff} (VO_4^{3-}) should be the weakest oxidizing agent because of the lower attraction for electrons.

20.4   (a) $Cr_2O_7^{2-}$ (b) Cr^{3+} (c) Cr^{2+} (d) Fe^{2+} (e) Cu^{2+}

20.5   (a) $Cr(OH)_2$ (b) $Cr(OH)_4^-$ (c) CrO_4^{2-} (d) $Fe(OH)_2$ (e) $Fe(OH)_3$

20.6 $[Cr(NH_3)_2(SCN)_4]^-$

20.7   In $Na_4[Fe(CN)_6]$ each sodium is in the +1 oxidation state (+4 total); each cyanide (CN^-) has a –1 charge (–6 total). The compound is neutral; therefore, the oxidation state of the iron is +2.

20.8   (a)

(b) Cr^{3+} is the Lewis acid. The glycinate ligand is the Lewis base. Nitrogen and oxygen are the ligand donor atoms. The chelate rings are identified in the drawing.
(c) The coordination number is 6. The coordination geometry is octahedral. The chromium is in the +3 oxidation state.

20.9   (a) tetraamminecopper(II) sulfate   (b) sodium tetrahydroxochromate(III)
       (c) triglycinatocobalt(III)   (d) pentaaquaisothiocyanatoiron(III) ion

20.10 (a) $[Zn(NH_3)_4](NO_3)_2$ (b) $Ni(CO)_4$ (c) $K[Pt(NH_3)Cl_3]$ (d) $[Au(CN)_2]^-$

20.11 (a) Two diastereoisomers are possible.

cis   trans

(b) No isomers are possible for a tetrahedral complex of the type MA_2B_2.

(c) Two diastereoisomers are possible.

(d) No isomers are possible for a complex of this type.

(e) Two diastereoisomers are possible.

trans   cis

(f) No diastereoisomers are possible for a complex of this type.

20.12 (1) and (2) are the same. (3) and (4) are the same. (1) and (2) are different from (3) and (4).

20.13 (a) chair, no   (b) foot, yes   (c) pencil, no
       (d) corkscrew, yes   (e) banana, no   (f) football, no

20.14 (a) (2) and (3) are chiral and (1) and (4) are achiral.
       (b) enantiomer of (2)   enantiomer of (3)

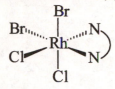 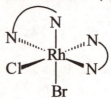

20.15 (a) $[Fe(C_2O_4)_3]^{3-}$ can exist as enantiomers.

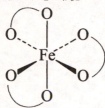 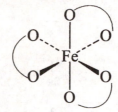

       (b) $[Co(NH_3)_4en]^{3+}$ cannot exist as enantiomers.
       (c) $[Co(NH_3)_2(en)_2]^{3+}$ can exist as enantiomers.

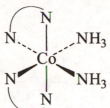 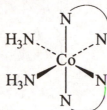

       (d) $[Cr(H_2O)_4Cl_2]^+$ cannot exist as enantiomers.

20.16 (a) The ion is absorbing in the red (625 nm), so the most likely color for the ion is blue.
       (b) 625 nm = 625 x 10^{-9} m

$$E = h\frac{c}{\lambda} = (6.626 \times 10^{-34}\ J \cdot s)\left(\frac{3.00 \times 10^8\ m/s}{625 \times 10^{-9}\ m}\right) = 3.18 \times 10^{-19}\ J$$

20.17 (a) Fe^{3+} [Ar]   ↑ ↑ ↑ ↑ ↑   —   — — —
                                         3d   4s   4p

       $[Fe(CN)_6]^{3-}$ [Ar]   ↑↓ ↑↓ ↑ ↑↓ ↑↓   ↑↓   ↑↓ ↑↓ ↑↓
                                         3d   4s   4p

                          d^2sp^3 1 unpaired e^-

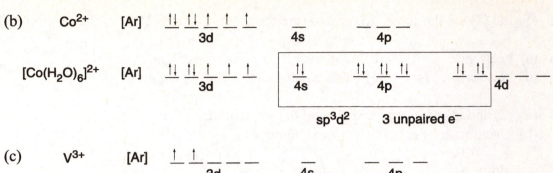

(b)   Co²⁺   [Ar]   3d   4s   4p

[Co(H₂O)₆]²⁺   [Ar]   3d   4s   4p   4d

sp³d²   3 unpaired e⁻

(c)   V³⁺   [Ar]   3d   4s   4p

[VCl₄]⁻   [Ar]   3d   4s   4p

sp³   2 unpaired e⁻

(d)   Pt²⁺   [Xe]   5d   6s   6p

[PtCl₄]²⁻   [Xe]   5d   6s   6p

dsp²   no unpaired e⁻

20.18

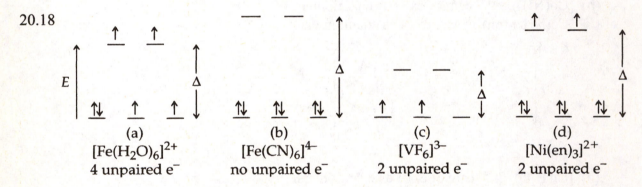

(a)   (b)   (c)   (d)
[Fe(H₂O)₆]²⁺   [Fe(CN)₆]⁴⁻   [VF₆]³⁻   [Ni(en)₃]²⁺
4 unpaired e⁻   no unpaired e⁻   2 unpaired e⁻   2 unpaired e⁻

20.19   Both [NiCl₄]²⁻ and [Ni(CN)₄]²⁻ contain Ni²⁺ with a [Ar] 3d⁸ electron configuration.

(a) [NiCl₄]²⁻ (tetrahedral)   (b) [Ni(CN)₄]²⁻ (square planar)

⇅ ↑ ↑
xy xz yz   x²–y²

⇅
xy
⇅
z²

⇅   ⇅   ⇅   ⇅
z² x²–y²   xz   yz

2 unpaired electrons   no unpaired electrons

20.20 $N_2(g) + H_2(g) \rightarrow HN=NH(g)$

$\Delta G^{\circ}_1 = [\Delta G^{\circ}_f(HN=NH)] - 0 = (1 \text{ mol})(243.8 \text{ kJ/mol}) - 0 = 243.8 \text{ kJ}$

$HN=NH(g) + H_2(g) \rightarrow H_2N-NH_2(g)$

$\Delta G^{\circ}_2 = [\Delta G^{\circ}_f(H_2N-NH_2)] - [\Delta G^{\circ}_f(HN=NH) + 0)]$

$\Delta G^{\circ}_2 = [(1 \text{ mol})(159.3 \text{ kJ/mol})] - [(1 \text{ mol})(243.8 \text{ kJ/mol}) + 0] = -84.5 \text{ kJ}$

$H_2N-NH_2(g) + H_2(g) \rightarrow 2 NH_3(g)$

$\Delta G^{\circ}_3 = [2 \Delta G^{\circ}_f(NH_3)] - [\Delta G^{\circ}_f(H_2N-NH_2) + 0)]$

$\Delta G^{\circ}_3 = [(2 \text{ mol})(-16.5 \text{ kJ/mol})] - [(1 \text{ mol})(159.3 \text{ kJ/mol}) + 0] = -192.3 \text{ kJ}$

The third step is the most favorable because it has the most negative ΔG.

20.21

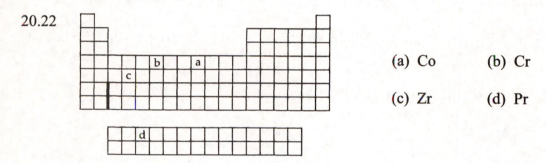

Conceptual Problems

20.22

(a) Co   (b) Cr

(c) Zr   (d) Pr

20.24 (a) The atomic radii decrease, at first markedly and then more gradually. Toward the end of the series, the radii increase again. The decrease in atomic radii is a result of an increase in Z_{eff}. The increase is due to electron-electron repulsions in doubly occupied d orbitals.

(b) The densities of the transition metals are inversely related to their atomic radii. The densities initially increase from left to right and then decrease toward the end of the series.

(c) Ionization energies generally increase from left to right across the series. The general trend correlates with an increase in Z_{eff} and a decrease in atomic radii.

(d) The standard oxidation potentials generally decrease from left to right across the first transition series. This correlates with the general trend in ionization energies.

20.26 (1) dichloroethylenediamineplatinum(II)
       (2) trans-diammineaquachloroplatinate(II) ion
       (3) amminepentachloroplatinate(IV) ion
       (4) cis-diaquabis(ethylenediamine)platinum(IV) ion

20.28. (a) (1) cis; (2) trans; (3) trans; (4) cis
       (b) (1) and (4) are the same.   (2) and (3) are the same.
       (c) None of the isomers exist as enantiomers because their mirror images are identical.

20.30

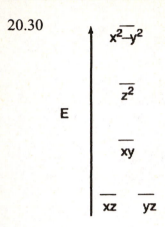

Section Problems
Electron Configurations and Properties of Transition Elements (Sections 20.1 and 20.2)

20.32 (a) Cr, [Ar] $3d^5 4s^1$ 　　(b) Zr, [Kr] $4d^2 5s^2$ 　　(c) Co^{2+}, [Ar] $3d^7$

　　(d) Fe^{3+}, [Ar] $3d^5$ 　　(e) Mo^{3+}, [Kr] $4d^3$ 　　(f) Cr(VI), [Ar] $3d^0$

20.34 (a) Cu^{2+}, 　[Ar] $3d^9$ ↿⇂ ↿⇂ ↿⇂ ↿⇂ ↿ 　　1 unpaired e^-
　　　　　　　　　　　　3d

　　(b) Ti^{2+}, 　[Ar] $3d^2$ ↿ ↿ _ _ _ 　　2 unpaired e^-
　　　　　　　　　　　　3d

　　(c) Zn^{2+}, 　[Ar] $3d^{10}$ ↿⇂ ↿⇂ ↿⇂ ↿⇂ ↿⇂ 　　0 unpaired e^-
　　　　　　　　　　　　3d

　　(d) Cr^{3+}, 　[Ar] $3d^3$ ↿ ↿ ↿ _ _ 　　3 unpaired e^-
　　　　　　　　　　　　3d

20.36 Ti is harder than K and Ca largely because the sharing of d, as well as s, electrons results in stronger metallic bonding.

20.38 (a) The decrease in radii with increasing atomic number is expected because the added d electrons only partially shield the added nuclear charge. As a result, Z_{eff} increases. With increasing Z_{eff}, the electrons are more strongly attracted to the nucleus, and atomic size decreases.
　　(b) The densities of the transition metals are inversely related to their atomic radii.

20.40 The smaller than expected sizes of the third-transition series atoms are associated with what is called the lanthanide contraction, the general decrease in atomic radii of the f-block lanthanide elements. The lanthanide contraction is due to the increase in Z_{eff} as the 4f subshell is filled.

20.42 Sc $(631 + 1235) = 1866$ kJ/mol
      Ti $(659 + 1310) = 1969$ kJ/mol
      V $(651 + 1410) = 2061$ kJ/mol
      Cr $(653 + 1591) = 2224$ kJ/mol
      Mn $(717 + 1509) = 2226$ kJ/mol
      Fe $(762 + 1562) = 2324$ kJ/mol
      Co $(760 + 1648) = 2408$ kJ/mol
      Ni $(737 + 1753) = 2490$ kJ/mol
      Cu $(745 + 1958) = 2703$ kJ/mol
      Zn $(906 + 1733) = 2639$ kJ/mol

Across the first transition element series, Z_{eff} increases and there is an almost linear increase in the sum of the first two ionization energies. This is what is expected if the two electrons are removed from the 4s orbital. Higher than expected values for the sum of the first two ionization energies are observed for Cr and Cu because of their anomalous electron configurations (Cr $3d^5 4s^1$; Cu $3d^{10} 4s^1$). An increasing Z_{eff} affects 3d orbitals more than the 4s orbital and the second ionization energy for an electron from the 3d orbital is higher than expected.

20.44 (a) $Cr(s) + 2 H^+(aq) \rightarrow Cr^{2+}(aq) + H_2(g)$ (b) $Zn(s) + 2 H^+(aq) \rightarrow Zn^{2+}(aq) + H_2(g)$
      (c) N.R.   (d) $Fe(s) + 2 H^+(aq) \rightarrow Fe^{2+}(aq) + H_2(g)$

Oxidation States (Section 20.3)

20.46 (b) Mn   (d) Cu

20.48 Sc(III), Ti(IV), V(V), Cr(VI), Mn(VII), Fe(VI), Co(III), Ni(II), Cu(II), Zn(II)

20.50 Cu^{2+} is a stronger oxidizing agent than Cr^{2+} because of a higher Z_{eff}.

20.52 A compound with vanadium in the +2 oxidation state is expected to be a reducing agent, because early transition metal atoms have a relatively low effective nuclear charge and are easily oxidized to higher oxidation states.

20.54 $Mn^{2+} < MnO_2 < MnO_4^-$ because of increasing oxidation state of the Mn.

Chemistry of Selected Transition Elements (Section 20.4)

20.56 (a) $Cr_2O_3(s) + 2 Al(s) \rightarrow 2 Cr(s) + Al_2O_3(s)$
      (b) $Cu_2S(l) + O_2(g) \rightarrow 2 Cu(l) + SO_2(g)$

20.58 $Cr(OH)_3(s) + OH^-(aq) \rightarrow Cr(OH)_4^-(aq)$
      The Cr in $Cr(OH)_4^-$ is in the +3 oxidation state. $Cr(OH)_4^-$ is deep green.

20.60 (c) $Cr(OH)_3$

20.62   (a) Add excess KOH(aq) and Fe^{3+} will precipitate as $Fe(OH)_3$(s). Na^+(aq) will remain in solution.

(b) Add excess NaOH(aq) and Fe^{3+} will precipitate as $Fe(OH)_3$(s). $Cr(OH)_4^-$(aq) will remain in solution.

(c) Add excess NH_3(aq) and Fe^{3+} will precipitate as $Fe(OH)_3$(s). $Cu(NH_3)_4^{2+}$(aq) will remain in solution.

20.64   (a) $Cr_2O_7^{2-}$(aq) + 6 Fe^{2+}(aq) + 14 H^+(aq) $\rightarrow$ 2 Cr^{3+}(aq) + 6 Fe^{3+}(aq) + 7 H_2O(l)

(b) 4 Fe^{2+}(aq) + O_2(g) + 4 H^+(aq) $\rightarrow$ 4 Fe^{3+}(aq) + 2 H_2O(l)

(c) Cu_2O(s) + 2 H^+(aq) $\rightarrow$ Cu(s) + Cu^{2+}(aq) + H_2O(l)

(d) Fe(s) + 2 H^+(aq) $\rightarrow$ Fe^{2+}(aq) + H_2(g)

20.66   (a) 2 CrO_4^{2-}(aq) + 2 H_3O^+(aq) $\rightarrow$ $Cr_2O_7^{2-}$(aq) + 3 H_2O(l)
              (yellow)   (orange)

(b) $[Fe(H_2O)_6]^{3+}$(aq) + SCN^-(aq) $\rightarrow$ $[Fe(H_2O)_5(SCN)]^{2+}$(aq) + H_2O(l)
                                                              (red)

(c) 3 Cu(s) + 2 NO_3^-(aq) + 8 H^+(aq) $\rightarrow$ 3 Cu^{2+}(aq) + 2 NO(g) + 4 H_2O(l)
                                                  (blue)

(d) $Cr(OH)_3$(s) + OH^-(aq) $\rightarrow$ $Cr(OH)_4^-$(aq)
     2 $Cr(OH)_4^-$(aq) + 3 HO_2^-(aq) $\rightarrow$ 2 CrO_4^{2-}(aq) + 5 H_2O(l) + OH^-(aq)
                                                 (yellow)

Coordination Compounds; Ligands (Sections 20.5 and 20.6)

20.68 Ni^{2+} accepts six pairs of electrons, two each from the three ethylenediamine ligands. Ni^{2+} is an electron pair acceptor, a Lewis acid. The two nitrogens in each ethylenediamine donate a pair of electrons to the Ni^{2+}. The ethylenediamine is an electron pair donor, a Lewis base. The formation of $[Ni(en)_3]^{2+}$ is a Lewis acid-base reaction.

20.70   (a) $[Ag(NH_3)_2]^+$ (b) $[Ni(CN)_4]^{2-}$ (c) $[Cr(H_2O)_6]^{3+}$

20.72   (a) $AgCl_2^-$
        2 Cl^-
        The oxidation state of the Ag is +1.
        (b) $[Cr(H_2O)_5Cl]^{2+}$
        4 H_2O (no charge)   1 Cl^-
        The oxidation state of the Cr is +3.
        (c) $[Co(NCS)_4]^{2-}$
        4 NCS^-
        The oxidation state of the Co is +2.
        (d) $[ZrF_8]^{4-}$
        8 F^-
        The oxidation state of the Zr is +4.

(e) $Co(NH_3)_3(NO_2)_3$

3 NH_3 (no charge)   3 NO_2^-

The oxidation state of the Co is +3.

20.74 (a) $Ni(CO)_4$ (b) $[Ag(NH_3)_2]^+$ (c) $[Fe(CN)_6]^{3-}$ (d) $[Ni(CN)_4]^{2-}$

20.76

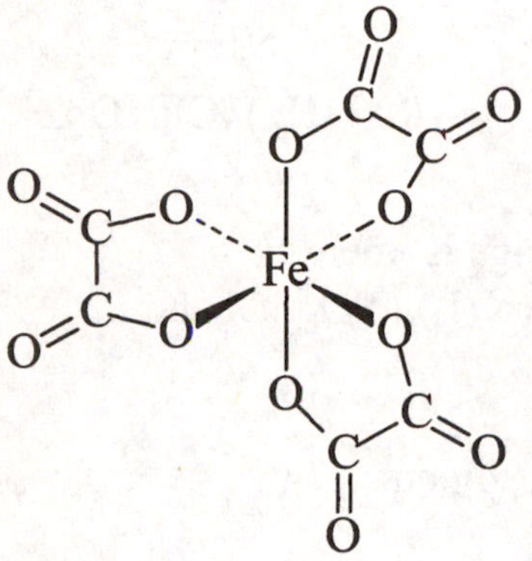

The iron is in the +3 oxidation state, and the coordination number is six. The geometry about the Fe is octahedral. The oxalate ligand is behaving as a bidentate chelating ligand. There are three chelate rings, one formed by each oxalate ligand.

20.78 (a) $Co(NH_3)_3(NO_2)_3$

3 NH_3 (no charge)   3 NO_2^-

The oxidation state of the Co is +3.

(b) $[Ag(NH_3)_2]NO_3$

2 NH_3 (no charge)   1 NO_3^-

The oxidation state of the Ag is +1.

(c) $K_3[Cr(C_2O_4)_2Cl_2]$

3 K^+ 2 $C_2O_4^{2-}$ 2 Cl^-

The oxidation state of the Cr is +3.

(d) $Cs[CuCl_2]$

1 Cs^+ 2 Cl^-

The oxidation state of the Cu is +1.

Naming Coordination Compounds (Section 20.7)

20.80 (a) tetrachloromanganate(II)

(b) hexaamminenickel(II)

(c) tricarbonatocobaltate(III)

(d) bis(ethylenediamine)dithiocyanatoplatinum(IV)

20.82 (a) cesium tetrachloroferrate(III)

(b) hexaaquavanadium(III) nitrate

(c) tetraamminedibromocobalt(III) bromide

(d) diglycinatocopper(II)

20.84 (a) $[Pt(NH_3)_4]Cl_2$

(b) $Na_3[Fe(CN)_6]$

(c) $[Pt(en)_3](SO_4)_2$

(d) $Rh(NH_3)_3(SCN)_3$

Isomers (Sections 20.8 and 20.9)

20.86

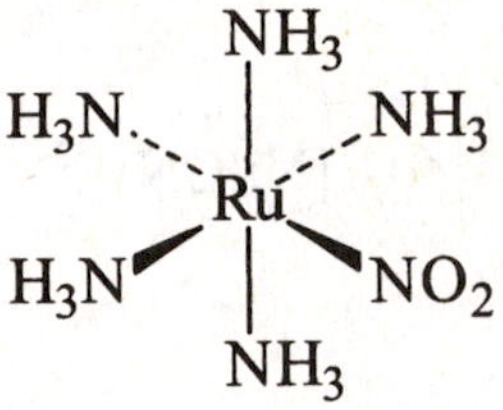

$[Ru(NH_3)_5(NO_2)]Cl$ 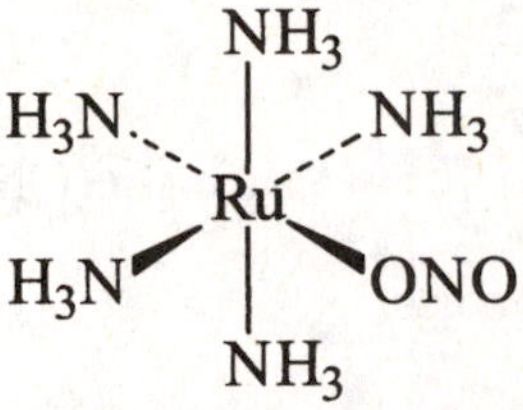 $[Ru(NH_3)_5(ONO)]Cl$ 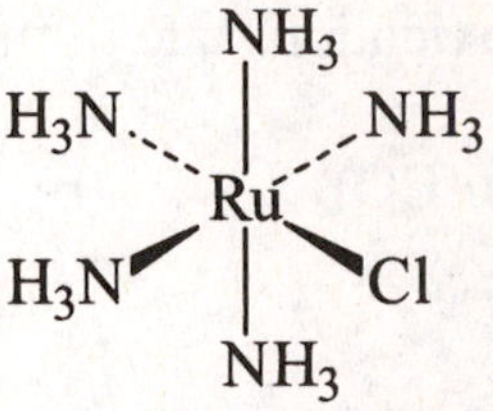 $[Ru(NH_3)_5Cl]NO_2$

$[Ru(NH_3)_5(NO_2)]Cl$ and $[Ru(NH_3)_5(ONO)]Cl$ are linkage isomers.
$[Ru(NH_3)_5Cl]NO_2$ is an ionization isomer of both $[Ru(NH_3)_5(NO_2)]Cl$ and
$[Ru(NH_3)_5(ONO)]Cl$.

20.88 (a) $[Cr(NH_3)_2Cl_4]^-$ can exist as cis and trans diastereoisomers.

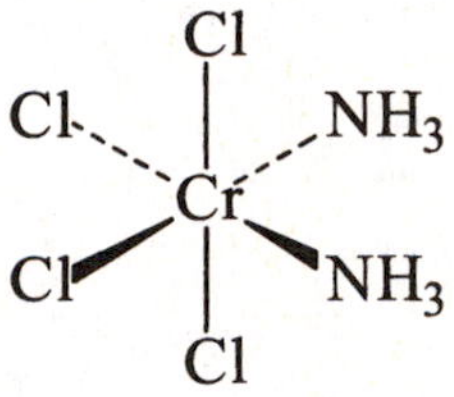

cis 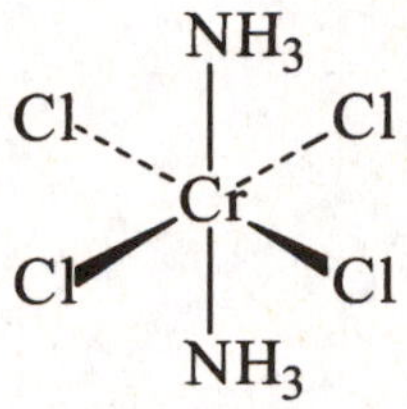   trans

(b) $[Co(NH_3)_5Br]^{2+}$ cannot exist as diastereoisomers.
(c) $[FeCl_2(NCS)_2]^{2-}$ (tetrahedral) cannot exist as diastereoisomers.
(d) $[PtCl_2Br_2]^{2-}$ (square planar) can exist as cis and trans diastereoisomers.

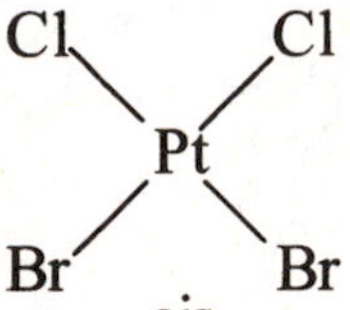

cis 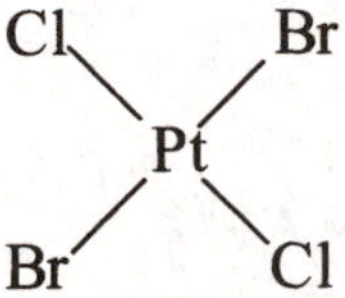   trans

20.90 (c) cis-$[Cr(en)_2(H_2O)_2]^{3+}$ (d) $[Cr(C_2O_4)_3]^{3-}$

20.92

enantiomers

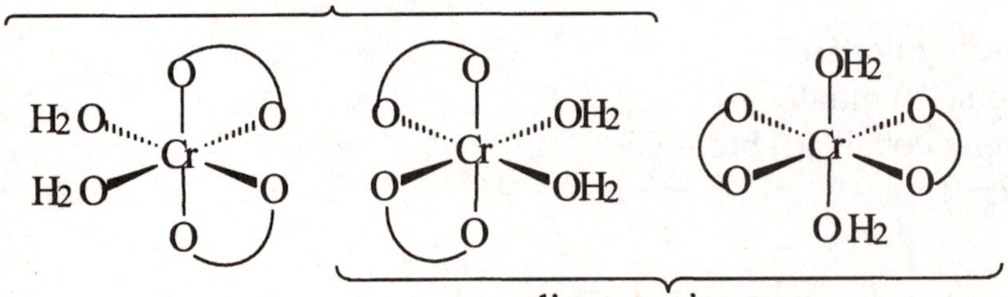

diastereoisomers

20.94 Plane-polarized light is light in which the electric vibrations of the light wave are restricted to a single plane. The following chromium complex can rotate the plane of plane-polarized light.

$[Cr(en)_3]^{3+}$

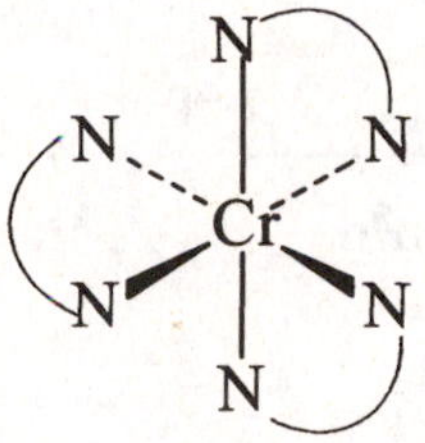
N
N
N
Cr
N
N
N

Color of Complexes; Valence Bond and Crystal Field Theories (Sections 20.10–20.12)

20.96 The measure of the amount of light absorbed by a substance is called the absorbance, and a graph of absorbance versus wavelength is called an absorption spectrum. If a complex absorbs at 455 nm, its color is orange (use the color wheel in Figure 20.26).

20.98 (a) $[Ti(H_2O)_6]^{3+}$

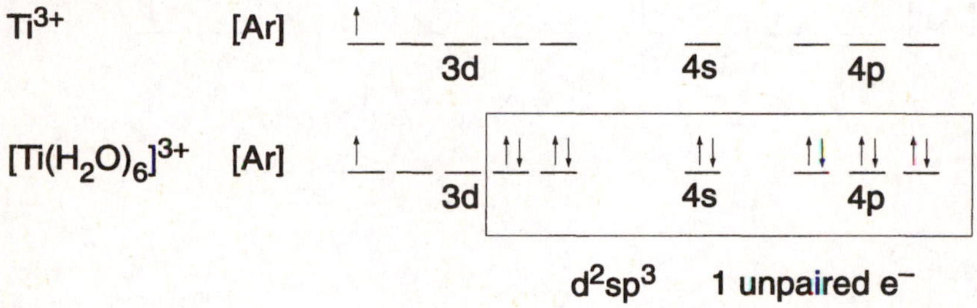
Ti3+ [Ar]
3d
4s
4p
[Ti(H2O)6]3+ [Ar]
3d
4s
4p
d2sp3 1 unpaired e−

(b) $[NiBr_4]^{2-}$

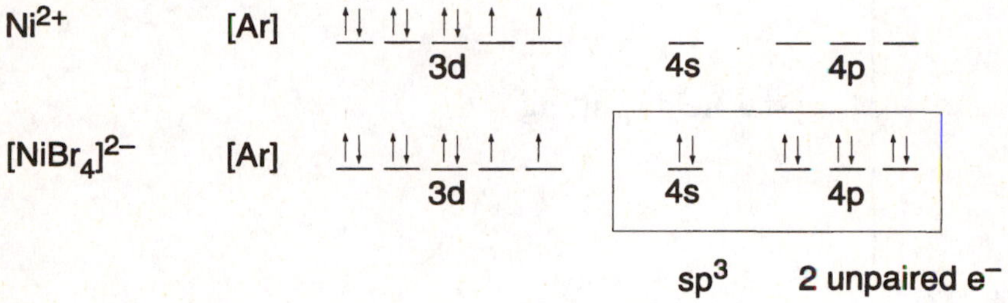
Ni2+ [Ar]
3d
4s
4p
[NiBr4]2− [Ar]
3d
4s
4p
sp3 2 unpaired e−

(c) $[Fe(CN)_6]^{3-}$ (low-spin)

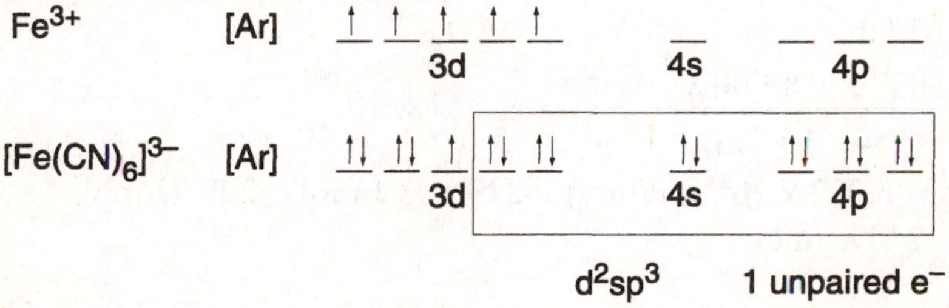
Fe3+ [Ar]
3d
4s
4p
[Fe(CN)6]3− [Ar]
3d
4s
4p
d2sp3 1 unpaired e−

(d) [MnCl$_6$]$^{3-}$ (high-spin)

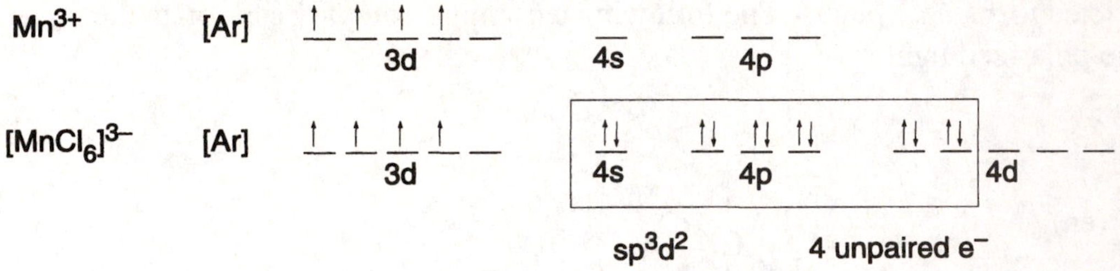

Mn^{3+} [Ar] ↑ ↑ ↑ ↑ _ _
3d

4s _ _ _ _
4p

[MnCl$_6$]$^{3-}$ [Ar] ↑ ↑ ↑ ↑ _
3d

↑↓ ↑↓ ↑↓ ↑↓ ↑↓ ↑↓ _ _ _
4s 4p 4d

sp^{3}d^2 4 unpaired e$^-$

20.100 (a) +3, M = Cr or Ni

(b)

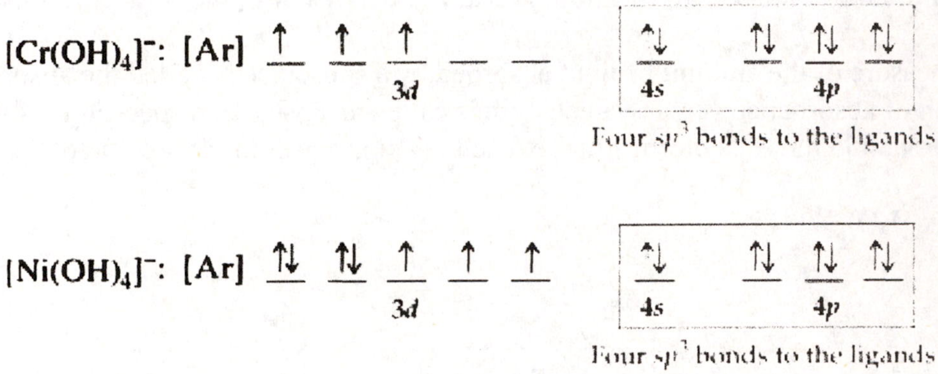

[Cr(OH)$_4$]$^-$: [Ar] ↑ ↑ ↑ _ _
3d

↑↓ ↑↓ ↑↓ ↑↓
4s 4p

Four sp^3 bonds to the ligands

[Ni(OH)$_4$]$^-$: [Ar] ↑↓ ↑↓ ↑ ↑ ↑
3d

↑↓ ↑↓ ↑↓ ↑↓
4s 4p

Four sp^3 bonds to the ligands

(c) [Cr(OH)$_4$]$^-$

20.102 [Ti(H$_2$O)$_6$]$^{3+}$ Ti^{3+} 3d^1

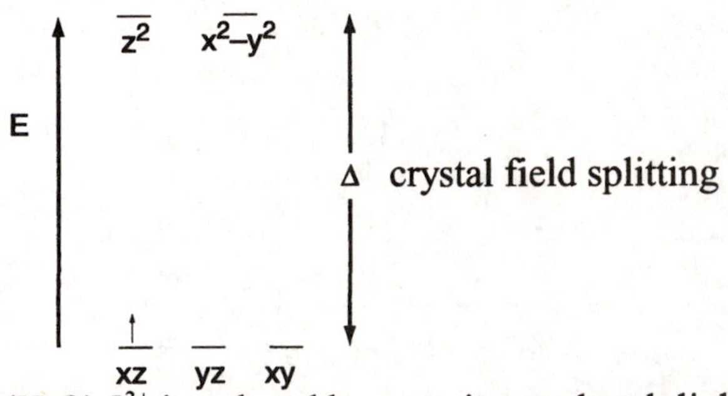

E

z^2 x^2–y^2
_ _

Δ crystal field splitting

↑ _ _
xz yz xy

[Ti(H$_2$O)$_6$]$^{3+}$ is colored because it can absorb light in the visible region, exciting the electron to the higher-energy set of orbitals.

20.104 λ = 544 nm = 544 x 10^{-9} m

$$\Delta = \frac{hc}{\lambda} = \frac{(6.626 \times 10^{-34} \text{ J} \cdot \text{s})(3.00 \times 10^8 \text{ m/s})}{(544 \times 10^{-9} \text{ m})} = 3.65 \times 10^{-19} \text{ J}$$

Δ = (3.65 x 10^{-19} J/ion)(6.022 x 10^{23} ion/mol) = 219,803 J/mol = 220 kJ/mol
For [Ti(H$_2$O)$_6$]$^{3+}$, Δ = 240 kJ/mol

Because $\Delta_{NCS^-} < \Delta_{H_2O}$ for the Ti complex, NCS^- is a weaker-field ligand than H_2O. If $[Ti(NCS)_6]^{3-}$ absorbs at 544 nm, its color should be red (use the color wheel in Figure 20.26).

20.106   (a) $[CrF_6]^{3-}$ (b) $[V(H_2O)_6]^{3+}$ (c) $[Fe(CN)_6]^{3-}$

       — —   — —   — —

     ↿ ↿ ↿
     3 unpaired e$^-$

                   ↿ ↿ —
                  2 unpaired e$^-$

                                            ⇅ ⇅ ↿
                                        1 unpaired e$^-$

20.108 $Ni^{2+}(aq)$ $Zn^{2+}(aq)$

     ↿ ↿   ⇅ ⇅

     ⇅ ⇅ ⇅   ⇅ ⇅ ⇅

$Ni^{2+}(aq)$ is green because the Ni^{2+} ion can absorb light, which promotes electrons from the filled d orbitals to the higher energy half-filled d orbitals. $Zn^{2+}(aq)$ is colorless because the d orbitals are completely filled and no electrons can be promoted, so no light is absorbed.

20.110   Weak-field ligands produce a small Δ. Strong-field ligands produce a large Δ. For a metal complex with weak-field ligands, $\Delta < P$, where P is the pairing energy, and it is easier to place an electron in either d_{z^2} or $d_{x^2-y^2}$ than to pair up electrons; high-spin complexes result. For a metal complex with strong-field ligands, $\Delta > P$ and it is easier to pair up electrons than to place them in either d_{z^2} or $d_{x^2-y^2}$; low-spin complexes result.

20.112   — x^2-y^2

     ⇅ xy

     ⇅ z^2

     ⇅ ⇅ xz, yz

Square planar geometry is most common for metal ions with d^8 configurations because this configuration favors low-spin complexes in which all four lower energy d orbitals are filled, and the higher energy $d_{x^2-y^2}$ orbital is vacant.

Chapter Problems

20.114 (a) $[Mn(CN)_6]^{3-}$ Mn^{3+} [Ar] $3d^4$
CN$^-$ is a strong-field ligand. The Mn^{3+} complex is low-spin.

— —

$\uparrow\downarrow$ $\uparrow$ $\uparrow$ 2 unpaired e$^-$, paramagnetic

(b) $[Zn(NH_3)_4]^{2+}$ Zn^{2+} [Ar] $3d^{10}$
$[Zn(NH_3)_4]^{2+}$ is tetrahedral.

$\uparrow\downarrow$ $\uparrow\downarrow$ $\uparrow\downarrow$

$\uparrow\downarrow$ $\uparrow\downarrow$ no unpaired e$^-$, diamagnetic

(c) $[Fe(CN)_6]^{4-}$ Fe^{2+} [Ar] $3d^6$
CN$^-$ is a strong-field ligand. The Fe^{2+} complex is low-spin.

— —

$\uparrow\downarrow$ $\uparrow\downarrow$ $\uparrow\downarrow$ no unpaired e$^-$, diamagnetic

(d) $[FeF_6]^{4-}$ Fe^{2+} [Ar] $3d^6$
F$^-$ is a weak-field ligand. The Fe^{2+} complex is high-spin.

$\uparrow$ $\uparrow$

$\uparrow\downarrow$ $\uparrow$ $\uparrow$ 4 unpaired e$^-$, paramagnetic

20.116 (a) $4\,[Co^{3+}(aq) + e^- \rightarrow Co^{2+}(aq)]$
$\underline{2\,H_2O(l) \rightarrow O_2(g) + 4\,H^+(aq) + 4\,e^-}$
$4\,Co^{3+}(aq) + 2\,H_2O(l) \rightarrow 4\,Co^{2+}(aq) + O_2(g) + 4\,H^+(aq)$

(b) $4\,Cr^{2+}(aq) + O_2(g) + 4\,H^+(aq) \rightarrow 4\,Cr^{3+}(aq) + 2\,H_2O(l)$

(c) $3\,[Cu(s) \rightarrow Cu^{2+}(aq) + 2\,e^-]$
$\underline{Cr_2O_7^{2-}(aq) + 14\,H^+(aq) + 6\,e^- \rightarrow 2\,Cr^{3+}(aq) + 7\,H_2O(l)}$
$3\,Cu(s) + Cr_2O_7^{2-}(aq) + 14\,H^+(aq) \rightarrow 3\,Cu^{2+}(aq) + 2\,Cr^{3+}(aq) + 7\,H_2O(l)$

(d) $2\,CrO_4^{2-}(aq) + 2\,H^+(aq) \rightarrow Cr_2O_7^{2-}(aq) + H_2O(l)$

20.118 EDTA^{4-} in mayonnaise will complex any metal cations that are present in trace amounts. Free metal ions can catalyze the oxidation of oils, causing the mayonnaise to become rancid. The bidentate ligand $H_2NCH_2CO_2^-$ will not bind to metal ions as strongly as does the hexadentate EDTA^{4-} and so would not be an effective substitute for EDTA^{4-}.

20.120 (a)

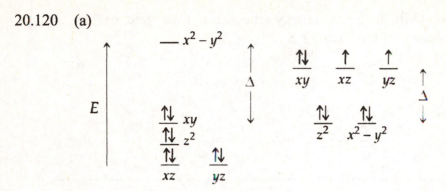

square planar nickel(II)   tetrahedral nickel(II)

(b) $NiCl_2L_2$ is tetrahedral, $Ni(NCS)_2L_2$ is square planar.

$$
\underset{SCN}{\overset{L}{\diagdown}}Ni\underset{L}{\overset{NCS}{\diagup}}
\qquad
\underset{L}{\overset{L}{\diagdown}}Ni\underset{NCS}{\overset{NCS}{\diagup}}
\qquad
\underset{L}{\overset{L}{|}}Ni\underset{Cl}{\overset{Cl}{\diagup}}
$$

(c) Square planar cis-$Ni(NCS)_2L_2$ and tetrahedral $NiCl_2L_2$ have a dipole moment.

20.122 Cl^- is a weak-field ligand, whereas CN^- is a strong-field ligand. Δ for $[Fe(CN)_6]^{3-}$ is larger than the pairing energy P; Δ for $[FeCl_6]^{3-}$ is smaller than P. Fe^{3+} has a $3d^5$ electron configuration.

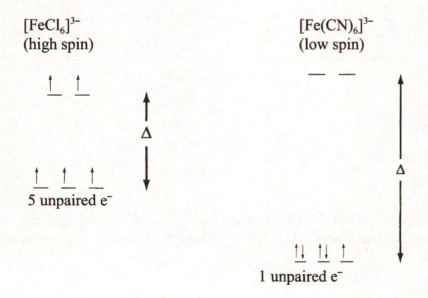

$[FeCl_6]^{3-}$
(high spin)

$[Fe(CN)_6]^{3-}$
(low spin)

5 unpaired e^-

1 unpaired e^-

Because of the difference in Δ, $[FeCl_6]^{3-}$ is high-spin with five unpaired electrons, whereas $[Fe(CN)_6]^{3-}$ is low-spin with only one unpaired electron.

20.124 A choice between high-spin and low-spin electron configurations arises only for complexes of metal ions with four to seven d electrons, the so-called d^4-d^7 complexes. For d^1-d^3 and d^8-d^{10} complexes, only one ground-state electron configuration is possible. In d^1-d^3 complexes, all the electrons occupy the lower-energy d orbitals, independent of the value of Δ. In d^8-d^{10} complexes, the lower-energy set of d orbitals is filled with three pairs of electrons, while the higher-energy set contains two, three, or four electrons, again independent of the value of Δ.

$$\frac{\uparrow\ \ -}{d^1}\ -\qquad\qquad \frac{\uparrow\ \uparrow}{d^2}\ -\qquad\qquad \frac{\uparrow\ \uparrow\ \uparrow}{d^3}$$

$$\underline{\uparrow}\ -\qquad\qquad\quad -\ -\qquad\qquad \underline{\uparrow}\ \underline{\uparrow}\qquad\qquad -\ -$$

$$\frac{\uparrow\ \uparrow\ \uparrow}{d^4\ \text{high-spin}}\quad \frac{\uparrow\downarrow\ \uparrow\ \uparrow}{d^4\ \text{low-spin}}\qquad \frac{\uparrow\ \uparrow\ \uparrow}{d^5\ \text{high-spin}}\quad \frac{\uparrow\downarrow\ \uparrow\downarrow\ \uparrow}{d^5\ \text{low-spin}}$$

$$\underline{\uparrow}\ \underline{\uparrow}\qquad\qquad -\ -\qquad\qquad \underline{\uparrow}\ \underline{\uparrow}\qquad\qquad \underline{\uparrow}\ -$$

$$\frac{\uparrow\downarrow\ \uparrow\ \uparrow}{d^6\ \text{high-spin}}\quad \frac{\uparrow\downarrow\ \uparrow\downarrow\ \uparrow\downarrow}{d^6\ \text{low-spin}}\qquad \frac{\uparrow\downarrow\ \uparrow\downarrow\ \uparrow}{d^7\ \text{high-spin}}\quad \frac{\uparrow\downarrow\ \uparrow\downarrow\ \uparrow\downarrow}{d^7\ \text{low-spin}}$$

$$\underline{\uparrow}\ \underline{\uparrow}\qquad\qquad \underline{\uparrow\downarrow}\ \underline{\uparrow}\qquad\qquad \underline{\uparrow\downarrow}\ \underline{\uparrow\downarrow}$$

$$\frac{\uparrow\downarrow\ \uparrow\downarrow\ \uparrow\downarrow}{d^8}\qquad\qquad \frac{\uparrow\downarrow\ \uparrow\downarrow\ \uparrow\downarrow}{d^9}\qquad\qquad \frac{\uparrow\downarrow\ \uparrow\downarrow\ \uparrow\downarrow}{d^{10}}$$

20.126 $[CoCl_4]^{2-}$ is tetrahedral. $[Co(H_2O)_6]^{2+}$ is octahedral. Because $\Delta_{tet} < \Delta_{oct}$, these complexes have different colors. $[CoCl_4]^{2-}$ has absorption bands at longer wavelengths.

20.128 $Co(gly)_3$

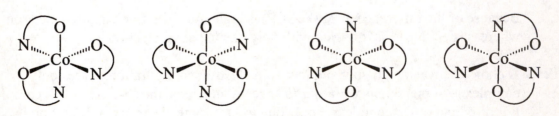

20.130

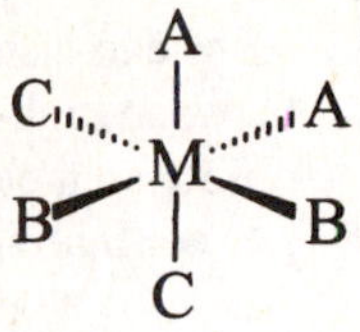

1

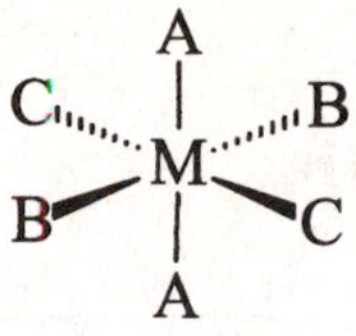

2

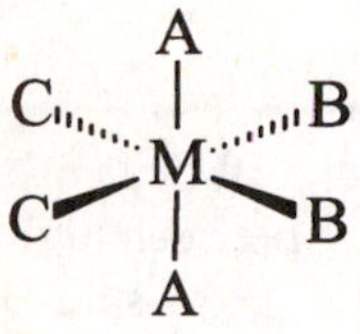

3

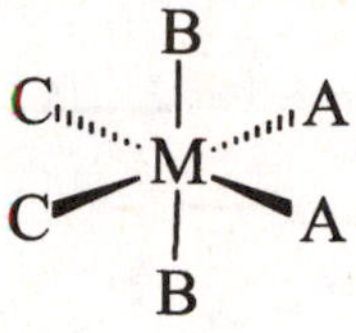

4

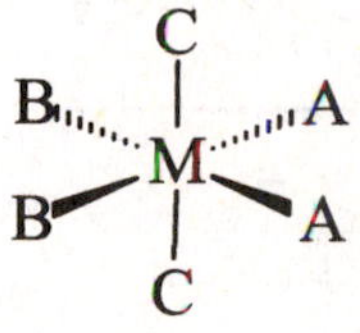

5

1 can exist as enantiomers.

20.132

	weak-field ligands	strong-field ligands	
Ti^{2+} [Ar] $3d^2$	— — ↑ ↑ —	— — ↑ ↑ —	BM cannot distinguish between high-spin and low-spin electron configurations
	$BM = \sqrt{2(2+2)} = 2.83$	$BM = \sqrt{2(2+2)} = 2.83$	
V^{2+} [Ar] $3d^3$	— — ↑ ↑ ↑	— — ↑ ↑ ↑	BM cannot distinguish between high-spin and low-spin electron configurations
	$BM = \sqrt{3(3+2)} = 3.87$	$BM = \sqrt{3(3+2)} = 3.87$	
Cr^{2+} [Ar] $3d^4$	↑ — ↑ ↑ ↑	— — ↑↓ ↑ ↑	BM can distinguish between high-spin and low-spin electron configurations
	$BM = \sqrt{4(4+2)} = 4.90$	$BM = \sqrt{2(2+2)} = 2.83$	
Mn^{2+} [Ar] $3d^5$	↑ ↑ ↑ ↑ ↑	— — ↑↓ ↑↓ ↑	BM can distinguish between high-spin and low-spin electron configurations
	$BM = \sqrt{5(5+2)} = 5.92$	$BM = \sqrt{1(1+2)} = 1.73$	

Fe^{2+} [Ar] $3d^6$	↑ ↑ ↑↓ ↑ ↑	— — ↑↓ ↑↓ ↑↓	BM can distinguish between high-spin and low-spin electron configurations
	$BM = \sqrt{4(4+2)} = 4.90$	$BM = 0$	
Co^{2+} [Ar] $3d^7$	↑ ↑ ↑↓ ↑↓ ↑	↑ — ↑↓ ↑↓ ↑↓	BM can distinguish between high-spin and low-spin electron configurations
	$BM = \sqrt{3(3+2)} = 3.87$	$BM = \sqrt{1(1+2)} = 1.73$	
Ni^{2+} [Ar] $3d^8$	↑ ↑ ↑↓ ↑↓ ↑↓	↑ ↑ ↑↓ ↑↓ ↑↓	BM cannot distinguish between high-spin and low-spin electron configurations
	$BM = \sqrt{2(2+2)} = 2.83$	$BM = \sqrt{2(2+2)} = 2.83$	
Cu^{2+} [Ar] $3d^9$	↑↓ ↑ ↑↓ ↑↓ ↑↓	↑↓ ↑ ↑↓ ↑↓ ↑↓	BM cannot distinguish between high-spin and low-spin electron configurations
	$BM = \sqrt{1(1+2)} = 1.73$	$BM = \sqrt{1(1+2)} = 1.73$	
Zn^{2+} [Ar] $3d^{10}$	↑↓ ↑↓ ↑↓ ↑↓ ↑↓	↑↓ ↑↓ ↑↓ ↑↓ ↑↓	BM cannot distinguish between high-spin and low-spin electron configurations
	$BM = 0$	$BM = 0$	

20.134 (a)

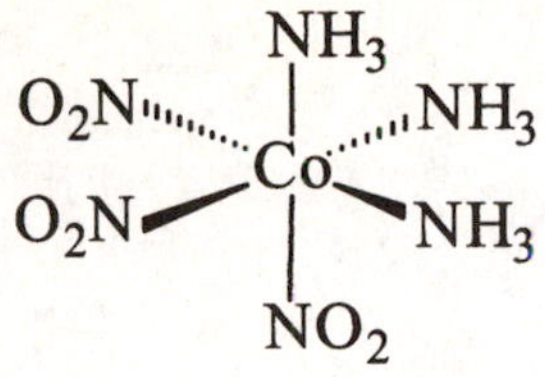

$$\begin{array}{cc} \textbf{1} & \textbf{2} \end{array}$$

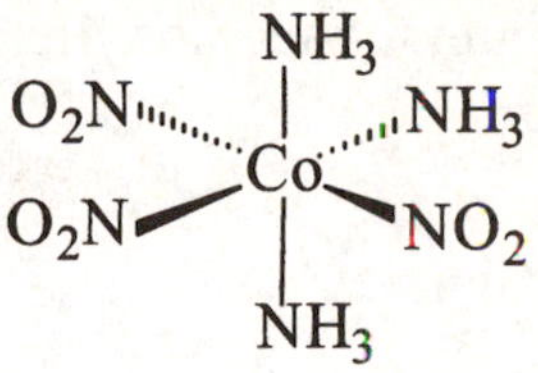

(b) Isomer **2** would give rise to the desired product because it has two trans NO_2 groups.

20.136 (a) $(NH_4)[Cr(H_2O)_6](SO_4)_2$, ammonium hexaaquachromium(III) sulfate

Cr^{3+} — —

$$\underline{\uparrow}\ \underline{\uparrow}\ \underline{\uparrow}$$
3 unpaired e⁻

(b) $Mo(CO)_6$, hexacarbonylmolybdenum(0)

Mo^0 — —

$$\underline{\uparrow\downarrow}\ \underline{\uparrow\downarrow}\ \underline{\uparrow\downarrow}$$
low-spin, no unpaired e⁻

(c) $[Ni(NH_3)_4(H_2O)_2](NO_3)_2$, tetraamminediaquanickel(II) nitrate

Ni^{2+} $\quad\underline{\uparrow}\ \underline{\uparrow}$

$$\underline{\uparrow\downarrow}\ \underline{\uparrow\downarrow}\ \underline{\uparrow\downarrow}$$
2 unpaired e⁻

(d) $K_4[Os(CN)_6]$, potassium hexacyanoosmate(II)

Os^{2+} — —

$$\underline{\uparrow\downarrow}\ \underline{\uparrow\downarrow}\ \underline{\uparrow\downarrow}$$
low-spin, no unpaired e⁻

(e) $[Pt(NH_3)_4](ClO_4)_2$, tetraammineplatinum(II) perchlorate

Pt^{2+}

$$\underline{\quad}$$
$$\underline{\uparrow\downarrow}$$
$$\underline{\uparrow\downarrow}$$
$$\underline{\uparrow\downarrow}\ \underline{\uparrow\downarrow}$$
low spin, no unpaired e⁻

(f) $Na_2[Fe(CO)_4]$, sodium tetracarbonylferrate(–II)

Fe^{2-} ⇅ ⇅ ⇅

⇅ ⇅
no unpaired e^-

20.138 For transition metal complexes, observed colors and absorbed colors are generally complementary. Using the color wheel (Figure 20.26), the absorbed colors in the table are complementary colors to those observed.

	Observed Color	Absorbed Color	Approximate λ (nm)
$Cr(acac)_3$	red	green	530
$[Cr(H_2O)_6]^{3+}$	violet	yellow	580
$[CrCl_2(H_2O)_4]^+$	green	red	700
$[Cr(urea)_6]^{3+}$	green	red	700
$[Cr(NH_3)_6]^{3+}$	yellow	violet	420
$Cr(acetate)_3(H_2O)_3$	blue-violet	orange-yellow	600

The magnitude of Δ is comparable to the energy of the absorbed light from the low energy red end to the high energy violet end (ROYGBIV). The red of $[CrCl_2(H_2O)_4]^+$ is lower energy than the yellow of $[Cr(H_2O)_6]^{3+}$, so $Cl^- < H_2O$. Because $[CrCl_2(H_2O)_4]^+$ and $[Cr(urea)_6]^{3+}$ are both red, Δ for 6 ureas is approximately equal to Δ for 2 Cl^-'s and 4 H_2O's. Therefore, urea is between Cl^- and H_2O.
The spectrochemical series is: $Cl^- < urea < acetate < H_2O < acac < NH_3$

Multiconcept Problems

20.140 (1) $Ni(H_2O)_6^{2+}(aq) + 6\,NH_3(aq) \rightleftharpoons Ni(NH_3)_6^{2+}(aq) + 6\,H_2O(l)$ $K_f = 2.0 \times 10^8$
(2) $Ni(H_2O)_6^{2+}(aq) + 3\,en(aq) \rightleftharpoons Ni(en)_3^{2+}(aq) + 6\,H_2O(l)$ $K_f = 4 \times 10^{17}$
(a) Reaction (2) should have the larger entropy change because three bidentate en ligands displace six water molecules.
(b) $\Delta G° = \Delta H° - T\Delta S°$
Because $\Delta H°_1$ and $\Delta H°_2$ are almost the same, the difference in $\Delta G°$ is determined by the difference in $\Delta S°$. Because $\Delta S°_2$ is larger than $\Delta S°_1$, $\Delta G°_2$ is more negative than $\Delta G°_1$ which is consistent with the greater stability of $Ni(en)_3^{2+}$.
(c) $\Delta H° - T\Delta S° = \Delta G° = -RT \ln K_f$
$\Delta H°_1 - T\Delta S°_1 - (\Delta H°_2 - T\Delta S°_2) = -RT \ln K_f(1) - [-RT \ln K_f(2)]$

$T\Delta S°_2 - T\Delta S°_1 = RT \ln K_f(2) - RT \ln K_f(1) = RT \ln \dfrac{K_f(2)}{K_f(1)}$

$\Delta S°_2 - \Delta S°_1 = R \ln \dfrac{K_f(2)}{K_f(1)} = [8.314\ J/(K \cdot mol)] \ln \dfrac{4 \times 10^{17}}{2.0 \times 10^8}$

$\Delta S°_2 - \Delta S°_1 = 178\ J/(K \cdot mol)$ or $180\ J/(K \cdot mol)$

20.142   (a) $Cr(s) + 2 H^+(aq) \rightarrow Cr^{2+}(aq) + H_2(g)$

(b) $mol\ Cr = 2.60\ g\ Cr \times \dfrac{1\ mol\ Cr}{52.00\ g\ Cr} = 0.0500\ mol\ Cr$

$mol\ H_2SO_4 = (0.050\ 00\ L)(1.200\ mol/L) = 0.060\ 00\ mol\ H_2SO_4$

The stoichiometry between Cr and H_2SO_4 is one to one, therefore Cr is the limiting reagent because of the smaller number of moles.

$mol\ H_2 = 0.0500\ mol\ Cr \times \dfrac{1\ mol\ H_2}{1\ mol\ Cr} = 0.0500\ mol\ H_2$

25 °C = 298 K

$PV = nRT; \qquad V = \dfrac{nRT}{P} = \dfrac{(0.0500\ mol)\left(0.082\ 06\ \dfrac{L \cdot atm}{K \cdot mol}\right)(298\ K)}{\left(735\ mm\ Hg \times \dfrac{1.00\ atm}{760\ mm\ Hg}\right)} = 1.26\ L\ of\ H_2$

(c) 0.060 00 mol H_2SO_4 can provide 0.1200 mol H^+. 0.0500 mol Cr reacts with 2 x (0.0500 mol H^+) = 0.100 mol H^+. This leaves 0.0200 mol H^+ and 0.0600 mol SO_4^{2-}, which will give, after neutralization, 0.0200 mol HSO_4^- and 0.0400 mol SO_4^{2-}.

$[HSO_4^-] = 0.0200\ mol/0.050\ 00\ L = 0.400\ M$

$[SO_4^{2-}] = 0.0400\ mol/0.050\ 00\ L = 0.800\ M$

The pH of this solution can be determined from the following equilibrium:

	$HSO_4^-(aq)$	+ $H_2O(l)$	$\rightleftharpoons$	$H_3O^+(aq)$	+ $SO_4^{2-}(aq)$
initial (M)	0.400			0	0.800
change (M)	−x			+x	+x
equil (M)	0.400 − x			x	0.800 + x

$K_{a2} = \dfrac{[H_3O^+][SO_4^{2-}]}{[HSO_4^-]} = 1.2 \times 10^{-2} = \dfrac{(x)(0.800 + x)}{0.400 - x}$

$x^2 + 0.812x - 0.0048 = 0$

Use the quadratic formula to solve for x.

$x = \dfrac{-(0.812) \pm \sqrt{(0.812)^2 - 4(1)(-0.0048)}}{2(1)} = \dfrac{-0.812 \pm 0.8237}{2}$

x = 0.005 85 and −0.818

Of the two solutions for x, only the positive value of x has physical meaning, because x is the $[H_3O^+]$.

$[H_3O^+] = x = 0.005\ 85\ M$

$pH = -\log[H_3O^+] = -\log(0.005\ 85) = 2.23$

(d) Crystal field d-orbital energy level diagram

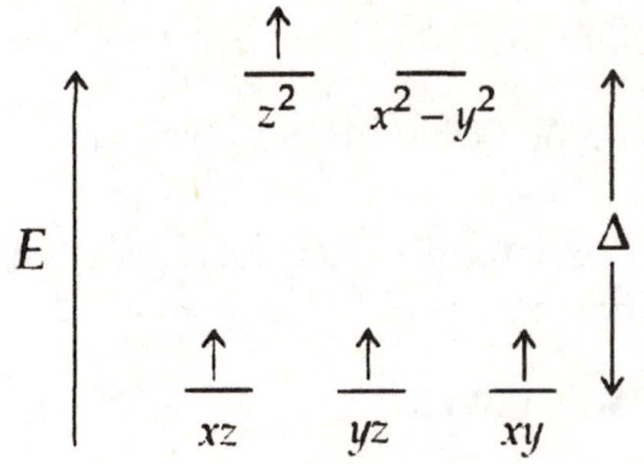

Valence bond orbital diagram

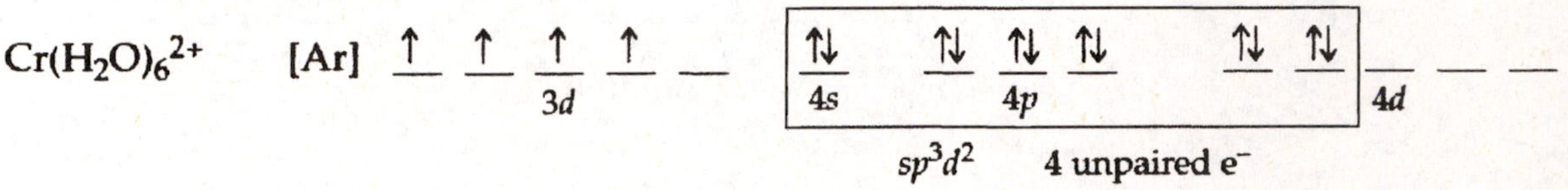

$Cr(H_2O)_6^{2+}$ [Ar]

sp^3d^2 4 unpaired e^-

(e) The addition of excess KCN converts $Cr(H_2O)_6^{2+}$(aq) to $Cr(CN)_6^{4-}$(aq). CN^- is a strong field ligand and increases Δ changing the chromium complex from high spin, with 4 unpaired electrons, to low spin, with only 2 unpaired electrons.

20.144 (a) Assume a 100.0 g sample of the chromium compound.

$$19.52 \text{ g Cr x } \frac{1 \text{ mol Cr}}{51.996 \text{ g Cr}} = 0.3754 \text{ mol Cr}$$

$$39.91 \text{ g Cl x } \frac{1 \text{ mol Cl}}{35.453 \text{ g Cl}} = 1.126 \text{ mol Cl}$$

$$40.57 \text{ g H}_2\text{O x } \frac{1 \text{ mol H}_2\text{O}}{18.015 \text{ g H}_2\text{O}} = 2.252 \text{ mol H}_2\text{O}$$

$Cr_{0.3754}Cl_{1.126}(H_2O)_{2.252}$, divide each subscript by the smallest, 0.3754.
$Cr_{0.3754/0.3754}Cl_{1.126/0.3754}(H_2O)_{2.252/0.3754}$
$CrCl_3(H_2O)_6$

(b) $Cr(H_2O)_6Cl_3$, 266.45; AgCl, 143.32
For **A**:

$$\text{mol Cr complex} = \text{mol Cr} = 0.225 \text{ g Cr complex x } \frac{1 \text{ mol Cr complex}}{266.45 \text{ g Cr complex}} = 8.44 \times 10^{-4} \text{ mol Cr}$$

$$\text{mol Cl} = \text{mol AgCl} = 0.363 \text{ g AgCl x } \frac{1 \text{ mol AgCl}}{143.32 \text{ g AgCl}} = 2.53 \times 10^{-3} \text{ mol Cl}$$

$$\frac{\text{mol Cl}}{\text{mol Cr}} = \frac{2.53 \times 10^{-3} \text{ mol Cl}}{8.44 \times 10^{-4} \text{ mol Cr}} = 3 \text{ Cl/Cr}$$

For **B**:

$$\text{mol Cr complex} = \text{mol Cr} = 0.263 \text{ g Cr complex} \times \frac{1 \text{ mol Cr complex}}{266.45 \text{ g Cr complex}} = 9.87 \times 10^{-4} \text{ mol Cr}$$

$$\text{mol Cl} = \text{mol AgCl} = 0.283 \text{ g AgCl} \times \frac{1 \text{ mol AgCl}}{143.32 \text{ g AgCl}} = 1.97 \times 10^{-3} \text{ mol Cl}$$

$$\frac{\text{mol Cl}}{\text{mol Cr}} = \frac{1.97 \times 10^{-3} \text{ mol Cl}}{9.87 \times 10^{-4} \text{ mol Cr}} = 2 \text{ Cl/Cr}$$

For **C**:

$$\text{mol Cr complex} = \text{mol Cr} = 0.358 \text{ g Cr complex} \times \frac{1 \text{ mol Cr complex}}{266.45 \text{ g Cr complex}} = 1.34 \times 10^{-3} \text{ mol Cr}$$

$$\text{mol Cl} = \text{mol AgCl} = 0.193 \text{ g AgCl} \times \frac{1 \text{ mol AgCl}}{143.32 \text{ g AgCl}} = 1.35 \times 10^{-3} \text{ mol Cl}$$

$$\frac{\text{mol Cl}}{\text{mol Cr}} = \frac{1.35 \times 10^{-3} \text{ mol Cl}}{1.34 \times 10^{-3} \text{ mol Cr}} = 1 \text{ Cl/Cr}$$

Because only the free Cl^- ions (those not bonded to the Cr^{3+}) give an immediate precipitate of AgCl, the probable structural formulas are:

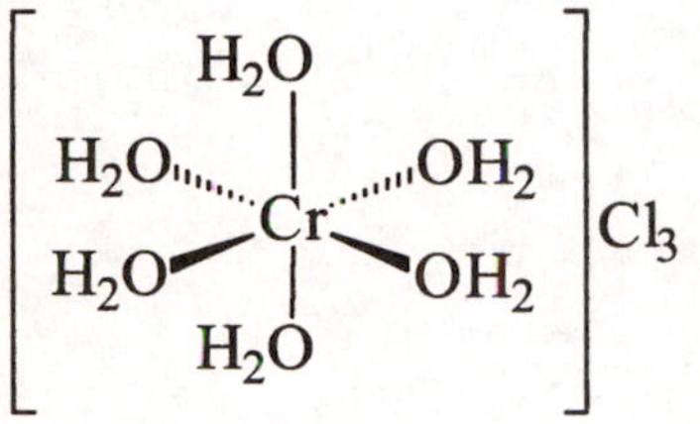

A

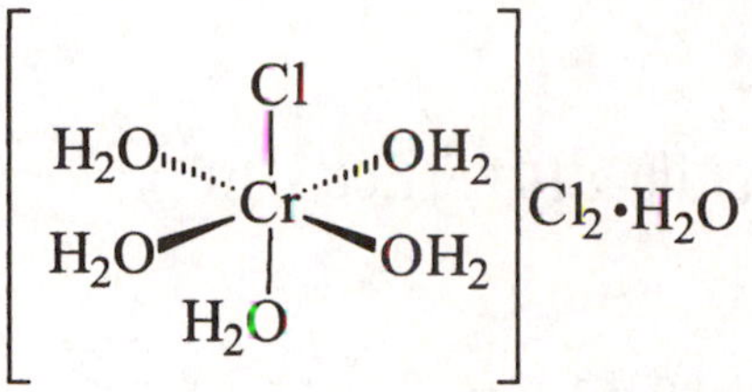

B

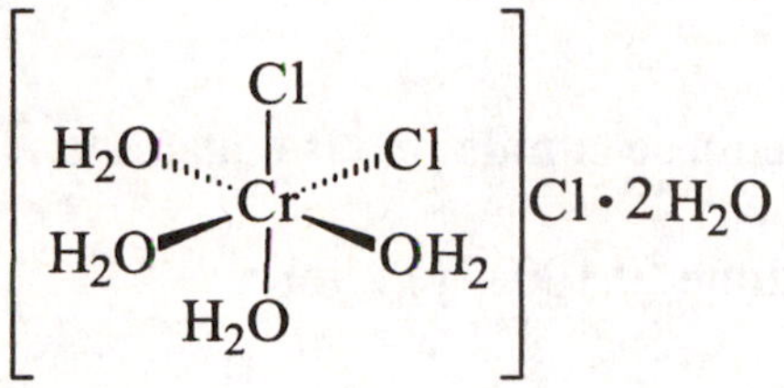

C

Structure **C** can exist as either cis or trans diastereoisomers.

(c) H_2O is a stronger field ligand than Cl^-. Compound **A** is likely to be violet absorbing in the yellow. Compounds **B** and **C** have weaker field ligands and would appear blue or green absorbing in the orange or red, respectively.

(d) $\Delta T = K_f \cdot m \cdot i$
For **A**, i = 4; for **B**, i = 3; and for **C**, i = 2.

For **A**, $\Delta T = K_f \cdot m \cdot i = (1.86\ °C/m)(0.25\ m)(4) = 1.9\ °C$
freezing point $= 0\ °C - \Delta T = 0\ °C - 1.9\ °C = -1.9\ °C$
For **B**, $\Delta T = K_f \cdot m \cdot i = (1.86\ °C/m)(0.25\ m)(3) = 1.4\ °C$
freezing point $= 0\ °C - \Delta T = 0\ °C - 1.4\ °C = -1.4\ °C$
For **C**, $\Delta T = K_f \cdot m \cdot i = (1.86\ °C/m)(0.25\ m)(2) = 0.93\ °C$
freezing point $= 0\ °C - \Delta T = 0\ °C - 0.93\ °C = -0.93\ °C$

20.146 (a) $K = \dfrac{[Cr_2O_7^{2-}]}{[CrO_4^{2-}]^2[H^+]^2} = 1.00 \times 10^{14}$

$[Cr_2O_7^{2-}]/[CrO_4^{2-}]^2 = 1.00 \times 10^{14}\ [H^+]^2$
In neutral solution, $[H^+] = 1.0 \times 10^{-7}$ and $[Cr_2O_7^{2-}]/[CrO_4^{2-}]^2 = 1$, so $[Cr_2O_7^{2-}]$ and $[CrO_4^{2-}]$ are comparable.
In basic solution, $[H^+] < 1.0 \times 10^{-7}$ and $[Cr_2O_7^{2-}]/[CrO_4^{2-}]^2 < 1$, so $[CrO_4^{2-}]$ predominates.
In acidic solution, $[H^+] > 1.0 \times 10^{-7}$ and $[Cr_2O_7^{2-}]/[CrO_4^{2-}]^2 > 1$, so $[Cr_2O_7^{2-}]$ predominates.

(b) At pH $= 4.000$, the $[H^+] = 1.00 \times 10^{-4}$ M
Let $x = [Cr_2O_7^{2-}]$ and $y = [CrO_4^{2-}]$

$$\dfrac{[Cr_2O_7^{2-}]}{[CrO_4^{2-}]^2} = [H^+]^2(1.00 \times 10^{14})$$

$$\dfrac{[Cr_2O_7^{2-}]}{[CrO_4^{2-}]^2} = (1.00 \times 10^{-4})^2(1.00 \times 10^{14})$$

$$\dfrac{[Cr_2O_7^{2-}]}{[CrO_4^{2-}]^2} = 1.00 \times 10^6 = \dfrac{x}{y^2}$$

Because there are 2 Cr atoms per $Cr_2O_7^{2-}$, the total Cr concentration is $2[Cr_2O_7^{2-}] + [CrO_4^{2-}]$, and therefore $2x + y = 0.100$.

$\dfrac{x}{y^2} = 1.00 \times 10^6$ and $2x + y = 0.100$ M; solve these simultaneous equations.

$x = (1.00 \times 10^6)y^2$ and $x = (0.100 - y)/2$; substitute $(0.100 - y)/2$ for x
$(0.100 - y)/2 = (1.00 \times 10^6)y^2$
$(2.00 \times 10^6)y^2 + y - 0.100 = 0$
Use the quadratic formula to solve for y.

$$y = \dfrac{-(1) \pm \sqrt{(1)^2 - 4(2.00 \times 10^6)(-0.100)}}{2(2.00 \times 10^6)} = \dfrac{(-1) \pm (894.2)}{4.00 \times 10^6}$$

$y = -2.24 \times 10^{-4}$ and $2.233 \times 10^{-4} = 2.23 \times 10^{-4}$
Of the two solutions for y, only the positive value of y has physical meaning because y is the $[CrO_4^{2-}]$.
$[CrO_4^{2-}] = 2.23 \times 10^{-4}$ M
$[Cr_2O_7^{2-}] = x = (1.00 \times 10^6)y^2 = (1.00 \times 10^6)(2.233 \times 10^{-4}\ M)^2 = 4.99 \times 10^{-2}$ M

(c) At pH = 2.000, the $[H^+] = 1.00 \times 10^{-2}$ M

Let $x = [Cr_2O_7^{2-}]$ and $y = [CrO_4^{2-}]$

$$\frac{[Cr_2O_7^{2-}]}{[CrO_4^{2-}]^2} = [H^+]^2(1.00 \times 10^{14})$$

$$\frac{[Cr_2O_7^{2-}]}{[CrO_4^{2-}]^2} = (1.00 \times 10^{-2})^2(1.00 \times 10^{14})$$

$$\frac{[Cr_2O_7^{2-}]}{[CrO_4^{2-}]^2} = 1.00 \times 10^{10} = \frac{x}{y^2}$$

Because there are 2 Cr atoms per $Cr_2O_7^{2-}$, the total Cr concentration is $2[Cr_2O_7^{2-}] + [CrO_4^{2-}]$, and therefore $2x + y = 0.100$.

$\dfrac{x}{y^2} = 1.00 \times 10^{10}$ and $2x + y = 0.100$ M; solve these simultaneous equations.

$x = (1.00 \times 10^{10})y^2$ and $x = (0.100 - y)/2$; substitute $(0.100 - y)/2$ for x

$(0.100 - y)/2 = (1.00 \times 10^{10})y^2$

$(2.00 \times 10^{10})y^2 + y - 0.100 = 0$

Use the quadratic formula to solve for y.

$$y = \frac{-(1) \pm \sqrt{(1)^2 - 4(2.00 \times 10^{10})(-0.100)}}{2(2.00 \times 10^{10})} = \frac{(-1) \pm (8.944 \times 10^4)}{4.00 \times 10^{10}}$$

$y = -2.24 \times 10^{-6}$ and $2.236 \times 10^{-6} = 2.24 \times 10^{-6}$

Of the two solutions for y, only the positive value of y has physical meaning because y is the $[CrO_4^{2-}]$.

$[CrO_4^{2-}] = 2.24 \times 10^{-6}$ M

$[Cr_2O_7^{2-}] = x = (1.00 \times 10^{10})y^2 = (1.00 \times 10^{10})(2.236 \times 10^{-6} \text{ M})^2 = 5.00 \times 10^{-2}$ M

21 Metals and Solid-State Materials

21.1   The most common oxidation state for the 3B transition metals (Sc, Y, and La) is 3+. The 3+ oxidation state of the cations conveniently matches the 3– charge of the phosphate ion resulting in a large lattice energy and corresponding insolubility for MPO_4 compounds.

21.2   (a) $Cr_2O_3(s) + 2\,Al(s) \rightarrow 2\,Cr(s) + Al_2O_3(s)$
        (b) $Cu_2S(s) + O_2(g) \rightarrow 2\,Cu(s) + SO_2(g)$
        (c) $PbO(s) + C(s) \rightarrow Pb(s) + CO(g)$
        (d) $2\,K^+(l) + 2\,Cl^-(l) \xrightarrow{\text{electrolysis}} 2\,K(l) + Cl_2(g)$

21.3 $CaO(s) + SiO_2(s) \rightarrow CaSiO_3(l)$ (slag)
        The O^{2-} in CaO behaves as a Lewis base and SiO_2 is the Lewis acid. They react with each other in a Lewis acid-base reaction to yield $CaSiO_3$ (Ca^{2+} and SiO_3^{2-}).

21.4   The electron configuration for Hg is [Xe] $4f^{14}\,5d^{10}\,6s^2$. Assuming the 5d and 6s bands overlap, the composite band can accomodate 12 valence electrons per metal atom. Weak bonding and a low melting point are expected for Hg because both the bonding and antibonding MOs are occupied.

21.5   (a)   The composite s-d band can accomodate 12 valence electrons per metal atom.
        Hf [Xe] $6s^2\,4f^{14}\,5d^2$, 4 valence electrons (4 bonding, 0 antibonding)
        The s-d band is 1/4 full, so Hf is picture (1).
        Pt [Xe] $6s^2\,4f^{14}\,5d^8$, 10 valence electrons (6 bonding, 4 antibonding)
        The s-d band is 5/6 full, so Pt is picture (2).
        Re [Xe] $6s^2\,4f^{14}\,5d^5$, 7 valence electrons (6 bonding, 1 antibonding)
        The s-d band is 7/12 full, so Re is picture (3).
        (b) Re has an excess of 5 bonding electrons and it has the highest melting point and is the hardest of the three.
        (c) Pt has an excess of only 2 bonding electrons and it has the lowest melting point and is the softest of the three.

21.6   Ge doped with As is an n-type semiconductor because As has an additional valence electron. The extra electrons are in the conduction band. The number of electrons in the conduction band of the doped Ge is much higher than for pure Ge, and the conductivity of the doped semiconductor is higher.

21.7   (a) (1), silicon; (2), white tin; (3), diamond; (4), silicon doped with aluminum
        (b) (3) < (1) < (4) < (2)
        Diamond (3) is an insulator with a large band gap. Silicon (1) is a semiconductor with a band gap smaller than diamond. The conduction band is partially occupied with a few

electrons and the valence band is partially empty. Silicon doped with aluminum (4) is a p-type semiconductor that has fewer electrons than needed for bonding and has vacancies (positive holes) in the valence band. White tin (2) has a partially filled s-p composite band and is a metallic conductor.

21.8 $E = 222 \text{ kJ/mol} \times \dfrac{1000 \text{ J}}{1 \text{ kJ}} \times \dfrac{1 \text{ mol}}{6.02 \times 10^{23}} = 3.69 \times 10^{-19} \text{ J}$

$\nu = \dfrac{E}{h} = \dfrac{3.69 \times 10^{-19} \text{ J}}{6.626 \times 10^{-34} \text{ J} \cdot \text{s}} = 5.57 \times 10^{14} \text{ s}^{-1}$

$\lambda = \dfrac{c}{\nu} = \dfrac{3.00 \times 10^{8} \text{ m/s}}{5.57 \times 10^{14} \text{ s}^{-1}} = 5.39 \times 10^{-7} \text{ m} = 539 \times 10^{-9} \text{ m} = 539 \text{ nm}$

21.9   8 Cu at corners $8 \times 1/8 = 1 \text{ Cu}$
       8 Cu on edges $8 \times 1/4 = \underline{2 \text{ Cu}}$
                              Total $= 3 \text{ Cu}$

       12 O on edges $12 \times 1/4 = 3 \text{ O}$
       8 O on faces $8 \times 1/2 = \underline{4 \text{ O}}$
                              Total $= 7 \text{ O}$

21.10 $Si(OCH_3)_4 + 4 H_2O \rightarrow Si(OH)_4 + 4 HOCH_3$

21.11 $Ba[OCH(CH_3)_2]_2 + Ti[OCH(CH_3)_2]_4 + 6 H_2O \rightarrow BaTi(OH)_6(s) + 6 HOCH(CH_3)_2$

$BaTi(OH)_6(s) \xrightarrow{\text{heat}} BaTiO_3(s) + 3 H_2O(g)$

21.12 (a) cobalt/tungsten carbide is a ceramic-metal composite.
       (b) silicon carbide/zirconia is a ceramic-ceramic composite.
       (c) boron nitride/epoxy is a ceramic-polymer composite.
       (d) boron carbide/titanium is a ceramic-metal composite.

21.13 The smaller the particle, the larger the band gap and the greater the shift in the color of the emitted light from the red to the violet. The yellow quantum dot is larger because yellow is closer to the red than is the blue.

21.14 atom diameter $= 250 \text{ pm} = 250 \times 10^{-12} \text{ m} = 0.25 \times 10^{-9} \text{ m} = 0.25 \text{ nm}$
       (a) $5.0 \text{ nm}/0.25 \text{ nm} = 20$ atoms on an edge of nanoparticle
       total atoms in nanoparticle $= 20^3 = 8,000$
       interior atoms in nanoparticle $= (20 - 2)^3 = 18^3 = 5,832$
       surface atoms in nanoparticle $=$ total atoms $-$ interior atoms $= 8,000 - 5,832 = 2,168$
       % of atoms on nanoparticle surface $= (2,168/8,000) \times 100\% = 27\%$
       (b) $10 \text{ nm}/0.25 \text{ nm} = 40$ atoms on an edge of nanoparticle
       total atoms in nanoparticle $= 40^3 = 64,000$
       interior atoms in nanoparticle $= (40 - 2)^3 = 38^3 = 54,872$

surface atoms in nanoparticle = total atoms – interior atoms = 64,000 – 54,872 = 9,128

% of atoms on nanoparticle surface = (9,128/64,000) x 100% = 14%

Conceptual Problems

21.16 (a) electrolysis   (b) roasting a metal sulfide

    (c)   A = Li, electrolysis.

            B = Hg, roasting of the metal sulfide.

            C = Mn, reduction of the metal oxide.

            D = Ca, reduction of the metal oxide.

21.18 (a) (2), bonding MO's are filled.

    (b) (3), bonding and antibonding MO's are filled.

    (c) (3) < (1) < (2). Hardness increases with increasing MO bond order.

21.20

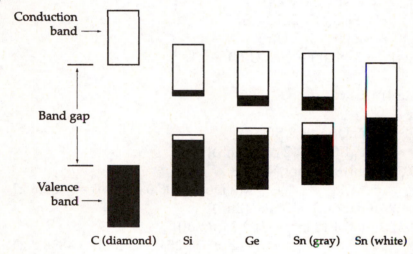

Section Problems
Sources of the Metallic Elements (Section 21.1)

21.22 TiO_2, MnO_2, and Fe_2O_3

21.24 (a) Cu is found in nature as a sulfide.   (b) Zr is found in nature as an oxide.

    (c) Pd is found in nature uncombined.   (d) Bi is found in nature as a sulfide.

21.26 The less electronegative early transition metals tend to form ionic compounds by losing electrons to highly electronegative nonmetals such as oxygen. The more electronegative late transition metals tend to form compounds with more covalent character by bonding to the less electronegative nonmetals such as sulfur.

21.28 (a) Fe_2O_3, hematite   (b) PbS, galena

    (c) TiO_2, rutile   (d) $CuFeS_2$, chalcopyrite

Metallurgy (Section 21.2)

21.30   The flotation process exploits the differences in the ability of water and oil to wet the surfaces of the mineral and the gangue. The gangue, which contains ionic silicates, is moistened by the polar water molecules and sinks to the bottom of the tank. The mineral particles, which contain the less polar metal sulfide, are coated by the oil and become attached to the soapy air bubbles created by the detergent. The metal sulfide particles are carried to the surface in the soapy froth, which is skimmed off at the top of the tank. This process would not work well for a metal oxide because it is too polar and will be wet by the water and sink with the gangue.

21.32   Because $E° < 0$ for Zn^{2+}, the reduction of Zn^{2+} is not favored.
Because $E° > 0$ for Hg^{2+}, the reduction of Hg^{2+} is favored.
The roasting of CdS should yield CdO because, like Zn^{2+}, $E° < 0$ for the reduction of Cd^{2+}.

21.34   (a) $V_2O_5(s) + 5\ Ca(s)\ \rightarrow\ 2\ V(s) + 5\ CaO(s)$
(b) $2\ PbS(s) + 3\ O_2(g)\ \rightarrow\ 2\ PbO(s) + 2\ SO_2(g)$
(c) $MoO_3(s) + 3\ H_2(g)\ \rightarrow\ Mo(s) + 3\ H_2O(g)$
(d) $3\ MnO_2(s) + 4\ Al(s)\ \rightarrow\ 3\ Mn(s) + 2\ Al_2O_3(s)$
(e) $MgCl_2(l)\ \xrightarrow{electrolysis}\ Mg(l) + Cl_2(g)$

21.36 $2\ ZnS(s)\ +\ 3\ O_2(g)\ \rightarrow\ 2\ ZnO(s)\ +\ 2\ SO_2(g)$
$\Delta H° = [2\ \Delta H°_f(ZnO) + 2\ \Delta H°_f(SO_2)] - [2\ \Delta H°_f(ZnS)]$
$\Delta H° = [(2\ mol)(-350.5\ kJ/mol) + (2\ mol)(-296.8\ kJ/mol)]$
$$- (2\ mol)(-206.0\ kJ/mol) = -882.6\ kJ$$
$\Delta G° = [2\ \Delta G°_f(ZnO) + 2\ \Delta G°_f(SO_2)] - [2\ \Delta G°_f(ZnS)]$
$\Delta G° = [(2\ mol)(-320.5\ kJ/mol) + (2\ mol)(-300.2\ kJ/mol)]$
$$- (2\ mol)(-201.3\ kJ/mol) = -838.8\ kJ$$

$\Delta H°$ and $\Delta G°$ are different because of the entropy change associated with the reaction. The minus sign for $(\Delta H° - \Delta G°)$ indicates that the entropy is negative, which is consistent with a decrease in the number of moles of gas from 3 mol to 2 mol.

21.38 $FeCr_2O_4(s)\ +\ 4\ C(s)\ \rightarrow\ Fe(s) + 2\ Cr(s)\ +\ 4\ CO(g)$
ferrochrome

(a) $FeCr_2O_4$, 223.84;   Cr, 52.00;   236 kg = 236 x 10^3 g
$$\text{mass Cr} = 236 \times 10^3\ g \times \frac{1\ mol\ FeCr_2O_4}{223.84\ g} \times \frac{2\ mol\ Cr}{1\ mol\ FeCr_2O_4} \times \frac{52.00\ g\ Cr}{1\ mol\ Cr} \times \frac{1.00\ kg}{1000\ g} = 110\ kg\ Cr$$

(b) $\text{mol CO} = 236 \times 10^3\ g \times \frac{1\ mol\ FeCr_2O_4}{223.84\ g} \times \frac{4\ mol\ CO}{1\ mol\ FeCr_2O_4} = 4217.3\ mol\ CO$

$$PV = nRT; \quad V = \frac{nRT}{P} = \frac{(4217.3 \text{ mol})\left(0.082\ 06\ \dfrac{L \cdot atm}{K \cdot mol}\right)(298\ K)}{\left(740 \text{ mm Hg} \times \dfrac{1.00 \text{ atm}}{760 \text{ mm Hg}}\right)} = 1.06 \times 10^5 \text{ L CO}$$

21.40 $Ni^{2+}(aq) + 2\ e^- \rightarrow Ni(s);$ $1\ A = 1\ C/s$

mass Ni = $52.5\ \dfrac{C}{s} \times 8.00\ h \times \dfrac{3600\ s}{1\ h} \times \dfrac{1\ mol\ e^-}{96,500\ C} \times \dfrac{1\ mol\ Ni}{2\ mol\ e^-} \times \dfrac{58.69\ g\ Ni}{1\ mol\ Ni} \times \dfrac{1.00\ kg}{1000\ g}$

mass Ni = 0.460 kg Ni

Iron and Steel (Section 21.3)

21.42 $Fe_2O_3(s) + 3\ CO(g) \rightarrow 2\ Fe(l) + 3\ CO_2(g)$
Fe_2O_3 is the oxidizing agent. CO is the reducing agent.

21.44 Slag is a byproduct of iron production, consisting mainly of $CaSiO_3$. It is produced from the gangue in iron ore.

21.46 Molten iron from a blast furnace is exposed to a jet of pure oxygen gas for about 20 minutes. The impurities are oxidized to yield a molten slag that can be poured off.
$P_4(l) + 5\ O_2(g) \rightarrow P_4O_{10}(l)$
$6\ CaO(s) + P_4O_{10}(l) \rightarrow 2\ Ca_3(PO_4)_2(l)$ (slag)

$2\ Mn(l) + O_2(g) \rightarrow 2\ MnO(s)$
$MnO(s) + SiO_2(s) \rightarrow MnSiO_3(l)$ (slag)

21.48 $SiO_2(s) + 2\ C(s) \rightarrow Si(s) + 2\ CO(g)$
$Si(s) + O_2(g) \rightarrow SiO_2(s)$
$CaO(s) + SiO_2(s) \rightarrow CaSiO_3(l)$ (slag)

21.50 No. In a blast furnace tungsten carbide (WC) would be formed.

Bonding in Metals (Section 21.4)

21.52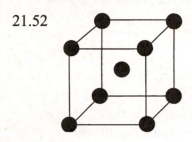

Each K has a single valence electron and has eight nearest neighbor K atoms. The valence electrons can't be localized in an electron-pair bond between any particular pair of K atoms.

21.54 Malleability and ductility of metals follow from the fact that the delocalized bonding extends in all directions. When a metallic crystal is deformed, no localized bonds are broken. Instead, the electron sea simply adjusts to the new distribution of cations, and the energy of the deformed structure is similar to that of the original. Thus, the energy required to deform a metal is relatively small.

21.56 The energy required to deform a transition metal like W is greater than that for Cs because W has more valence electrons and hence more electrostatic "glue."

21.58 The difference in energy between successive MOs in a metal decreases as the number of metal atoms increases so that the MOs merge into an almost continuous band of energy levels. Consequently, MO theory for metals is often called band theory.

21.60 The energy levels within a band occur in degenerate pairs; one set of energy levels applies to electrons moving to the right, and the other set applies to electrons moving to the left. In the absence of an electrical potential, the two sets of levels are equally populated. As a result there is no net electric current. In the presence of an electrical potential those electrons moving to the right are accelerated, those moving to the left are slowed down, and some change direction. Thus, the two sets of energy levels are now unequally populated. The number of electrons moving to the right is now greater than the number moving to the left, and so there is a net electric current.

21.62 (a)   (b)

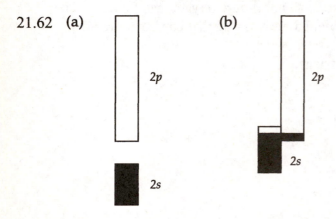

Diagram (b) shows the 2s and 2p bands overlapping in energy and the resulting composite band is only partially filled. Thus, Be is a good electrical conductor.

21.64 Transition metals have a d band that can overlap the s band to give a composite band consisting of six MOs per metal atom. Half of the MOs are bonding and half are antibonding, and thus one expects maximum bonding for metals that have six valence electrons per metal atom. Accordingly, the melting points of the transition metals go through a maximum at or near group 6B.

Semiconductors and Semiconductor Applications (Sections 21.5 and 21.6)

21.66 A semiconductor is a material that has an electrical conductivity intermediate between that of a metal and that of an insulator. Si, Ge, and Sn (gray) are semiconductors.

21.68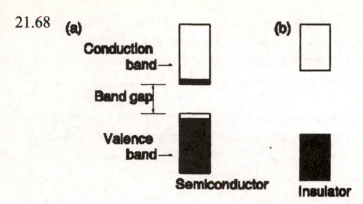

The MOs of a semiconductor are similar to those of an insulator, but the band gap in a semiconductor is smaller. As a result, a few electrons have enough energy to jump the gap and occupy the higher-energy, conduction band. The conduction band is thus partially filled, and the valence band is partially empty. When an electrical potential is applied to a semiconductor, it conducts a small amount of current because the potential can accelerate the electrons in the partially filled bands.

21.70 As the band gap increases, the number of electrons able to jump the gap and occupy the higher-energy conduction band decreases, and thus the conductivity decreases.

21.72 An n-type semiconductor is a semiconductor doped with a substance with more valence electrons than the semiconductor itself. Si doped with P is an example.

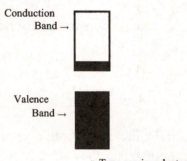

n-Type semiconductor

21.74 In the MO picture, the extra electrons occupy the conduction band. The number of electrons in the conduction band of the doped Ge is much greater than for pure Ge, and the conductivity of the doped semiconductor is correspondingly higher.

21.76 (a) p-type (In is electron deficient with respect to Si)
(b) n-type (Sb is electron rich with respect to Ge)
(c) n-type (As is electron rich with respect to gray Sn)

21.78 $Cd(CH_3)_2(g) + H_2Se(g) \rightarrow CdSe(s) + 2 CH_4(g)$

21.80 Al_2O_3 < Ge < Ge doped with In < Fe < Cu

21.82   In a diode, current flows only when the junction is under a forward bias (negative battery terminal on the n-type side). A p-n junction that is part of a circuit and subjected to an alternating potential acts as a rectifier, allowing current to flow in only one direction, thereby converting alternating current to direct current.

21.84   Both an LED and a photovoltaic cell contain p-n junctions, but the two devices involve opposite processes. An LED converts electrical energy to light; a photovoltaic, or solar, cell converts light to electricity.

21.86 $E = 193 \text{ kJ/mol} \times \dfrac{1000 \text{ J}}{1 \text{ kJ}} \times \dfrac{1 \text{ mol}}{6.02 \times 10^{23}} = 3.21 \times 10^{-19} \text{ J}$

$\nu = \dfrac{E}{h} = \dfrac{3.21 \times 10^{-19} \text{ J}}{6.626 \times 10^{-34} \text{ J·s}} = 4.84 \times 10^{14} \text{ s}^{-1}$

$\lambda = \dfrac{c}{\nu} = \dfrac{3.00 \times 10^8 \text{ m/s}}{4.84 \times 10^{14} \text{ s}^{-1}} = 6.20 \times 10^{-7} \text{ m} = 620 \times 10^{-9} \text{ m} = 620 \text{ nm, orange light}$

Superconductors (Section 21.7)

21.88   (1) A superconductor is able to levitate a magnet.
(2) In a superconductor, once an electric current is started, it flows indefinitely without loss of energy. A superconductor has no electrical resistance.

21.90   Some K^+ ions are surrounded octahedrally by six C_{60}^{3-} ions; others are surrounded tetrahedrally by four C_{60}^{3-} ions.

Ceramics and Composites (Sections 21.8 and 21.9)

21.92   Ceramics are inorganic, nonmetallic, nonmolecular solids, including both crystalline and amorphous materials. Ceramics have higher melting points, and they are stiffer, harder, and more resistant to wear and corrosion than are metals.

21.94   Ceramics have higher melting points, and they are stiffer, harder, and more wear resistant than metals because they have stronger bonding. They maintain much of their strength at high temperatures, where metals either melt or corrode because of oxidation.

21.96   The brittleness of ceramics is due to strong chemical bonding. In silicon nitride each Si atom is bonded to four N atoms and each N atom is bonded to three Si atoms. The strong, highly directional covalent bonds prevent the planes of atoms from sliding over one another when the solid is subjected to a stress. As a result, the solid can't deform to relieve the stress. It maintains its shape up to a point, but then the bonds give way suddenly and the material fails catastrophically when the stress exceeds a certain threshold value. By contrast, metals are able to deform under stress because their planes of metal cations can slide easily in the electron sea.

21.98   Ceramic processing is the series of steps that leads from raw material to the finished ceramic object.

21.100 $Zr[OCH(CH_3)_2]_4 + 4\ H_2O \rightarrow Zr(OH)_4 + 4\ HOCH(CH_3)_2$

21.102 $(HO)_3Si{-}O{-}H + H{-}O{-}Si(OH)_3 \rightarrow (HO)_3Si{-}O{-}Si(OH)_3 + H_2O$
         Further reactions of this sort give a three-dimensional network of Si–O–Si bridges. On heating, SiO_2 is obtained.

21.104 $2\ Ti(BH_4)_3(soln) \rightarrow 2\ TiB_2(s) + B_2H_6(g) + 9\ H_2(g)$

21.106 $3\ SiCl_4(g) + 4\ NH_3(g) \rightarrow Si_3N_4(s) + 12\ HCl(g)$

21.108   Graphite/epoxy composites are good materials for making tennis rackets and golf clubs because of their high strength-to-weight ratios.

Chapter Problems

21.110 $\Delta G = \Delta H - T\Delta S = -160.8\ kJ - (298\ K)(-0.410\ kJ/K) = -38.6\ kJ$
         Because $\Delta G < 0$, the reaction is spontaneous at 25 °C (298 K).
         Set $\Delta G = 0$ and solve for T to find the temperature at which the reaction becomes nonspontaneous.
$$0 = \Delta H - T\Delta S; \qquad T = \frac{\Delta H}{\Delta S} = \frac{-160.8\ kJ}{-0.410\ kJ/K} = 392\ K = 119\ °C$$

21.112 $2\ Eu^{3+}(aq) + Zn(s) \rightarrow 2\ Eu^{2+}(aq) + Zn^{2+}(aq)$
         $Eu^{2+}(aq) + SO_4^{2-}(aq) \rightarrow EuSO_4(s)$

21.114   The chemical composition of the alkaline earth minerals is that of metal sulfates and sulfites, MSO_4 and MSO_3.

21.116   Band theory better explains how the number of valence electrons affects properties such as melting point and hardness.

21.118   V   [Ar] $3d^3\ 4s^2$ Zn   [Ar] $3d^{10}\ 4s^2$
         Transition metals have a d band that can overlap the s band to give a composite band consisting of six MOs per metal atom. Half of the MOs are bonding and half are antibonding. Strong bonding and a high enthalpy of vaporization are expected for V because almost all of the bonding MOs are occupied and all of the antibonding MOs are empty. Weak bonding and a low enthalpy of vaporization are expected for Zn because both the bonding and the antibonding MOs are occupied.

21.120 With a band gap of 130 kJ/mol, GaAs is a semiconductor. Because Ge lies between Ga and As in the periodic table, GaAs is isoelectronic with Ge.

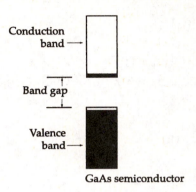

GaAs semiconductor

21.122 $YBa_2Cu_3O_7$, 666.20; $Cu(OCH_2CH_3)_2$, 153.67
$Y(OCH_2CH_3)_3$, 224.09; $Ba(OCH_2CH_3)_2$, 227.45

$$\text{mol } Cu(OCH_2CH_3)_2 = 75.4 \text{ g} \times \frac{1 \text{ mol}}{153.67 \text{ g}} = 0.4907 \text{ mol } Cu(OCH_2CH_3)_2$$

mass $Y(OCH_2CH_3)_3$ = 0.4907 mol $Cu(OCH_2CH_3)_2$ x

$$\frac{1 \text{ mol } Y(OCH_2CH_3)_3}{3 \text{ mol } Cu(OCH_2CH_3)_2} \text{ x } \frac{224.09 \text{ g } Y(OCH_2CH_3)_3}{1 \text{ mol } Y(OCH_2CH_3)_3} = 36.7 \text{ g } Y(OCH_2CH_3)_3$$

mass $Ba(OCH_2CH_3)_2$ = 0.4907 mol $Cu(OCH_2CH_3)_2$ x

$$\frac{2 \text{ mol } Ba(OCH_2CH_3)_2}{3 \text{ mol } Cu(OCH_2CH_3)_2} \text{ x } \frac{227.45 \text{ g } Ba(OCH_2CH_3)_2}{1 \text{ mol } Ba(OCH_2CH_3)_2} = 74.4 \text{ g } Ba(OCH_2CH_3)_2$$

mass $YBa_2Cu_3O_7$ = 0.4907 mol $Cu(OCH_2CH_3)_2$ x

$$\frac{1 \text{ mol } YBa_2Cu_3O_7}{3 \text{ mol } Cu(OCH_2CH_3)_2} \text{ x } \frac{666.20 \text{ g } YBa_2Cu_3O_7}{1 \text{ mol } YBa_2Cu_3O_7} = 109 \text{ g } YBa_2Cu_3O_7$$

21.124 (a) 6 $Al(OCH_2CH_3)_3$ + 2 $Si(OCH_2CH_3)_4$ + 26 H_2O →
    6 $Al(OH)_3(s)$ + 2 $Si(OH)_4(s)$ + 26 $HOCH_2CH_3$
        sol

(b) H_2O is eliminated from the sol through a series of reactions linking the sol particles together through a three-dimensional network of O bridges to form the gel.
$(HO)_2Al–O–H + H–O–Si(OH)_3 \rightarrow (HO)_2Al–O–Si(OH)_3 + H_2O$

(c) The remaining H_2O and solvent are removed from the gel by heating to produce the ceramic, 3 $Al_2O_3 \cdot 2 SiO_2$.

21.126 (a)

$$H_2C\!\!=\!\!CH \quad + \quad H_2C\!\!=\!\!CH \quad \longrightarrow \quad -CH_2-CH-CH_2-CH-$$
$$\overset{|}{CN} \qquad\qquad \overset{|}{CN} \qquad\qquad\qquad\qquad \overset{|}{CN} \qquad \overset{|}{CN}$$

(b) $\Delta H° = D_{C=C} - 2 D_{C-C} = 611$ kJ $- 2(350)$ kJ $= -89$ kJ/unit; exothermic

21.128 (a)

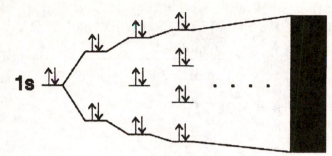

This material is an insulator because all MOs are filled, preventing the movement of electrons.

(b)

Neutral hydrogen atoms have only 1 valence electron, compared with 2 in H⁻. Partially empty antibonding MOs will allow the movement of electrons, so the doped material will be a conductor.

(c) The missing electrons in the doped material create "holes" which are positive charge carriers. This type of doped material is a p-type semiconductor.

Multiconcept Problems

21.130 (a) $8.894 \text{ g Al} \times \dfrac{1 \text{ cm}^3}{2.699 \text{ g Al}} = 3.295 \text{ cm}^3$

$3.295 \text{ cm}^3 = (36.5 \text{ cm})^2 \times \text{h}$

$\text{h} = \dfrac{3.295 \text{ cm}^3}{(36.5 \text{ cm})^2} = 2.47 \times 10^{-3} \text{ cm}$

$\text{h} = 2.47 \times 10^{-3} \text{ cm} \times \dfrac{1 \text{ m}}{100 \text{ cm}} = 2.47 \times 10^{-5} \text{ m}$

(b) $\text{d} = 143 \text{ pm} = 143 \times 10^{-12} \text{ m}$

For a face-centered cube, the unit cell face diagonal = $4\text{r} = \sqrt{2} \cdot \text{d}$

$\text{unit cell edge} = \text{d} = \dfrac{4\text{r}}{\sqrt{2}} = \dfrac{(4)(143 \times 10^{-12} \text{ m})}{\sqrt{2}} = 4.04 \times 10^{-10} \text{ m}$

$\dfrac{2.47 \times 10^{-5} \text{ m}}{4.04 \times 10^{-10} \text{ m/unit cell}} = 6.11 \times 10^4 \text{ unit cells thick}$

21.132 $Cr_2O_7^{2-}(aq) + 6 Fe^{2+}(aq) + 14 H^+(aq) \rightarrow 6 Fe^{3+}(aq) + 2 Cr^{3+}(aq) + 7 H_2O(l)$
mol $Cr_2O_7^{2-}$ = (0.038 89 L)(0.018 54 mol/L) = 7.210 x 10^{-4} mol $Cr_2O_7^{2-}$

mass Fe = 7.210 x 10^{-4} mol $Cr_2O_7^{2-}$ x $\dfrac{6 \text{ mol Fe}^{2+}}{1 \text{ mol Cr}_2O_7^{2-}}$ x $\dfrac{55.847 \text{ g Fe}^{2+}}{1 \text{ mol Fe}^{2+}}$ = 0.2416 g Fe^{2+}

mass % Fe = $\dfrac{0.2416 \text{ g}}{0.3249 \text{ g}}$ x 100% = 74.36% Fe

21.134   (a) 431 pm = 431 x 10^{-12} m
There are 4 oxygen atoms in the face-centered cubic unit cell.
mass of unit cell = (5.75 g/cm^3)(431 x 10^{-12} m)^3(100 cm/1 m)^3 = 4.604 x 10^{-22} g

mass of Fe in unit cell = (4.604 x 10^{-22} g) – 4 O atoms x $\dfrac{15.9994 \text{ g O}}{6.022 \text{ x } 10^{23} \text{ O atoms}}$

= 3.541 x 10^{-22} g Fe

number of Fe atoms in unit cell = 3.541 x 10^{-22} g Fe x $\dfrac{6.022 \text{ x } 10^{23} \text{ Fe atoms}}{55.847 \text{ g Fe}}$

= 3.818 Fe atoms

For Fe_xO, x = $\dfrac{3.818 \text{ Fe atoms}}{4 \text{ O atoms}}$ = 0.955

(b) The average oxidation state of Fe = $\dfrac{+2}{0.955}$ = 2.094

(c) Let X equal the fraction of Fe^{3+} and Y equal the fraction of Fe^{2+} in wustite.
So, X + Y = 1 and 3X + 2Y = 2.094
Y = 1 – X
3X + 2(1 –X) = 2.094
3X + 2 – 2X = 2.094
X + 2 = 2.094
X = 2.094 – 2 = 0.094
9.4% of the Fe in wustite is Fe^{3+}.

(d) d = 431 pm
d = $\dfrac{n\lambda}{2 \sin\theta}$ = $\dfrac{3 \cdot 70.93 \text{ pm}}{2 \sin\theta}$ = 431 pm

$\sin\theta$ = $\dfrac{3 \cdot 70.93 \text{ pm}}{2 \cdot 431 \text{ pm}}$ = 0.247 and θ = 14.3 °

(e) The presence of Fe^{3+} in the semiconductor leads to missing electrons that create "holes" which are positive charge carriers. This type of doped material is a p-type semiconductor.

21.136 $Ni(s) + 4 CO(g) \rightleftharpoons Ni(CO)_4(g)$
$\Delta H° = -160.8$ kJ; $\Delta S° = -410$ J/K = -410 x 10^{-3} kJ/K
(a) 150 °C = 423 K
$\Delta G° = \Delta H° - T\Delta S° = -160.8$ kJ $-(423$ K$)(-410$ x 10^{-3} kJ/K) = +12.6 kJ

$\Delta G° = - RT \ln K$

$\ln K = \dfrac{-\Delta G°}{RT} = \dfrac{-12.6 \text{ kJ/mol}}{[8.314 \times 10^{-3} \text{ kJ/(K} \cdot \text{mol)}](423 \text{ K})} = -3.58$

$K = K_p = e^{-3.58} = 0.028$

(b) 230 °C = 503 K

$\Delta G° = \Delta H° - T\Delta S° = -160.8 \text{ kJ} - (503 \text{ K})(-410 \times 10^{-3} \text{ kJ/K}) = +45.4 \text{ kJ}$

$\Delta G° = - RT \ln K$

$\ln K = \dfrac{-\Delta G°}{RT} = \dfrac{-45.4 \text{ kJ/mol}}{[8.314 \times 10^{-3} \text{ kJ/(K} \cdot \text{mol)}](503 \text{ K})} = -10.86$

$K = K_p = e^{-10.86} = 1.9 \times 10^{-5}$

(c) $\Delta S°$ is large and negative because as the reaction proceeds in the forward direction, the number of moles of gas decrease from four to one.

Because $\Delta S°$ is negative, $-T\Delta S°$ is positive, and as T increases, $\Delta G°$ becomes more positive because $\Delta G° = \Delta H° - T\Delta S°$.

(d) The reaction is exothermic because $\Delta H°$ is negative.

$Ni(s) + 4 CO(g) \rightleftharpoons Ni(CO)_4(g) + \text{heat}$

Heat is added as the temperature is raised and the reaction proceeds in the reverse direction to relieve this stress, as predicted by Le Châtelier's principle. As the reverse reaction proceeds, the partial pressure of CO increases and the partial pressure of $Ni(CO)_4$ decreases. K_p decreases as calculated because $K_p = \dfrac{P_{Ni(CO)_4}}{(P_{CO})^4}$.

21.138 (a) $(NH_4)_2Zn(CrO_4)_2(s) \rightarrow ZnCr_2O_4(s) + N_2(g) + 4 H_2O(g)$

(b) mol $(NH_4)_2Zn(CrO_4)_2 = 10.36 \text{ g } (NH_4)_2Zn(CrO_4)_2 \times \dfrac{1 \text{ mol } (NH_4)_2Zn(CrO_4)_2}{333.45 \text{ g } (NH_4)_2Zn(CrO_4)_2}$

$= 0.03107 \text{ mol } (NH_4)_2Zn(CrO_4)_2$

mass $ZnCr_2O_4 = 0.03107 \text{ mol } (NH_4)_2Zn(CrO_4)_2 \times \dfrac{1 \text{ mol } ZnCr_2O_4}{1 \text{ mol } (NH_4)_2Zn(CrO_4)_2}$

$\times \dfrac{233.38 \text{ g } ZnCr_2O_4}{1 \text{ mol } ZnCr_2O_4} = 7.251 \text{ g } ZnCr_2O_4$

(c) mol N_2 + mol $H_2O = 0.03107 \text{ mol } (NH_4)_2Zn(CrO_4)_2 \times \dfrac{5 \text{ mol gas}}{1 \text{ mol } (NH_4)_2Zn(CrO_4)_2}$

$= 0.1554 \text{ mol gaseous by-products}$

292 °C = 565 K;

$PV = nRT; \quad V = \dfrac{nRT}{P} = \dfrac{(0.1554 \text{ mol})\left(0.082\ 06 \dfrac{L \cdot atm}{K \cdot mol}\right)(565 \text{ K})}{\left(745 \text{ mm Hg} \times \dfrac{1.00 \text{ atm}}{760 \text{ mm Hg}}\right)} = 7.35 \text{ L}$

(d) A face-centered cubic unit cell has four octahedral holes and eight tetrahedral holes. This unit cell contains one Zn^{2+} ion in a tetrahedral hole and two Cr^{3+} ions in octahedral holes, therefore 1/8 of the tetrahedral holes and 1/2 of the octahedral holes are filled.

(e)

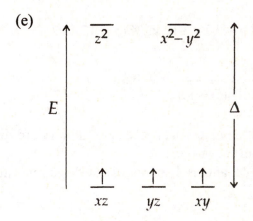

Octahedral Cr^{3+} has three unpaired electrons in the lower-energy d orbitals (xy, xz, yz). Cr^{3+} can absorb visible light to promote one of these d electrons to one of the higher-energy d orbitals making this compound colored. All of the d orbitals in Zn^{2+} are filled and no d electrons can be promoted, consequently the Zn^{2+} ion does not contribute to the color.

21.140

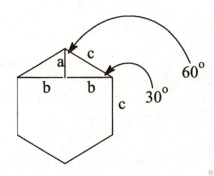

$c = 141.5 \text{ pm} = 141.5 \times 10^{-12} \text{ m}$
$\cos(60) = a/c$ and $\sin(60) = b/c$
$a = \cos(60) \cdot c = (0.5)(141.5 \times 10^{-12} \text{ m}) = 7.075 \times 10^{-11} \text{ m}$
$b = \sin(60) \cdot c = (0.866)(141.5 \times 10^{-12} \text{ m}) = 1.225 \times 10^{-10} \text{ m}$
diameter $= 1.08 \text{ nm} = 1.08 \times 10^{-9} \text{ m} = 1080 \times 10^{-12} \text{ m} = 1080 \text{ pm}$

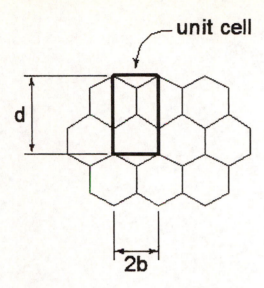

$d = 2a + 2c = 2(7.075 \times 10^{-11} \text{ m}) + 2(141.5 \times 10^{-12} \text{ m}) = 4.245 \times 10^{-10}$ m

$d = 424.5 \times 10^{-12}$ m $= 424.5$ pm

$2b = 2(1.225 \times 10^{-10} \text{ m}) = 2.450 \times 10^{-10}$ m $= 245.0 \times 10^{-12}$ m $= 245.0$ pm

Area of unit cell $= (424.5 \text{ pm})(245.0 \text{ pm}) = 1.04 \times 10^5$ pm^2/cell

No. of C atoms/cell $= (4)(1/4) + (2)(1/2) + 2 = 4$ C atoms/cell

Surface area of nanotube $= \pi d l = \pi(1080 \text{ pm})(1.0 \times 10^9 \text{ pm}) = 3.39 \times 10^{12}$ pm^2

No. of C atoms $= (3.39 \times 10^{12} \text{ pm}^2)\left(\dfrac{1 \text{ cell}}{1.04 \times 10^5 \text{ pm}^2} \right)\left(\dfrac{4 \text{ C atoms}}{\text{cell}} \right) = 1.3 \times 10^8$ C atoms

22

Nuclear Chemistry

22.1   (a) ^{199}Au has a higher neutron/proton ratio and decays by beta emission. ^{173}Au has a lower neutron/proton ratio and decays by alpha emission.

        (b) ^{196}Pb has a lower neutron/proton ratio and decays by positron emission. ^{206}Pb is nonradioactive.

22.2   For $^{16}_{8}$O:

First, calculate the total mass of the nucleons (8 n + 8 p)

Mass of 8 neutrons = (8)(1.008 66) = 8.069 28

Mass of 8 protons = (8)(1.007 28) = 8.058 24

Mass of 8 n + 8 p   = 16.127 52

Next, calculate the mass of a ^{16}O nucleus by subtracting the mass of 8 electrons from the mass of a ^{16}O atom.

Mass of ^{16}O atom   = 15.994 91

−Mass of 8 electrons = −(8)(5.486 x 10^{-4})   = −0.004 39

Mass of ^{16}O nucleus   = 15.990 52

Then subtract the mass of the ^{16}O nucleus from the mass of the nucleons to find the mass defect:

Mass defect = mass of nucleons − mass of nucleus

        = (16.127 52) − (15.990 52) = 0.137 00 u

Mass defect in grams = (0.137 00 u)(1.660 54 x 10^{-24} g/u) = 2.2749 x 10^{-25} g

Mass defect in g/mol = (2.2749 x 10^{-25} g)(6.022 x 10^{23} mol^{-1}) = 0.136 99 g/mol

Now, use the Einstein equation to convert the mass defect into the binding energy.

$\Delta E = \Delta mc^2 = (0.136\ 99\ \text{g/mol})(10^{-3}\ \text{kg/g})(3.00 \times 10^8\ \text{m/s})^2$

$\Delta E = 1.233 \times 10^{13}\ \text{J/mol} = 1.233 \times 10^{10}\ \text{kJ/mol}$

$$\Delta E = \frac{1.233 \times 10^{13}\ \text{J/mol}}{6.022 \times 10^{23}\ \text{nuclei/mol}} \times \frac{1\ \text{MeV}}{1.60 \times 10^{-13}\ \text{J}} \times \frac{1\ \text{nucleus}}{16\ \text{nucleons}} = 8.00\ \frac{\text{MeV}}{\text{nucleon}}$$

22.3 $\Delta m = \dfrac{\Delta E}{c^2} = \dfrac{-820 \times 10^3\ \text{J}}{(3.00 \times 10^8\ \text{m/s})^2} = \dfrac{-820 \times 10^3\ \text{kg·m}^2/\text{s}^2}{(3.00 \times 10^8\ \text{m/s})^2} = -9.11 \times 10^{-12}\ \text{kg}$

        $\Delta m = -9.11 \times 10^{-9}\ \text{g}$

22.4 $\quad {}^{1}_{0}n + {}^{235}_{92}U \rightarrow {}^{137}_{52}Te + {}^{97}_{40}Zr + 2\,{}^{1}_{0}n$

mass ${}^{235}_{92}U$	235.0439
mass ${}^{1}_{0}n$	1.008 66
−mass ${}^{137}_{52}Te$	−136.9254
−mass ${}^{97}_{40}Zr$	−96.9110
−mass 2 ${}^{1}_{0}n$	−(2)(1.008 66)

mass change $\qquad$ 0.1988 u

$(0.1988\text{ u})(1.660\,54 \times 10^{-24}\text{ g/u})(6.022 \times 10^{23}\text{ mol}^{-1}) = 0.1988\text{ g/mol}$
$\Delta E = \Delta mc^2 = (0.1988\text{ g/mol})(10^{-3}\text{ kg/g})(3.00 \times 10^8\text{ m/s})^2$
$\Delta E = 1.79 \times 10^{13}\text{ J/mol} = 1.79 \times 10^{10}\text{ kJ/mol}$

22.5 $\quad {}^{1}_{1}H + {}^{2}_{1}H \rightarrow {}^{3}_{2}He$

mass ${}^{1}H$	1.007 83
mass ${}^{2}H$	2.014 10
−mass ${}^{3}He$	−3.016 03

mass change $\qquad$ 0.005 90 u

$(0.005\,90\text{ u})(1.660\,54 \times 10^{-24}\text{ g/u})(6.022 \times 10^{23}\text{ mol}^{-1}) = 0.005\,90\text{ g/mol}$
$\Delta E = \Delta mc^2 = (0.005\,90\text{ g/mol})(10^{-3}\text{ kg/g})(3.00 \times 10^8\text{ m/s})^2$
$\Delta E = 5.31 \times 10^{11}\text{ J/mol} = 5.31 \times 10^8\text{ kJ/mol}$

22.6 $\quad {}^{40}_{18}Ar + {}^{1}_{1}p \rightarrow {}^{40}_{19}K + {}^{1}_{0}n$

22.7 $\quad {}^{238}_{92}U + {}^{2}_{1}H \rightarrow {}^{238}_{93}Np + 2\,{}^{1}_{0}n$

22.8 $\quad \ln\left(\dfrac{N}{N_0}\right) = (-0.693)\left(\dfrac{t}{t_{1/2}}\right); \qquad \dfrac{N}{N_0} = \dfrac{\text{Decay rate at time t}}{\text{Decay rate at time t} = 0}$

$\qquad \ln\left(\dfrac{2.4}{15.3}\right) = (-0.693)\left(\dfrac{t}{5730\text{ y}}\right); \qquad t = 1.53 \times 10^4\text{ y}$

22.9 $\quad$ 2 billion years ago unusually rich uranium deposits, with the fissionable ${}^{235}U$ isotope at about 3% abundance, were flooded by groundwater, which acted as a moderator to slow the neutrons released by fission of ${}^{235}U$, thereby allowing a nuclear chain reaction to take place. Today, however, the natural abundance of ${}^{235}U$ is only about 0.7% because of nuclear decay over the past few billion years. Because a chain-reaction is no longer self-sustaining at the present 0.7% level of ${}^{235}U$, the conditions needed for natural reactors are no longer present on Earth.

Section Problems
Nuclear Stability (Section 22.1)

22.10 ^{160}W is neutron poor and decays by alpha emission. ^{185}W is neutron rich and decays by beta emission.

22.12   "Neutron rich" nuclides emit beta particles to decrease the number of neutrons and increase the number of protons in the nucleus.

Energy Changes during Nuclear Reactions (Section 22.2)

22.14   The loss in mass that occurs when protons and neutrons combine to form a nucleus is called the mass defect. The lost mass is converted into the binding energy that is used to hold the nucleons together.

22.16 $E = (1.50 \text{ MeV})\left(\dfrac{1.60 \times 10^{-13} \text{ J}}{1 \text{ MeV}}\right) = 2.40 \times 10^{-13} \text{ J}$

$\lambda = \dfrac{hc}{E} = \dfrac{(6.626 \times 10^{-34} \text{ J·s})(3.00 \times 10^{8} \text{ m/s})}{2.40 \times 10^{-13} \text{ J}} = 8.28 \times 10^{-13} \text{ m} = 0.000\ 828 \text{ nm}$

22.18   (a) For $^{52}_{26}$Fe:
First, calculate the total mass of the nucleons (26 n + 26 p)
Mass of 26 neutrons = (26)(1.008 66) = 26.225 16
Mass of 26 protons = (26)(1.007 28) = 26.189 28
Mass of 26 n + 26 p   = 52.414 44
Next, calculate the mass of a ^{52}Fe nucleus by subtracting the mass of 26 electrons from the mass of a ^{52}Fe atom.
Mass of ^{52}Fe atom   = 51.948 11
−Mass of 26 electrons = −(26)(5.486 x 10^{-4}) = −0.014 26
Mass of ^{52}Fe nucleus   = 51.933 85
Then subtract the mass of the ^{52}Fe nucleus from the mass of the nucleons to find the mass defect:
Mass defect = mass of nucleons − mass of nucleus
           = (52.414 44) − (51.933 85) = 0.480 59 u
Mass defect in g/mol:
(0.480 59 u)(1.660 54 x 10^{-24} g/u)(6.022 x 10^{23} mol^{-1}) = 0.480 58 g/mol

(b) For $^{92}_{42}$Mo:
First, calculate the total mass of the nucleons (50 n + 42 p)
Mass of 50 neutrons = (50)(1.008 66) = 50.433 00
Mass of 42 protons = (42)(1.007 28) = 42.305 76
Mass of 50 n + 42 p   = 92.738 76
Next, calculate the mass of a ^{92}Mo nucleus by subtracting the mass of 42 electrons from the mass of a ^{92}Mo atom.

Mass of ^{92}Mo atom	= 91.906 81
−Mass of 42 electrons = −(42)(5.486 x 10^{-4})	= −0.023 04
Mass of ^{92}Mo nucleus	= 91.883 77

Then subtract the mass of the ^{92}Mo nucleus from the mass of the nucleons to find the mass defect:

Mass defect = mass of nucleons − mass of nucleus
$$= (92.738 \ 76) - (91.883 \ 77) = 0.854 \ 97 \text{ u}$$

Mass defect in g/mol:
$(0.854 \ 99 \text{ u})(1.660 \ 54 \text{ x } 10^{-24} \text{ g/u})(6.022 \text{ x } 10^{23} \text{ mol}^{-1}) = 0.854 \ 99 \text{ g/mol}$

22.20   (a) For $^{58}_{28}$Ni:

First, calculate the total mass of the nucleons (30 n + 28 p)

Mass of 30 neutrons = (30)(1.008 66) = 30.259 80	
Mass of 28 protons = (28)(1.007 28) = 28.203 84	
Mass of 30 n + 28 p	= 58.463 64

Next, calculate the mass of a ^{58}Ni nucleus by subtracting the mass of 28 electrons from the mass of a ^{58}Ni atom.

Mass of ^{58}Ni atom	= 57.935 35
−Mass of 28 electrons = −(28)(5.486 x 10^{-4})	= −0.015 36
Mass of ^{58}Ni nucleus	= 57.919 99

Then subtract the mass of the ^{58}Ni nucleus from the mass of the nucleons to find the mass defect:

Mass defect = mass of nucleons − mass of nucleus
$$= (58.463 \ 64) - (57.919 \ 99) = 0.543 \ 65 \text{ u}$$

Mass defect in g/mol:
$(0.543 \ 65 \text{ u})(1.660 \ 54 \text{ x } 10^{-24} \text{ g/u})(6.022 \text{ x } 10^{23} \text{ mol}^{-1}) = 0.543 \ 64 \text{ g/mol}$

Now, use the Einstein equation to convert the mass defect into the binding energy.

$\Delta E = \Delta mc^2 = (0.543 \ 64 \text{ g/mol})(10^{-3} \text{ kg/g})(3.00 \text{ x } 10^8 \text{ m/s})^2$

$\Delta E = 4.893 \text{ x } 10^{13} \text{ J/mol} = 4.893 \text{ x } 10^{10} \text{ kJ/mol}$

$$\Delta E = \frac{4.893 \text{ x } 10^{13} \text{ J/mol}}{6.022 \text{ x } 10^{23} \text{ nuclei/mol}} \text{ x } \frac{1 \text{ MeV}}{1.60 \text{ x } 10^{-13} \text{ J}} \text{ x } \frac{1 \text{ nucleus}}{58 \text{ nucleons}} = 8.76 \text{ MeV/nucleon}$$

(b) For $^{84}_{36}$Kr:

First, calculate the total mass of the nucleons (48 n + 36 p)

Mass of 48 neutrons = (48)(1.008 66) = 48.415 68	
Mass of 36 protons = (36)(1.007 28) = 36.262 08	
Mass of 48 n + 36 p	= 84.677 76

Next, calculate the mass of a ^{84}Kr nucleus by subtracting the mass of 36 electrons from the mass of a ^{84}Kr atom.

Mass of ^{84}Kr atom	= 83.911 51
−Mass of 36 electrons = −(36)(5.486 x 10^{-4})	= −0.019 75
Mass of ^{84}Kr nucleus	= 83.891 76

Then subtract the mass of the ^{84}Kr nucleus from the mass of the nucleons to find the mass defect:

Mass defect = mass of nucleons – mass of nucleus
$$= (84.677\ 76) - (83.891\ 76) = 0.786\ 00\ u$$

Mass defect in g/mol:
$$(0.786\ 00\ u)(1.660\ 54\ x\ 10^{-24}\ g/u)(6.022\ x\ 10^{23}\ mol^{-1}) = 0.785\ 98\ g/mol$$

Now, use the Einstein equation to convert the mass defect into the binding energy.

$$\Delta E = \Delta mc^2 = (0.785\ 98\ g/mol)(10^{-3}\ kg/g)(3.00\ x\ 10^8\ m/s)^2$$

$$\Delta E = 7.074\ x\ 10^{13}\ J/mol = 7.074\ x\ 10^{10}\ kJ/mol$$

$$\Delta E = \frac{7.074\ x\ 10^{13}\ J/mol}{6.022\ x\ 10^{23}\ nuclei/mol}\ x\ \frac{1\ MeV}{1.60\ x\ 10^{-13}\ J}\ x\ \frac{1\ nucleus}{84\ nucleons} = 8.74\ MeV/nucleon$$

22.22 $^{174}_{77}Ir \rightarrow ^{170}_{75}Re + ^{4}_{2}He$

mass $^{174}_{77}Ir$	173.966 66
–mass $^{170}_{75}Re$	–169.958 04
–mass $^{4}_{2}He$	– 4.002 60
mass change	0.006 02 u

$$(0.006\ 02\ u)(1.660\ 54\ x\ 10^{-24}\ g/u)(6.022\ x\ 10^{23}\ mol^{-1}) = 0.006\ 02\ g/mol$$

$$\Delta E = \Delta mc^2 = (0.006\ 02\ g/mol)(10^{-3}\ kg/g)(3.00\ x\ 10^8\ m/s)^2$$

$$\Delta E = 5.42\ x\ 10^{11}\ J/mol = 5.42\ x\ 10^8\ kJ/mol$$

22.24 $$\Delta m = \frac{\Delta E}{c^2} = \frac{92.2\ x\ 10^3\ J}{(3.00\ x\ 10^8\ m/s)^2} = \frac{92.2\ x\ 10^3\ kg \cdot m^2/s^2}{(3.00\ x\ 10^8\ m/s)^2} = 1.02\ x\ 10^{-12}\ kg$$

$$\Delta m = 1.02\ x\ 10^{-9}\ g$$

22.26   Mass of positron and electron
$$= 2(9.109\ x\ 10^{-31}\ kg)(6.022\ x\ 10^{23}\ mol^{-1}) = 1.097\ x\ 10^{-6}\ kg/mol$$

$$\Delta E = \Delta mc^2 = (1.097\ x\ 10^{-6}\ kg/mol)(3.00\ x\ 10^8\ m/s)^2$$

$$\Delta E = 9.87\ x\ 10^{10}\ J/mol = 9.87\ x\ 10^7\ kJ/mol$$

Nuclear Transmutation (Section 22.4)

22.28   (a) $^{109}_{47}Ag + ^{4}_{2}He \rightarrow ^{113}_{49}In$

(b) $^{10}_{5}B + ^{4}_{2}He \rightarrow ^{13}_{7}N + ^{1}_{0}n$

22.30 $^{209}_{83}Bi + ^{58}_{26}Fe \rightarrow ^{266}_{109}Mt + ^{1}_{0}n$

22.32 $^{238}_{92}U + ^{12}_{6}C \rightarrow ^{246}_{98}Cf + 4\ ^{1}_{0}n$

Chapter Problems

22.34 $\ln\left(\dfrac{N}{N_0}\right) = -0.693\left(\dfrac{t}{t_{1/2}}\right) = -0.693\left(\dfrac{9.5\ \text{y}}{87.7\ \text{y}}\right) = -0.0751$

$\dfrac{N}{N_0} = e^{-0.0751} = 0.928$

power output = 240 W x 0.928 = 223 W = 220 W

22.36 $^{232}_{90}\text{Th} \rightarrow {}^{208}_{82}\text{Pb} + 6\ {}^{4}_{2}\text{He} + 4\ {}^{0}_{-1}\text{e}$

Reactant: $^{232}_{90}\text{Th}$ nucleus = $^{232}_{90}\text{Th}$ atom – 90 e⁻

Product: $^{208}_{82}\text{Pb}$ nucleus + (6)($^{4}_{2}$He nucleus) + 4 e⁻

$\qquad = ({}^{208}_{82}\text{Pb}$ atom – 82 e⁻) + (6)($^{4}_{2}$He atom – 2 e⁻) + 4 e⁻

$\qquad = {}^{208}_{82}\text{Pb}$ atom + (6)($^{4}_{2}$He atom) – 90 e⁻

Change: $({}^{232}_{90}\text{Th}$ atom – 90 e⁻) – [$^{208}_{82}\text{Pb}$ atom + (6)($^{4}_{2}$He atom) – 90 e⁻]

$\qquad = {}^{232}_{90}\text{Th}$ atom – [$^{208}_{82}\text{Pb}$ atom + (6)($^{4}_{2}$He atom)] (electrons cancel)

Mass change = (232.038 054) – [(207.976 627) + (6)(4.002 603)]
$\qquad\qquad = 0.045\ 809$ u

(0.045 809 u)(1.660 54 x 10⁻²⁴ g/u)(6.022 x 10²³ mol⁻¹) = 0.045 808 g/mol

$\Delta E = \Delta mc^2 = $ (0.045 808 g/mol)(10⁻³ kg/g)(3.00 x 10⁸ m/s)²

$\Delta E = 4.12$ x 10¹² J/mol = 4.12 x 10⁹ kJ/mol

22.38 (a) For $^{50}_{24}\text{Cr}$:

First, calculate the total mass of the nucleons (26 n + 24 p)

Mass of 26 neutrons = (26)(1.008 66) = 26.225 16

Mass of 24 protons = (24)(1.007 28) = 24.174 72

Mass of 26 n + 24 p $\qquad\qquad$ = 50.399 88

Next, calculate the mass of a ⁵⁰Cr nucleus by subtracting the mass of 24 electrons from the mass of a ⁵⁰Cr atom.

Mass of ⁵⁰Cr atom $\qquad\qquad\qquad$ = 49.946 05

–Mass of 24 electrons = –(24)(5.486 x 10⁻⁴) = –0.013 17

Mass of ⁵⁰Cr nucleus $\qquad\qquad\quad$ = 49.932 88

Then subtract the mass of the ⁵⁰Cr nucleus from the mass of the nucleons to find the mass defect:

Mass defect = mass of nucleons – mass of nucleus
$\qquad\qquad$ = (50.399 88) – (49.932 88) = 0.467 00 u

Mass defect in g/mol:

(0.467 00 u)(1.660 54 x 10⁻²⁴ g/u)(6.022 x 10²³ mol⁻¹) = 0.466 99 g/mol

Now, use the Einstein equation to convert the mass defect into the binding energy.

$\Delta E = \Delta mc^2 = (0.466\ 99\ \text{g/mol})(10^{-3}\ \text{kg/g})(3.00 \times 10^8\ \text{m/s})^2$

$\Delta E = 4.203 \times 10^{13}\ \text{J/mol} = 4.203 \times 10^{10}\ \text{kJ/mol}$

$\Delta E = \dfrac{4.203 \times 10^{13}\ \text{J/mol}}{6.022 \times 10^{23}\ \text{nuclei/mol}} \times \dfrac{1\ \text{MeV}}{1.60 \times 10^{-13}\ \text{J}} \times \dfrac{1\ \text{nucleus}}{50\ \text{nucleons}} = 8.72\ \text{MeV/nucleon}$

(b) For $^{64}_{30}\text{Zn}$:

First, calculate the total mass of the nucleons (34 n + 30 p)

Mass of 34 neutrons = (34)(1.008 66) = 34.294 44

Mass of 30 protons = (30)(1.007 28) = 30.218 40

Mass of 34 n + 30 p   = 64.512 84

Next, calculate the mass of a ^{64}Zn nucleus by subtracting the mass of 30 electrons from the mass of a ^{64}Zn atom.

Mass of ^{64}Zn atom   = 63.929 15

−Mass of 30 electrons = −(30)(5.486 × 10⁻⁴) = −0.016 46

Mass of ^{64}Zn nucleus   = 63.912 69

Then subtract the mass of the ^{64}Zn nucleus from the mass of the nucleons to find the mass defect:

Mass defect = mass of nucleons − mass of nucleus

          = (64.512 84) − (63.912 69) = 0.600 15 u

Mass defect in g/mol:

$(0.600\ 15\ \text{u})(1.660\ 54 \times 10^{-24}\ \text{g/u})(6.022 \times 10^{23}\ \text{mol}^{-1}) = 0.600\ 14\ \text{g/mol}$

Now, use the Einstein equation to convert the mass defect into the binding energy.

$\Delta E = \Delta mc^2 = (0.600\ 14\ \text{g/mol})(10^{-3}\ \text{kg/g})(3.00 \times 10^8\ \text{m/s})^2$

$\Delta E = 5.401 \times 10^{13}\ \text{J/mol} = 5.401 \times 10^{10}\ \text{kJ/mol}$

$\Delta E = \dfrac{5.401 \times 10^{13}\ \text{J/mol}}{6.022 \times 10^{23}\ \text{nuclei/mol}} \times \dfrac{1\ \text{MeV}}{1.60 \times 10^{-13}\ \text{J}} \times \dfrac{1\ \text{nucleus}}{64\ \text{nucleons}} = 8.76\ \text{MeV/nucleon}$

The ^{64}Zn is more stable.

22.40 $^2_1\text{H} + {}^3_2\text{He} \rightarrow {}^4_2\text{He} + {}^1_1\text{H}$

mass ^2_1H	2.0141
mass ^3_2He	3.0160
−mass ^4_2He	− 4.0026
−mass ^1_1H	−1.0078
mass change	0.0197 u

$(0.0197\ \text{u})(1.660\ 54 \times 10^{-24}\ \text{g/u})(6.022 \times 10^{23}\ \text{mol}^{-1}) = 0.0197\ \text{g/mol}$

$\Delta E = \Delta mc^2 = (0.0197\ \text{g/mol})(10^{-3}\ \text{kg/g})(3.00 \times 10^8\ \text{m/s})^2$

$\Delta E = 1.77 \times 10^{12}\ \text{J/mol} = 1.77 \times 10^9\ \text{kJ/mol}$

22.42 $^{238}_{92}\text{U} + {}^1_0\text{n} \rightarrow {}^{239}_{94}\text{Pu} + 2\,{}^{\ 0}_{-1}\text{e}$

22.44 $^{10}B + {}^{1}n \rightarrow {}^{4}He + {}^{7}Li + \gamma$

mass ^{10}B	10.012 937
mass ^{1}n	1.008 665
–mass ^{4}He	– 4.002 603
–mass ^{7}Li	–7.016 004
mass change	0.002 995 u

$(0.002\ 995\ u)(1.660\ 54 \times 10^{-24}\ g/u) = 4.973 \times 10^{-27}\ g$

$\Delta E = \Delta mc^2 = (4.973 \times 10^{-27}\ g)(10^{-3}\ kg/g)(3.00 \times 10^8\ m/s)^2 = 4.476 \times 10^{-13}\ J$

Kinetic energy = 2.31 MeV x $\dfrac{1.60 \times 10^{-13}\ J}{1\ MeV}$ = $3.696 \times 10^{-13}\ J$

γ photon energy = ΔE – KE = $4.476 \times 10^{-13}\ J$ – $3.696 \times 10^{-13}\ J$ = $7.80 \times 10^{-14}\ J$

$\qquad\qquad = 7.80 \times 10^{-14}\ J \times \dfrac{1\ MeV}{1.60 \times 10^{-13}\ J} = 0.488\ MeV$

22.46   (a) $^{100}_{43}Tc \rightarrow {}^{0}_{1}e + {}^{100}_{42}Mo$ (positron emission)

$\qquad$ $^{100}_{43}Tc + {}^{0}_{-1}e \rightarrow {}^{100}_{42}Mo$ (electron capture)

(b) Positron emission

Reactant: $^{100}_{43}Tc$ nucleus = $^{100}_{43}Tc$ atom – 43 e$^-$

Product: $^{100}_{42}Mo$ nucleus + e$^+$ = $^{100}_{42}Mo$ atom – 42 e$^-$ + 1 e$^+$

___

Change: ($^{100}_{43}Tc$ atom – 43 e$^-$) – ($^{100}_{42}Mo$ atom – 42 e$^-$ + 1 e$^+$)

$\qquad$ = $^{100}_{43}Tc$ atom – $^{100}_{42}Mo$ atom – 2 e$^-$

Mass change = (99.907 657) – (99.907 48) – (2)(0.000 5486)

$\qquad\qquad$ = –0.000 92 u (energy is absorbed)

$(-0.000\ 92\ u)(1.660\ 54 \times 10^{-24}\ g/u)(6.022 \times 10^{23}\ mol^{-1}) = -0.000\ 92\ g/mol$

$\Delta E = \Delta mc^2 = (-0.000\ 92\ g/mol)(10^{-3}\ kg/g)(3.00 \times 10^8\ m/s)^2$

$\Delta E = -8.3 \times 10^{10}\ J/mol = -8.3 \times 10^7\ kJ/mol$

Electron Capture

Reactant: $^{100}_{43}Tc$ nucleus + e$^-$ = $^{100}_{43}Tc$ atom – 42 e$^-$

Product: $^{100}_{42}Mo$ nucleus = $^{100}_{42}Mo$ atom – 42 e$^-$

___

Change: ($^{100}_{43}Tc$ atom – 42 e$^-$) – ($^{100}_{42}Mo$ atom – 42 e$^-$)

$\qquad$ = $^{100}_{43}Tc$ atom – $^{100}_{42}Mo$ atom   (electrons cancel)

Mass change = (99.907 657) – (99.907 48) = 0.000 177 u

$(0.000\ 177\ u)(1.660\ 54 \times 10^{-24}\ g/u)(6.022 \times 10^{23}\ mol^{-1}) = 0.000\ 177\ g/mol$

$\Delta E = \Delta mc^2 = (0.000\ 177\ \text{g/mol})(10^{-3}\ \text{kg/g})(3.00 \times 10^8\ \text{m/s})^2$

$\Delta E = 1.6 \times 10^{10}\ \text{J/mol} = 1.6 \times 10^7\ \text{kJ/mol}$

Only electron capture is observed because only this process involves a mass decrease and a release of energy.

Multiconcept Problems

22.48 $BaCO_3$, 197.34

$$1.000\ \text{g}\ BaCO_3 \times \frac{1\ \text{mol}\ BaCO_3}{197.34\ \text{g}\ BaCO_3} \times \frac{1\ \text{mol}\ C}{1\ \text{mol}\ BaCO_3} \times \frac{12.011\ \text{g}\ C}{1\ \text{mol}\ C} = 0.060\ 86\ \text{g}\ C$$

$4.0 \times 10^{-3}\ \text{Bq} = 4.0 \times 10^{-3}\ \text{disintegrations/s}$

$(4.0 \times 10^{-3}\ \text{Bq} = 4.0 \times 10^{-3}\ \text{disintegrations/s})(60\ \text{s/min}) = 0.24\ \text{disintegrations/min}$

$$\text{sample radioactivity} = \frac{0.24\ \text{disintegrations/min}}{0.060\ 86\ \text{g}\ C} = 3.94\ \text{disintegrations/min per gram of C}$$

$$\ln\left(\frac{N}{N_0}\right) = (-0.693)\left(\frac{t}{t_{1/2}}\right); \qquad \frac{N}{N_0} = \frac{\text{Decay rate at time}\ t}{\text{Decay rate at time}\ t = 0}$$

$$\ln\left(\frac{3.94}{15.3}\right) = (-0.693)\left(\frac{t}{5730\ \text{y}}\right); \qquad t = 11{,}000\ \text{y}$$

22.50   First find the activity of the ^{51}Cr after 17.0 days.

$$\ln\left(\frac{N}{N_0}\right) = (-0.693)\left(\frac{t}{t_{1/2}}\right); \qquad \frac{N}{N_0} = \frac{\text{Decay rate at time}\ t}{\text{Decay rate at time}\ t = 0}$$

$$\ln\left(\frac{N}{4.10}\right) = (-0.693)\left(\frac{17.0\ \text{d}}{27.7\ \text{d}}\right)$$

$\ln N - \ln(4.10) = -0.4253$

$\ln N = -0.4253 + \ln(4.10) = 0.9857$

$N = e^{0.9857} = 2.68\ \mu\text{Ci/mL}$

$(20.0\ \text{mL})(2.68\ \mu\text{Ci/mL}) = (\text{total blood volume})(0.009\ 35\ \mu\text{Ci/mL})$

total blood volume = 5732 mL = 5.73 L

430

23 Organic and Biological Chemistry

23.1

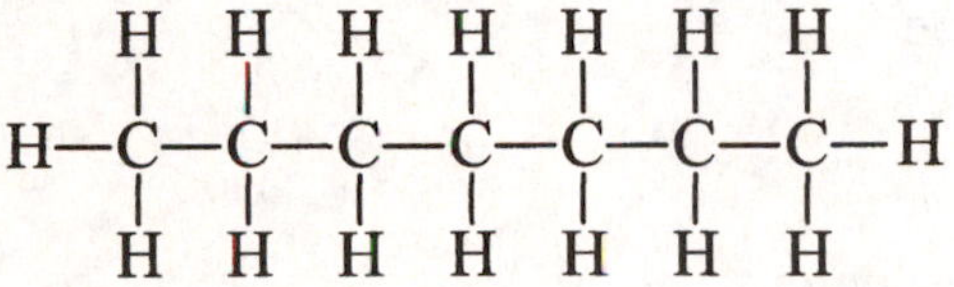

$$H-\underset{\underset{H}{|}}{\overset{\overset{H}{|}}{C}}-\underset{\underset{H}{|}}{\overset{\overset{H}{|}}{C}}-\underset{\underset{H}{|}}{\overset{\overset{H}{|}}{C}}-\underset{\underset{H}{|}}{\overset{\overset{H}{|}}{C}}-\underset{\underset{H}{|}}{\overset{\overset{H}{|}}{C}}-\underset{\underset{H}{|}}{\overset{\overset{H}{|}}{C}}-\underset{\underset{H}{|}}{\overset{\overset{H}{|}}{C}}-H$$

23.2

(Structures drawn — isomers of the carbon chain.)

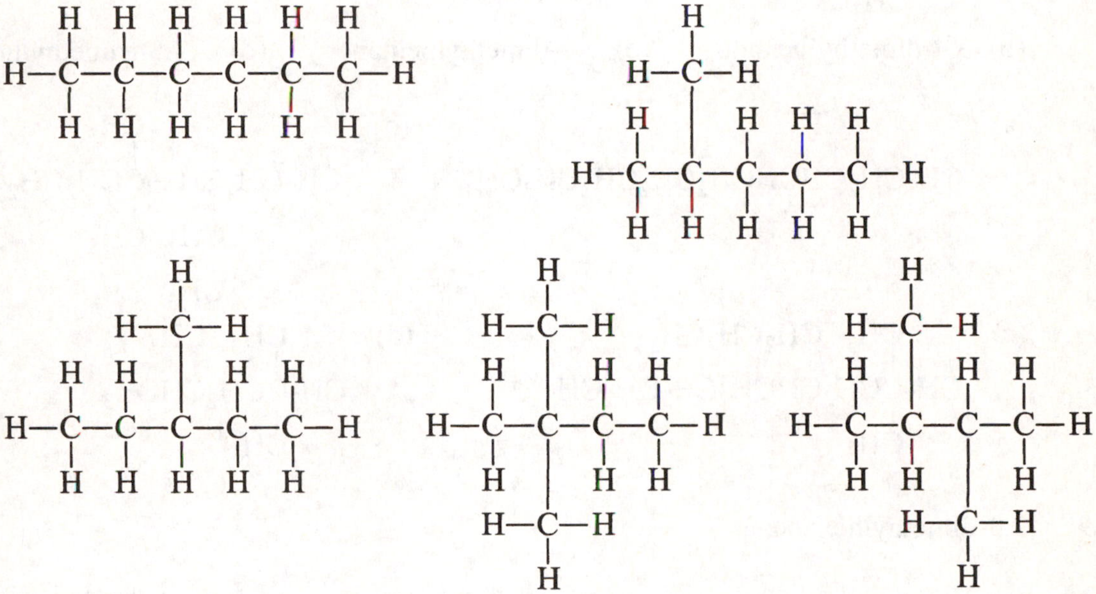

23.3 Structures (a) and (c) are identical. They both contain a chain of six carbons with two –CH$_3$ branches at the fourth carbon and one –CH$_3$ branch at the second carbon. Structure (b) is different, having a chain of seven carbons.

23.4 C$_7$H$_{16}$

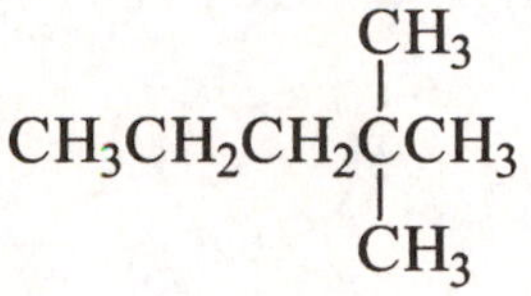

$$CH_3CH_2CH_2\underset{\underset{CH_3}{|}}{\overset{\overset{CH_3}{|}}{C}}CH_3$$

23.5 (a) carboxylic acid

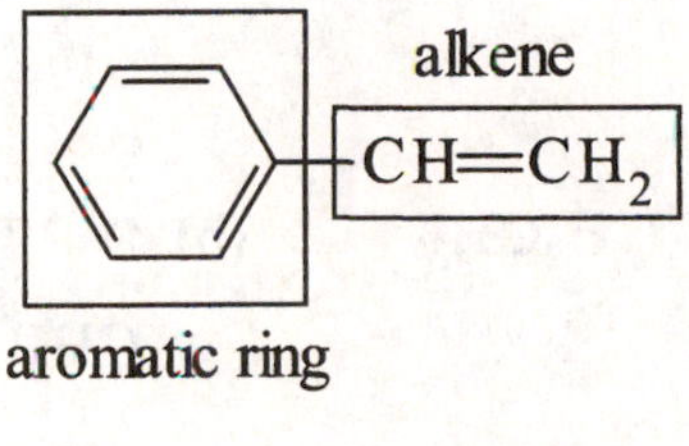

$$CH_3-\underset{\underset{\boxed{OH}}{|}}{\overset{\overset{H}{|}}{C}}-\boxed{\overset{O}{\overset{\|}{C}}-OH}$$

alcohol

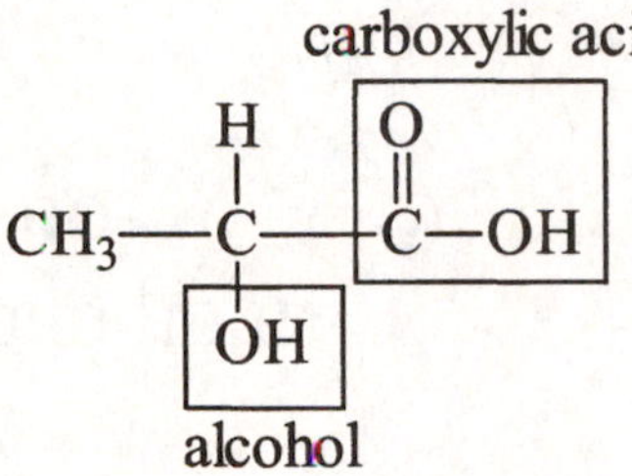

(b) alkene

aromatic ring — $\boxed{CH=CH_2}$

23.6   (a)

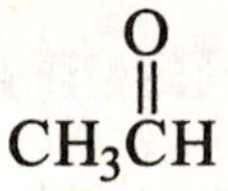

$$\underset{\displaystyle CH_3CH}{\overset{\displaystyle O}{\,}}$$

(b)

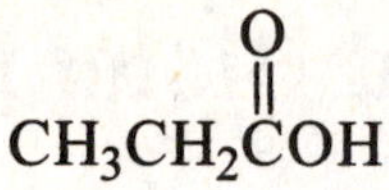

$$\underset{\displaystyle CH_3CH_2COH}{\overset{\displaystyle O}{\,}}$$

(c)

$$\underset{\displaystyle CH_3COCH_2CH_3}{\overset{\displaystyle O}{\,}} \qquad \underset{\displaystyle CH_3CH_2COCH_3}{\overset{\displaystyle O}{\,}}$$

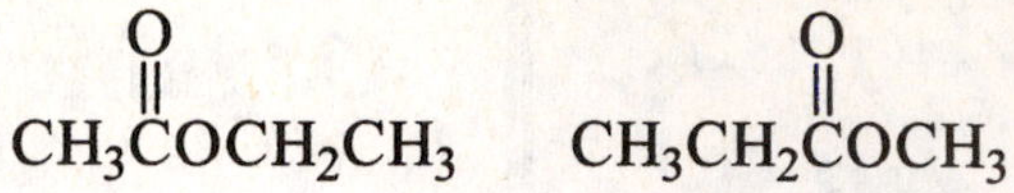

23.7   (a) $CH_3CH_2CH_2CH_2CH_3$ pentane

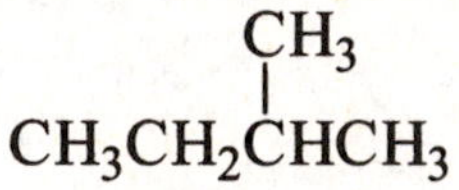

$$\underset{\displaystyle CH_3CH_2\overset{\displaystyle |}{C}HCH_3}{\overset{\displaystyle CH_3}{\,}} \qquad \text{2-methylbutane}$$

$$CH_3\overset{\displaystyle CH_3}{\underset{\displaystyle CH_3}{\overset{\displaystyle |}{\underset{\displaystyle |}{C}}}}CH_3 \qquad \text{2,2-dimethylpropane}$$

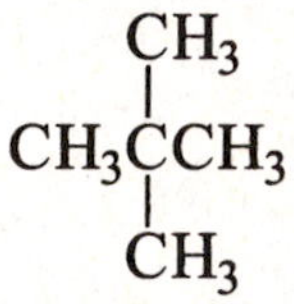

(b) 3,4-dimethylhexane   (c) 2,4-dimethylpentane   (d) 2,2,5-trimethylheptane

23.8   (a)

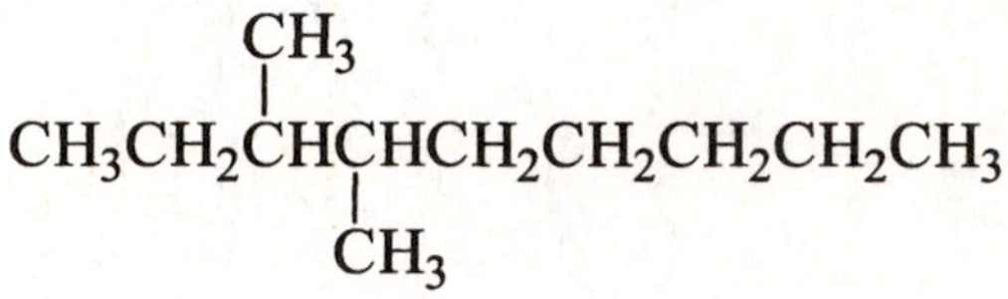

$$\underset{\displaystyle \qquad\qquad\; CH_3}{CH_3CH_2\overset{\displaystyle CH_3}{\overset{\displaystyle |}{C}}HCHCH_2CH_2CH_2CH_2CH_3}$$

(b)

$$CH_3CH_2CH{-}\overset{\displaystyle CH_3}{\underset{\displaystyle \overset{\displaystyle |}{CH_2}}{\overset{\displaystyle |}{C}}}CH_2CH_2CH_3 \quad CH_3$$
(with CH_2 branch continuing to CH_3)

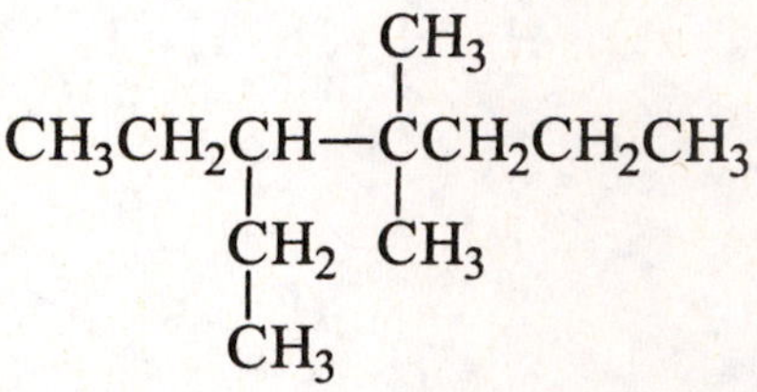

(c)

$$CH_3\overset{\displaystyle CH_3}{\underset{\displaystyle CH_3}{\overset{\displaystyle |}{\underset{\displaystyle |}{C}}}}CH_2\overset{\displaystyle CH_2CH_2CH_3}{\overset{\displaystyle |}{C}}HCH_2CH_2CH_2CH_3$$

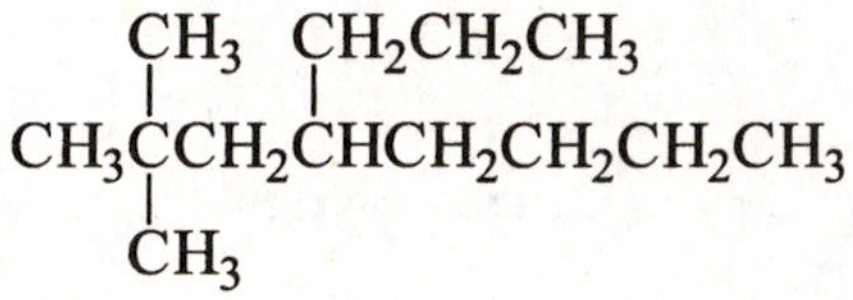

(d)

$$CH_3\overset{\displaystyle CH_3}{\underset{\displaystyle CH_3}{\overset{\displaystyle |}{\underset{\displaystyle |}{C}}}}CH_2\overset{\displaystyle CH_3}{\overset{\displaystyle |}{C}}HCH_3$$

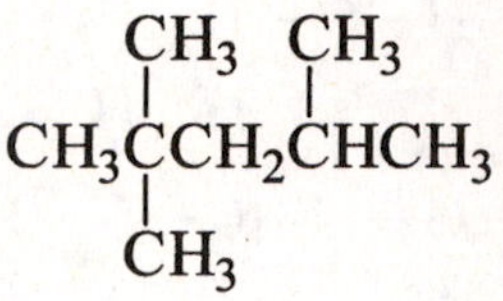

23.9   2,3-dimethylhexane

23.10   (a) 3-methyl-1-butene   (b) 4-methyl-3-heptene   (c) 3-ethyl-1-hexyne

23.11   (a)

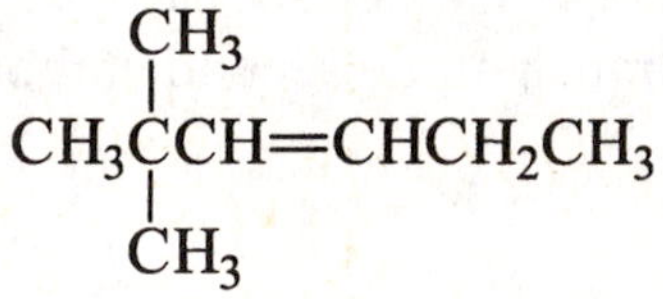

$$CH_3\overset{\displaystyle CH_3}{\underset{\displaystyle CH_3}{\overset{\displaystyle |}{\underset{\displaystyle |}{C}}}}CH{=}CHCH_2CH_3$$

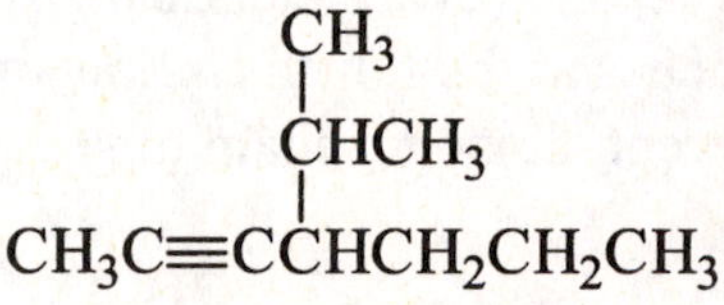

(b)

$$\underset{\displaystyle CH_3C\equiv CCHCH_2CH_2CH_3}{\overset{\displaystyle \overset{\displaystyle CH_3}{\overset{\displaystyle |}{CHCH_3}}}{\,}}$$

(c)

$$\overset{\displaystyle CH_3CH_2}{\underset{\displaystyle H}{\,}}C{=}C\overset{\displaystyle H}{\underset{\displaystyle CH_2CH_2CH_3}{\,}}$$

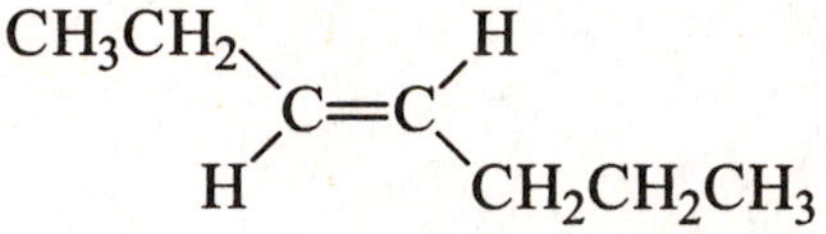

(d)

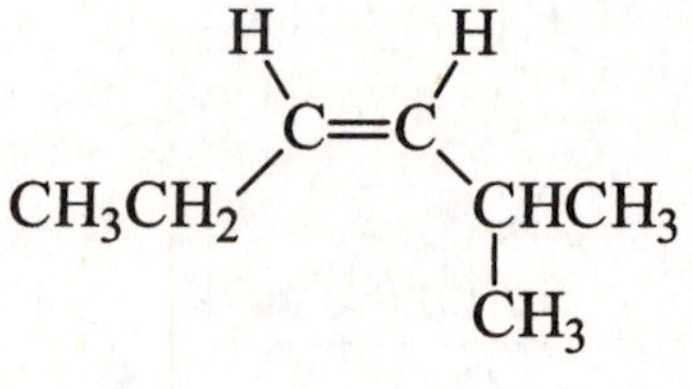

$$\overset{\displaystyle H}{\underset{\displaystyle CH_3CH_2}{\,}}C{=}C\overset{\displaystyle H}{\underset{\displaystyle \underset{\displaystyle CH_3}{\overset{\displaystyle |}{CHCH_3}}}{\,}}$$

23.12   (a) $CH_3CH_2CH_2CH_3$ (b)

$$\underset{\displaystyle CH_3CHCHCH_3}{\overset{\displaystyle Br\;\;Br}{\overset{\displaystyle |\;\;\;|}{\,}}}$$

(c)

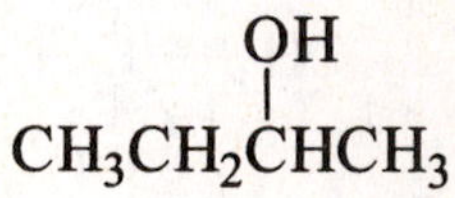

$$\underset{\displaystyle CH_3CH_2CHCH_3}{\overset{\displaystyle OH}{\overset{\displaystyle |}{\,}}}$$

432

23.13

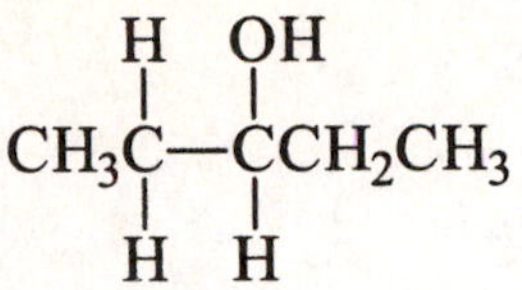

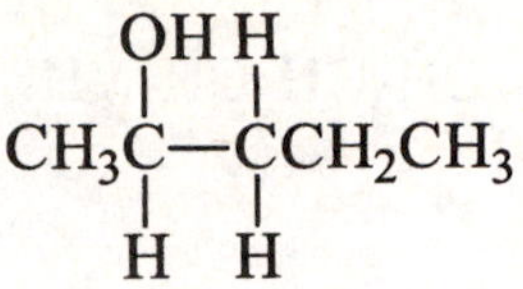

$$CH_3\overset{\displaystyle H}{\underset{\displaystyle H}{C}}-\overset{\displaystyle OH}{\underset{\displaystyle H}{C}}CH_2CH_3 \qquad\qquad CH_3\overset{\displaystyle OH}{\underset{\displaystyle H}{C}}-\overset{\displaystyle H}{\underset{\displaystyle H}{C}}CH_2CH_3$$

23.14　(a) 1,4-dimethylcyclohexane
　　　 (b) 1-ethyl-3-methylcyclopentane
　　　 (c) isopropylcyclobutane

23.15　(a) 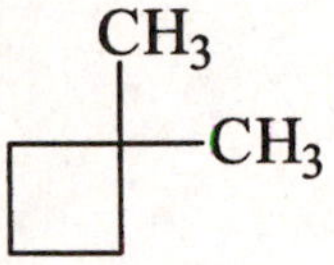 　(b) 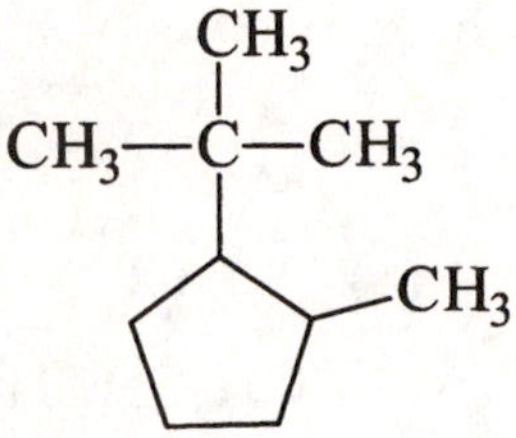 　(c)

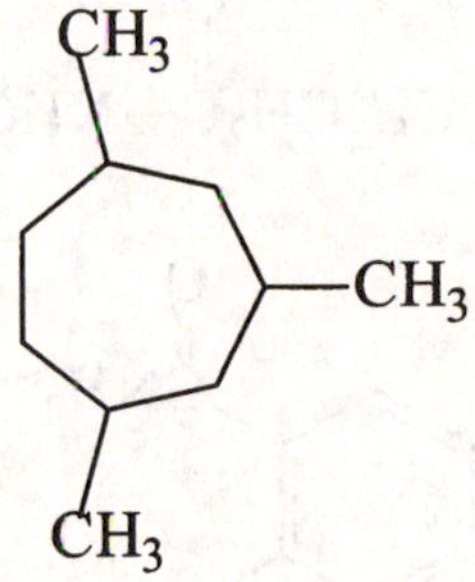

　　　 (d)

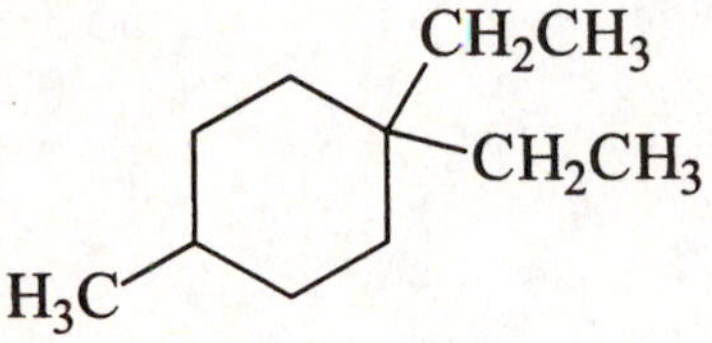

23.16　(a) 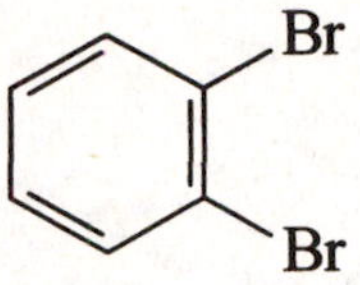 　(b)  　(c)

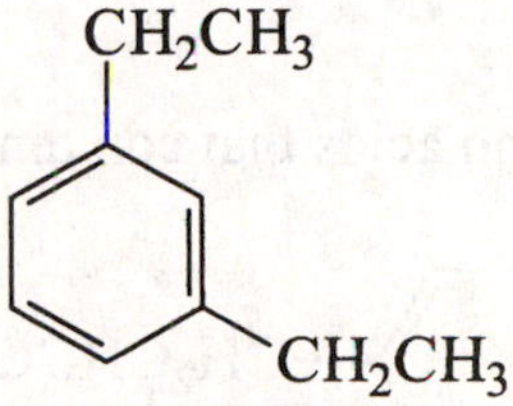

23.17　(a) 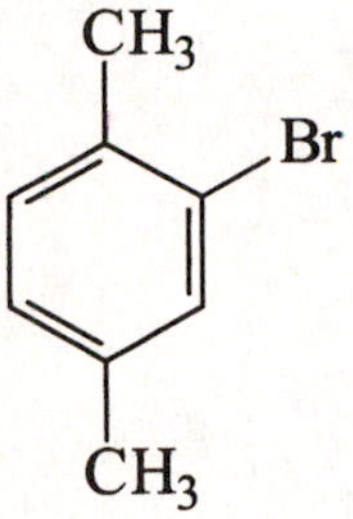 　(b)

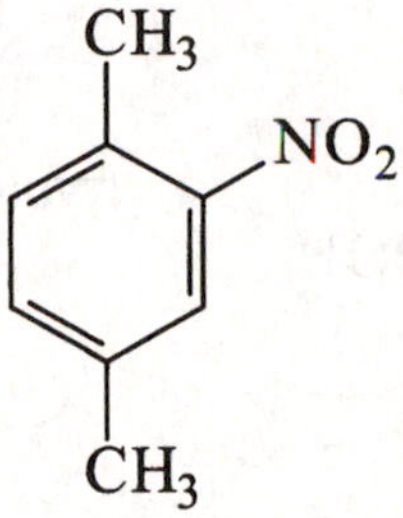

23.18

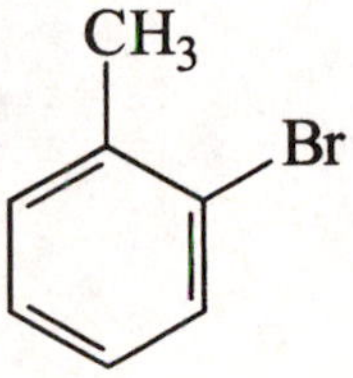

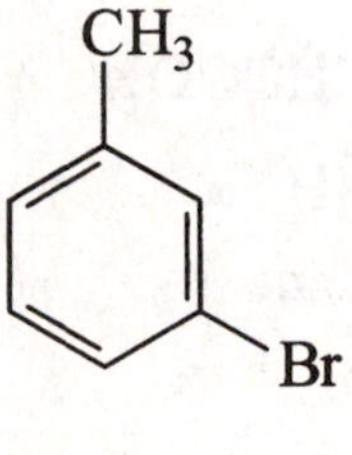

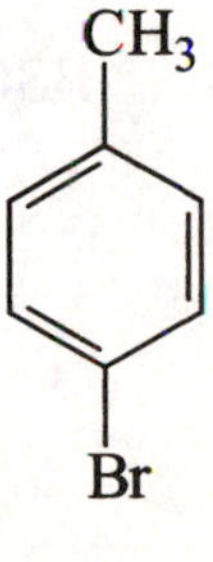

23.19 (a)

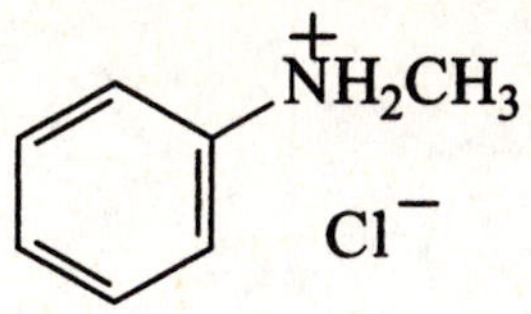

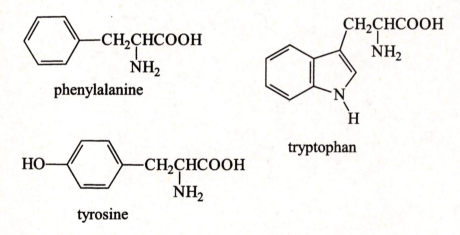

(b) $CH_3CH_2CH_2\overset{+}{N}H_3 \quad Cl^-$

23.20 (a)

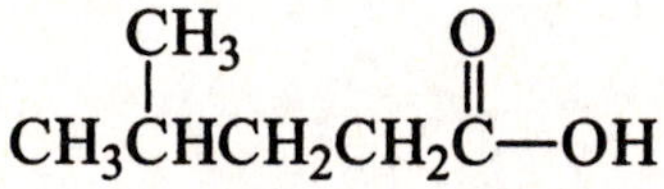

(b)

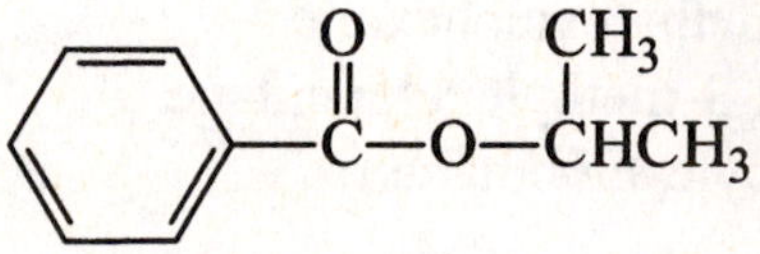

(c)

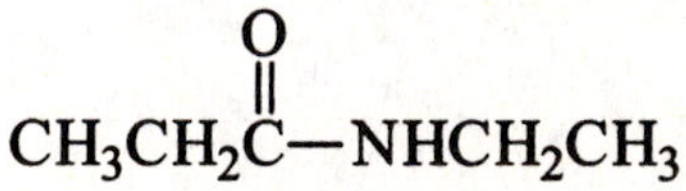

23.21 (a)

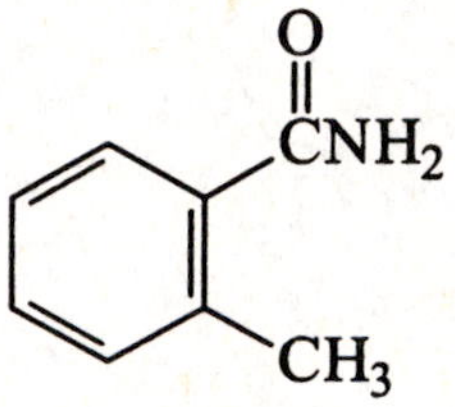

(b)

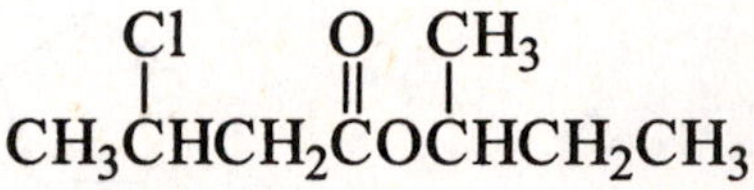

23.22

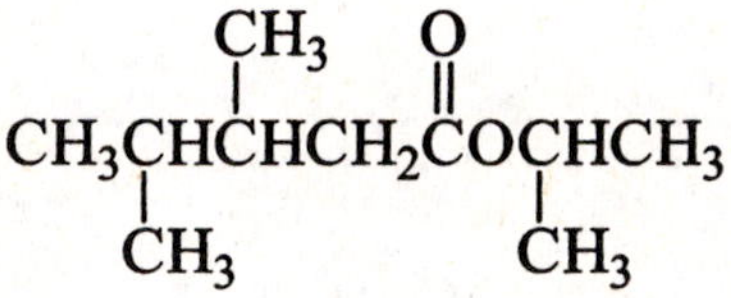

23.23 Amino acids that contain an aromatic ring:

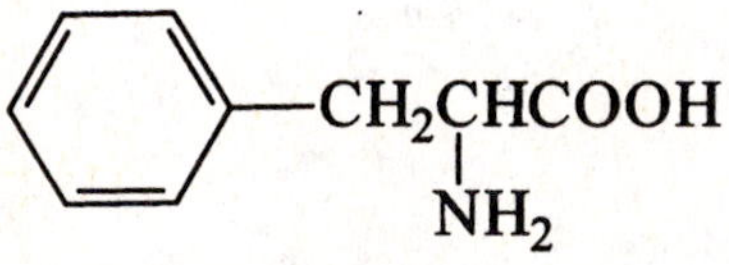

phenylalanine

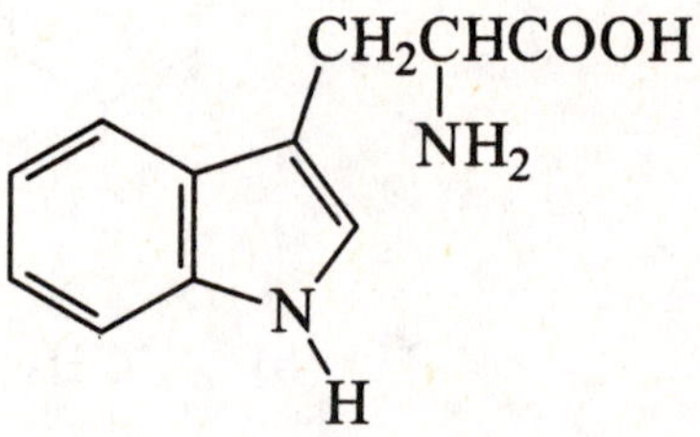

tryptophan

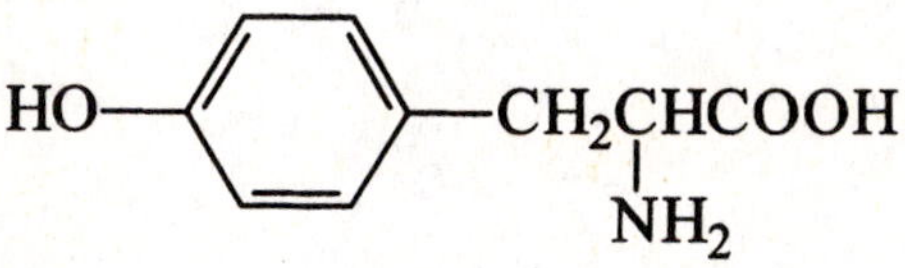

tyrosine

Amino acids that contain sulfur:

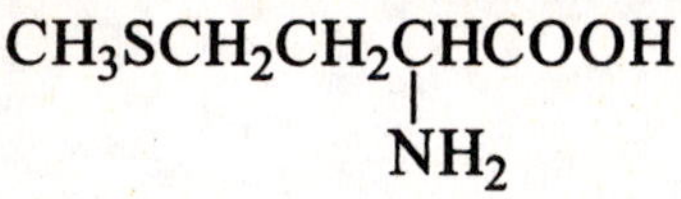

methionine

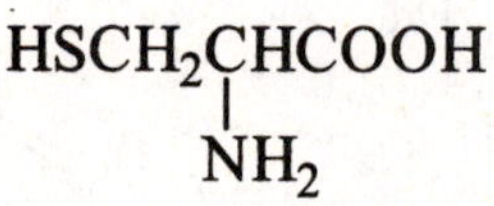

cysteine

Amino acids that are alcohols:

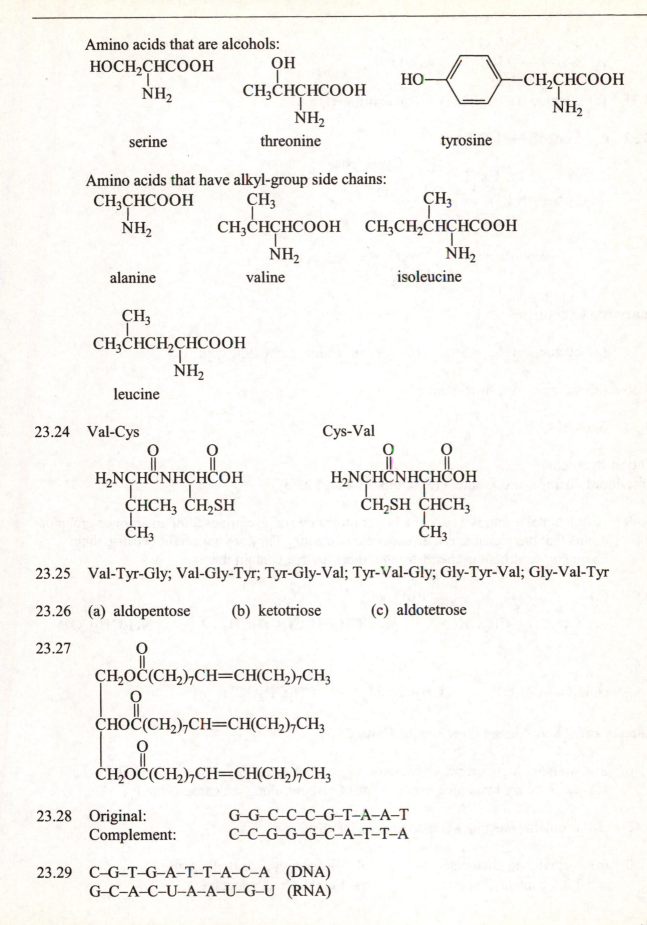

HOCH₂CHCOOH
|
NH₂

serine

OH
|
CH₃CHCHCOOH
|
NH₂

threonine

HO—⟨benzene⟩—CH₂CHCOOH
|
NH₂

tyrosine

Amino acids that have alkyl-group side chains:

CH₃CHCOOH
|
NH₂

alanine

CH₃
|
CH₃CHCHCOOH
|
NH₂

valine

CH₃
|
CH₃CH₂CHCHCOOH
|
NH₂

isoleucine

CH₃
|
CH₃CHCH₂CHCOOH
|
NH₂

leucine

23.24   Val-Cys

O   O
‖   ‖
H₂NCHCNHCHCOH
| |
CHCH₃ CH₂SH
|
CH₃

Cys-Val

O   O
‖   ‖
H₂NCHCNHCHCOH
| |
CH₂SH CHCH₃
|
CH₃

23.25   Val-Tyr-Gly; Val-Gly-Tyr; Tyr-Gly-Val; Tyr-Val-Gly; Gly-Tyr-Val; Gly-Val-Tyr

23.26   (a) aldopentose   (b) ketotriose   (c) aldotetrose

23.27
O
‖
CH₂OC(CH₂)₇CH═CH(CH₂)₇CH₃
| O
| ‖
CHOC(CH₂)₇CH═CH(CH₂)₇CH₃
| O
| ‖
CH₂OC(CH₂)₇CH═CH(CH₂)₇CH₃

23.28   Original: G–G–C–C–C–G–T–A–A–T
        Complement: C–C–G–G–G–C–A–T–T–A

23.29   C–G–T–G–A–T–T–A–C–A   (DNA)
        G–C–A–C–U–A–A–U–G–U   (RNA)

435

23.30   U–G–C–A–U–C–G–A–G–U   (RNA)
        A–C–G–T–A–G–C–T–C–A   (DNA)

23.31   (a) adenine (DNA, RNA)   (b) thymine (DNA)

23.32

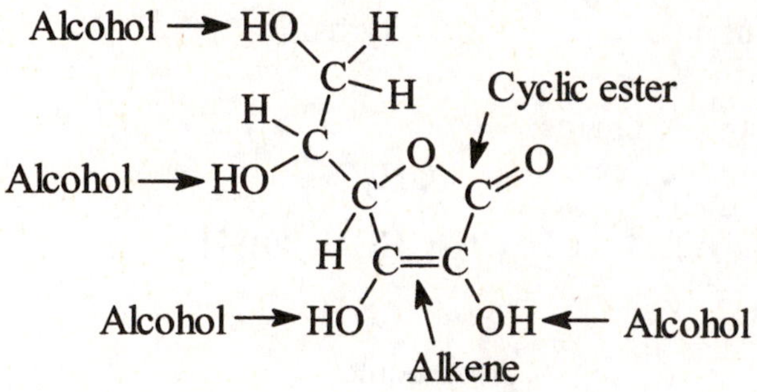

Conceptual Problems

23.34   (a) alkene, ketone, ether   (b) alkene, amine, carboxylic acid

23.36   (a) serine   (b) methionine

23.38   Ser-Val

Section Problems
Functional Groups and Isomers (Sections 23.2 and 23.3)

23.40   A functional group is a part of a larger molecule and is composed of an atom or group of atoms that has a characteristic chemical behavior. They are important because their chemistry controls the chemistry in molecules that contain them.

23.42   (a)

$$CH_3CH_2\overset{\overset{\textstyle O}{\|}}{C}CH_2CH_3$$

(b)

$$CH_3CH_2CH_2\overset{\overset{\textstyle O}{\|}}{C}OCH_2CH_3$$

(c)

$$NH_2CH_2\overset{\overset{\textstyle O}{\|}}{C}OH$$

23.44

$$CH_3CH_2CH_2OH \qquad CH_3\overset{\overset{\textstyle OH}{|}}{C}HCH_3 \qquad CH_3CH_2OCH_3$$

Alkanes and Their Names (Sections 23.1 and 23.3)

23.46   In a straight-chain alkane, all the carbons are connected in a row. In a branched-chain alkane, there are branching connections of carbons along the carbon chain.

23.48 C_3H_9 contains one more H than needed for an alkane.

23.50   (a) 4-ethyl-3-methyloctane   (b) 4-isopropyl-2-methylheptane
        (c) 2,2,6-trimethylheptane   (d) 4-ethyl-4-methyloctane

23.52 (a)

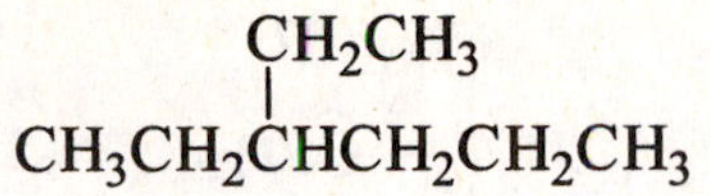

CH₂CH₃
|
CH₃CH₂CHCH₂CH₂CH₃

(b)

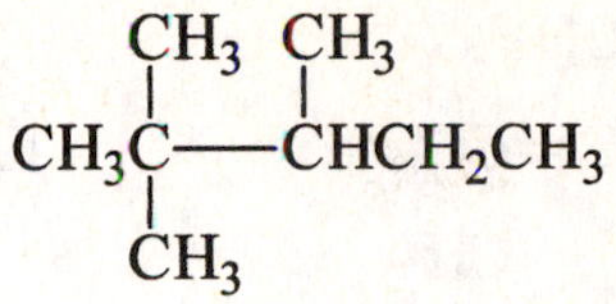

CH₃ CH₃
| |
CH₃C——CHCH₂CH₃
|
CH₃

(c)

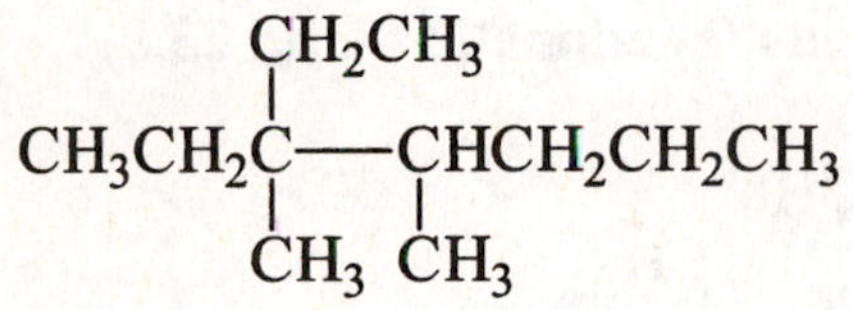

CH₂CH₃
|
CH₃CH₂C——CHCH₂CH₂CH₃
| |
CH₃ CH₃

(d)

CH₃
|
          CHCH₃
CH₃ |
| |
CH₃CHCH₂CH₂CHCH₂CH₂CH₃

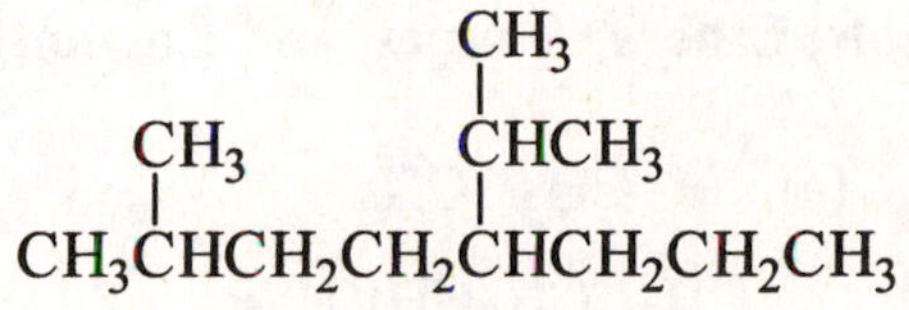

Alkenes, Alkynes, and Aromatic Compounds (Sections 23.4 and 23.6)

23.54   Today the term "aromatic" refers to the class of compounds containing a six-membered ring with three double bonds, not to the fragrance of a compound.

23.56   (a) CH₃CH=CHCH₂CH₃   (b) HC≡CCH₂CH₃   (c)

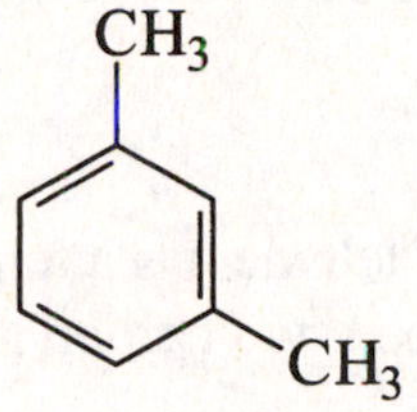

23.58   (a) 4-methyl-2-pentene
        (b) 3-methyl-1-pentene
        (c) 1,2-dichlorobenzene, or o-dichlorobenzene
        (d) 2-methyl-2-butene
        (e) 7-methyl-3-octyne

23.60   (a) CH₂=CHCH₂CH₂CH₂CH₃   This compound cannot form cis-trans isomers.

        (b) CH₃CH=CHCH₂CH₂CH₃   This compound can form cis-trans isomers because of the different groups on each double bond C.

        (c) CH₃CH₂CH=CHCH₂CH₃   This compound can form cis-trans isomers because of the different groups on each double bond C.

23.62   (a)

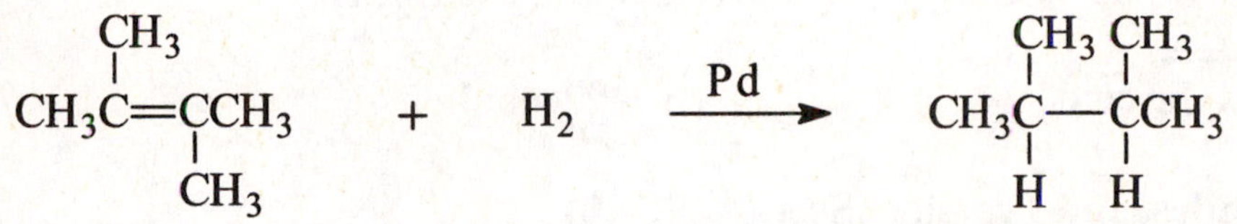

CH₃
|
CH₃C=CCH₃   +   H₂   ——Pd——→
|
CH₃

CH₃ CH₃
| |
CH₃C—CCH₃
| |
H   H

(b)

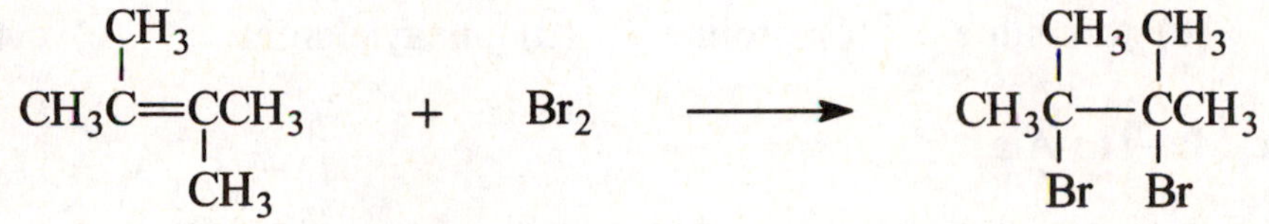

CH₃
|
CH₃C=CCH₃   +   Br₂   ——————→
|
CH₃

CH₃ CH₃
| |
CH₃C—CCH₃
| |
Br Br

(c)

$$CH_3\overset{\overset{\displaystyle CH_3}{|}}{C}{=}CCH_3 \quad + \quad H_2O \quad \xrightarrow{H_2SO_4} \quad CH_3\overset{\overset{\displaystyle CH_3}{|}}{\underset{\underset{\displaystyle OH}{|}}{C}}{-}\overset{\overset{\displaystyle CH_3}{|}}{\underset{\underset{\displaystyle H}{|}}{C}}CH_3$$

with CH_3 on the lower left carbon

Alcohols, Ethers, Amines, and Carbonyl Compounds (Sections 23.7 and 23.8)

23.64 (a)

$$CH_3\overset{\overset{\displaystyle CH_3}{|}}{\underset{\underset{\displaystyle OH}{|}}{C}}CH_2\overset{\overset{\displaystyle CH_3}{|}}{C}HCH_3$$

(b)

cyclohexane ring with OH and two CH_3 groups

(c)

$$HOCH_2CH_2CH_2CH_2\overset{\overset{\displaystyle CH_2CH_3}{|}}{\underset{\underset{\displaystyle CH_2CH_3}{|}}{C}}CH_2CH_3$$

(d)

$$CH_3CH_2\overset{\overset{\displaystyle CH_2CH_3}{|}}{\underset{\underset{\displaystyle OH}{|}}{C}}CH_2CH_2CH_3$$

23.66 An aldehyde has a terminal carbonyl group. A ketone has the carbonyl group located between two carbon atoms.

23.68 (a) ketone   (b) aldehyde   (c) ketone   (d) amide   (e) ester

23.70 (a)

$$CH_3CH_2CH_2CH_2\overset{\overset{\displaystyle O}{\|}}{C}{-}OCH_3$$

(b)

$$CH_3CH_2\overset{\overset{\displaystyle O}{\|}}{\underset{\underset{\displaystyle CH_3}{|}}{CHC}}{-}OC\overset{\overset{\displaystyle CH_3}{|}}{\underset{\underset{\displaystyle CH_3}{|}}{H}}$$

(c)

$$CH_3\overset{\overset{\displaystyle O}{\|}}{C}{-}O{-}\text{cyclohexyl}$$

23.72 amine, aromatic ring, and ester

$$H_2N{-}\boxed{}{-}\overset{\overset{\displaystyle O}{\|}}{C}OH$$

carboxylic acid

$$HOCH_2CH_2\overset{\overset{\displaystyle CH_2CH_3}{|}}{N}CH_2CH_3$$

alcohol

Amino Acids, Peptides, and Proteins (Sections 23.10)

23.74 (a) serine   (b) threonine   (c) proline   (d) phenylalanine   (e) cysteine

23.76 Val-Ser-Phe-Met-Thr-Ala

23.78 Met-Ile-Lys, Met-Lys-Ile, Ile-Met-Lys, Ile-Lys-Met, Lys-Met-Ile, Lys-Ile-Met

Carbohydrates (Section 23.11)

23.80   An aldose contains the aldehyde functional group while a ketose contains the ketone functional group.

23.82

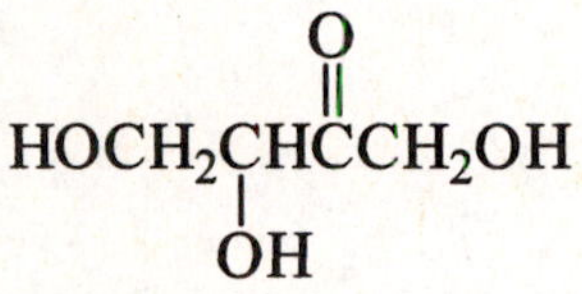

$$
\underset{\underset{OH}{|}}{HOCH_2CHC}\overset{\overset{O}{\|}}{C}CH_2OH
$$

Lipids (Section 23.12)

23.84   Long-chain carboxylic acids are called fatty acids. Fatty acids are usually unbranched and have an even number of carbon atoms in the range of 12 –22.

23.86

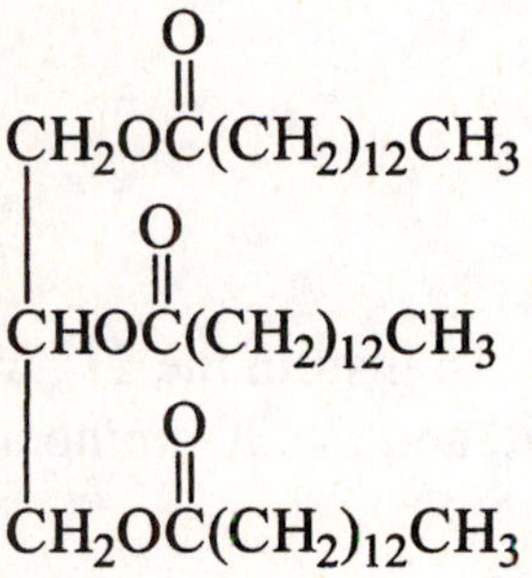

$$
\begin{aligned}
&CH_2O\overset{\overset{O}{\|}}{C}(CH_2)_{12}CH_3\\
&|\\
&CHO\overset{\overset{O}{\|}}{C}(CH_2)_{12}CH_3\\
&|\\
&CH_2O\overset{\overset{O}{\|}}{C}(CH_2)_{12}CH_3
\end{aligned}
$$

23.88

$$
\begin{aligned}
&CH_2O\overset{\overset{O}{\|}}{C}(CH_2)_{16}CH_3\\
&|\\
&CHO\overset{\overset{O}{\|}}{C}(CH_2)_{16}CH_3\\
&|\\
&CH_2O\overset{\overset{O}{\|}}{C}(CH_2)_{14}CH_3
\end{aligned}
\qquad
\begin{aligned}
&CH_2O\overset{\overset{O}{\|}}{C}(CH_2)_{16}CH_3\\
&|\\
&CHO\overset{\overset{O}{\|}}{C}(CH_2)_{14}CH_3\\
&|\\
&CH_2O\overset{\overset{O}{\|}}{C}(CH_2)_{16}CH_3
\end{aligned}
$$

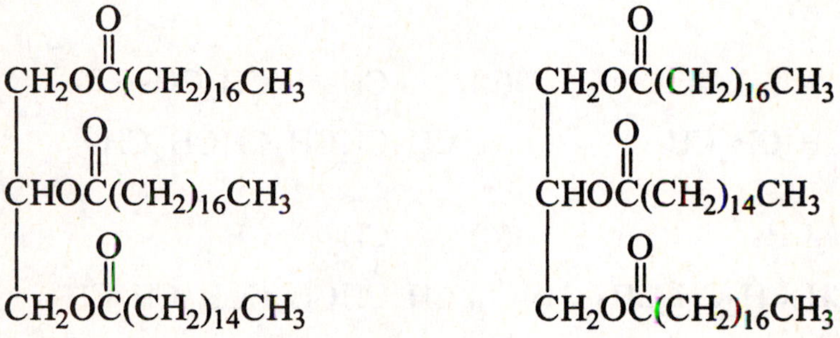

The two fat molecules differ from each other depending on where the palmitic acid chain is located. In the first fat molecule, palmitic acid is on an end and in the second it is in the middle.

Nucleic Acids (Sections 23.13)

23.90   Just as proteins are polymers made of amino acid units, nucleic acids are polymers made up of nucleotide units linked together to form a long chain. Each nucleotide contains a phosphate group, an aldopentose sugar, and an amine base.

23.92

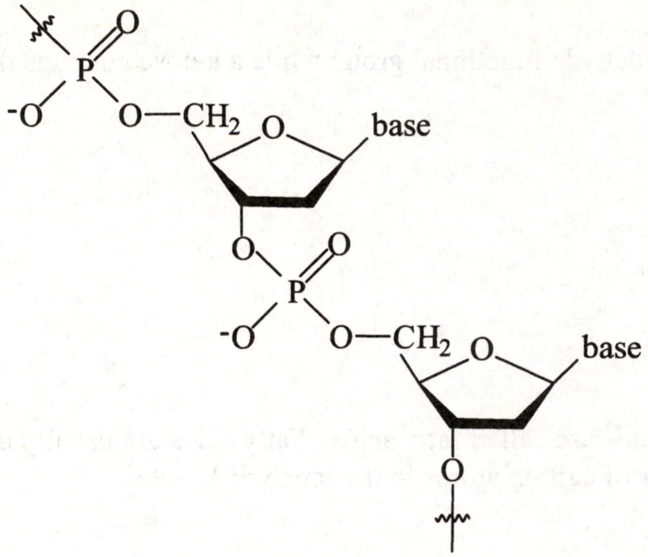

23.94 Original: T–A–C–C–G–A
       Complement: A–T–G–G–C–T

23.96 It takes three nucleotides to code for a specific amino acid. In insulin the 21 amino acid chain would require (3 x 21) = 63 nucleotides to code for it, and the 30 amino acid chain would require (3 x 30) = 90 nucleotides to code for it.

Chapter Problems

23.98 (a)

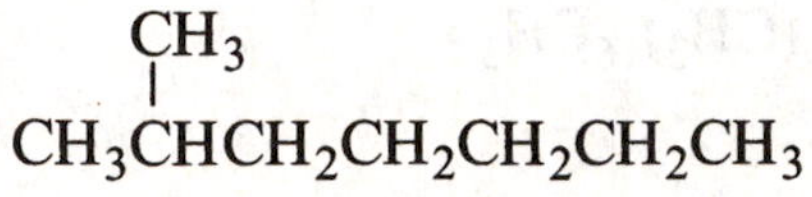

$$CH_3CHCH_2CH_2CH_2CH_2CH_3$$
with CH_3 branch

(b)

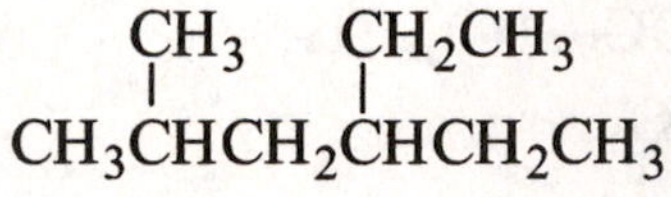

$$CH_3CHCH_2CHCH_2CH_3$$
with CH_3 and CH_2CH_3 branches

(c)

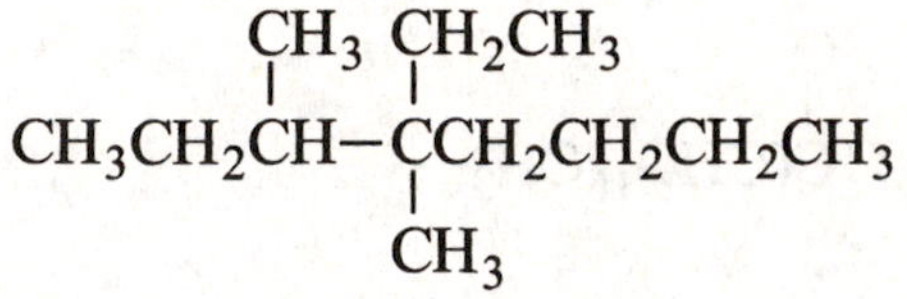

$$CH_3CH_2CH-CCH_2CH_2CH_2CH_3$$
with CH_3, CH_2CH_3, and CH_3 branches

(d)

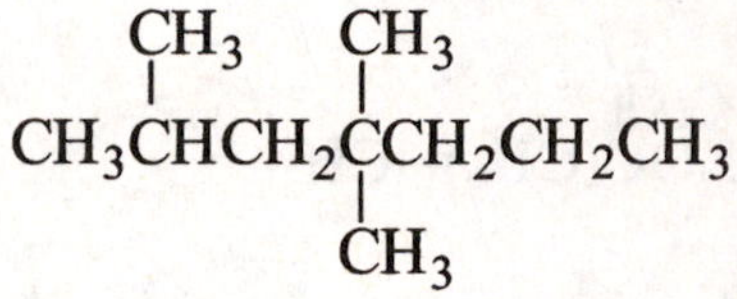

$$CH_3CHCH_2CCH_2CH_2CH_3$$
with CH_3, CH_3, and CH_3 branches

(e) CH_3 CH_3 on cyclopentane

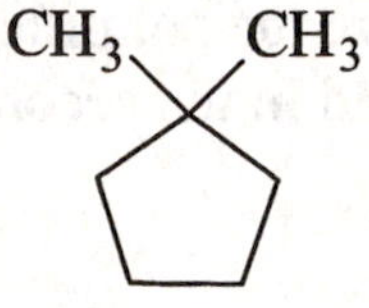

(f)

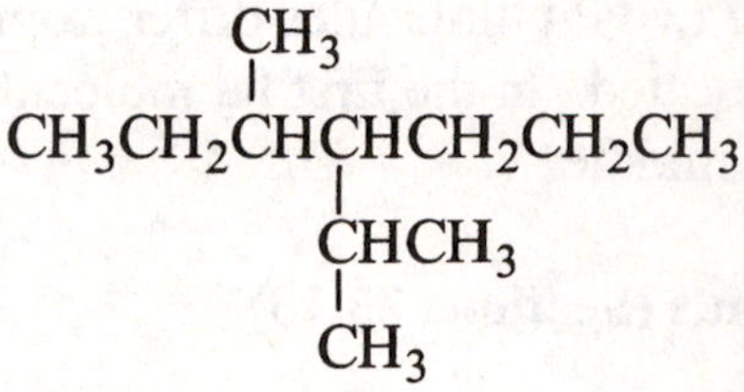

$$CH_3CH_2CHCHCH_2CH_2CH_3$$
with CH_3 branch, $CHCH_3$ and CH_3

23.100

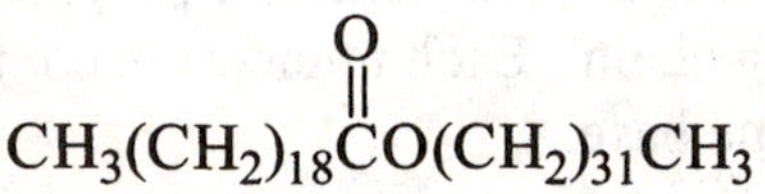

$$CH_3(CH_2)_{18}\overset{O}{\underset{\|}{C}}O(CH_2)_{31}CH_3$$

23.102 (a)

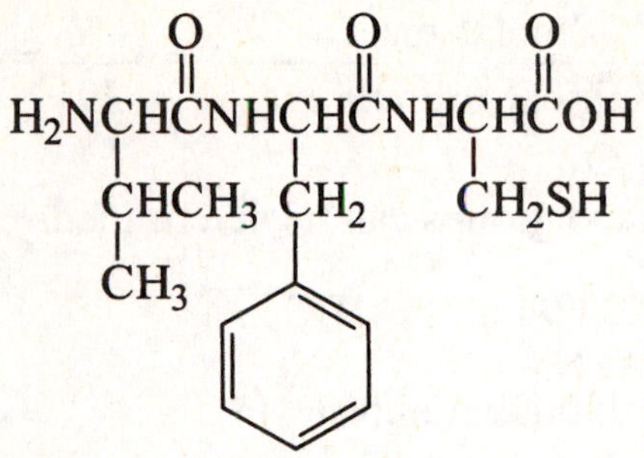

$$H_2NCHCNHCHCNHCHCOH$$

with three C=O groups; substituents CHCH₃, CH₂, CH₂SH; CHCH₃ with CH₃; CH₂ with phenyl ring

(b)

$$H_2NCHC-N-CHCNHCHCNHCHCOH$$

with four C=O groups; substituents CH₂, pyrrolidine ring; CH₂COOH; CHCH₃, CH₂CHCH₃; CH₂CH₃, CH₃

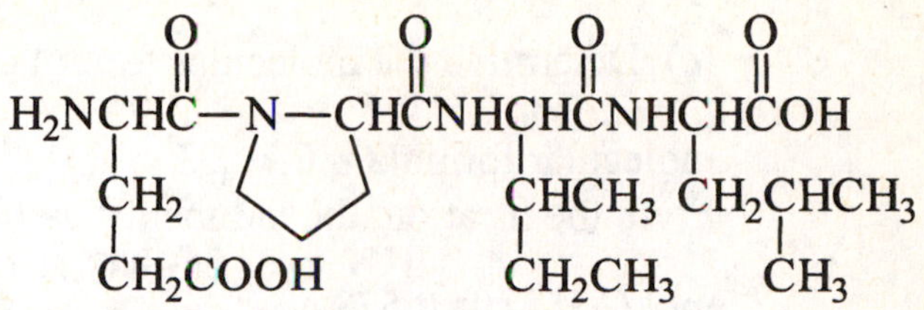

23.104 Original: A–G–T–T–C–A–T–C–G
       Complement: T–C–A–A–G–T–A–G–C

23.106 (a) Amine, ester, and arene

(b)

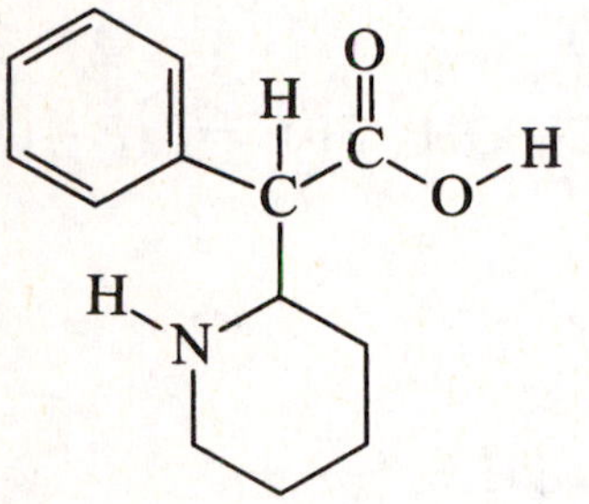

and CH_3OH

Multiconcept Problems

23.108 (a) Calculate the empirical formula. Assume a 100.0 g sample of fumaric acid.

$$41.4 \text{ g C} \times \frac{1 \text{ mol C}}{12.01 \text{ g C}} = 3.45 \text{ mol C}$$

$$3.5 \text{ g H} \times \frac{1 \text{ mol H}}{1.008 \text{ g H}} = 3.47 \text{ mol H}$$

$$55.1 \text{ g O} \times \frac{1 \text{ mol O}}{16.00 \text{ g O}} = 3.44 \text{ mol O}$$

Because the mol amounts for the three elements are essentially the same, the empirical formula is CHO (29).

(b) Calculate the molar mass from the osmotic pressure.

$$\Pi = MRT; \quad M = \frac{\Pi}{RT} = \frac{\left(240.3 \text{ mm Hg} \times \dfrac{1.00 \text{ atm}}{760 \text{ mm Hg}}\right)}{\left(0.082\,06 \dfrac{L \cdot atm}{K \cdot mol}\right)(298 \text{ K})} = 0.0129 \text{ M}$$

$$(0.1000 \text{ L})(0.0129 \text{ mol/L}) = 1.29 \times 10^{-3} \text{ mol fumaric acid}$$

$$\text{fumaric acid molar mass} = \frac{0.1500 \text{ g}}{1.29 \times 10^{-3} \text{ mol}} = 116 \text{ g/mol}$$

molecular mass = 116

(c) Determine the molecular formula.

$$\frac{\text{molar mass}}{\text{empirical formula mass}} = \frac{116}{29} = 4$$

molecular formula $= C_{(1 \times 4)}H_{(1 \times 4)}O_{(1 \times 4)} = C_4H_4O_4$

From the titration, the number of carboxylic acid groups can be determined.

$$\text{mol } C_4H_4O_4 = 0.573 \text{ g} \times \frac{1 \text{ mol } C_4H_4O_4}{116 \text{ g}} = 0.004\ 94 \text{ mol } C_4H_4O_4$$

mol NaOH used $= (0.0941 \text{ L})(0.105 \text{ mol/L}) = 0.0099 \text{ mol NaOH}$

$$\frac{\text{mol NaOH}}{\text{mol } C_4H_4O_4} = \frac{0.0099 \text{ mol}}{0.004\ 94 \text{ mol}} = 2$$

Because 2 mol of NaOH are required to titrate 1 mol $C_4H_4O_4$, $C_4H_4O_4$ is a diprotic acid. Because $C_4H_4O_4$ gives an addition product with HCl and a reduction product with H_2, it contains a double bond.

(d) The correct structure is

23.110 (a) grams $I_2 = (0.0250 \text{ L})(0.200 \text{ mol/L})\left(\dfrac{253.81 \text{ g } I_2}{1 \text{ mol } I_2}\right) = 1.27 \text{ g } I_2$

(b) mol $Na_2S_2O_3 = (0.08199 \text{ L})(0.100 \text{ mol/L}) = 8.20 \times 10^{-3} \text{ mol}$

grams excess $I_2 = 8.20 \times 10^{-3} \text{ mol } Na_2S_2O_3 \times \dfrac{1 \text{ mol } I_2}{2 \text{ mol } Na_2S_2O_3} \times \dfrac{253.81 \text{ g } I_2}{1 \text{ mol } I_2} = 1.04 \text{ g } I_2$

grams I_2 reacted $= 1.27 \text{ g} - 1.04 \text{ g} = 0.23 \text{ g } I_2$

(c) iodine number $= \dfrac{0.23 \text{ g } I_2}{0.500 \text{ g milkfat}} \times 100 = 46$

(d) mol milkfat $= 0.500 \text{ g milkfat} \times \dfrac{1 \text{ mol milkfat}}{800 \text{ g milkfat}} = 6.25 \times 10^{-4} \text{ mol}$

mol I_2 reacted $= 0.23 \text{ g } I_2 \times \dfrac{1 \text{ mol } I_2}{253.81 \text{ g } I_2} = 9.06 \times 10^{-4} \text{ mol}$

number of double bonds per molecule $= \dfrac{9.06 \times 10^{-4} \text{ mol } I_2}{6.25 \times 10^{-4} \text{ mol milkfat}} = 1.4 \text{ double bonds}$